DISCARDED

An Introduction to Phytoplanktons: Diversity and Ecology

Ruma Pal • Avik Kumar Choudhury

An Introduction to Phytoplanktons: Diversity and Ecology

Ruma Pal
Avik Kumar Choudhury
Department of Botany
University of Calcutta
Kolkata, West Bengal, India

ISBN 978-81-322-1837-1   ISBN 978-81-322-1838-8 (eBook)
DOI 10.1007/978-81-322-1838-8
Springer New Delhi Heidelberg New York Dordrecht London

Library of Congress Control Number: 2014939609

© Springer India 2014
This work is subject to copyright. All rights are reserved by the Publisher, whether the whole or part of the material is concerned, specifically the rights of translation, reprinting, reuse of illustrations, recitation, broadcasting, reproduction on microfilms or in any other physical way, and transmission or information storage and retrieval, electronic adaptation, computer software, or by similar or dissimilar methodology now known or hereafter developed. Exempted from this legal reservation are brief excerpts in connection with reviews or scholarly analysis or material supplied specifically for the purpose of being entered and executed on a computer system, for exclusive use by the purchaser of the work. Duplication of this publication or parts thereof is permitted only under the provisions of the Copyright Law of the Publisher's location, in its current version, and permission for use must always be obtained from Springer. Permissions for use may be obtained through RightsLink at the Copyright Clearance Center. Violations are liable to prosecution under the respective Copyright Law.
The use of general descriptive names, registered names, trademarks, service marks, etc. in this publication does not imply, even in the absence of a specific statement, that such names are exempt from the relevant protective laws and regulations and therefore free for general use.
While the advice and information in this book are believed to be true and accurate at the date of publication, neither the authors nor the editors nor the publisher can accept any legal responsibility for any errors or omissions that may be made. The publisher makes no warranty, express or implied, with respect to the material contained herein.

Printed on acid-free paper

Springer is part of Springer Science+Business Media (www.springer.com)

Preface

Phytoplankton community in water bodies support the base of the natural food chain depending on which the natural fauna including the fish populations can survive. At the same time they produce almost 70 % of world's atmospheric oxygen. On the other hand, excessive phytoplankton production in the water bodies causes extensive problems like fish poisoning and deterioration of water quality in drinking water management, swimming pools and other water-based recreations. Therefore, it is dire necessity to study the different factors controlling phytoplankton growth or in other words phytoplankton ecology.

The microscopic phytoplanktonic genera are represented by different algal groups like Cyanobacteria, Chlorophyta, Bacillariophyta, Euglenophyta, Prymnesiophyta, etc. Therefore, identification of planktonic genera requires a good knowledge in algal taxonomy. Algal taxonomy is nowadays based on different characters like ultrastructure of algal chloroplast and flagella, cellular biochemistry, molecular characterization, etc., rather than morphotaxonomy of early days. With the advent of different sophisticated instruments like TEM, SEM, AFM, HPLC, HPTLC, etc., change in algal taxonomy is a regular phenomenon. For this reason, an attempt has been taken to discuss the chronological changes in algal taxonomy. Pigment composition is nowadays an important characteristic for classifying the algal kingdom, especially the marine phytoplanktons; therefore, we have attempted to discuss pigment composition of different groups of algae in detail.

To correlate the phytoplankton productivity together with varied physicochemical parameters, different statistical methods are to be employed for data analysis, based upon which different ecological models are proposed. To study the phytoplankton diversity and ecology, sampling is an important factor to get the actual results. Different methods of samplings have also been discussed here. Therefore, phytoplankton ecology is a complex science, as it includes the interactions between biogeochemical cycling and environmental parameters. This phenomenon can be explained by the simple question, 'What lives where and why?', as stated by the famous phytoplankton ecologist Reynolds.

In general, there is scarcity of books on phytoplanktons related to diversity and ecology for students as well as researchers, which I experienced during my research work for the last 20 years on phytoplankton ecology. Therefore, presently for the benefit of students and phytoplankton researchers, we have tried to compile a general account of phytoplanktons, their physical and chemical environments, sampling methods and the statistical analysis together with similar case studies on phytoplankton ecology. The results of the case studies are the doctoral work of my student Dr. Avik Kumar Choudhury and research findings of my other students, Sri Nirupam Barman, Sri Gour Gopal Satpati and Ms Anindita Singha Roy. I think this book will help the students of botany, zoology, microbiology and environmental biology together with the plankton researchers.

Kolkata, India   Ruma Pal
17 Sep 2013

Acknowledgement

The development of this book was made possible by the financial support from University Grants Commission (UGC), Govt. of India, for two research projects on phytoplankton dynamics of Eastern Indian coast and other ecological niche. Dr. Avik Choudhury and Ms. Anindita Singha Roy worked as Project Fellows, and their research findings are represented as Case Study in Chap. 5.

Financial support from Council of Scientific and Industrial Research (CSIR), Govt. of India, is also acknowledged for survey work at Indian Sunderbans area under NIMTLI program. Mr. Nirupam Barman and Sri Gour Gopal Satpati surveyed different parts of Sunderbans, and their work is also included in this book.

20 Jan 2014   Ruma Pal

Contents

1	**A Brief Introduction to Phytoplanktons**		1
	1.1	General	1
	1.2	Historical Perspective	1
	1.3	Algal Classification	2
		1.3.1 Old Classical Taxonomy	3
		1.3.2 Modern System of Algal Classification	8
	1.4	Classification of Phytoplanktons	15
		1.4.1 On the Basis of Size Variations	15
		1.4.2 On the Basis of Habitat	16
	1.5	Algal Pigments	18
	1.6	Ecological Significance	23
		1.6.1 Phytoplankton Bloom	23
		1.6.2 Cyanobacterial Bloom	24
		1.6.3 Dinoflagellate Bloom	25
		1.6.4 Algal Toxins	27
	1.7	Algae in Wetlands	28
		1.7.1 Role of Phytoplanktons in Wetlands	29
		1.7.2 Nutrient Status of Wetland Ecosystem	29
		1.7.3 Types of Wetlands	30
	1.8	Key to Identification of Common Phytoplankton Genera	30
		1.8.1 Division: Cyanobacteria	30
		1.8.2 Division: Chlorophyta	31
		1.8.3 Order: Volvocales	32
		1.8.4 Order: Chlorococcales	32
		1.8.5 Division: Bacillariophyta (Diatoms)	33
	References		39
2	**Physicochemical Environment of Aquatic Ecosystem**		43
	2.1	Physical Factors	43
		2.1.1 Light and Temperature	43
		2.1.2 Turbulence	45
	2.2	Major Nutrients	46
		2.2.1 Redfield Ratio	47
		2.2.2 Carbon	47
		2.2.3 Nitrogen	49
		2.2.4 Phosphorus	50

		2.2.5	Silicon	51
		2.2.6	Nutrient Uptake Model	52
	References			52
3	**Phytoplanktons and Primary Productivity**			55
	References			57
4	**Community Pattern Analysis**			59
	4.1	Phytoplankton Sampling		59
		4.1.1	Bottle Samplers	59
		4.1.2	Plankton Pumps	60
		4.1.3	Plankton Nets	61
	4.2	Biomass Estimation		62
		4.2.1	Cell Counts	63
		4.2.2	Utermohl Sedimentation Method for Cell Counts	63
		4.2.3	Biovolume	64
		4.2.4	Chlorophyll and Photopigments	65
	4.3	Species Diversity Index		68
		4.3.1	Species Evenness	69
		4.3.2	Species Richness	69
	4.4	Multivariate Analysis		70
		4.4.1	Preparation of Data Sets	70
		4.4.2	Data Transformations	70
		4.4.3	Exploratory Analysis	71
	References			73
5	**Case Study**			75
	5.1	Studies from Eastern Mediterranean Region: A Review		75
	5.2	Case Study I: Phytoplankton Diversity of East Calcutta Wetland: A Ramsar Site		78
		5.2.1	Study Area	78
		5.2.2	Results	79
	5.3	Case Study II: Phytoplankton Diversity of Indian Sunderbans		88
		5.3.1	List of Phytoplankton Genera Recorded from Sunderbans	89
		5.3.2	Taxonomic Account of a Few Phytoplankton Taxa Recorded from Sunderbans (Rest Are in Case Studies I and III)	89
	5.4	Case Study III: Phytoplankton Dynamics of Eastern Indian Coast		92
		5.4.1	Study Area	92
		5.4.2	Nutrient and Phytoplankton Dynamics of Coastal West Bengal	92
	References			158
Glossary				163
Bibliography				167

About the Author

Dr. Ruma Pal did her M.Sc. and Ph.D. from University of Calcutta. Presently, she is Associate Professor in Botany, Department of Botany, University of Calcutta, India. Prior to this assignment she had served in Presidency College, Calcutta and Nara Sigh, Dutt College Howrah. She has more than 30 years of research experience in Phycology. Her research interest is related to various fields of algal biotechnology, like, Phycoremediation, aquaculture, biofuel production, nanotechnology, etc. Algal diversity study, phytoplankton dynamics and ecological modelling are also the areas of her interest. She has already conducted more than 15 research projects in the field of algal application and has published more than 50 papers in refereed journals.

Dr. Avik Kumar Choudhury completed his postgraduation from Banaras Hindu University in Botany in 2004. Subsequently, he worked as a UGC Project Fellow at Department of Botany, University of Calcutta. He completed his doctoral work from the same institute and received his Ph.D. degree in 2011 under the guidance of Dr. R. Pal. He also worked as a DBT (Department of Biotechnology, Govt. of India) Postdoctoral Research Associate at Department of Biological Sciences, Indian Institute of Science Education and Research. Presently he is in West Bengal School service.

A Brief Introduction to Phytoplanktons

1.1 General

'Phytoplanktons' are free-floating, photosynthetic, aquatic microorganisms, which move from one place to another, either actively by their locomotory organs (flagella) or passively by water currents. The name 'phytoplankton' came from the Greek words 'φυτόν' (phyton), meaning 'plant', and 'πλαγκτός' (planktos), meaning 'wanderer' or 'drifter'. The term 'plankton' was first used by the German biologist Victor Hensen in 1887. According to Hensen, 'plankton included all organic particles which float freely and involuntarily in open water, independent of shores and bottom (Ruttner 1940; Hutchinson 1957)'.

Most of the phytoplanktons survive on the open surface waters of lakes, rivers and oceans. The phytoplankton community is mainly represented by algal representatives including both prokaryotes and eukaryotic genera. Plankton populations are mostly represented by members of Cyanobacteria, Chlorophyta, Dinophyta, Euglenophyta, Haptophyta, Chrysophyta, Cryptophyta and Bacillariophyta. Planktonic representative taxa are absent in other algal divisions like Phaeophyta and Rhodophyta.

1.2 Historical Perspective

It has been proposed that studies related to plankton are included under 'limnology' where the name comes from the Greek word 'limnos' meaning 'pool' or 'lake' or 'swamp'. The study of biological limnology originated in 1674 with the first microscopic description of *Spirogyra* from Berkelse Lake, Netherlands, by Leeuwenhoek. The first work on limnology was probably published in the USA by Louis Agassiz (1850) entitled *Lake Superior: Its Physical Character, Vegetation and Animals*. Professor Forel of the University of Lausanne, who is considered as the 'father of limnology', published the first textbook on limnology in 1869 on the bottom fauna of Geneva Lake entitled *Introduction a letude de la faune profonde du Lac Leman* and for the first time he used the term 'limnology' in his book.

The study of lotic estuarine systems was initiated by J.R. Lorenz in the Elbe, Germany, in the 1860s. At the same time, pollution research started in the Thames in England, and the realization of the problems of survival in brackish waters was initiated with the biological approaches (Meyer and Möbius 1865–72). The importance of ecological concepts in limnology was first established by the English botanist Tansley which was later popularized by G.E. Hutchinson and R.L. Lindenman in the latter's paper entitled *The Trophic Dynamic Aspect of Ecology (1942)*. Professor Birge of the University of Wisconsin contributed to limnology in the USA through his study of the plankton of Lake Mendota (1917). Other eminent scientists like C.A. Kofoid, J.G. Needham and C. Juday also worked on different rivers and lakes. In the early part of the twentieth century, Professor P.S. Welch wrote the first American textbook on limnology. Similar books were also written like *Fundamentals of Limnology* by Franz Ruttner (1940) and Professor G.E. Hutchinson published the book entitled *Treatise on Limnology* (1957, 1967) which is considered as standard reference work throughout the world.

In the nineteenth century, it was felt to publish the journals on limnology that would pull together the increasing volume of limnological informations that were developing every day. Thus, on January 1, 1936, the Limnological Society of America was established, which was later (1948) named as the 'American Society of Limnology and Oceanography' and used to publish till date the most well-circulated and popular journal titled *Limnology and Oceanography*. The year 1948 is also marked for the formation of the 'Freshwater Biological Association' in Britain. This association maintains a continuous record of physical, chemical and biological informations of 17 lakes in the Lake District of northwest England. Similarly 'Istituto Italiano di Idrobiologia' in Italy carried out intensive studies on northern Italian lakes, and the science of limnology in Europe flourished. Similar other renowned institutes developed around this time in Europe, for example, at Plön, Germany; Uppsala and Lund, Sweden; Copenhagen, Denmark; and on Lake Constance, Germany.

1.3 Algal Classification

Algae are considered as the most primitive plants with well diversifications. They all have a chlorophyll-bearing thalloid plant body and their reproductive structures are without sterile jackets. They have variations in morphology and cellular biochemistry together with reproductive behaviour and life cycle patterns. On the basis of these characters, algologists classified the entire algal kingdom into different divisions and classes from time to time.

The systematics or classification of algae has been changing dramatically through ages. In early times, the classification was mainly based on the morphotaxonomy. But very recently the situation is marked by the quest of a compromise between the conventional (artificial) system and the phylogenetic system together with the molecular genetics. The classical approaches using morphological characters do not reflect the phylogenetic relationships, e.g. molecular data revealed that among the green algae, the genera with spherical ball-type thallus evolved independently in different lineages of algae. On the other hand, highly diverse morphotypes can belong to one and the same phylogenetic lineage.

In algal systematics nowadays, a polyphasic approach is widely suggested. Therefore, phylogeneticists suggest the following criteria for proper identification and phylogenetic placement of any algal groups:

1. Conventional morphological and ecophysiological study of algae under field condition
2. Isolation of unialgal culture for morphological, ontogenetic, biochemical and physiological studies under laboratory conditions
3. Ultrastructural studies of different cell organelles of algal cell, especially the chloroplast ultrastructure
4. Use of molecular markers (conserved sequences) for molecular phylogenetic analysis

The most commonly used sequences are the 16S rRNA or the small subunit (SSU) of ribosomal RNA (rRNA) for prokaryotic cells and 18S rRNA for eukaryotic cells. Besides some other markers are also used for phylogenetic analysis, e.g. *rbcL* and *tufA* gene.

Therefore, for convenience we can consider two phases of algal systematics. In the first phase, starting from placement of algae in plant kingdom by Eichler (1886), followed by algal classification by Fritsch (1935) and Smith (1950), up to Bold and Wynne's (1985) system, only morphological characters were considered for taxonomic purpose. Therefore, these classifications can be considered under 'old classical taxonomy'. In another approach or in the second phase, other parameters including

1.3 Algal Classification

ultrastructural, biochemical and molecular characters are also considered – reflecting the evolution and phylogeny of and can be designated as 'modern system of classification'.

1.3.1 Old Classical Taxonomy

Eichler (1886) mentioned the phylogenetic position of algae in relation to other plant groups. He divided the entire plant kingdom into two groups: Cryptogamia (spore-producing plants) and Phanerogamia (seed plants). These two groups are again divided into three divisions and four classes. Algae are placed in Class I of division Thallophyta under Cryptogamia in close association with fungi. The only difference between algae and fungi was recognized as the presence or absence of chlorophyll.

This system of classification is considered as incorrect by most of the scientists now. Indeed, we think it is still needed to be taught to a botanist to understand the phylogenetic position of algae and other plant groups which also indicate the evolutionary tendency among different plant groups in a broader sense.

In the old classical system of algal classification, famous algologists like Smith, Fritsch, Prescott, Bold and Wynne and Chapman and Chapman have classified the algal kingdom on the basis of variation in cell structure (prokaryotic and eukaryotic), flagellar position, number and structure, pigment composition, reserve food matters, mode of reproduction, etc.

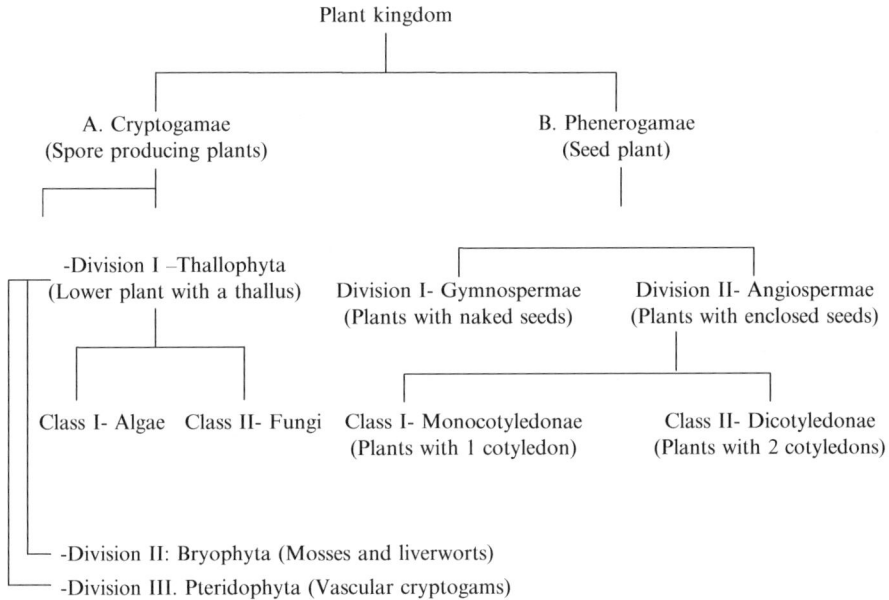

Eichler's system of classification (1886)

Afterwords, Fritsch (1935) divided the algal kingdom into 11 classes:
Class
  I. Chlorophyceae
  II. Xanthophyceae
  III. Chrysophyceae
  IV. Bacillariophyceae
  V. Cryptophyceae
  VI. Dinophyceae
  VII. Chloromonadineae
  VIII. Euglenineae
  IX. Phaeophyceae
  X. Rhodophyceae
  XI. Myxophyceae

On the basis of the morphological characters, Fritsch distinguished clearly 11 classes of the algae. The termination 'phyceae' has been adopted wherever the class includes forms with an algal organization, while for flagellates the old designation is retained. Therefore, he designated the different groups of algae as 'classes' – rather than 'divisions'.

The 11 classes were characterized as follows:
1. Chlorophyceae (Isokontae) – Members with grass-green-coloured chromatophores and contain the same pigment compositions and approximately in the same proportions as in higher plants. Starch is the customary form of storage products of photosynthesis, often (especially in resting stages) accompanied by oil, and pyrenoids commonly surrounded by a starch sheath are frequently present inside the chloroplasts. Cells are surrounded by a cell wall in which cellulose is the constituent. The motile cells are with equal whiplash type of flagella (commonly two or four) which arise from the front end of swarmers. In many members the cells contain only one or few chromatophores. The members of Chlorophyceae exhibit sexual reproduction – ranging from isogamy to advanced oogamy. Most of the taxa are haploid with zygote representing the only diploid phase, but some exhibit a regular alternation of generation with similar haploid and diploid individuals. The class is more widely represented by freshwater members than in salt water, and there is a marked terrestrial tendency. Examples are *Chlamydomonas* and *Spirogyra*.
2. Xanthophyceae (Heterokontae) – The members are with yellow-green chromatophores due to presence of xanthophylls as major pigment. Starch is absent and storage product is oil. The algal members have a cell wall which is often rich in pectic compounds. The motile cells possess two very unequal flagella (or sometimes only one) arising from the front end. As a general rule the cells contain a number of discoid chromatophores. Sexual reproduction is always isogamous. The most advanced forms have a simple filamentous habit. All are probably haploid. The class is more widely distributed in freshwater than in the sea. An example is *Vaucheria*.
3. Chrysophyceae – Members are with brown- or orange-coloured chromatophores containing one or more accessory pigments. Starch is absent, but naked pyrenoid-like bodies are occasionally present. Fat and a compound leucosin are found in the form of rounded whitish opaque lumps as the food storage. A large proportion of the members are flagellate and devoid of a special cell membrane. The motile cells possess one or two (rarely three) flagella attached at the front end. The cells typically contained one or two parietal chromatophores. The most advanced habit is that of a branched filament. Sexual reproduction is extremely rare and not yet quite clearly established in any one case; the existing records point only to isogamy. The class is widely distributed in freshwaters, but a few are marine. An example is *Chromulina*.
4. Bacillariophyceae (diatoms) – Unicellular members with yellow or golden-brown chromatophores containing accessory brown pigments. Pyrenoid-like bodies are often present and the products of photosynthesis are fat and volutin. All the members are unicellular or colonial. A cell wall is always present and is composed of mainly silica and partly of pectic substances. The cell consists of two halves, each composed of two or more pieces, and is commonly richly ornamented. One set of forms (Centrales) is radially and the other (Pinales) bilaterally symmetrical. The diatoms produce a special type of spore – the auxospore. The Pinales show a special type of sexual fusion between the protoplasts of the ordinary individuals. The members of this class are probably diploid. Diatoms are very widely distributed in the sea and in all kinds of freshwaters, as well as in the soil and in other terrestrial habitat. Examples are *Navicula* and *Chaetoceros*.
5. Cryptophyceae – The members are of flagellate organization and with oogamous type of reproduction. The cells are with usually two large parietal chromatophores showing very diverse pigmentation. Pyrenoid-like bodies occur, but appear often to be independent of the chromatophores; the products of photosynthesis are

solid carbohydrates, in some cases starch, in others a compound akin to it. The motile cells are pronouncedly dorsiventral, have two slightly unequal flagella and possess a very specialized and characteristic structure. There is often a complex vacuolar system. The class is relatively small and appears to be equally scantily represented in the sea and freshwater. An example is *Cryptomonas*.

6. Dinophyceae (Peridiniidae) – The majority of the members are motile unicells and many possess a very elaborate cellulose envelope composed of a large number of often richly sculptured plates; some are with a branched filament. The members usually have numerous discoid chromatophores which are dark yellow, brown, etc., and contain a number of special pigments. The products of photosynthesis are starch and oil (fat). Many species are colourless saprophytes or exhibit holozoic nutrition; one extensive series is parasitic. The motile cells have two furrows, the transverse one harbouring the transverse flagellum which usually encircles the body and the other longitudinal constituting the starting point for the longitudinal flagellum which is directed backwards. Resting cysts of characteristic form are often produced. Isogamous sexual reproduction is certainly of rare occurrence and not yet clearly established. A class of mainly plankton organisms is more widely represented in the sea than in freshwaters. An example is *Gymnodinium*.

7. Chloromonadineae – The members of this class are motile flagellates, with two almost equal flagella. Cells are with numerous discoid chromatophores having a bright green tint and containing an excess of xanthophylls. Pyrenoids are lacking and oil is the assimilatory product. Although superficially like Xanthophyceae the detailed structure of the cells is altogether different (complex vacuolar apparatus, etc.). The class is only recorded from freshwaters. An example is *Gonyostomum* (raphidophytes).

8. Euglenineae – Unicellular thallus with pure green chromatophores, each cell usually with several pyrenoid-like bodies. The product of photosynthesis is a polysaccharide, paramylum, which occurs in the form of solid grains of diverse and often very distinctive shape. Only flagellate members are known and the majority is motile with the help of one or two flagella which arise from the base of a canal-like invagination at the front end. There is a complex vacuolar system and a large and prominent nucleus. Only few cases of sexuality (isogamous) are known and these are not quiet fully substantiated. The bulk of the members of this class probably inhabits freshwaters. The class is highly specialized and no really simple form is known. An example is *Euglena*.

9. Phaeophyceae – The majority of cells are with brown chromatophores containing, apart from the usual pigments, the yellow fucoxanthin. Naked pyrenoid-like bodies occur in some of the lower forms. The assimilatory products are sugar alcohols (mannitol) with only traces of sugars, as well as polysaccharides (laminarin) and fats. Characteristic fucosan vesicles are present which probably represent waste products. The motile reproductive cells have two laterally attached flagella, of which one is directed forwards and the other backwards. These swarmers are always formed in special organs which are either unilocular or septate with numerous small compartments (plurilocular sporangia). Sexual reproduction is of wide occurrence and ranges from isogamy to oogamy of a primitive type, with liberation of the ovum prior to fertilization. The zygote exhibits no resting period. The life cycle is very diverse, with varied types of alternation of generations. An example is *Laminaria*.

10. Rhodophyceae – The majority attains to a considerable complexity of structure, though the simplest forms are filamentous. Cells are with chromatophores containing, apart from the usual pigments, others like the red phycoerythrin and blue phycocyanin. Pyrenoid-like bodies are found in the lower groups and the product of assimilation is a solid polysaccharide similar to starch (floridean starch). Neither motile reproductive stages

nor flagellate members are known. Evident protoplasmic connections are the rule between the cells of the majority of forms. Most of the Rhodophyceae are marine. All exhibit sexual reproduction of an advanced oogamous type, the female organ having a long receptive neck and the antheridium producing but a single motionless male cell. As a result of fertilization, special spores (carpospores) are produced from bunches of threadlike structures that arise from the female organ after fertilization. The Rhodophyceae are either haploid or exhibit a regular alternation of similar haploid and diploid individuals, the latter bearing characteristic sporangia (tetrasporangia), each producing four spores. An example is *Polysiphonia*.

11. Myxophyceae (Cyanophyceae) – Members with a simple type of cell, containing at the best only a very rudimentary nucleus (central body) and without a proper chromatophore, the photosynthetic pigments being diffused through the peripheral cytoplasm. The pigments present are chlorophyll, carotene, phycocyanin and phycoerythrin, the last two being in varying proportions and the colour of the cells being very commonly blue green. The products of photosynthesis are sugars and glycogen. No motile stages are known and all the members have a membrane around the cell. There is no sexual reproduction. The members of this class are of simple organization and many propagate entirely by simple division or by vegetative means. Most types are filamentous, many of them with a peculiar 'false' branching. They occur very abundantly in freshwaters and in terrestrial habitats and are not common in the sea. An example is *Spirulina*.

But in 1950, Smith proposed seven divisions of algae in his system of classification as follows:
1. Chlorophyta
2. Euglenophyta
3. Chrysophyta
4. Phaeophyta
5. Pyrrophyta
6. Cyanophyta
7. Rhodophyta

The divisions were characterized by the following basic characteristics:
1. *Chlorophyta* – The grass-green algae with the major pigments chlorophyll a and b together with carotenes and xanthophylls. Photosynthetic reserves are usually stored in the form of starch. It is always in association with pyrenoid. Motile stages have flagella of equal length with a few exceptions. The zoospores and motile gametes have two or four flagella. Sexual reproduction is a phenomenon of wide occurrence within the group and in various orders it ranges all the way from isogamy to oogamy. This group also shows a wide range in vegetative structures (*Chlorella*).
2. *Euglenophyta* – All the members are unicellular and most of them are naked free-swimming cells with one, two or three flagella. Many of the genera have grass-green, discoid band-shaped or stellate chloroplasts, with or without pyrenoids. The chloroplasts contain the same chlorophylls as Chlorophyceae, beta-carotene and xanthophylls unlike Chlorophyta. Food reserve is paramylum; nutrition may be holophytic, holozoic or saprophytic. There are one or two contractile vacuoles at the anterior end of the cell, which are connected with a reservoir, which in turn is connected with the cell's exterior by a narrow gullet (*Euglena*).
3. *Chrysophyta* – The members have yellowish-green to golden-brown pigment because of the predominance of carotenes and xanthophylls. The food reserves include a complex carbohydrate leucosin and oils. Some members are with cell wall made up of silica and are composed of two overlapping halves. Cells may be flagellated or non-flagellated, solitary or united in colonies of definite or indefinite shape. Sexual reproduction is usually isogamous by a union of flagellated and non-flagellated gametes, e.g. members of present-day Xanthophyta (*Vaucheria*) and Bacillariophyta (*Nitzschia*).
4. *Phaeophyta* – The Phaeophyta or brown algae have many celled complex type of plant body that is usually of macroscopic size and distinctive shape. The chromatophore or pigment-bearing organelles are yellowish brown in

colour due to the presence of xanthophylls in greater amount than that of chlorophyll and carotenes. The two principle reserve foods are laminarin, a polysaccharide, and mannitol. Zoospores or gametes are pyriform with two laterally inserted flagella of unequal length. Reproductive organs are of two kinds – the one celled or unilocular reproductive organ is always a sporangium and born on a diploid thallus. The other kind of reproductive organ is many celled and with each cell containing a single gamete or single zoospore. Most of the members of Phaeophyta have a life cycle in which there is alternation of two independent multicellular generations, haploid and the other is diploid. The two generations may be identical in size and structure, in others they are dissimilar in both size and structure (*Ectocarpus*).

5. *Pyrrophyta* – Members of this division are greenish to golden brown. The pigments are Chl *a*, Chl *c*, beta-carotene and four xanthophylls. Photosynthetic compounds are reserved as starch and also oil. The nucleus is distinctive in which chromatin lies in numerous bead-like structures on thread. Cell wall when present contains cellulose.

6. *Cyanophyta* – The Cyanophyta or blue-green algae are a distinctive group sharply delimited from other algae (prokaryotic algae). Their pigments are not localized on chromatophores; rather they are present in the peripheral portion of the protoplast and include chlorophyll *a*, carotenes and distinctive xanthophylls. In addition, there is a blue pigment (C-phycocyanin) and a red pigment (C-phycoerythrin). The unique feature of Cyanophyta is the presence of primitive type of nucleus within the cell (central body) which lacks a nucleolus and a nuclear membrane. They lack any flagellated structure and devoid of sexual reproduction (*Oscillatoria*).

7. *Rhodophyta* – The Rhodophyta or red algae have multicellular thalli of microscopic or macroscopic size and often of distinctive shape. Red algae differ from all other algae in structure of their sexual organs, in mode of fertilization followed by formation of a spore-producing structure, the so-called cystocarp. Pigments are localized in chromatophores. In addition to chlorophylls, carotene, xanthophyll and R-phycoerythrin and R-phycocyanin are present.

Prescott (1984) first classified the algal kingdom into seven phyla like other groups of biological kingdom. Each phylum is again divided into different classes and orders as follows:

I. Phylum Chlorophyta
   (i) Class – Chlorophyceae (17 orders)
   (ii) Class – Charophyceae (1 order)
II. Phylum Euglenophyta
III. Phylum Chrysophyta
   (i) Chrysophyceae
   (ii) Bacillariophyceae
   (iii) Heterokontae
IV. Phylum Pyrrophyta
   (i) Desmokontae
   (ii) Dinokontae
V. Phylum Phaeophyta
   (i) Isogeneratae
   (ii) Heterogeneratae
VI. Phylum Rhodophyta
   Subphylum – Bangioideae (4 orders)
   Subphylum – Florideae (6 orders)
VII. Phylum Cyanophyta
   Subphylum – Coccogoneae
   Subphylum – Hormogoneae

Among the old classical algal taxonomy, *Bold and Wynne's* (1985) system of classification is the most well-accepted one. He divided the algal kingdom into nine divisions and introduced a new division Charophyta, which is considered as progenitor of land plants.

Divisions

I. Cyanophyta and Prochlorophyta
II. Chlorophyta (16 orders)
III. Charophyta
IV. Euglenophyta
V. Phaeophyta (13 orders)
VI. Chrysophyta (6 class)
   Chrysophyceae
   Prymnesiophyceae
   Xanthophyceae
   Eustigmatophyceae
   Raphidophyceae
   Bacillariophyceae

VII. Pyrrophyta
VIII. Rhodophyta
IX. Cryptophyta

Charophyta

Commonly known as stoneworts, small plantlike, basal rhizoidal part rootlike and upper region differentiated into nodes and internodes. At the nodal region primary and secondary laterals are present. Branching is also prominent. Small leaflike stipules and bracts are present. Reproduction is oogamous type. Male reproductive structure antheridia and female reproductive structure oogonia are covered by sterile jacket, shield cell and tube cell, respectively. They share many characters with land plants. They possess unilateral type of flagellar root – in contrast to cruciate types in the members of Chlorophyta. Mitosis – open type, i.e. they lack nuclear membrane in late prophase, whereas it is closed types in other members of Chlorophyta (*Chara, Nitella*).

Bold and Wynne (1985) mentioned about Prochlorophyta together with Cyanophyta in their system of classification. Prochlorophyta were considered as a new division by Lee (1980, 1989, 1999) with the prokaryotic members having Chl *b* (*Prochloron*). But later on, other characters are also considered including molecular markers and it was found that those are nothing but Chl *b*-bearing cyanobacteria.

1.3.2 Modern System of Algal Classification

In modern approach, the classification is based on the evolutionary process of the biological kingdom. The entire kingdom or individual groups are divided into several kingdoms or classes considering different characters, reflecting the evolution and phylogeny. Together with morphological parameters, other characters, like biochemical and molecular data, are also considered.

Whittaker's (1969) system of biome classification is one of such example, where the entire biome is divided into five kingdoms, viz. Monera, Protista, Mycota, Metaphyta and Metazoa, as follows:

1. Kingdom: Monera (prokaryotic organisms)
2. Kingdom: Protista (primitive eukaryotic organisms)
3. Kingdom: Mycota (exclusively fungi)
4. Kingdom: Metaphyta (advanced eukaryotic plants)
5. Kingdom: Metazoa (all multicellular animals)

In this classification, all the cyanobacterial genera or prokaryotic algae are placed in 'Monera', together with prokaryotic bacteria. All unicellular members are included in 'Protista', including protozoa of animal kingdom and unicellular algae of plant kingdom. Multicellular higher plants including algae are placed in the kingdom 'Plantae' together with bryophytes and pteridophytes.

According to this classification, Monera represent the most primitive group of organisms. The Monera are thought to have given rise to Protista from which the three other kingdoms of organisms, namely, the fungi, plants and animals, evolved along separate lines. Fungi are thought to first to appear from Protista. Later, about a billion years ago, some protists must have evolved into primitive multicellular animals. Still later, probably about 350 million years ago, some protists must have evolved into higher forms of plants.

Like all systems of classification, the five-kingdom classification has also certain merits and demerits. However, it is largely the most accepted system of modern classification mainly because of the phylogenetic placing of different groups of living organisms.

This system of classification looks more scientific and natural because of the following considerations:

1. All the prokaryotes are placed into an independent kingdom. It is justifiable because they differ from all other organisms in their general organization.
2. The kingdom Protista contains all the unicellular eukaryotes. It solved many problems, particularly related to the position of some unusual organisms like *Euglena*.
3. Separation of the group fungi to a separate kingdom is justifiable since fungi totally differ

from other primitive eukaryotes like algae and protozoans.
4. The five-kingdom classification gives a clear indication of cellular organization and modes of nutrition.

However, the five-kingdom classification has some demerits also, particularly with reference to the lower forms of life:
1. The kingdoms Monera and Protista include both photosynthetic (autotrophic) as well as non-photosynthetic (heterotrophic) organisms and organisms which have cells with cell wall as well as without cell wall.
2. The three higher kingdoms or multicellular lines have originated from Protista several times (polyphyletic).
3. Unicellular or colonial green algae like *Chlamydomonas* and *Volvox* have not been included under Protista because of their resemblance to other green algae.
4. Slime moulds are different from other members of Protista.
5. Viruses have not been given proper place in this system of classification.

Nevertheless, the five-kingdom classification has found a wide acceptance with biologists all over the world.

Status of Viruses

The position of viruses in the biological kingdom is one of the unsolved mysteries. Due to the absence of a cellular organization, viruses cannot be placed with either prokaryotes or eukaryotes. They are considered as intermediate between living and nonliving systems. Viruses are active and show reproduction only inside the host cell. In the free state, they are totally inactive. They may even be purified and crystallized like chemical substances. Viruses have a genetic material represented by either DNA or RNA, surrounded by a protein sheath. Viruses reproduce by using the metabolic machinery and raw materials of the host cell. Because of these peculiarities, viruses do not fit into any of the five kingdoms of life.

From 1975 onwards, with the advent of electron microscopy and other sophisticated instruments like high-resolution SEM, TEM, HPLC and HPTLC, fine characters of algal cells like ultrastructural and biochemical and more recently the molecular data are also considered for algal classification together with the morphological characters of algal thallus. Since then the science of phycology has sustained major conceptual changes. The increased resolution of the electron microscope revealed the presence and structure of flagella, flagellar hairs, flagellar roots, eyespots, chloroplast, endoplasmic reticulum, etc., which were found to be important in basic systematics of algae especially the green algae. Important observations were made by Pickett-Heaps (1967, 1969, 1972a, b, 1975), which revealed that different types of microtubular arrangements are involved in cytokinesis of green algae. In some orders of Chlorophyta, cell divisions are characterized by the collapse of the interzonal spindle apparatus after mitosis, which give rise to a 'phycoplast' with the microtubules oriented in the plane of cell division, viz. Volvocales, Tetrasporales, Chlorococcales, Oedogoniales and Ulotrichales. On the other hand, some members together with land plants have a persistent spindle and develop a cleavage and furrow (Klebsormidiales) or a 'phragmoplast' in which the microtubules are oriented perpendicular to the plane of cytoplasmic division (Coleochaetales, Charales and Conjugales). At the same time, biochemical analysis also clarified the presence and structure of algal pigments, storage products and cell wall constituents. Considering all these parameters, Stewart and Mattox (1975) classified the green algal division Chlorophyta into five classes, considering comparative cytology, viz. Chlorokybales, Zygnematales, Klebsormidiales, Coleochaetales and Charales.

1.3.2.1 Algal Chloroplast in Classification

Mereschkowski (1905) first studied about the nature and origin of chloroplast in eukaryotic algae and the evolutionary pattern among them. In 1905, he published the most extensive paper on the origin of chloroplast based on the idea that eukaryotic algal cell originated by the process of endosymbiotic events between the cyanobacterial cell and the primitive eukaryotic phagocytotic protozoa. He was with the opinion, as he observed many characters of unicellular photosynthesizing motile algae similar to that of heterotrophic protozoa like, cell structure, their movement, flagellar structure. On the other hand, chloroplast also has its own DNA and self-replication process. But at that time this hypothesis was not accepted. Later on, with the advent of electron microscope and other sophisticated instruments, some evidences, especially the peptidoglycan wall surrounding the glaucophycean chloroplast, proved the endosymbiotic events in chloroplast evolution. The hypothesis was accepted almost after 75 years of Mereschkowski's original hypothesis, and Lee (1980) first accepted this hypothesis in his system of classification.

Lee (1980) also adopted the endosymbiotic theory and proposed that chloroplast evolution took place in three lines of evolution.

In the first line of evolution for chloroplast development, a prokaryotic algal cell was captured in a food vesicle by a phagocytotic non-photosynthetic protozoan. The protozoans instead of digesting the algal cells started it maintaining as endosymbiont. The endosymbionts are used to get shelter from the host, whereas the host is used to get the supply of photosynthetic products from the endosymbiont. In the course of evolution, the endosymbiont turned into the photosynthetic cell organelle or chloroplast – modifying the photosynthetic protozoa to eukaryotic algal cell. Eventually, in the process of evolution, the plasma membrane of the endosymbiont became the inner membrane of the chloroplast and the food vesicle membrane of the host became the outer membrane of the chloroplast envelope (chloroplast of Glaucophyta represents the intermediate stage). This process is termed as primary endosymbiosis.

In the second line of evolution, a chloroplast of a eukaryotic alga was taken up into a food vesicle by a phagocytotic protozoa. Eventually, the food vesicle membrane of the host became the single membrane of chloroplast endoplasmic reticulum surrounding the two chloroplastic envelope and the process is known as secondary endosymbiosis, where the two-membrane chloroplast evolved by primary endosymbiosis and the three-membrane algal chloroplast by secondary events (Group III of Lee's classification).

In the third evolutionary line, the phagocytotic protozoan took up a red alga into a food vesicle. The nucleus of red alga reduced to a nucleomorph. This protozoan along with its algal symbiont was taken up by a second phagocytotic protozoan into a food vesicle. The nucleus of the first protozoan took over the functioning of the cellular apparatus, and the nucleus of the second protozoan was lost. Also the food vesicle membrane of the second protozoa and the outer nuclear envelope of the first protozoa were also lost. Ultimately the four-membrane chloroplast evolved. The outer membrane of the chloroplast endoplasmic reticulum was the plasmalemma of the first protozoa and the inner membrane of chloroplastic endoplasmic reticulum was derived from the food vesicle membrane of the first protozoan.

Therefore, Lee (1980) first brought the revolutionary changes in algal classification and proposed four distinct evolutionary groups within algae on the basis of cell structure (prokaryotes and eukaryotes) and chloroplast ultrastructure, and it was accepted by all. Lee's classification was further revised on 1989, 1999 and 2008. According to this classification there are four distinct evolutionary groups.

According to Lee's classification the groupings are mainly done on the basis of evolutionary pattern of algal chloroplast and flagellar ultrastructure and mainly divided into four groups as follows:

The members of the first group are prokaryotic in nature and lack membrane-bound chloroplast having only free thylakoids embedded in cytoplasm. All the cyanobacteria are included in this group.

In the second group, the chloroplasts with basic structure, i.e. double membrane-bound (chloroplast envelope) organelle enclosing the ground substance stroma and the membrane-bound saclike structure thylakoids embedded in it, are present in three divisions of algae, viz. Glaucophyta, Rhodophyta and Chlorophyta (group II). These algae evolved through primary endosymbiosis.

Among these three algal divisions, in chloroplast of Glaucophyta, thylakoids are arranged equidistantly at peripheral region of chloroplast (like cyanobacterial cell). There are similarities between chloroplast of Glaucophyta and cyanobacterial cell as follows:
- They are about the same size.
- They evolve oxygen in photosynthesis.
- They have 70S ribosomes.
- They have circular prokaryotic DNA without basic proteins.
- They have peptidoglycan wall surrounding the chloroplast envelope.

For this reason chloroplast of Glaucophyta is termed as 'cyanelles' or the incipient chloroplast.

The chloroplast of Rhodophyta (rhodoplast) is also surrounded by two chloroplastic membranes with no chloroplast ER. Inside the chloroplast, thylakoids occur singly and the DNA molecule occurs as microfibrils and the phycobilin pigments are localized into phycobilisomes on the surface of the thylakoids (similar to Cyanophyceae). Chl a and d are present inside the thylakoids. Among the carotenoids zeaxanthin is found in the greatest quantities. Phycobiliproteins include R-phycocyanin, allophycocyanin and three forms of phycoerythrin in maximum amounts.

In Chlorophyta, chloroplasts are highly evolved like higher-plant chloroplast, with two membranes of chloroplast envelope; thylakoids are stacked to form grana which are embedded in the matrix called stroma. Chloroplast pigments are also similar to higher plants. Chlorophylls a and b are present and the main carotenoid is lutein. Extraplastidic carotenoids are sometimes present (haematochrome).

The chloroplastic DNA is partially looped and ribosome is of 70S type. Starch is formed in the chloroplast in association with pyrenoid. The starch is similar to that of higher plant and is composed of amylose and amylopectin. A pyrenoid is a differentiated region within the chloroplast that is composed of polypeptides with enzymatic properties of ribulose-1,5-bisphosphate carboxylase that are capable of fixing carbon dioxide. Storage products are frequently associated with pyrenoids. The pyrenoid is denser than the surrounding stroma and may or may not be traversed by thylakoids.

The chloroplast of group III algal divisions, viz. Euglenophyta, Dinophyta and Apicomplexa, is surrounded by two membranes of the chloroplast envelope and one membrane of chloroplast endoplasmic reticulum, which is not continuous with nuclear membrane. The thylakoids are grouped in bands of three. Major pigments present are Chl a and b (Euglenophyta); Chl a and $c2$ with peridinin as major xanthophylls (Dinophyta); and rudimentary chloroplast (Apicomplexa).

In addition, some accessory chloroplasts are present in Dinophyta (originated from further endosymbiotic process between dinoflagellate cell and Cyanophyta or Bacillariophyta or Rhodophyta) giving characteristic colour.

The membranes of the fourth evolutionary group Heterokontophyta contain chloroplast having four membranes (two chloroplast envelopes and two chloroplastic endoplasmic reticula).

The chloroplast of Phaeophyta is known as phaeoplast. Phaeoplast is a four-membrane-bound structure having three thylakoids per band. Membrane-bound structures are also present between chloroplast envelope and chloroplast ER. The chloroplast contains Chl a, $c1$ and $c2$ with major carotenoid, fucoxanthin. Pyrenoids of Phaeophyceae are stalklike structures which set off from the main body of chloroplast containing granular substances not traversed by thylakoids. Surrounding the pyrenoid is a saclike structure containing the reserve food matters, generally present (laminarin as major component).

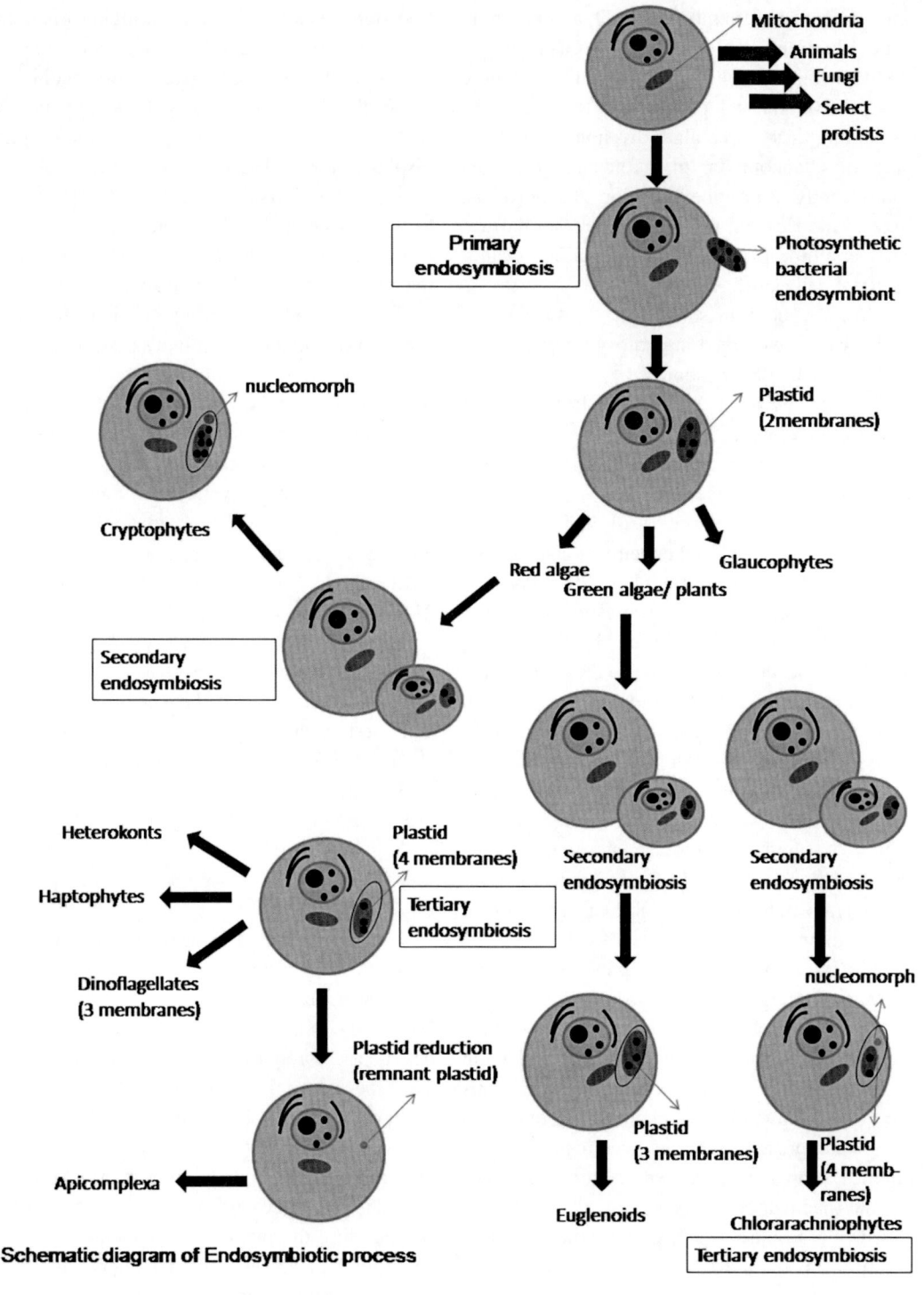

Schematic diagram of Endosymbiotic process

1.3.2.2 Outline of Lee's (2008) Classification with Basic Characters of Different Groups

In this system of algal classification, the entire algal kingdom is classified into four distinct groups:

I. Prokaryotes
II. Eukaryotic algae with chloroplast with two membranes
III. Eukaryotic algae with chloroplast with one chloroplast ER (total three membranes)
IV. Eukaryotic algae with chloroplast with two chloroplast ER (total four membranes)

Group I:
 1. Cyanobacteria

Group II:
 1. Glaucophyta
 2. Rhodophyta
 3. Chlorophyta

Group III:
 1. Euglenophyta
 2. Dinophyta
 3. Apicomplexa

Group IV:
 1. Cryptophyta
 2. Heterokontophyta
 3. Prymnesiophyta

Cyanobacteria: Prokaryotic members containing chlorophyll *a* and phycobiliproteins. Some members contain chlorophyll *b* and are known as green cyanobacteria (Prochlorophyta). Cell wall is similar to Gram-negative bacteria containing peptidoglycan layer outside the cell membrane. Naked circular DNA present at the central portion of the cell. Cyanophycin, the polymer of arginine and aspartic acid; carboxysome; polyphosphate bodies; and polyglucan granules are the other cellular inclusions. Gas vacuoles, containing a large number of gas vesicles, are present.

Glaucophyta: Members unicellular with primitive type of chloroplast, showing many characters similar to cyanobacterial ancestor and are known as cyanelle. Pigments are similar to cyanobacteria. Members are unicellular flagellates.

Rhodophyta: Both marine and freshwater in habitat. Unicellular to huge thallus, especially for marine seaweeds. Chloroplast with one thylakoid per band and no chloroplast ER, floridean starch grain synthesized in the cytoplasm, no flagella, pit connections between the cells. Reproductive unit spermatia and carpogonia. Post fertilization change prominent.

Single class – Rhodophyceae

Chlorophyta: Members have chlorophylls *a* and *b* and starch is the reserve food matter formed within the chloroplast. Chloroplast with two membranes only, thylakoids form the grana. Both freshwater and marine in habitat with wide range in morphology. Flagella isokont type with fine hairs if present. Flagellar root cruciate type or unilateral type. Eyespot present.

Four classes: Prasinophyceae, Charophyceae, Ulvophyceae and Chlorophyceae

Euglenophyta: Euglenoid flagellate, unicellular, surrounded by pellicle. Chloroplast with two chloroplastic membranes and one chloroplastic endoplasmic reticulum containing Chl *a* and *b*. Marine or freshwater in habitat. Cytosome or gullet-like structures are present at the anterior region of the cell.

Single class: Euglenophyceae

Dinophyta: Unicellular with two halves, epicone and hypocone, made up of thecal plates. Chloroplast with two chloroplastic membranes and one chloroplastic endoplasmic reticulum with Chl *a* and *c*2 and the carotenoids peridinin and neoperidinin as major pigments. Storage product is starch. Two flagella – one longitudinal and one transverse.

Single class: Dinophyceae

Apicomplexa

Unicellular having reduced colourless plastid called *apicoplast*. Apicoplast and dinoflagellate plastids originated from red algae by a single endosymbiotic event. The apical complex consists of a *polar ring* and a *conoid* formed of spirally coiled microtubules. Apicomplexa are *endoparasites* that cause some of the most significant tropical diseases like malaria or diarrhoea. The parasite attaches to the host cell with the conoid protruding to produce a *stylet* that forms a tight junction with the host cell. The apicomplexan cell is taken up into the host cell in the *parasitophorous* vacuole.

Cryptophyta

Cells with chlorophylls a and c, phycobiliproteins and nucleomorph present between inner and outer membranes of chloroplast ER starch grains stored between inner membrane of chloroplast ER and chloroplast envelope, periplast inside plasma membrane and tripartite hairs on flagella.

Single class: Cryptophyceae

Heterokontophyta

Cells are with tripartite hairs; anterior tinsel flagellum and posterior whiplash flagellum. Pigments are chlorophylls a and c and fucoxanthin. Storage product usually chrysolaminarin present in vesicles of cytoplasm.

These divisions include 12 classes:
1. Chrysophyceae
2. Synurophyceae
3. Eustigmatophyceae
4. Pinguiophyceae
5. Dictyochophyceae
6. Pelagophyceae
7. Bolidophyceae
8. Bacillariophyceae
9. Raphidophyceae
10. Xanthophyceae
11. Phaeothamniophyceae
12. Phaeophyceae

Prymnesiophyta (Haptophytes)

Two whiplash flagella, haptonema present, chlorophylls a and c, fucoxanthin, scales common outside cell, storage product usually chrysolaminarin in vesicles in cytoplasm.

Single class: Prymnesiophyceae

McFadden (2001) introduced another division of algae in his classification, where the members possess four-membrane chloroplast with Chl a and b as the major pigments.

1.3.2.3 Other divisions of Algae

Chloroplast derived from a green alga, chlorophylls a and b present, nucleomorph between inner and outer membrane of chloroplast ER. Chloroplast with four membranes – two chloroplastic membranes and two chloroplastic endoplasmic reticula. Vegetative cells are naked, uninucleate with amoeboid projections. An example is *Chlorarachnion*.

Bremer (1985) combined charophytes and embryophytes into a single division *Streptophyta*, based on ultrastructure and molecular characters as they share many similar characters and also consider charophytes as progenitor of land plants.

Cavalier (1981) proposed the term *Viridiplantae* to comprise the true green plants including green algae and higher plants as an arguably monophyletic group based on ultrastructure of flagella (stellate structure in flagellar transition region) and plastid characters (Chl a, Chl b).

Members of phytoplankton belonging to different groups of algae and cyanobacteria contain different types of pigments in different compositions, which are important in algal taxonomy. These pigments play an important role in oceanographic research as they act as tracers to elucidate the composition and fate of phytoplankton in the world's ocean and are also associated with important biogeochemical cycles like carbon dynamics. Pigments of phytoplanktons often change the colour of the water indicating the biomass growth rate, which are also used in their estimation in situ *and in remote-sensing application*. By analysing these pigments using HPLC or HPTLC, presence or absence of picoplanktons can be determined. These techniques have allowed identification of many new pigments. Therefore, pigment analysis of phytoplanktons is very important in oceanographic research especially for chemotaxonomic purpose. From 1997 onwards HPLC-linked mass spectrometry is used for marine plankton analysis.

Phytoplankton biomass of the ocean is generally determined by measuring the ocean colour from space; therefore, in situ pigment measurements have become high-priority areas for oceanographic research (Jeffrey et al. 1997a; Nair et al. 2008).

In recent years to develop the map of microalgal population, the major pigments of a particular group are considered for field study. Many new algal genera have been identified on the basis of pigment composition (e.g. members of Herptophyton) (Zapata et al. 2004). They recovered many new members of chlorophyll c and fucoxanthin families from different phytoplankton genera. Jeffrey et al. (1997b) published a very important book, *Phytoplankton Pigments in Oceanography: Guidelines to Modern Methods* (SOR-UNESCO volume). They listed 12 microalgal classes as common members of phytoplanktons, viz. diatoms, dinoflagellates, haptophytes, chrysophytes, rhodophytes, raphidophytes, cryptomonads, chlorophytes, euglenophytes, eustigmatophytes, cyanobacteria and prochlorophytes.

The use of advanced HPLC and ultrahigh-performance liquid chromatography (UPLC) helps to identify different phytoplankton taxa in addition to it considering culture code and gene bank references.

A new system of classification of eukaryotes (including only photosynthetic microalgal groups) from a protistan perspective has been proposed by Adl et al. (2005), which is mainly used in oceanographic literature. According to them the traditional 'kingdoms', such as Metazoa, Fungi and Plantae, are recognized as deriving from monophyletic protist lineages. The authors grouped the molecular phytogenies into six clusters as follows:

1. Opisthokonta: animals, fungi, choanoflagellates and Mesomycetozoea
2. Amoebozoa: traditional amoebae, slime moulds, etc.
3. Rhizaria: foraminifera, radiolarian, heterotrophic flagellates, etc.
4. Archaeplastida: red algae, green algae, Glaucophyta and Plantae

Classification scheme of Adl et al. (2005)

Super groups	First rank	Second rank (examples of photosynthetic eukaryotes)
Rhizaria	Cercozoa	Chlorarachniophyta, Paulinella
Archaeplastida	Glaucophyta	Glaucophyceae
	Rhodophyceae	Subdivisions uncertain according to Adl et al. (2005)
	Chloroplastida	Charophyta[a], Chlorophyta, Mesostigma, Prasinophyta
Chromalveolata	Cryptophyceae	Cryptomonadales
	Haptophyta	Pavlovophyceae, Prymnesiophyceae
	Stramenopiles	Bacillariophyta, Bolidomonas, Chrysophyceae, Dictyochophyceae, Eustigmatales, Pelagophyceae, Phaeophyceae[a], Phaeothamniophyceae, Pinguiochrysidales, Raphidophyceae, Synurales, Xanthophyceae
	Alveolata	Apicomplexa, Dinozoa
Excavata	Euglenozoa	Euglenida

[a]Clades with multicellular groups

1.4 Classification of Phytoplanktons

1.4.1 On the Basis of Size Variations

Phytoplanktons show a wide range of size variations. The earliest work on phytoplankton classification on the basis of size variations was proposed by Schütt (1892). During the 1950s, development of different types of nets, filters and screens with different pore sizes leads to identification of micro-, nano- and ultraplanktons. Picoplanktons are the smallest in cell size with a diameter of 0.2–2.0 µm, followed by nanoplankton (2.0–20 µm), microplankton (20–200 µm), mesoplankton (0.2–2.0 mm) and macroplankton (2–20 mm).

Interestingly, in earlier studies picoplanktons were referred only to 'heterotrophic picoplanktons' that are presently known as 'bacterioplankton' (Sieburth 1978). In later periods, chroococcoid cyanobacteria were observed in oceanic samples along with other eukaryotic taxa that are presently referred as 'picoplankton' (Johnson and Sieburth 1979; Li et al. 1983). Several other terms have also been introduced like 'ultrananoplankton' that referred to algae that were less than 2 µm (Dussart and Roger 1966) and 'ultraplankton'. The use of this term was mainly dependent on the discretion of the author and the size which ranged from 0.5 to 15 µm (Hutchinson 1967; Reynolds 1984). Some scientists used the term 'net plankton' for those planktons that were >45 µm in size (Throndsen 1978).

Picoplanktons have gained considerable interests among plankton biologists in the later part of the twentieth century. Several studies have indicated that in oligotrophic waters, picoplanktons are the primary population that constitutes 50–70 % of the total productivity of oceanic ecosystems (Caron et al. 1985). Thus, based upon their pigment composition and genetic diversity, picoplanktons have been categorized as 'prokaryotic picocyanobacteria' and 'eukaryotic phototrophs'. Unfortunately taxonomic identification of picoplanktons on the basis of morphological variations is very difficult to enumerate. Thus, pigment composition and analysis by molecular methods are increasingly been used as possible tools to understand picoplankton populations.

Table showing the classification of planktons on the basis of size variations

Group	Size range	Examples
Megaplankton	(20+ mm)	Metazoans, e.g. jellyfish, ctenophores, salps and pyrosomes (pelagic Tunicata), Cephalopoda
Macroplankton	(2–20 mm)	Copepods, amphipod, polychaete
Mesoplankton	(0.2–2 mm)	Protozoa, Foraminifera, Hydrozoa
Microplankton	(20–200 µm)	Dinoflagellates (e.g. *Dinophysis, Gymnodinium, Ceratium*), diatoms (e.g. *Biddulphia, Thalassiosira, Coscinodiscus*)
Nanoplankton	(2–20 µm)	Flagellates (*Distephanus, Thalassomonas, Tetraselmis*)
Picoplankton	(0.2–2 µm)	Cyanobacteria (*Synechococcus*)
Femtoplankton	(<0.2 µm)	Mostly viruses

1.4.2 On the Basis of Habitat

Phytoplanktons are widespread in their distribution in different aquatic habitats around the world. They live in freshwater, brackish and salt water, ice, moist soil and other damp places.

The freshwater lentic ecosystems like lakes and ponds can be divided into different zones (Fig. 1.1):

1. The *supralittoral zone* which is above the edge of the standing water that gets wet only due to wave actions and splash during windy seasons. Here, algal populations remain sparse due to the drying nature of the habitat along with other deterrents like the abrasive effects of sand and gravel.
2. The *littoral zone* which extends to about 6 m in depth from the water edge into the water body. This zone is a highly productive area in lotic systems in terms of productivity with dominance of periphytic algae like diatoms and desmids.
3. The *sublittoral zone* which extends down to the compensation point.
4. The *profundal zone* which occurs below the compensation depth where the phytoplankton population is mainly composed of colourless heterotrophic algal cells and resting spores of photosynthetic algae.

In lotic ecosystems like oceans and seas, there are two main divisions:

1. *Pelagic* species are those that live near the surface of the ocean during all or most part of their life cycle. The pelagic diatoms are further divided into *oceanic* and *neritic* species.
    (a) *Oceanic* species are those capable of living and reproducing entirely in the open ocean. Oceanic taxa are mostly reported from depth in excess of 200 m. Thus, the oceanic province can be subdivided into two distinct zones:

1.4 Classification of Phytoplanktons

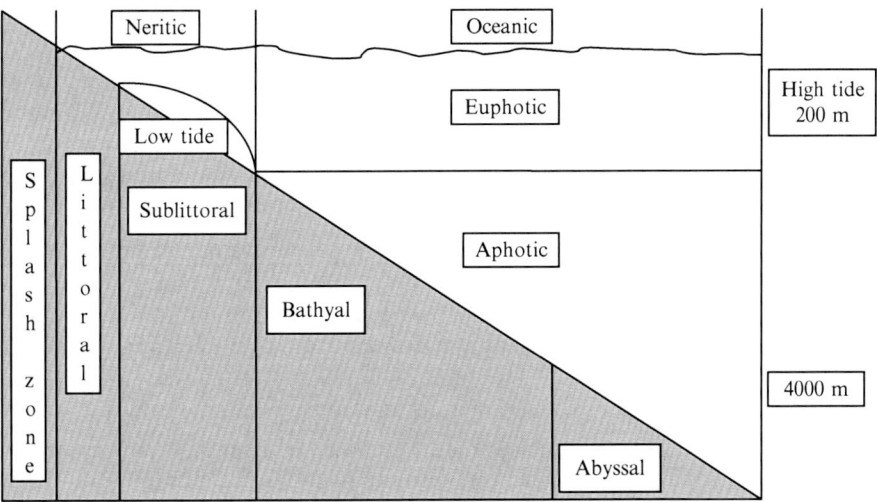

Fig. 1.1 Zonation of marine ecosystem

(i) *Euphotic* or *epipelagic zone* which extends from the surface to a depth of 200 m. Photosynthetic activities occur mainly in the upper portion of this zone.

(ii) *Aphotic zone* which extends from a depth of 200 m. Here light is insufficient to support planktonic photosynthetic activity.

(b) *Neritic* species are those which have their origin near the coast and reproduce most efficiently under coastal conditions. It is not easy to clearly distinct between neritic and oceanic species.

2. *Benthic* environment that represents the ocean bottom. This environment can be further categorized as:

(a) *Littoral* region which extends from high tide to low tide zone

(b) *Deep-sea* region which includes the entire ocean beneath the low tide mark. This region can be further subdivided into three distinct zones:

(i) *Sublittoral* zone which extends from the low tide mark to a depth of 200 m. This zone mainly consists of large seaweeds belonging to Rhodophyta and Phaeophyta. The availability of these seaweeds at different depth depends on the turbidity and light penetration along the water column.

(ii) *Bathyal* zone which extends from a depth of 200–4,000 m, corresponding to the continental slope, the geomorphic province beyond the continental shelf.

(iii) *Abyssal* zone which represents depths beyond 4,000 m.

Both oceanic and neritic species may be further divided into three main groups according to the latitude in which they are most commonly found or have had their origin. Thus, we speak of arctic, temperate and tropical oceanic species and arctic, temperate and tropical neritic species. Subdividing these again, we have boreal (northern but not arctic species), north temperate, south temperate and subtropical species. Some species are ubiquitous, others restricted within rather definite regions. There are a number of species which it is still impossible to place in any definite group. The boundaries of the regions are variable and the flora of each locality is continually changing with seasonal changes and movements of the water.

Most of the coastal or neritic species have a special adaptation, the resting spores, which serve as protection against changing conditions (e.g. diatoms). The spores sink into deeper water and may be found there for several months after the species has disappeared from the surface. The majority remains on the bottom in shallow coastal water until conditions favour their germination (Gross 1937). Resting spores must be the means by which many species continue in coastal waters in spite of the fact that conditions are more variable there than in the open ocean and may be favourable to diatoms for only a limited part of each year (members of Biddulphiales).

1.5 Algal Pigments

Algae are photosynthetic organisms and they possess chlorophyll in their chloroplasts. The primary photosynthetic pigment of algae is chlorophyll and is the light receptor in photosystem I of light reaction. There are different types of chlorophyll like Chl *a*, *b*, *c* (*c*1, *c*2), *d* and *e* present in algal cells. Among them Chl *a* is universal and present in all members of autotrophic algae (with a few exception of heterotrophic algae). Due to presence of Chl *a*, members of cyanobacteria are also considered as prokaryotic algae by algologist. Chlorophyll a is soluble in alcohol, diethyl ether, benzene and acetone but insoluble in water. Chlorophyll is composed of a porphyrin ring system that is very similar to that of haemoglobin but has a magnesium atom at the centre instead of an iron atom. Chlorophyll content ranges from 0.3 % of the dry weight among different algal genera of different classes. Chlorophyll has two main absorption bands in the red light region at 663 nm and the other at 430 nm.

Unlike chlorophyll *a*, other types of chlorophylls have a more limited distribution and function as accessory photosynthetic pigments. Chlorophyll *b* is found in the Euglenophyta, Chlorophyta and Chlorarachniophyta. Chlorophyll *b* functions as a light-harvesting pigment transferring absorbed light energy to chlorophyll *a*. Chlorophyll *b* has two main absorption maxima in acetone or methanol, one at 645 nm and the other at 435 nm.

Chlorophyll *c* has two components which are spectrally different, viz. chlorophyll *c*1 and *c*2. The ratio of chlorophyll *a* to chlorophyll *c* ranges from 1.2:2 to 5:1. Chlorophyll *c* probably functions as an accessory pigment to photosystem II. Chl *c* is soluble in ether, acetone, methanol and ethyl acetate but is insoluble in water and petroleum ether. Chlorophyll *c*1 has main absorption maxima at 634, 583 and 440 nm in methanol, whereas chlorophyll *c*2 has maxima at 635, 586 and 452 nm.

Chlorophyll *d* is a minor component present in many members of Rhodophyta. It is soluble in ether, acetone, alcohol and benzene and very slightly soluble in petroleum ether showing three main absorption bands at 696, 456 and 400 nm.

Carotenoids are yellow, orange or red hydrocarbons present in algal members as accessory pigments and usually occur inside the plastid but may be outside in certain cases. Carotenoids can be divided into two classes: (1) oxygen-free hydrocarbons, the carotenoids, and (2) their oxygenated derivatives, the xanthophylls. The most common algal carotene is the β-carotene.

1.5 Algal Pigments

Chla

Chlb

Chld

Chl c_1

Chl c_2

Chemical configuration of algal pigments

β-carotene

Phycoerythrobilin   Phycocyanobilin

Chemical configuration of few major pigments of algae

There are a large number of different xanthophylls present in different groups of algae. Fucoxanthin is the principal xanthophylls in the golden-brown algae (Chrysophyta, Bacillariophyta, Prymnesiophyta and Phaeophyta), giving these algae their characteristic colour. Like the chlorophyll, the carotenoids are also fat-soluble pigment being soluble in alcohols, benzene and acetone but insoluble in water.

Among all the algal pigments, phycobiliproteins – present in Cyanobacteria, Rhodophyta and Cryptophyta – are water-soluble blue or red pigments. They are located on (Cyanophyta, Rhodophyta) or inside (Cryptophyta) the thylakoids of algal chloroplasts. They are associated with macromolecular protein and are known as chromoproteins (coloured proteins). They have the prosthetic group (nonprotein part of the molecule) or chromophore. Chromophore is a tetrapyrrole (bile pigment) known as phycobilin and is tightly bound by covalent linkages to its apoprotein (protein part of the molecule). As the pigment is tightly bound to the apoprotein, the term phycobiliprotein is used. The apoproteins are again of two types, α and β, which together form the basic unit of the phycobiliproteins. The major 'blue' chromophore is called the phycocyanobilin and is present in phycocyanin and allophycocyanin, and the major 'red' chromophore phycoerythrobilin is present dominantly in phycoerythrin.

1.5 Algal Pigments

Each phycobiliprotein usually consists of a basic aggregate of three molecules of α- apoprotein and three molecules of β-apoprotein and the chromophores attached to the apoproteins. The basic aggregate is designated as (αβ)$_3$. In some cases the basic aggregate is double of this along with a linker polypeptide and can be designated as (αβ)$_6$. Different linker polypeptides interact with the same phycobiliprotein to give complexes with physical properties determined by the linker polypeptides (Yu et al. 1981). A third apoprotein, γ, is present in B- and R-phycoerythrins. The core of the phycobilisome is probably composed of allophycocyanin, with the peripheral rods containing phycocyanin and phycoerythrin. The length of the peripheral rods varies, being dependent on the wavelength of light under which the cells are grown. This phenomenon is a form of chromatic adaptation (Bryant 1994).

Two classes of phycobilisomes exist: (1) the hemi-ellipsoidal phycobilisomes and the hemi-discoidal phycobilisomes in the red algae, blue-green algae and the cyanelles (endosymbiotic blue-green algae).

It is believed that the light energy transfers in the following pathway during photosynthesis:

Phycoerythrin → phycocyanin → allophycocyanin → allophycocyanin B → chlorophyll a
(λ max = 565) (λ max = 620 – 638) (λ max 650) (λ max = 670)
or
Phycoerythrocyanin
(λ max 568)
(Glazer et al. 1985)

The efficiency of energy transfer from the phycobilisome to chlorophyll a in the thylakoids exceeds more than 90 % in intact cells (Porter et al. 1978).

Table showing different pigment compositions of various groups of algae with thallus organization

Division/class	Thallus organization	Pigments
Cyanobacteria	Members of Cyanobacteria are in the form of unicells, or colonies, up to 2 mm in size. Picoplanktonic forms are 1–2 μm in diameter; filamentous forms are unbranched, pseudobranched or branched	Chl a, Chl b (in Prochlorophyta, now called green cyanobacteria) and traces of MgDVP[a]
		Phycobilins: phycocyanin, phycoerythrin, allophycocyanin.
		Carotenoids: β-carotene, myxoxanthophyll, zeaxanthin, echinenone, canthaxanthin, oscillaxanthin, nostoxanthin, aphanizophyll, 4-keto-myxoxanthophyll. Jeffrey and Wright (2006) categorized five different types of pigment composition, cyano-1–5
Glaucophyta	Biflagellated unicells, oblong or coccoid, dorsiventral, 10–30 μm diameter; also in palmelloid stage	Chl a
		Carotenoids: β-carotene, zeaxanthin, β-cryptoxanthin
		Biliproteins: phycocyanin, allophycocyanin
Rhodophyta	Coccoid unicells, filamentous, simple or polysiphonous, thalloid leaflike	Chl a
		Carotenoids: β-carotene, zeaxanthin;
		Biliproteins: phycoerythrin, phycocyanin (less amount, allophycocyanin)
Chlorophyta	Small green flagellates, naked, or ovoid in form, 10–40 mm in diameter, unicellular, coccoid, colonial, filamentous – simple uniseriate to branched form	Chlorophylls a and b, MgDVP; Carotenoids: lutein, violaxanthin, neoxanthin, antheraxanthin, β-carotene, β,ε-carotene, zeaxanthin, β,ψ-carotene (trace), astaxanthin and two pigment types are recognized: CHLORO-1 and CHLORO-2 (Jeffrey and Egeland 2008)

(continued)

(continued)

Division/class	Thallus organization	Pigments
Euglenophyta	Unicellular, ovoid or fusiform; most species have a flexible pellicle, which allows movement by deformation; other species have a rigid lorica	Chl a and b, MgDVP; Carotenoids: eutreptiellanone, diadinoxanthin, diatoxanthin, 9′-cis-neoxanthin, β,β-carotene, β,ε-carotene (Jeffrey and Wright 1997)
Dinophyta	Mostly unicellular (5–200 μm), with a transverse girdle groove into an upper epicingulum and a lower hypocingulum; unarmoured or armoured with cellulose plates	Chlorophylls: Chl a, $c2$, MgDVP; Carotenoids: peridinin, diadinoxanthin, diatoxanthin, dinoxanthin, peridininol, pyrrhoxanthin and β,β-carotene
		Five combinations of pigments: the major dinoflagellate carotenoid peridinin containing DINO-1 or with their endosymbiont pigments, e.g. haptophytes (DINO-2), diatoms (DINO-3), cryptophytes (DINO-4) or prasinophytes (DINO-5) (Jeffrey and Wright 2006)
Heterokontophyta (Bacillariophyceae – diatoms)	Unicellular or colonial forms. Cells are covered by characteristic siliceous frustules with two overlapping halves.	Chlorophylls: Three combinations of pigment are distinguished on the basis of Chl c derivatives (Zeffry et al. 2011)
		DIATOM-1: Chl a, Chl $c1$, Chl $c2$, MgDVP
	Morphology of diatoms is based on radial (centric) or bilateral symmetry (pennate)	DIATOM-2: Chl a, Chl $c2$, Chl $c3$, MgDVP
		DIATOM-3: Chl a, Chl $c1$, Chl $c2$ $c2$, Chl $c3$, MgDVP(trace)
		Carotenoids: fucoxanthin, diadinoxanthin, diatoxanthin, β,β-carotene, 19′-butanoyloxyfucoxanthin, violaxanthin, antheraxanthin, zeaxanthin
Bolidophyceae	Picoplanktonic cells: round or heart shaped	Chl a, $c3$
Chrysophyceae	Colonial with coccoid or ovoid cell, flagellated or amoeboid	Chl a, $c1$, $c2$
		Carotenoids: β,β-carotene, fucoxanthin, violaxanthin, antheraxanthin, zeaxanthin
Dictyochophyceae	Picoplanktonic or larger, unicellular, naked or inside a siliceous skeleton	Chl a, $c2$, $c3$
		Carotenoid: β-carotene, fucoxanthin, diatoxanthin, diadinoxanthin
Eustigmatophyceae	Unicellular, coccoid or ovoid, flagellated	Chl a, MgDVP
		Carotenoids: β,β-carotene, antheraxanthin, vaucheriaxanthin, violaxanthin, zeaxanthin
Pelagophyceae	Unicellular or colonial. Cells coccoid or ovoid, filamentous, palmelloid	Chl a, $c2$
		Carotenoids: diadinoxanthin, diatoxanthin, fucoxanthin, beta-carotene, gyroxanthin diester
Haptophyta	Unicellular, 5–20 μm in diameter; almost exclusively flagellates, cells elongated	HAPTO-1
		Chlorophylls: Chl a, $c1$, $c2$, MgDVP
		Carotenoids: fucoxanthin, diadinoxanthin, diatoxanthin, β-carotene
		HAPTO-2
		Chlorophylls: Chl a, $c1$, $c2$, MgDVP
		Carotenoids: fucoxanthin, diadinoxanthin, diatoxanthin, β-carotene
Cryptophyta	Ovoid asymmetrical unicells (6–20 μm), often flattened A unique cell covering or pellicle is made up of a ridged periplast superimposed on an inner layer of thin proteinaceous plates; no microtubular cytoskeleton	Chlorophylls: Chl a and $c2$, MgDVP
		Carotenoids: alloxanthin, crocoxanthin, monadoxanthin, β,ε-carotene lipoproteins, red or blue phycobiliproteins

[a]MgDVP – Mg-3,8-divinyl-pheoporphyrin a5 monomethyl ester

1.6 Ecological Significance

Phytoplanktons play an important role in aquatic ecosystems, both in freshwater and in marine environment. They are the primary producer organisms, therefore, supporting zooplanktons, fishes and other members of aquatic fauna. Thus, they are placed at the base of the trophic strata or at the bottom of the aquatic food web. Phytoplanktons also play a major role in global carbon dioxide fixation. In marine environment they fix almost 48 Pg C. year^{-1} [1 Pg = 1 × 10^{15} g], which is almost 48 % of the total fixed carbon on the earth's surface (Geider et al. 1997). Phytoplanktons also maintain the oxygen level of the water body, which is designated as dissolved oxygen or DO.

The phytoplankton population, controlling the life cycle of each species, is again controlled by several factors, like the availability of nutrients, degree of thermal stratification, algal movements relative to the water current, zooplankton grazing, intra-algal competition and parasitism by protozoans, fungi, bacteria or viruses. It is observed in many places that the chlorophyte, chrysophyte, cryptophyte and euglenophyte algae often dominate and form in the summer peak due to an ability to take up nutrients at low level and to maintain their positions by swimming. In some lakes, they also exist below the thermocline and at greater depths.

Different genera of phytoplanktons evolved various strategy to overcome the nutrient depletion and grazing. They generally produce different types of enzymes, some of which are directly involved in nutrient uptake and others responsible to secrete some chelators or siderophores which form complex with the nutrients and make them available for uptake. Due to diatom bloom the nutrient level becomes depleted especially silica (SiO_2) – required for diatom frustule growth. Phytoplanktons can move directly or float through water current. Movement by swimming or a change in cell density may allow them to reach new sources of nutrients actively or passively. In adverse environmental conditions, some algae produce resting spores, remain active at the junction of epilimnion and hypolimnion region of thermocline. To protect themselves from grazing also, many species produce protective spores, gelatinous coats, or grow fast to produce a large population. Grazing rate of zooplanktons may be reduced due to thick growth of algae or sometimes due to production of unpalatable species such as blue-green algae.

1.6.1 Phytoplankton Bloom

A rapid increase or accumulation in the population of algae (typically microscopic) in an aquatic system is regarded as an 'algal bloom'. Many of the planktonic genera form the blooms.

Phytoplankton population generally grows in a series of pulses or blooms. Blooms occur when cell numbers exceed their annual average or background concentration manifolds or when a certain high cell number is reached, for example, 5×10^6 cells/L. A bloom can colour the water.

In temperate region, the first bloom is initiated in spring due to increased sunlight; subsequently, the growth is terminated in autumn when light decreases in water. On the other hand, in tropical regions growth may be nearly continuous when sufficient nutrients are available. In polar region only a short period of growth can be observed mainly of diatoms as the sunlight and ice-free periods are very brief.

In productive lakes of temperate regions, holoplanktons, i.e. the algae which always remain in planktonic form, flourish more during spring time, like the diatom genera – *Asterionella, Fragilaria* and *Tabellaria* – forming the spring blooms. Excessive growth of chytrid fungi, zooplankton grazing or protozoans interference may affect the algal bloom.

Meroplanktonic genus *Melosira,* which is only sometimes in planktonic form, appears in large number in winter. Actually due to resuspension of live cells of *Melosira*, i.e. germination of resting spores from the sediments in favourable season, resulted in changes in population pattern and the bloom appeared. *Melosira* has a slow growth rate, but they are quite able to take the advantage of the high nutrient levels and benefits from low levels of competition and grazing.

The other meroplanktonic cyanobacterial taxa like *Aphanizomenon*, *Anabaena* and *Microcystis* flourish more in warm lakes in summer and falls, but the diatoms grow at a faster rate in winter of a tropical country. They produce the resting spore or thick-walled cell during winter to withstand the unfavourable season. Moreover, cyanobacteria can regulate their buoyancy by their minute gas vesicles and are able to adopt themselves in thermocline of different seasons at different depths. Many cyanobacteria and eukaryotic algae have an ability to take up nutrients like phosphate and ammonia at low levels. A few of them can also fix atmospheric nitrogen gas. All other algal groups as well as many other cyanobacteria lack this ability. Therefore, cyanobacterial blooms appear in comparatively low nutrient level even at nitrogen-depleted condition. Among the algal divisions cyanobacteria and dinoflagellate members produce bloom very frequently in different ecological conditions and are also ecologically significant.

1.6.2 Cyanobacterial Bloom

Cyanobacteria grow profusely, congregate and make the lake water coloured soup. This phenomenon is called 'cyanobacterial bloom'. Due to huge growth of buoyant cyanobacteria, the lake and ocean water turned coloured and the transparent water suddenly became soupy in appearance often turned into bright blue, grey, tan or even red in colour.

Generally chlorophyll production of bloom is 10 mg μm^{-3} (ca 20,000 cells mL^{-1}). Bloom formation is mainly the function of the environment; both freshwater ecosystem and the oceanic surface are suitable for cyanobacterial bloom formation.

Marine planktonic cyanobacteria control the biogeochemical cycle and the trophic status of marine environment. They control the marine production and nutrient cycling in two main ways – firstly they fix CO_2 to a greater extent at blooming condition, therefore controlling the C-cycling, and secondly some of them can fix atmospheric N_2; as a result the level of soluble bioavailable nitrogen in ocean water is increased. Nitrogen-fixing cyanobacteria generally form the surface bloom and supply the nitrogen to N-depleted ocean water. Ocean water is generally unproductive and can be called as oligotrophic in nature. In the early times, it was believed that marine cyanobacteria are very few in numbers. But since the 1970s with the advent of high-resolution fluorescence microscopy, electron microscopy, HPTLC, etc., it is proved that the oceanic primary productivity is mainly maintained by pico (≤ 5 μm)- and nanoplanktonic (5–20 μm) cyanobacteria together with comparatively less microplanktonic (≥ 20 μm) cyanobacterial members. Marine planktonic cyanobacteria are represented by unicellular (*Synechococcus*, *Prochlorococcus*, *Synechocystis*, *Aphanothece*), colonial (*Merismopedia*), filamentous non-heterocystous (*Lyngbya*, *Oscillatoria*, *Phormidium*, *Spirulina*, *Trichodesmium*) and heterocystous taxa (*Anabaena*, *Aphanizomenon*, *Nodularia*, *Richelia*). Picoplanktonic genera of *Synechococcus*, *Prochlorococcus* and *Synechocystis* constitute 30–50 % of phytoplankton biomass and therefore control the primary productivity by fixing CO_2. Cyanobacterial taxa grow in different ecological conditions forming bloom. The free-floating *Trichodesmium erythraeum*, the non-heterocystous taxa, forms a dense colony, ensheathed with polysaccharidic sheath, therefore producing an O_2-free microenvironment on sea surface and fixing atmospheric N_2. Another N_2-fixing filamentous genus *Richelia* grows inside the diatom *Rhizosolenia* and fixes atmospheric N_2 in ocean water. Heterocystous planktonic genera, *Nodularia*, *Aphanizomenon*, etc., grow in estuarine entropic environment and fix considerable amount of atmospheric nitrogen.

The cyanobacterial taxa having gas vesicles form the surface bloom like planktonic filamentous *Anabaena* and *Aphanizomenon* or colonial form *Microcystis*, etc. Cyanobacterial blooms form the surface scum at early morning, as the internal carbohydrate is generally consumed at night. During daytime when photosynthetic rate is increased and carbohydrate is accumulated within the cell, the gas vesicles collapsed due to the increased turgor pressure. As a result the organisms descend out of the surface scum and the bloom disappeared. Afterwards they reappear the next morning, when the stored carbohydrate is consumed. Due to some physiological damages like photoinhibition, photo-oxidation and dehydration, persistent surface blooms can appear which indicate the failure of buoyancy regulation. According to Pearl and Ustach (1982), the surface blooms are a part of an ecological strategy for optional use of photosynthetically active radiation and atmospheric carbon dioxide. Surface bloom also affects light penetration through the water column, not only by covering the water surface but also by scattering the light by the gas vesicles. Due to light scattering by cyanobacterial gas vesicles, the planktons of euphotic zone have an advantage of intercepting more light.

1.6.2.1 Cyanobacterial Bloom Control

Bloom formation can be controlled by nutrient concentration and/or controlling the light and temperature. Mixing is another process by which the turbulent environment of lakes or streams can be initiated which affect the natural phenomenon of bloom formation. The following are few practiced process by which bloom can be controlled:

1. By collapsing gas vesicles: To prevent buoyancy, if the cyanobacterial population is exposed to a hydrostatic pressure of 0.4 MPa, a significant proportion of gas vesicles would be collapsed; therefore, the bloom would disappear. Circulation of water containing surface scum of cyanobacteria up to a depth of 90 m can destroy the gas vesicles – therefore, the surface water would be free of cyanobacterial scum.
2. Artificial mixing: Cyanobacterial members prefer stable water column together with high nitrate and phosphate level for bloom formation. Artificial mixing of water column is the effective process for controlling cyanobacterial bloom. After the removal of the cyanobacteria, other members of diatoms or dinoflagellates may appear in blooming condition in turbulent environment with high nutrient levels.
3. Application of biological control: Inoculation of cyanophage (virus that infects cyanobacterial cell) for controlling cyanobacterial bloom is a common phenomenon.
4. Use of different algicides like copper sulphate for bloom controlling is a regular phenomenon for common people. But the toxic effects of chemicals also hamper the total ecological balance. Application of $CaCO_3$ in fish pond to control algal bloom is a common practice.
5. Light shielding: Light is the major factor for bloom formation. Light shielding is therefore an easy process for bloom control. Shielding can be done by covering with some dark sheet or by floating angiospermic plants like *Eischornia* and *Lemna*.

1.6.3 Dinoflagellate Bloom

Occasionally large populations of dinoflagellate, i.e. dinoflagellate blooms, appear on the surface of lakes, estuaries and ocean as red covering of the water body and are visible in the naked eye. Dinoflagellate blooms are usually known as 'red tide', e.g. bloom of *Peridinium* in alpine lake. Most of the members of dinoflagellates can quickly regulate their position by swimming. They are comparatively larger and can move fast and are phototactic in nature. They generally swim to the surface of water in the morning for photosynthesis and then swim down again in late afternoon.

Dinoflagellate members like *Peridinium* and *Ceratium* generally grow fast in summer and autumn and are quite able to swim actively to get the positions of favourable light and nutrients. The algae are positively phototactic and may form reddish-brown surface patches, called red tides. Nutrient requirements of dinoflagellate members are complex and may include organic substrate due to their heterotrophic nutrition. Dinoflagellate's population may decline due to heavy zooplankton grazing, competition from other algae and possibly nutrient depletion. The growth cycles of different algae actually depend on both physical and chemical factors and therefore is a complex phenomenon. Large colonies rise or sink faster than unicellular or filamentous form. Large size and unpalatability prevent serious loss to zooplankton grazing.

Recently, harmful algal blooms (HABs) have become a cause of concern due to their toxic nature mainly composed of dinoflagellate members and a few other algae including cyanobacteria. A *harmful algal bloom* (HAB) is an algal bloom that causes negative impacts to other organisms via production of natural toxins above the permissible level, mechanical damage to other organisms or by other means. HABs are often associated with large-scale marine mortality events and have been associated with various types of shellfish poisonings. Harmful algal blooms have been observed to cause adverse effects to varying species of marine mammals and sea turtles, with each presenting specific toxicity-induced reductions in developmental, immunological, neurological and reproductive capacities.

Dinoflagellates constitute approximately 50 % of all red tide species and 75 % of all HAB species (Sournia 1973; Smayda 1997); therefore, research related to dinoflagellate ecology is essential for understanding the red tide and HAB phenomena. The well-known 'Florida red tide' that occurs in the Gulf of Mexico is an HAB caused by *Karenia brevis* that produces brevetoxin, causing neurotoxic shellfish poisoning. Bloom of *Karenia brevis* is a recurrent phenomenon in the West Florida Shelf as well. This increase in the population mainly originates in the oligotrophic offshore water which is subsequently transported to the West Florida Shelf via winds and tidal currents (Steidinger and Haddad 1981). These thermal fonts not only act as a barrier but also as transport mechanism to concentrate *Karenia brevis* population in the shelf waters. Thus, the resultant bloom is not due to the excessive productivity but the concentration process. A study on this phenomenon from October 1998 to January 2002 showed that the cell density ranged from 1 to 5.4 million cells/L under an N-limited but P-enriched condition (Vargo et al. 2001). It was reported that during these periods, wind mixing as well as upwelling favouring winds allowed onshore transport of water. This accounted for transportation of estuarine waters into the coastal zone that led to the breakdown of vertical stratification and promoted horizontal stratification with the formation of thermal and salinity fonts. Thus, it can be said that development of 'blooms' is not only a nutrient-regulated phenomenon but also involves physical processes as well.

Similar algal blooms occur in freshwater as well as in marine environments. Typically, only one or a small number of phytoplankton species are involved in bloom formation, and some blooms may cause characteristic discolouration of the water resulting from the high density of pigmented cells. If the algal bloom results in a high enough concentration of algae, the water may become discoloured, varying in colour from purple to almost pink, normally being red or green (Fig. 1.2).

HABs occur in many regions of the world, and in the USA, these are recurring phenomena in multiple geographical regions. The Gulf of Maine frequently experiences blooms of the dinoflagellate *Alexandrium fundyense* that produces saxitoxin, the neurotoxin responsible for paralytic shellfish poisoning. California coastal waters also experience seasonal blooms of *Pseudonitzschia*, a diatom known to produce domoic acid, responsible for amnesic shellfish poisoning. Off the west coast of South Africa, HABs caused by *Alexandrium catanella* occur in every spring. These blooms used to cause severe intoxication of filter-feeding shellfish which affects the entire food chain (Fig. 1.3).

At the end of the growing season, the dead and decomposed planktonic biomass depletes the oxygen from the water together with the increased loading of organic matter into the water body. This causes massive death of aquatic animals due to suffocation and intoxication, which ultimately produce the dead zone. Dead zones are hypoxic (low-oxygen) areas in the world's oceans – the observed incidences of which have been increasing as oceanographers recorded since the 1970s. These occur near inhabited coastlines, where aquatic life is most

1.6 Ecological Significance

Fig. 1.2 A red tide off the coast of La Jolla, San Diego, California (Source: www.wikipedia.com)

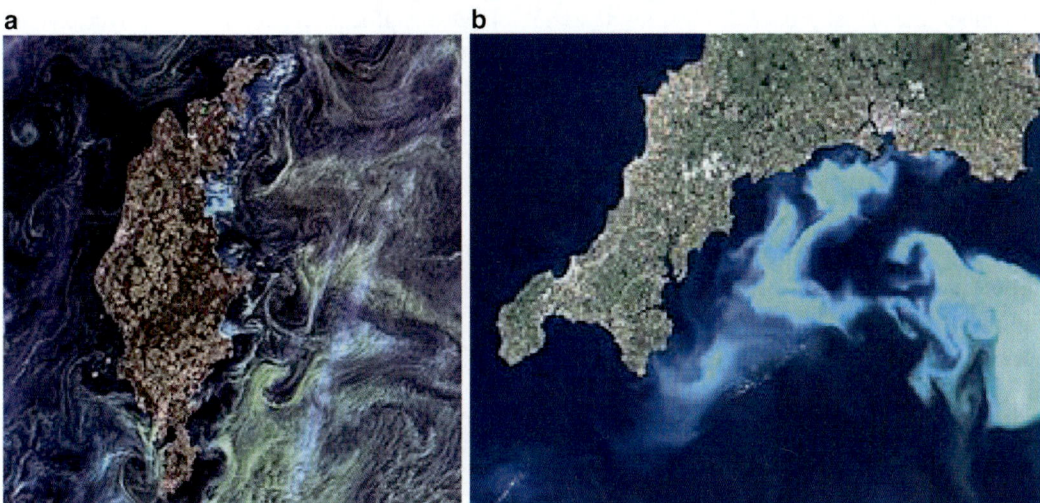

Fig. 1.3 (a) Satellite image of phytoplankton swirling around the Swedish island of Gotland in the Baltic Sea, in 2005, and (b) an algae bloom off the southern coast of Devon and Cornwall in England, in 1999 (Source: www.wikipedia.com)

concentrated. In March 2004, when the recently established UN Environment Programme published its first Global Environment Outlook Year Book (*GEO Year Book 2003*), it reported 146 dead zones in the world's oceans where marine life could not be supported due to depleted oxygen levels. Some of these were as small as a square kilometre (0.4 mi^2), but the largest dead zone covered 70,000 km^2 (27,000 mi^2). A recent study counted 405 dead zones worldwide.

1.6.4 Algal Toxins

The toxicity of some algal genera including cyanobacteria has been known since early times. Most of these taxa are planktonic causing deleterious effect on marine and freshwater fauna and flora. Generally the concentration of these toxins, viz. cyanotoxin, paralytic shellfish poison (PSP) or domoic acid (DA), increase when the secreting taxa appear in blooming condition, causing intoxication or death of other organisms especially the mammals.

1.6.4.1 Cyanotoxin

The toxins secreted by cyanobacteria or the cyanotoxins are the most common toxins of both freshwater and marine habitat. Different types of cyanotoxins have been recorded with varied chemical structure. The most commonly occurring toxins are anatoxin-a, saxitoxin, microcystin, scytophycin, cyanobactrin, lyngbyatoxin, etc. Depending on the toxic effect of the cyanotoxins, it can be categorized into different types like neurotoxins, hepatotoxin and cytotoxins.

Neurotoxins are chemically nitrogen-containing compounds of low molecular weight, for example, anatoxin and saxitoxin. They generally block the signal transmission from neurone to neurone and neurone to muscle. The neurotoxic amino acid L-beta-N-methylamino-l-alanine was shown to be produced by diverse cyanobacterial taxa. In acute intoxication it may be fatal due to respiratory arrest. Cyanobacterial taxa like *Anabaena flos-aquae*, *Aphanizomenon flos-aquae*, *Oscillatoria* and *Trichodesmium* produce neurotoxins.

The most common hepatotoxin is microcystin secreted by *Microcystis aeruginosa*. It is a chemically cyclic hepatic peptide. A total of seven peptides are involved to form the structure of a microcystin, and the molecular formula is (2S,3S, 8S,9S,4E,6E)-3-amino-9-methoxy-2,6,8-trimethyl-10-phenyl-4,6-decadienoic acid. It contains D-alanine (D-ala) and D-METHYL-ASPARTATE (Masp). *Microcystis* poisoning was first reported about 1,000 years ago in southern China, when a Chinese general reported the death of his troops who drank green-coloured water from a river. Another hepatotoxin is nodularin, secreted by other cyanobacterial taxa including *Nodularia*.

A large number of cyanobacterial taxa produce cytotoxic compounds which are active against cell tissue line and have antitumor activities.

1.6.4.2 Dinoflagellate Toxins

Toxins produced by marine dinoflagellates are among the most potent nonproteinaceous lethal materials known. The most notorious of the dinoflagellate toxins are neurotoxins. These toxins act either as depolarizing agents or as non-depolarizing agents in membranes of excitable cells. Dinoflagellate toxins are a matter of public health concern as they infect many marine products through the food chain especially during red tide formation.

The water-soluble toxins, saxitoxins, are produced by members of *Gonyaulax*, a tetrahydropurine composed of two guanidinium moieties. These toxins are categorized as follows:
1. Neosaxitoxin – 1(*N*)hydroxysaxitoxin (class I paralytic shellfish poison)
2. Sulfated 11 hydroxyl substitution (class II) – gonyautoxins 1,2,3 and 4
3. *N*-sulfoconjugation on the carbamoyl position (class III) – gonyautoxins 5 and 6
4. Dual sulfoconjugation at the 11 hydroxyl and carbamoyl positions (class IV)

Lipid-soluble toxins (Florida red tide toxin) are produced by:
1. *Ptychodiscus brevis* – brevetoxins
2. *Gymnothora javanicus* – ciguatoxin
3. *Prorocentrum lima* – okadaic acid
4. *Dinophysis fortii* – methyl okadaic acid

The dinoflagellate toxins act predominantly by altering transmembrane fluxes of cations, primarily sodium. Only maitotoxin acts in a sodium-independent manner. They block the inward flow of sodium ions without affecting the potassium channel. Ciguatoxin enhances the inward flow of sodium ions. Saxitoxin binds to specific receptors in the nerve membrane, whereas maitotoxin possesses a specific calcium-dependent transport.

1.7 Algae in Wetlands

Wetlands are waterlogged ecosystem with a periodic fluctuation in water depth. Most of the wetlands are with a depth of less than 2 m (<2). Wetlands occur in every continent except Antarctica. According to Mitsch and Gosselink (1993), wetlands comprise as much as 6 % of global land area. It is also reported that 14 % land area of Canada is wetland – most of which are confined to northern peatlands.

1.7 Algae in Wetlands

Wetlands are important for their high productivity as they are considered as nutrient sinks (Dolan et al. 1981; Reeder 1994), flood control buffers and the breeding ground of different aquatic animals like waterfowl and others (Batt et al. 1989). Wetland research is carried out in different countries by various scientists (Barica et al. 1980; Hanson and Butler 1994; Moss 1983).

1.7.1 Role of Phytoplanktons in Wetlands

In wetland ecosystem, phytoplanktons are one of the fundamental players of physical, chemical and biological processes that characterize the wetland ecosystems. They mostly act as primary producer, therefore placed in the wetland food web. Pinckney and Zingmark (1993) calculated the relative primary production of different types of algal community of wetlands and obtained as much as 22–38 % of epipelagic algae, 10–17 % for macroalgae and 30–59 % for vascular plants. The average C fixation rate of freshwater marsh phytoplanktons varied from 3 to 16 μ L^{-1} h^{-1} as estimated by different authors (Kotak and Robinson 1991; Mitsch and Reeder 1991). Whereas in peat bogs, it was 5–32 µg L^{-1} h^{-1} (Henebry and Cairna 1984); in salt marsh it was recorded as 100 g m^{-2} $year^{-1}$ (Vernberg 1993), and in tundra ponds, it was at minimum level of 0.6–0.9 g $^{-2}$ $year^{-1}$ (Stanley 1976).

The importance of wetland phytoplanktons as a resource of food to herbivores depends on their availability throughout the year. Their variation in size, species composition and productivity support the zooplankton and other aquatic fauna as continuous food source.

Generally the levels of phytoplankton chlorophyll in wetlands varied from 50 to >200 µg L^{-1} during periodic cyanobacterial bloom (Barica 1975).

1.7.2 Nutrient Status of Wetland Ecosystem

In a wetland ecosystem, nutrient efflux from different sources enriches the nutrient level; therefore, high entropic condition resulted in periodical phytoplankton bloom. Efflux from the sediments is a major source of nutrients to the water column. Significant water input from the adjacent lake, river or streams is the main source of nutrients and algal inocula for lakeshore and riverside wetlands. Nutrient efflux due to various anthropogenic factors is another source of nutrients for wetlands; waste water from different sources like industrial waste and municipal waste also increases the nutrient level in wetland ecosystem, especially in stagnant wetlands.

Limnologists are of the opinion that freshwater wetlands are generally P limited, though there are exceptions also. In north temperate wetland of North America, the N–P ratio in the water column is mostly >10 (Barica 1990). The nutrient deficiency of northern wetlands is due to the variation in phytoplankton population there. Murkin et al. (1991) estimated the alkaline phosphatase activity, ammonia uptake rate and the ratios of P–Cl, N–P, N–C and chlorophyll–C composition ratio and found severe N and P deficiency in summer due to insufficient nutrient loading. But other wetlands enriched by cattle feedlot effluent did not have any nutrient-deficient symptoms. Campeau et al. (1994) studied the ecological effect of the addition of N and P to a nutrient-poor marsh and observed the stimulation in biomass production of phytoplanktons together with epiphyton and metaphyton but not the epipelon. Generally shallow wetlands are dominated by phytoplanktons (Barica et al. 1980), but some wetlands with deepwater levels have abundant macrophytes and associated epiphytes. Physicochemical parameters and allelopathic effects of different organisms also play a major role in wetland ecology and phytoplankton abundance together with water depth.

Sometimes there is a direct role of fluctuating water depth on wetland nutrient status. It is already reported that exposure of sediments during draughts promotes the decomposition of organic matter and the nutrients are liberated during subsequent reflooding (Kadlec 1986; Schoenberg and Oliver 1988). van der Valk (1994) reported that in late stage of flooding, macrophytes are killed by flood stress and their decomposed biomass releases N and P (Murkin et al. 1989). Moreover, both algae and higher plants of wetland ecosystem release a diverse type of soluble organic substances into the water column during their growth and decompositions (Hall and Fisher 1985; Briggs et al. 1993). Therefore, due to abundance of organic N and P species, the algal community may also continue their heterotrophic nutrition. Pip and Robinson (1982) experimented with radiolabelled mixture of glucose, fructose and sucrose and observed that most of the algal genera including cyanobacteria can also utilize the organic carbon sources.

1.7.2.1 Physical Factors Affecting Wetland Ecosystem

I. Light: Light penetration in water column of shallow wetlands is affected greatly by wind-driven sediment resuspension which also increases the productivity of shallow wetlands (Klarer and Millie 1992). The bottom sediments are also subjected to regular disturbances by the benthivorous carp fishes (*Cyprinus carpio*) and other macrophytes (Meijer et al. 1990). Sometimes floating algal mats or *Lemna* mats disrupt light penetration; therefore, the population productivity of wetland flora is greatly affected, where it becomes light limited rather than nutrient limited.

II. Temperature: In wetland ecosystem differences in temperature with depth are minor, as the wetlands are shallow and the wind-driven mixing minimizes the temperature gradient of wetland water column. Sometimes the thick algal mats or floating mat of duckweed may increase the temperature variation in water column reducing the temperature.

1.7.3 Types of Wetlands

Based on the depth and duration of the water-logged condition of the wetland and their respective algal flora, the limnologists classified wetlands into different types.

1.7.3.1 Dry State

In dry state wetland, water levels occur at very low level after a long period of draught, i.e. draw down in existing wetland or the low level of water by early flooding of a new wetland. Epipelic algae are predominant in this type forming a crust on the bottom sediment.

1.7.3.2 Open State

An open state wetland is with deep and turbulent water column. This type of wetland may arise by two ways: by gradual filling of dry state wetland or by biomanipulation of phytoplankton-dominated lake state wetland. Open state wetlands are dominated by macrophytes, and with abundant epiphyton biomass on them, comparatively phytoplanktons and epipelic algae are less.

1.7.3.3 Sheltered State

In some wetland areas, nutrient load is very high, but the area is shady due to bordering vegetations and excessive growth of macrophytes. These metaphytes remain covered by epiphytic growth. These epiphytic algal mats sometimes detach from their substratum and float on the water surface as metaphytons. Whitton (1970) recorded high nutrient status, high irradiance, alkaline pH, high calcium and low N–P ratios as the main contributors for metaphyton growth.

1.7.3.4 Lake State

Lake state wetlands are characterized by high water column and nutrient levels, therefore ideal for phytoplankton growth. Due to high depth of water, macrophyte growth is checked together with the epiphytic and metaphyton growth. The water column is turbid due to phytoplankton growth. Lake state wetland may develop by rapid water input to other types of wetlands, by natural process or by anthropogenic manipulation. This leads to the death of macrophytes and epiphytons, leading to a growth of lake state wetlands.

1.8 Key to Identification of Common Phytoplankton Genera

1.8.1 Division: Cyanobacteria

Cells smaller, blue green in colour, cells without membrane-bound cell organelles; pigments diffused throughout peripheral portion of the protoplast; water-soluble pigments like phycocyanin and phycoerythrin present; photosynthetic storage products are glycogen and glycoproteins; definite nucleus and chromosome absent, mucilage conspicuous, hydrostatic gas vacuoles present, reproduction by fission and by fragmentation, sexual reproduction absent.

1.8.1.1 Class: Cyanophyceae

Unicellular/multicellular without true nucleus or chromatophore; chlorophyll *a* and phycobiliproteins are main pigments that are composed of allophycocyanin, phycocyanin and phycoerythrin; photosynthetic product glycogen or glycosides, no starch; reproduction by division or through fragmentation of thallus, endospores, exospores, hormogones, motile flagellate cells absent; sexual reproduction almost absent.

1.8.1.2 Key to Orders

Thallus unicellular or colonial, sometimes forming pseudofilamentous colony, no trichome organization, no differentiation into base and apex, endospores not formed in sporangia, no exospores, nanocysts present – Chroococcales

1.8.1.3 Key to Family

Cells unicellular or forming colonies, not forming filament-like growth – Chroococcaceae

1.8.1.4 Key to Genera

1. Cells single or few together in a shapeless colony	2
1. Cells generally many in a single colony	8
2. Cells spherical	3
2. Cells elongated	5
3. Without individual mucilage envelopes	*Synechocystis*
3. With distinct envelope	4
4. Vesicular sheath	*Gloeocapsa*
5. Cell division transverse, with a firm vesicular sheath	*Gloeothece*
5. Cell division transverse, without such a sheath	6
6. Cells ellipsoidal or cylindrical with round ends	7
7. Cells single or 2–4 together, erect without a common mucilage	*Synechococcus*
8. Cells without any regular or definite arrangement	9
8. Cells with definite arrangement in distinct colonies	14
9. Cells in a general amorphous mucilage, without or with a few distinct sheaths round the individual cells	10
9. Cells with distinct individual envelope or sheaths, colonial mucilage not homogenous	12
10. Cells typically well packed into microscopic colonies of definite shapes, mostly planktonic	*Microcystis*
10. Cells loosely arranged, mostly not planktonic, forming macroscopic colonies	11
11. Cells spherical	*Aphanocapsa*
11. Cells ellipsoidal to cylindrical	*Aphanothece*
12. Individual sheaths vesicular and broad and formed one in another	13
12. Individual sheaths not vesicular, cells spherical	*Chroococcus*
13. Cells spherical	*Gloeocapsa*
13. Cells ellipsoidal to cylindrical	*Gloeothece*
14. Colony with cells arranged in a tabular or cubical or 3D colony	15
14. Colony a hollow sphere with cells arranged along the margin uniformly	16
15. Cells in regular transverse and longitudinal rows, tabular or flat colonies	*Merismopedia*
16. Cells spherical, colonial mucilage homogenous	*Coelosphaerium*
16. Cells pear shaped or weakly spherical, colonial mucilage not homogenous, cells with distinct mucilage sheaths	*Gomphosphaeria*

1.8.2 Division: Chlorophyta

Grass-green chloroplasts, chloroplasts often with pyrenoids, food reserve starch, cell wall composed of cellulose and pectic compounds

1.8.2.1 Class: Chlorophyceae

Unicells (sometimes motile), simple or well-organized colonies, simple or branched filaments, partitioned coenocytes and true coenocytes (filaments without cross walls). Reproduction both sexual and asexual

1.8.2.2 Key to Orders

1. Motile in the vegetative condition; flagella 2 or 4, rarely 8, equal in length; organism 1 celled or colonial	Volvocales
1. Not motile in the vegetative condition	2
2. Cells embedded in copious mucilage (homogenous or lamellated), united in colonies of indefinite shapes or forming gelatinous strands (pseudofilaments) or mucilage invested, some unicellular or forming dendroid colonies which are epiphytic or epizoic, cells often with false flagella (pseudociliates) returning to a motile condition without resorting to reproductive stage	Tetrasporales
2. Cells not embedded in mucilage	3
3. Thallus filamentous, composed of cells adjoined end to end in definite series, sometimes interrupted	4
3. Thallus not composed of cells arranged to form filaments; unicellular or colonial; or if filamentous, occurring as coenocytes without cross walls	15
4. Filaments unbranched; attached or free floating	5
4. Filaments with branches, the branches sometimes closely appressed, forming pseudoparenchymatous masses	13
5. Filaments composed of a single series of cells	6
5. Filaments composed of more than one series of cells; cells adjoined; thallus a hollow tube or ribbon/frond-like expansion	12
6. Chloroplasts 1 to several, large, in the form of spiral bands, stellate masses or broad plates; pyrenoid conspicuous; reproduction by conjugation	Zygnematales
7. Thallus a single cell or a colony of definite or indefinite form; cells with various shapes that range from spherical, pyramidal to polygonal; no vegetative reproduction, reproduction by autospores, zoospores or isogametes	Chlorococcales

1.8.3 Order: Volvocales

Both vegetative and reproductive cells are motile; 2, 4 or rarely 8 flagella, generally a conspicuous pigment spot, cup-shaped parietal chloroplasts; reproduction by cell division, by zoospores, by isogametes or by heterogametes.

1.8.3.1 Key to Families

1. Cells possessing a definite wall and sometimes a mucilaginous sheath; solitary or united in colonies	2
2. Cells solitary	3
2. Cells united in colonies	5
3. Cell wall not bivalved, not flattened	4
4. Cells with protoplasts located at some distance within the cell wall and connected to it by radiating cytoplasmic strands	Haematococcaceae (in part)
4. Cells without radiating cytoplasmic strands	Chlamydomonadaceae
5. Cells united to form flat or globular colonies, evenly dispersed, although sometimes closely arranged within colonial mucilage	Volvocaceae

1.8.4 Order: Chlorococcales

One celled or colonies of definite shapes; cells may be adjoined or merely enclosed by colonial mucilaginous envelope. No cell division in vegetative state. Asexual reproduction by zoospore formation, sexual reproduction isogamous

1.8.4.1 Key to Families

1. Unicellular or colonial; cells varied in shape but not irregular; wall of uniform thickness and not definitely lamellated; free living or attached, rarely subaerial	2
2. Free floating or adherent on soil	3
3. Cells cylindrical and forming a macroscopic network or triangular or polyhedral and united to form either a flat and circular or a globose coenobium (colony)	Hydrodictyaceae

3. Cells not cylindrical and not forming colonies as above	4
4. Unicellular, solitary or sometimes gregarious, free floating (usually on moist soil if adherent), reproduction by zoospores (rarely aplanospores) which do not adhere to one another but which are liberated separately from the parent cell	Chlorococcaceae
4. Colonial or solitary, not reproducing by zoospores	5
5. Thallus not a globose, hollow coenobium	6
6. Cells not in a peripheral arrangement with no coloured mucilage	7
7. Cells solitary or in colonies of definite or indefinite shape; cells variable in form (spherical, ovate, lunate, polyhedral, etc.), not adjoined to one another; reproduction by autospores	Oocystaceae
7. 2–8 cells adjoined together or adherent to form a pattern of definite shape (a linear series, stellate or cruciate); reproduction by the formation of autocolonies within the cells of the parent coenobium	Scenedesmaceae

1.8.5 Division: Bacillariophyta (Diatoms)

1.8.5.1 Order: Centrales

Valves with a concentric or radiating sculpture around a point or points, central or lateral. Without raphe or pseudoraphe. Cells circular, oval or elliptical, sometimes polygonal, rarely crescent shaped or spindle shaped. Processes common

I. Suborder: Discoideae
Cells disc shaped or cylindrical. Valves circular, surface flat or convex, sometimes hemispherical. Spines frequent. Without horns or knobs; when present, small
  1. Valves circular. Not divided into definite sectors by ribs, rays or undulating sectors. Sculpturing sometimes arranged in bundles, long spines often present.
     Family – Coscinodiscaceae
     (a) Cells lens shaped, round or cylindrical. Usually united into more or less long typical chains. Intercalary bands often sculptured. Valve mantle usually strongly developed.
         Subfamily – Melosirinae
     Common genera – *Melosira*, *Stephanopyxis*
     Key to Genera
     Cells globose, elliptical or cylindrical, closely united in straight, beadlike chains by the centres of the valves. Valves either simply punctate or punctate and areolate. Intercalary bands none or many and narrow.
                                              Melosira
     Cells oblong, oval or nearly circular, with hexagonal areolations. Usually in short chains.
                                              Stephanopyxis
     (b) Cells short or elongated–cylindrical, bound into close chains by delicate siliceous projections or gelatinous threads. Cell wall usually weakly siliceous.
         Subfamily – Skeletoneminae
     Common genera – *Skeletonema*, *Thalassiosira*
     Key to Genera
     Cells circular, lens-shaped, oblong, or cylindrical. Valves circular, somewhat arched, without distinct structure, with a row of fine spines at the edge of the valve parallel to longitudinal, pervalvar axis. Spines interlock midway between adjacent cells and unite cells into chains.
                                              Skeletonema
     Cells similar to those of *Coscinodiscus*, usually drum or disc shaped, united in flexible chains by a cytoplasmic or gelatinous thread or living in formless gelatinous masses or seldom solitary. One or more intercalary bands to each valve. Valves with areolae or delicate radial rows of punctations. Structure often difficult to see. Marginal capsule or little spines present, usually distinct.
                                              Thalassiosira

(c) Cells disc or drum shaped, solitary. Valve surface slightly convex, sometimes nearly flat or slightly concave, with prominent sculptures (commonly hexagonal). Intercalary bands hyaline or rarely very delicately sculptured. Valve mantle not particularly well developed.

                Subfamily – Coscinodiscinae

Common genera – *Coscinodiscus, Cyclotella, Planktoniella*

Key to Genera

Cells disc or box shaped, single or in twos immediately after cell division. Valves circular, without large knobs or processes, with hexagonal areolae arranged in various ways or fine round puncta.

                Coscinodiscus

Cells single, disc shaped, with a hyaline winglike expansion all around consisting of extracellular chambers strengthened by radial rays. The winglike expansion weakly siliceous, an organ of flotation. Valves areolated like those of *Coscinodiscus excentricus*.

                Planktoniella

2. Valves circular. Divided into distinct, complete or incomplete sectors by radial ribs or undulations or by wide hyaline rays from a characteristically constructed centrum. Without horns or prominent spines.

                Family – Actinodisceae

(a) Valves divided into sharply distinct sectors by radial ridges uniformly running from the margin to the hyaline central area. Small but distinct spines usually at the marginal ends of these ridges. Alternate sectors generally depressed.

                Subfamily – Actinoptychinae

Common Genus – *Actinoptychus*

Key to Genera

Cells disc shaped, single. Valves divided into sectors which are alternately raised and depressed. Smooth central area. No intercalary bands. Cell wall usually of several layers, the individual membranes punctuated; puncta in crossing lines and more or less strongly areolated.

                Actinoptychus

3. Valves usually radially waved, eyes or knobs on the elevations or valves flat and then with singly placed wartlike elevations or circle of needles.

                Family – Eupodisceae

(a) Valves with knobs. Only one genus.

                Subfamily – Aulacodiscinae

Common genus – *Aulacodiscus*

Key to Genera

Cells discoid or box shaped. Valves with a circular outline, flat or slightly lower in the middle or convex, with four or more conical processes symmetrically arranged near the margin. Areolated. Intercalary bands present.

                Aulacodiscus

II. Suborder: Solenoideae

Valves oval or circular in cross section. Cells elongated, cylindrical or subcylindrical, with numerous intercalary bands, without internal septa. Valve structure arranged in relation to an excentric pole. Cells united into chains by their valves

As above – Family: Solenieae

(a) Valves flat or raised, with or without marginal spines. Excentric process or asymmetrical spine absent.

                Subfamily – Lauderiinae

Common genera – *Corethron, Leptocylindrus*

Key to Genera

Cells living singly. Cylindrical with rounded valves having a crown of long thin spines or setae at the margin directed outwards at an angle. Numerous intercalary bands, scale-like, often very indistinct.

                Corethron

Cells long, cylindrical, united into chains by whole valve surface. Valves flat, without spines or processes.

                Leptocylindrus

(b) Valves with a single often very short spine or process, usually excentrically placed, thus destroying the symmetry of the cell. Valves flat or convex
> Subfamily – Rhizosoleniinae

Common genus: *Rhizosolenia*

Key to Genera

Cells cylindrical with greatly elongated pervalvar axis, living singly or in compact or loose chains. Cells usually straight or more rarely curved, forming spirally twisted chains. Cross section elliptical or circular.
> *Rhizosolenia*

III. Suborder: Biddulphioideae

Cells box shaped. Pervalvar axis generally shorter, sometimes slightly longer, than the valvar axis. Valves usually oval, sometimes polygonal, circular or semicircular; unipolar, bipolar or multipolar, each pole represented by an angle or by a horn or spine or by both angles and horns.

Valves with long setae, longer than the cells. Cells united into chains by basal part of the setae. Seldom living as single cells. Intercalary bands only seldom present. Valves circular or oval. All species pelagic.
> Family – Chaetocerotaceae

Common genera: *Bacteriastrum, Chaetoceros*

Key to Genera

Cells cylindrical, in cross section circular. Bound into loose chains by the fusion of the more or less numerous setae that are regularly arranged around the margin of the cells. Setae of two adjacent cells are fused for a certain distance beyond the base, farther out divided again.
> *Bacteriastrum*

Cells with oval section to almost or rarely completely circular in valve view; in broad girdle view quadrangular with straight sides and concave, flat or slightly convex ends. Valve with more or less flat end surface or valve surface and a cylindrical part or valve mantle which is bound together without a seam.
> *Chaetoceros*

Horns short and thick or with claws at the end of long horn. Cells united into chains by the ends of the horns. Valvar plane circular or elliptical, usually with one to several poles and multiangled. Intercalary bands and septa often present. Mainly planktonic, partly littoral. Majority of species marine.
> Family – Biddulphiaceae

(i) Knobs and horns without claws on the end Valves bipolar. Cell wall weakly siliceous. Plankton forms.
> Subfamily – Eucampiinae

Common genus – *Eucampia*

Key to Genera

Valves elliptical in surface view with two blunt processes, without spines or setae. Numerous intercalary bands difficult to see in water mounts. Chains spirally curved. Large apertures between the cells.
> *Eucampia*

Valves tripolar to multipolar, sometimes with bipolar varieties. Angles not bearing domelike protrusions or horns. Intercalary bands frequently present. Marine forms, usually littoral, a few pelagic.
> Subfamily – Triceratiinae

Common genera – *Ditylum, Lithodesmium, Triceratium*

Key to Genera

Cells elongated, prismatic to box shaped. Solitary except immediately after division. Valves three to four cornered, seldom bipolar, with a strong central siliceous hollow spine and a marginal ridge strengthened by ribs.
> *Ditylum*

Cells united in usually long, straight chains with concealed apertures. Valves three cornered. Valves with marginal pervalvar- directed membrane by which adjacent cells are joined. Long, thin, hollow spine in the centre of valve. Intercalary bands present, collar like.

Lithodesmium

Valves bipolar, tripolar or multipolar. Each angle with a domelike protrusion or a horn. Usually strongly siliceous and forms chains. Predominantly littoral, but sometimes pelagic. A few species are typically planktonic and are then more weakly siliceous.

Subfamily – Biddulphiinae

Common genus – *Biddulphia*

Key to Genera

Cells box shaped to cylindrical. Valves elliptical, with two poles or three- or four sided (rarely five sided). At the corners or at the ends of the apical axis, more or less strongly developed processes or horns may be present.

Biddulphia

Valves unipolar. Cells in broad girdle view rhombic or trapezoid. Large to very large, robust marine forms. Littoral.

Subfamily –Isthmiinae

Common genus – *Isthmia*

Key to Genera

Cells cylindrical–box shaped with elliptical valve surface and usually longer pervalvar axis. Intercalary bands absent.

Isthmia

(ii) Horns with claws on the end

Valves bipolar, tripolar or quadripolar. Each angle with a long vertical horn tipped with a claw.

Family – Hemiaulineae

Common genera – *Cerataulina, Hemiaulus*

Key to Genera

Cells cylindrical, usually in chains. Valves slightly arched, with two blunt projections or processes near their margin, attached to adjacent cell by means of a fine, small, curved, hairlike process which fits into the valve of the adjacent cell. Intercalary bands numerous, annular.

Cerataulina

Cells single or united in chains. Valves elliptical in section, with two narrow, pointed, more or less long processes at the ends of the apical axis, parallel to pervalvar axis.

Hemiaulus

Cells without horns. Valves without internal septa. Valves semicircular, broader than long. Intercalary bands very seldom present. Cells in girdle view cuneate.

Family – Euodieae

Common genus – *Hemidiscus*

Key to Genera

Cells shaped like a sector of a sphere, in girdle view wedge shaped, narrowing from the dorsal towards the ventral side. Valve semicircular to asymmetrically elliptical. Valves flat with short valve mantles. Intercalary bands and septa absent, pervalvar axis not particularly elongated.

Hemidiscus

1.8.5.2 Order: Pennales

Valves elongated, bilaterally symmetrical. Outline generally boat shaped or rod shaped, sometimes oval, cuneate, crescent shaped or sigmoid; markings generally pinnate or transverse. True raphe, or hyaline median line (pseudoraphe), or raphe obscured by lateral wings or keel (cryptoraphe) always present. Processes absent. Cell capable of spontaneous movement if a true raphe is present.

1. Raphe absent. Pseudoraphe usually present.

Suborder – Araphidineae

Cells in general rod shaped to tabular prism shaped, in valve view usually more or less linear, seldom club shaped. In girdle view linear to tabular rectangular. Intercalary bands and septa frequently present. Valves with transapical striae or ribs, sometimes areolated–punctated, often with mucilage pores. Without a true raphe, but usually with a median pseudoraphe. Chromatophores usually more or less numerous small platelets, seldom a single large plate.

Family – Fragilarioideae

Cells in valve view usually linear, more seldom wedge shaped, frequently with transapical inflations or constrictions. In girdle view usually linear to tabular, seldom wedge shaped. Intercalary bands and septa always present and distinct. Valvar plane not bent in the pervalvar direction, the two valves of a cell usually entirely alike. Cells usually united into bands. Marine and freshwater forms.

Subfamily – Tabellarieae

Common genera – *Rhabdonema, Grammatophora*

Key to Genera

Cells in girdle view rectangular with rounded corners, usually united into zigzag chains.

Grammatophora

Cells with poles of apical axis unlike. In girdle view, as in valve view, wedge shaped. Intercalary bands and septa present. Cells stalked and often united into strongly branched colonies. Marine.

Subfamily – Licmophorinae

Common genus – *Licomophora*

Key to Genera

Cells with wedge-shaped girdle band side and wedge- or club-shaped valve. Two intercalary bands in resting cells, with a more or less long penetrating septum on the head pole. Valves with transapical punctated striae, seldom with weak transapical ribs and extremely delicately punctated intercalary space.

Licomophora

Cells usually rod shaped. Usually linear in both valve and girdle view, seldom wedge shaped or with tabular girdle view. Intercalary bands sometimes present, but always without or with only very rudimentary septa.

Subfamily – Fragilariieae

Common genera – *Fragilaria, Synedra, Thalassionema, Thalassiothrix, Asterionella*

Key to Genera

Cells united into more or less long bands by the whole valve sides, occasionally with regular apertures between the cells. Connection of the cells with one another often assisted by tiny spines on the valve margin.

Fragilaria

Cells single or united into fanlike to clustered starlike colonies, seldom in short bands. In general with greatly elongated apical axis, rodlike, sometimes bent in the direction of the apical axis.

Synedra

Cells forming zigzag bands or star-shaped colonies, adjacent cells united to each other by small gelatinous cushions on one cell end. In girdle view linear.

Thalassionema

Cells living singly or forming star-shaped colonies, zigzag bands or bunches, united to one another by a gelatinous cushion on the end of the cell. In girdle view narrow linear.

Thalassiothrix

Cells united by one end (the larger end) into star-shaped colonies or spirally curved, sometimes straight, comb-shaped bands.

Asterionella

Valves with transapical ribs that are, however, sometimes limited to one valve of a cell or to one single middle rib. Freshwater or marine forms. Cells as a rule united into closed or zigzag bands.

Subfamily – Diatominae

Common genus – *Diatoma*

2. One valve of the cell always with *Navicula-like* raphe, the other without a raphe or with a rudimentary raphe knot.

Suborder – Monoraphideae

Cells with linear, lanceolate or elliptical valvar plane, more or less distinctly bent about the apical or transapical axis. Valves sometimes with polar pseudosepta. The membrane lying between the transapical rows stronger, often thickened, riblike. The two valves of a cell usually considerably differentiated in regard to the structure as well as to the development of the raphe.

Family – Achnanthoideae

Cells bent about the transapical axis, sometimes also about the apical axis; the valve with the raphe, concave; the one without a raphe, convex. One valve always with a developed raphe, the other without, seldom with a very short raphe knot or with a rudimentary raphe. Outline of valve usually linear–lanceolate, seldom elliptical. Intercalary bands sometimes present, true septa absent.

Subfamily – Achnanthaceae

Key to Genera

Cells single or united into ribbon-like bands, with or without a gelatinous stalk to hold the chains to the substrate. Seldom pelagic. Valves usually linear–lanceolate, seldom elliptical; in girdle view in general rectangular, but more or less strongly broken along the transapical axis.

Achnanthes

Cells elliptic to near round, rapheless valve with slightly radial transapical striae, pseudoraphe narrow, valve with raphe with radial punctuate striae.

Subfamily – Cocconeideae

3. Both valves with developed raphe.

Suborder – Biraphideae

Cells of various types as regards to structure of membrane, cell contents and general form. All forms with a similar characteristic raphe system: outer and inner fissures and accompanying end and central knots. Knots often greatly reduced or in many species only slightly developed. Inner and outer fissures often difficult to distinguish from each other. Raphe usually in the valvar plane, generally distinct, not developed as a canal raphe; usually without a keel or strongly developed wings, but when present always without marginal canal and keel puncta.

Family – Naviculoideae

Cells as a rule of symmetrical construction, transapical axis only seldom with unlike ends. In girdle view usually rectangular, valves elliptical, linear or lanceolate, often S shaped, seldom club shaped or crescent shaped. Raphe usually in the valvar plane. Cells usually solitary, sometimes in gelatinous tubes or on a gelatinous stalk.

Subfamily – Naviculeae

Key to Genera

Cells usually free, motile. In plankton species usually united into ribbon-like chains. Valves linear to elliptical, with rounded, capitate or rostrate ends.

Navicula

Valves linear or lanceolate. Usually sigmoid. Axial area very narrow. Central area small. Striae punctate, in transverse and longitudinal rows.

Gyrosigma

Valves linear to lanceolate, usually sigmoid. Raphe usually sigmoid, central or excentric. Striae finely punctate in oblique and transverse lines. Central nodule usually small and rounded.

Pleurosigma

Raphe on a keel or wing that usually lies in the midline of the valve. Cells usually twisted about the apical axis.

Subfamily – Amphiproroideae

Key to Genera

Cells single or in ribbon-like chains. Cells constricted in the middle. Valves lanceolate, convex, with raphe, central nodule and a sigmoid keel. One-half of keel lies on each side of the chain axis. Terminal nodules present.

Amphiprora

Keel with canal raphe lying in the valvar plane, often displaced transapically as far as to the valve margin. Both valves with canal raphe. Keel with puncta.

Family – Nitzschiaceae

Raphe obscured by punctate marginal keel. Markings always transverse.

Sub-family Nitzschieae

Key to Genera

Cells spindle shaped, single or united into colonies. Valves keeled, the keel including a concealed raphe, usually diagonally opposite, either central or excentric.

Nitzschia

References

Adl, S. M., Simpson, A. G. B., Farmer, M. A., Andersen, R. A., Anderson, O. R., Barta, J. R., Bowser, S. S., Brugerolle, G., Fensome, R. A., Fredericq, S., James, T. Y., Karpov, S., Kugreens, P., Krug, J., Lane, C. E., Lewis, L. A., Lodge, J., Lynn, D. H., Mann, D. G., McCourt, R. M., Mendoza, L., Moestrup, O., MozleStandridge, S. E., Nerad, T. A., Shearer, C. A., Smirnov, A. V., Spiegel, F. W., Max, F., & Taylor, J. R. (2005). The new higher level classification of eukaryotes with emphasis on the taxonomy of protists. *The Journal of Eukaryotic Microbiology, 52*, 399–451.

Agassiz, L. (1850). *Lake superior: Its physical character, vegetation and animals* (428 pp). Boston: Gould, Kendal and Lincoln.

Batt, B. D. J., Anderson, M. G., Anderson, C. D., & Caswell, F. D. (1989). The use of prairie potholes by North American ducks. In A. van der Valk (Ed.), *Northern Prairie wetlands* (204–227pp.). Ames: Iowa State University Press.

Barica, J. (1975). Collapses of algal blooms in prairie pothole lakes: Their mechanism and ecological impact. *Verh. – Int. Thoor. Angew, Limnol., 19*, 606–615.

Barica, J. (1990). Seasonal variability of N : P ratios in eutrophic lakes. *Hydrobiologia, 191*, 97–103.

Barica, J., Kling, H., & Gibson, J. (1980). Experimental manipulation of algal bloom composition by nitrogen addition. *Canadian Journal of Fisheries and Aquatic Sciences, 37*, 1175–1183.

Bold, H. C., & Wynne, M. J. (1985). *Introduction to the algae*. Englewood Cliffs: Prentice Hall.

Bremer, K. (1985). Summary of green plant phylogeny and classification. *Cladistics, 1*, 369–385.

Briggs, S. V., Maher, M. T., & Tongway, D. J. (1993). Dissolved and particulate organic carbon in two wetlands in southwestern New South Wales, Australia. *Hydrobiologia, 264*, 13–19.

Bryant, D. A. (1994). *The molecular biology of cyanobacteria*. Dordrecht: Kluwer Academic Publishers.

Campeau, S., Murkin, H. R., & Titman, R. D. (1994). Relative importance of algae and emergent plant litter to freshwater marsh invertebrates. *Canadian Journal of Fisheries and Aquatic Sciences, 51*, 681–692.

Caron, J. M., Jones, A. L., Rall, L. B., & Kirschner, M. W. (1985). Autoregulation of tubulin synthesis in enucleated cells. *Nature, 317*, 648–651.

Cavalier, S. T. (1981). Eukaryote kingdoms: Seven or nine? *Bio Systems, 14*, 461–481.

Dolan, T. J., Bayley, S. E., Zoltek, J., Jr., & Hermann, A. J. (1981). Phosphorus dynamics of a Florida freshwater marsh receiving treated wastewater. *Journal of Applied Ecology, 18*, 205–219.

Dussart, B., & Roger, G. (1966). Faune planctonique du lac Tchad: 1. Crustacés copepods. *Cashiers ORSTOM Serie Oceanoraphie, 4*(3), 77–91.

Eichler, A. N. (1886). *Syllabus der Vorlesungen über specielle and medicin isch pharma ceutische Botanik* (4th ed., 68 pp). Berlin.

Fritsch, F. E. (1935). *Structure and reproduction of the algae* (Vols. I and II). Cambridge: Cambridge University Press.

Geider, R. J., Mac Intyre, H. L., & Kana, T. M. (1997). Dynamic model of phytoplankton growth and acclimation: Responses of the balanced growth rate and the chlorophyll a: Carbon ratio to light, nutrient-limitation and temperature. *Marine Ecology Progress Series, 148*, 187–200.

Gross, F. (1937). Notes on the culture of some marine plankton organisms. *Journal of the Marine Biological Association of the United Kingdom (New Series), 21*(02), 753–768.

Glazer, A. N., Yeh, S. W., Webb, S. P., & Clark, J. H. (1985). Disk to disk transfer as the rate limiting step for energy flow in phycobilisomes. *Science, 227*, 419–423.

Hall, S. L., & Fisher, F. M. (1985). Annual productivity and extracellular release of dissolved organic compounds by the epibenthic algal community of a brackish marsh. *Journal of Phycology, 21*, 277–281.

Hanson, M. A., & Butler, M. G. (1994). Responses to food web manipulation in a shallow waterfowl lake. *Hydrobiologia, 279–280*, 457–466.

Henebry, M. S., & Cairna, J., Jr. (1984). Protozoan colonization rates and trophic status of some freshwater wetland lakes. *Journal of Protozoology, 31*, 456–467.

Hutchinson, G. E. (1957). *A treatise on limnology* (1015 pp). New York: Wiley.

Hutchinson, G. E. (1967). *A treatise on limnology* (Vol. II, 1115 pp). New York: Wiley.

Jeffrey, S. W., & Egeland, E. S. (2008). Pigments of green and red forms of Dunaliella, and related chlorophytes. In A. Ben-Amotz, J. E. W. Polle, & D. V. Subba Rao (Eds.), *The alga Dunaliella: Biodiversity, physiology, genomics and biotechnology* (pp. 111–145). Enfield: Science Publishers.

Jeffrey, S. W., & Wright, S. W. (1997). Qualitative and quantitative HPLC analysis of SCOR reference algal cultures. In S. W. Jeffrey, R. F. C. Mantoura, & S. W. Wright (Eds.), *Phytoplankton pigments in oceanography: Guidelines to modern methods* (pp. 343–360). Paris: UNESCO Publishing.

Jeffrey, S. W., & Wright, S. W. (2006). Photosynthetic pigments in marine microalgae: Insights from cultures and the sea. In D. V. Subba Rao (Ed.), *Algal cultures, analogues of blooms and applications* (pp. 33–90). Enfield: Science Publishers.

Jeffrey, S. W., Mantoura, R. F. C., & Bjørnland, T. (1997a). Data for the identification of 47 key pohytoplankton pigments. In S. W. Jeffrey, R. F. C. Mantoura, & S. W. Wright (Eds.), *Phytoplankton pigments in oceanography: Guidelines to modern methods* (pp. 449–559). Paris: UNESCO Publishing.

Jeffrey, S. W., Mantoura, R. F. C., & Wright, S. W. (Eds.). (1997b). *Phytoplankton pigments in oceanography: Guidelines to modern methods*. Paris: UNESCO Publishing.

Johnson, P. W., & Sieburth, J. M. N. (1979). Chroococcoid cyanobacteria in the sea: A ubiquitous and diverse phototrophic biomass. *Limnology and Oceanography, 24*, 928–935.

Lee, R. E. (1980, 1989, 1999, 2008). *Phycology*. Cambridge: Cambridge University Press.

Li, W. K. W., Subba Rao, D. V., Harrison, W. G., Smith, J. C., Cullen, J. J., Irwin, B., & Platt, T. (1983). Autotrophic pico- plankton in the tropical ocean. *Science, 219*, 292–295.

Kadlec, J. A. (1986). Effects of flooding on dissolved and suspended nutrients in small diked marshes. *Canadian Journal of Fisheries and Aquatic Sciences, 43*, 1999–2008.

Klarer, D. M., & Millie, D. F. (1992). Aquatic macrophytes and algae at Old Woman Creek estuary and other Great Lakes coastal wetlands. *Journal of Great Lakes Research, 18*, 622–633.

Kotak, B. G., & Robinson, G. G. C. (1991). Artificially-induced water turbulence and the physical and biological features within small enclosures. *Archives of Hydrobiology, 122*, 335–349.

McFadden, G. I. (2001). Primary and secondary endosymbiosis and the origin of plastids. *Journal of Phycology, 37*, 951–959.

Meijer, M. A., deHaan, M. W., Breukelaar, A. W., & Buiteveld, H. (1990). Is reduction of the benthivorous fish as important cause of high transparency following biomanipulation in shallow lakes? *Hydrobiologia, 200–201*, 303–315.

Mereschkowski, K. (1905). Über Natur und ursprung der chromatophoren im pflanzenreiche. *Biologisches Centralblatt, 25*, 593–604, 689–691.

Mitsch, W. J., & Gosselink, J. G. (1993). *Wetlands* (2nd ed.). New York: Van Nostrand-Reinhold.

Mitsch, W. J., & Reeder, B. C. (1991). Modelling nutrient retention of a freshwater coastal wetland: Estimating the roles of primary productivity, sedimentation resuspension and hydrology. *Ecological Modelling, 54*, 151–187.

Moss, B. (1983). The Norfolk Broadland: Experiments in the restoration of a complex wetland. *Biological Reviews of the Cambridge Philosophical Society, 58*, 521–561.

Murkin, H. R., van der Walk, A. G., & Davis, C. B. (1989). Decomposition of four dominant macrophytes in the Delta Marsh, Manitoba. *Wildlife Society Bulletin, 17*, 215–221.

Murkin, H. R., Stainton, M. P., Boughen, J. A., Pollard, J. B., & Titman, R. D. (1991). Nutrient status of wetlands in the interlake region of Manitoba, Canada. *Wetlands, 11*, 105–122.

Nair, A., Sathyendranath, S., Platt, T., Morales, J., Stuart, V., Forget, M. H., Devred, E., & Bouman, H. (2008). Remote sensing of phytoplankton functional types. *Remote Sensing of Environment, 112*, 3366–3375.

Pearl, H. W., & Ustach, J. F. (1982). Blue green algae scums: An explanation for their occurrence during freshwater blooms. *Limnology and Oceanography, 27*, 212–217.

Pickett-Heaps, J. D. (1967). Ultrastructure and differentiation in Chara sp. II. Mitosis. *Australian Journal of Biological Sciences, 20*, 883–894.

Pickett-Heaps, J. D. (1969). The evolution of the mitotic apparatus: An attempt at comparative ultrastructural cytology in dividing plant cells. *Cytobios, 1*, 257–280.

Pickett-Heaps, J. D. (1972a). Variation in mitosis and cytokinesis in plant cells: Its significance in the phylogeny and evolution of ultrastructural systems. *Cytobios, 5*, 59–77.

Pickett-Heaps, J. D. (1972b). Cell division in *Klebsormidium subtilissimum* (formerly *Ulothrix subtilissima*) and its possible phylogenetic significance. *Cytobios, 6*, 167–184.

Pickett-Heaps, J. D. (1975). *Green algae – Structure, reproduction and evolution in selected genera*. Sunderland: Sinauer.

Pinckney, J., & Zingmark, R. G. (1993). Photophysiological responses of intertidal benthic microalgal communities to in situ light environments: Methodological considerations. *Limnology and Oceanography, 38*, 1373–1383.

Porter, G., Tredwell, C. J., Searl, G. F. W., & Barber, I. (1978). Picosecond time – Resolved energy transfer in *Phormidium cruentum*. *Biochimica et Biophysica Acta, 501*, 232–245.

Prescott, G. W. (1984). *The algae – Review*. Koenigstein: Otto-Koeltz Science Publishers.

Reeder, B. C. (1994). Estimating the role of autotrophs in nonpoint source phosphorus retention in a Laurentian Great Lakes coastal wetland. *Ecological Engineering, 3*, 161–169.

Reynolds, C. S. (1984). *The ecology of freshwater phytoplankton* (pp. 1–396). Cambridge: Cambridge University Press.

Ruttner, F. (1953). *Fundamentals of limnology* (242p.). Toronto: University of Toronto Press.

Schütt, F. (1892). Das Pflanzenleben der Hochsee. *Ergebnisse der Plankton-Expedition der Humboldt-Stiftung, 1A*, 243–324.

Sieburth, J. (1978). Pelagic ecosystem structure: Heterotrophic compartments of the plankton and their relationship to plankton size fractions. *Limnology and Oceanography, 23*(6), 1256–1263.

Smayda, T. J. (1997). Harmful algal blooms: Their ecophysiology and general relevance to phytoplankton blooms in the sea. Part 2: The ecology and oceanography of harmful algal blooms. *Liminology and Oceanography, 42*(5), 1137–1153.

References

Smith, G. M. (1950). *Fresh water algae of the United States*. New York/Toronto/London: McGraw-Hill.

Sournia, A. (1973). La production primaire planctonique en Mediterranee, Essai de mise en jour. *Bulletin Etude en Commun de la Méditerranée, 5*(no sp.), 128pp.

Stanley, D. W. (1976). Productivity of epipelic algae in tundra ponds and a lake near Barrow, Alaska. *Ecology, 57*, 1015–1024.

Steidinger, K. A., & Haddad, K. D. (1981). Biologie and hydrographic aspects of red tides. *Bioscience, 31*(11), 814–819.

Stewart, K. D., & Mattox, K. R. (1975). Comparative cytology, evolution and classification of the green algae with some consideration of the origin of other organisms with chlorophylls a and b. *The Botanical Review, 41*, 104–135.

Throndsen, J. (1978). Productivity and abundance of ultra- and nanoplankton. *Oslofjorden, 63*(4), 273–284.

Van der Valk, A. G. (1994). Effects of prolonged flooding on the distribution and biomass of emergent species along a freshwater wetland coenocline. *Vegetatio, 110*, 185–196.

Vargo, G. A., Heil, C. A., Spence, D., Neely, M. B., Merkt, R., Lester, K., Weisberg, R. H., Walsh, J. J., & Fanning, K. (2001). In: G. M. Hallegraeff, S. I. Blackburn, C. Bolch, & R. J. Lewis (Eds.), *IOC of UNESCO* (pp. 157–160).

Vernberg, F. J. (1993). Salt-marsh processes: A review. *Environmental Toxicology and Chemistry, 12*, 2167–2165.

von Meyer, H. A., & Möbius, K. (1865–1872). *Fauna der Kieler bucht*. Leipzig: W. Engelmann.

Whittaker, R. H. (1969). New concepts of kingdoms or organisms. Evolutionary relations are better represented by new classifications than by the traditional two kingdoms. *Science, 163*(3863), 150–160.

Whitton, B. (1970). Biology of Cladophora in freshwaters. *Water Research, 4*, 457–476.

Yu, M. H., Glazer, A. N., & William, R. C. (1981). Cyanobacterial phycobilisome. Phycocyanin assembly in the rod substructure of *Anabaena variabilis* phycobilisomes. *Journal of Biological Chemistry, 256*, 13130–13136.

Zapata, M., Jeffrey, S. W., Wright, S. W., Rodriguez, F., Garrido, J. L., & Clementson, L. (2004). Photosynthetic pigments in 37 species (65 strains) of Haptophyta: Implications for oceanography and chemotaxonomy. *Marine Ecology Progress Series, 270*, 83–102.

Physicochemical Environment of Aquatic Ecosystem

Influence of physical and chemical environment of a water body together with the growth pattern of individuals plays important roles in phytoplankton dynamics.

2.1 Physical Factors

Among the physical factors light and temperature are the major ones which control the phytoplankton growth.

2.1.1 Light and Temperature

Pelagic ecosystems are dominated by phytoplankton populations that cover 70 % of the world's surface (Reynolds 2006). Falkowski (1995) also opined that 45 % of the earth's photosynthesis is accounted by phytoplankton populations around the world. Light absorption ability of natural phytoplankton populations is directly related to the spectral nature of the light-harvesting capabilities of the pigment molecules present in the phytoplankton population (Bergmann et al. 2004). It is well known among plant biologists that maximum light absorption of chlorophyll is achieved in blue-violet and red regions of the spectrum, while the accessory pigments like carotenoids and xanthophylls absorb mainly in the blue-green region with phycobiliproteins showing maximum absorption efficiency in yellow-red region of the spectrum. The exponential decrease in light intensity due to depth can be attributed to absorption and refraction, a phenomenon known as vertical light attenuation which can be mathematically expressed as

$$E_d(z) = E_d(0) \cdot e_d^{-k_d z} \quad [\text{Beer Lambert's Law}]$$

where $E_d(0)$ and $E_d(z)$ are light intensity at surface and at depth z respectively.

Light and temperature are the most widely studied environmental parameter that affects algal growth, both in in situ and in vitro studies. Surface waters of different ecosystems get light and heat from solar irradiation on earth surface, and as a result different distinct vertical zones are developed. Actually sunlight enters into the water of ocean or lake and is converted into heat, as a result of which euphotic zone develops. In an aquatic ecosystem, the depth up to which light penetrates is called 'euphotic zone'. It is considered as the depth up to which 1 % of the surface irradiance is reached in the water column. Turbidity of the water column can act as an important regulator of the light availability in the water column in an aquatic ecosystem. An increase in turbidity is mainly caused due to SPM (suspended particulate matter) load and colloidal matter in the water column, including both inorganic (e.g. dispersed sediment particles) and organic (e.g. phytoplankton, zooplankton). Thus, it can be said that there will be an exponential decrease in light availability with depth at a rate which is dependent on the particle content of the water column. Thus, the euphotic zone may reach from a few meters in coastal and estuarine waters

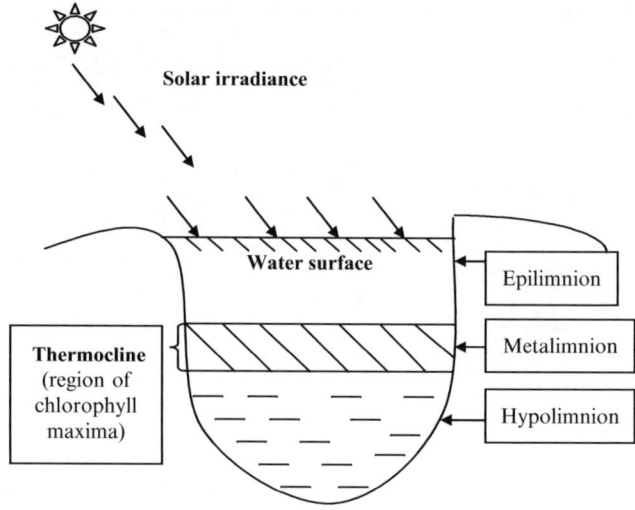

Fig. 2.1 Specific zonations on the basis of light and temperature variations at different depths of the water column

to more than 100 m in the Mediterranean Sea and the Pacific gyre. Many plankton biologists have considered this depth as the lower limit beyond which phytoplankton cells become incapable of photosynthetic activity. Exceptionally, there are reports that picoplanktonic form like *Prochlorococcus* can photosynthesize at 0.01 % of surface irradiance availability (Goericke et al. 2000). With an increase in depth, light availability decreases and the spectral range narrows to the blue region since the blue wavelength is least attenuated in a water column. So it can be said that phytoplankton cells have the ability to acclimatize to depth and light availability by increasing their pigment content and by shifting their pigment composition (e.g. *Prochlorococcus* has the ability to increase their Chl b:Chl a ratio since chlorophyll b absorbs optimally blue wavelength). Under high irradiance (top 20 m in oligotrophic waters), photosynthetic activity of phytoplanktons becomes photoinhibited due to damage of the photosystem core proteins by UV exposure. Cells counteract these detrimental effects by increasing the amount of photoprotective pigments such as zeaxanthin or diatoxanthin. Thus, the diversity in a water column is dependent on the vertical distribution of algal population. On the basis of light variation, vertical distribution can be separated into two different heads: high light, epilimnion, and low light, hypolimnion. Cells present in the hypolimnion undergo passive sedimentation due to cellular senescence or overwintering phase. Different planktonic species of the epilimnion like the dinoflagellates may migrate to the hypolimnion periodically for nutrient supplementation under nutrient replete conditions of the epilimnion. Thus, sampling by depth samplers and sedimentation chambers at different depths of the water column can provide important insights in the overall ecosystem functioning.

Temperature also follows the similar trend as was found for light, and the variation in temperature is directly related to solar irradiance. This process divides the water body into distinct layer or strata, called stratification. In lakes, the euphotic zone is divided into upper warm and less dense layer, termed epilimnion, and the lower cooler and denser layer is called the hypolimnion (Fig. 2.1). Between these two layers is an intermediate layer, the metalimnion, with a sharp decline in water temperature that gives rise to a prominent temperature gradient called the 'thermocline'. If metalimnion region lies in the euphotic zone, then maximum phytoplankton concentrates here, giving rise to the zone of 'chlorophyll maxima'.

A vertical gradient in ocean ecosystem due to differences in salinity or temperature is called 'pycnocline'. Thermal stratification in aquatic ecosystem has important consequences for phytoplankton growth and abundance. The upper warmer region with maximum light intensity is most suitable for plankton growth. Moreover, the nutrient mixing is predominant in this region due to various forms of water motion developed by interactions of air current and other forces.

In recent times, excessive fossil fuel burning has enhanced the emission of 'green house gases', mainly CO_2, that have become a major cause of concern among ecologists. This is mainly due to the high solubility of atmospheric CO_2 in oceanic water that results in an increase of sea surface temperature (SST). Increase in SST is responsible for thermal expansion of water that results in dissolution of more land mass along the low-lying coastal areas. Reports from the Sunderbans provide further evidence to this alarming issue where an average of 0.09 °C rise in sea surface temperature has been observed, much higher than the global average of 0.6 ± 0.2 °C (IPCC 2001). This has resulted in a sea level rise of 1.9 mm/year in the past 5 years around Sagar Island that have resulted in extinction of islands like Lohachara, Bedford, Kabasgadi and Mathabhanga. Moreover, such increase in temperature can alter the partial pressure in CO_2 as well as the mixed layer depth that causes drastic shifts in phytoplankton communities (Tortell et al. 2002; Kim et al. 2006). Other works have also suggested that similar increases in temperature may result in shifts in population from diatom dominance to diatom recedence due to decrease in nitrate reductase at elevated temperatures (Lomas and Gilbert 1999). Many workers have also opined that phytoplankton taxa may respond differently to temperature changes by expanding or contracting their ranges (Hays et al. 2005) or by shifting size and/or community composition, although these responses may not be consistent between different algal groups and trophic levels (Edwards and Richardson 2004).

2.1.2 Turbulence

A combination of forces from rotation of the earth, winds, solar irradiation and the tidal cycle generate different types of water motion which affect the

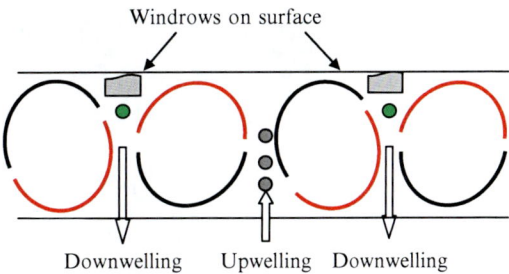

Fig. 2.2 Cross-sectional views of adjacent Langmuir cells which rotate in opposite directions, creating regions of downwelling (where the windrows occur) and upwelling. Accumulation of foam and positively buoyant algae (*green circles*) occurs in the downwelling region, whereas negatively buoyant algae (*grey circles*) accumulate in the upwelling regions beneath the surface

phytoplankton population to a greater extent. The buoyant genera with different floating devices like presence of vacuoles, flagella, other processes (setae, horns, etc.) and with less dense cell sap (with high concentration of K^+ ion instead of Na^+ ion) generally overcome the combined force and float on the water surface. Different planktons differ in their tolerance of turbulence due to structural and physiological variations (Fogg 1991), directly affecting the growth pattern and bloom formation. Following are the water motions that significantly affect the ecology of phytoplanktons:

When the wind speed of adjacent water bodies like lakes, etc., is 11 km·h^{-1} or more, then some elongated wind-driven surface rotations are formed and move spirally according to wind directions in opposite ways. These are marked by conspicuous lines of foams called 'windrows', developed from wave action. This type of wind-driven surface rotations is called convection cells or Langmuir cells. Adjacent convection cells rotate in opposite directions creating alternate upwelling and downwelling regions. Buoyant phytoplankton like *Microcystis* sp. (with cyanophycean vesicles) can survive in the region of downwelling. In the region of upwelling, negatively buoyant planktons concentrate (Fig. 2.2).

When wind blows in a particular direction, it generates the waves on the water body which move from one side to the other (north to south). But after crushing of the wave on the opposite shore, the force will return again to the north

Fig. 2.3 The convergence of the Oyashio and Kuroshio currents. When two currents with different temperatures and densities (cold, Arctic water is saltier and denser than subtropical waters) collide, they create eddies. Phytoplankton growing in the surface waters becomes concentrated along the boundaries of these eddies. The swirls of colour visible in the waters southeast of Hokkaido (*upper left*), show different kinds of phytoplankton that are using chlorophyll and other pigments to capture sunlight for photosynthesis (Source: www.wikipedia.com)

shore with a subsurface return force and act like a conveyor belt. Some fraction of this force also flows to east and west coast. Here also buoyant species like *Microcystis* sp. can survive against the wind force, but the dinoflagellate member like *Ceratium* sp. floats with the help of return force along the subsurface area.

In oceanic environment, sometimes by tremendous wind speed along with gravitational and other forces, a special type of water current develops. Due to this current, a portion of the entire ecosystem including phytoplanktons, zooplanktons, fishes and other aquatic organisms is pinched off from the whole system with varying diameters (100–200 km) and forms discrete ecosystems. These wind-driven water currents are called 'eddies' or 'loops'. Besides this, the coastal water is enriched by nutrients by upwelling water together with wave action towards the coast or away from the coast (Fig. 2.3). This is brought about by the combined action of wind and the Coriolis force of the earth's rotation. For these reasons, nutrient statuses of coastal waters vary from place to place, thereby controlling the phytoplankton diversity.

2.2 Major Nutrients

Four major nutrient elements like carbon, nitrogen, phosphorus and silica (C, N, P and Si) are regarded as the major chemical factors that control the phytoplankton productivity in any type of aquatic ecosystem. Thus, the spatial and temporal distributions of these elements play an important role in plankton dynamics. Other factors like uptake, growth, grazing and sedimentation also interact with the chemical nutrients. Many authors investigated the relationships between the distribution of phytoplankton and major nutrients. The distribution of individual species can sometimes be correlated with the concentration of the major ions. Thus, Harris and Vollenweider (1982) observed that increase in *Coscinodiscus* (estuarine inhabitant) cell counts in English Lakes was due to urban run-offs and the use of salts on the roads in winter. Light and nutrients are perhaps the two most important parameters that regulate the quantity, the distribution and structure of phytoplankton populations in natural aquatic habitats (Huisman and Weissing 1995; Diehl 2002; Hansen 2002). Unlike light and temperature which tend to have a unidirectional flow in a natural ecosystem, nutrients can be recycled. In natural aquatic habitats, nutrients can be well mixed in a water column with homogeneous distribution (mixing condition) or it can accumulate in deeper layers (under stratified conditions).

2.2.1 Redfield Ratio

The Redfield ratio has remained as one of the tenets among both biologists and geochemists with regard to aquatic biogeochemistry. Named in honour of Alfred Redfield, this concept attempts to establish the relationship between organism composition and water chemistry. Redfield (1958) opined that the elemental composition of plankton was 'uniform in a statistical sense' and that quantitative variations in inorganic C, N and P content in seawater were 'almost entirely as a result of the synthesis or decomposition of organic matter'. In this observation the C:N:P content in plankton was reported as 106:16:1.

With regard to Redfield ratio, biologists and geochemists interpret it differently. Geochemists use a C:N:P stoichiometry 105:15:1 based on the covariation of nitrate, phosphate and non-calcite contribution to total inorganic C in deep seawater, whereas biologists use a ratio of 106:16:1 based on Fleming's analysis of the average elemental composition of marine organisms (Goldman et al. 1979). The Redfield C:N has its main application in oceanography for calculation of export production and for nutrient-based productivity calculations as well as in models of ocean productivity. The Redfield N:P ratio of 16:1 is often regarded as a standard value to distinguish between N-limited and P-limited habitats among the different water bodies especially with reference to oceans (Behrenfeld and Falkowski 1997; Tyrell 1999). Accordingly, it has often been suggested that if N:P < 16, the system is N-limited, whereas it is P-limited when N:P > 16.

However, some works on oceans in periods of glacial maxima showed that the generally accepted ratio of 16:1 may not hold true and can reach as high as 25 mol N:mol P (Broecker and Henderson 1998). Keeping in view of these findings, Falkowski (2000) opined that "the upper bounds of N/P ratios in the dissolved inorganic phase in the oceans is almost certainly a consequence of the intrinsic chemical composition of marine phytoplankton" although no numerical value for this upper boundary was suggested by him. It is remarkable that most oceans around the world have a deep water N:P ratio of approximately 16, although several biochemist and physiologists have questioned the plasticity of the elemental composition of phytoplankton populations both in field and in laboratory cultures (Hecky et al. 1993). Redfield later acknowledged this fact and opined that the deep water constancy of N:P ratio is due to a complex balance between several biological processes including nitrogen fixation and denitrification (Redfield 1958).

2.2.2 Carbon

Carbon (C) is the main element that regulates the functioning of natural waters because of the intricate equilibrium that exists between CO_2, bicarbonate and carbonate that determine the acidity or alkalinity of natural waters. Moreover, this is the central element required in maximum quantity by photosynthetic organisms. The main source of carbon in water is the dissolved CO_2 from air. It is generally considered that about 90 % C in seawater exists as bicarbonate, 10 % as carbonate ion and approximately less than 1 % remains as unionized CO_2. The presence of bicarbonate as the main source of C results in slightly alkaline nature of habitat that is about 7.9 for seawater and 6–9 for freshwater respectively. In slightly acidic waters (pH $\leq$ 6), aerial CO_2 dissolved in water to form carbonic acid (H_2CO_3). For freshwater phytoplankton H_2CO_3 is available as the pH level of freshwater lakes and rivers remains slightly acidic (pH 6–6.5):

$$H_2CO_3 \rightleftarrows H_2O + CO_2;$$
$$H_2O + 2CO_2 \rightleftarrows 2HCO_3^{2-}$$

But in marine water, where pH ranges from 7.5 to 8.4, bicarbonate ions become predominant. But above pH 10, carbon exists in the form of insoluble carbonate. Thus, pH level regulates the carbonate–bicarbonate–CO_2/H_2CO_3 equilibrium in freshwater or marine habitats. Therefore, total carbon pool in aquatic ecosystems can be represented as follows:

$$TCO_2 = [CO_2] + [H_2CO_3] + [HCO_3^-] + [CO_3^{2-}].$$

As natural waters are open systems, addition or removal of CO_2 is a common feature of these habitats, especially in marine systems. The solubility of CO_2 is much higher in water than oxygen. Over the last century the burning of fossil fuels for energy generation has resulted in atmospheric CO_2 increase from 280 to 390 μ atm and a consequential decrease in surface ocean pH by 0.12 units. The current CO_2 emission scenario is predicted to raise CO_2 to 700 μ atm over the next 100 years, which will decrease seawater pH by a further 0.3 units, and raise the sea surface temperature (SST) by 2–6 °C (IPCC 2007). Climatic changes due to an increase in CO_2 concentrations can enhance algal growth (Wolf-Gladrow et al. 1999; Hutchins et al. 2007) which may also affect coccolithophores by reducing the calcification rates (Tortell et al. 2002).

Further experimental studies revealed that an excessive increase in pCO_2 may result in a shift towards diatom dominated population.

Phytoplankton abundance in relation to concentration of carbon depends upon the form it exists. Some taxa (dinoflagellate *Amphidinium carterae* and *Heterocapsa oceanica*) require CO_2 as their inorganic carbon source. Therefore, in the marine environment, they depend upon the *c*arbon *c*oncentrating *m*echanism (CCM) to get CO_2 from HCO_3^- or CO_3^{2-}. Almost all members of Heterokontophyta are with CCM present inside the cell and can thrive well exploiting HCO_3^- of marine environment.

2.2.2.1 CCM in Phytoplanktons

Algal cells primarily assimilate environmental CO_2 through C3 pathway (Calvin cycle) using RuBisCO. But very few algal forms also assimilate CO_2 by following alternative pathways in CO_2 starvation due to the weak binding affinity of RuBisCO as carboxylase. Thus, under these conditions, algal cells tend to produce phosphoglycolate as a by-product during RuBisCO performance as oxygenase. The phosphoglycolate thus produced inhibits RuBisCO activity which is further alleviated by formation of glycolate by the enzyme phosphoglycolate phosphatase. The glycolate thus produced can be excreted out by algal cells or can be further utilized as a substrate during photorespiration. In cyanobacteria CO_2 is converted to bicarbonate in the wall of the thylakoids. They also accumulate bicarbonate ion using plasma membrane transporter. This bicarbonate ion is converted to CO_2 by the enzyme carbonic anhydrase (CA) present in carboxysomes. In eukaryotic phytoplankton genera, diverse types of CCM are available. Among them, CA enzymes and membrane transporter are important. In some algal genera, pyrenoids also help in CCM. CCM that is operative in algae pumps HCO_3^- from outside the algal cell via the plasma membrane and the chloroplast membrane into the algal chloroplast. Within the chloroplast, HCO_3^- remains unchanged due to the alkaline nature of the stroma. A significant proportion of this HCO_3^- is subsequently passed into the thylakoid lumen. The enzyme carbonic anhydrase (CA) attached to the thylakoid membrane converts HCO_3^- in the lumen to CO_2 at a rate hundred times faster than nonenzymatic conversion of the same:

$$HCO_3^- + H^+ \rightleftarrows CO_2 + H_2O$$

This CO_2 formed by CA rapidly diffuses away from the thylakoid lumen to the chloroplast stroma that inhibits oxygenase activity of RuBisCO and promotes carboxylase activity of RuBisCO initiating the carbon reduction cycle.

Some algae have C_4-like photosynthesis where enzyme PEP carboxylase traps inorganic carbon. Among these CCM processes, CA is the most abundant among algal genera. CA may be secreted externally (*Skeletonema costatum*) or present at periplasmic space (*Chlamydomonas reinhardtii*). Some algae excrete protons (H^+) across the cell membrane to get CO_2 from external bicarbonate source, which is on the other hand related to $CaCO_3$ deposition (calcification). Presence of HCl in the intermembrane space between chloroplast membrane and chloroplast ER (endoplasmic reticulum) also converts CO_2 from HCO_3^- in the group Heterokontophyta that allows it to acclimatize in marine environment.

Organic carbon is also used by phytoplankton genera like *Chlorella* spp., *Cocconeis* spp., etc. Lewitus and Kana (1994) reported about the use of glucose as carbon source by *Closterium* at estuarine region.

2.2.3 Nitrogen

Phytoplankton productivity and their diversity are highly controlled by nitrogen. Presence and cycling of nitrogen in ocean and lakes is a complex phenomenon because it exists in four dissolved forms of inorganic nitrogen (dissolved N_2 gas, NO_3^-, NO_2^- and NH_4^+). Both atmospheric nitrogen and dissolved inorganic nitrogen (DIN) of water bodies play an important role in controlling phytoplankton population. Organic nitrogen sources like urea and amino acids are present in ocean surfaces. Nitrogen is primarily present in water as dissolved inert nitrogen gas. In ocean, almost 95 % of nitrogen occurs as N_2. Dissolved inorganic nitrogen includes the nitrate (NO_3^-) ion in oxygen-rich water, nitrite (NO_2^-) ion in moderate oxygen level and ammonium (NH_4^+) ion in water body with less oxygen with high BOD. Previously it was assumed that nitrogen was the universal limiting nutrient in marine waters. Although planktonic populations are capable of assimilating different forms of nitrogen, the more preferred form is ammonia as it directly gets utilized in the biosynthesis of amino acids. In cyanobacterial species ammonia is considered to be a complete inhibitor of nitrogen fixation and heterocyst formation. In contrast, conversion of nitrate to ammonia is a more energy-requiring process and requires the enzyme nitrate reductase. Thus, to minimize the energy currency, ammonia is often preferred by natural phytoplankton populations especially cyanobacteria. In natural habitat like oceans and lakes, the dynamics of nitrogen cycling is complex as it involves the interconversion between four dissolved inorganic forms which can be achieved only by bacterial action in the environment or within living cells. It is opined that more than 90 % of oceanic N presents as molecular N_2 and the rest in other forms. Thus, it is believed that the concentrations of these nutrients primarily depend on the degradation of biological material synthesized by the biotic components in the surface waters of these habitats. The supply of NO_3^- in the surface waters from the cooler hypolimnetic nutrient-rich waters depends upon vertical advection and eddy diffusion across the thermocline. Due to the high rate of nitrification and ubiquitous presence of oxygen in the surface waters of natural lotic systems, NO_2^- concentrations are much lower as compared to NO_3^- concentrations. The higher concentrations can only be expected at deeper waters beyond the thermocline where oxygen concentrations are relatively low as compared to that of surface waters.

In oligotrophic waters, NO_3^- is the dominant form of DIN (dissolved inorganic nitrogen) because of the high rate of nitrification and the even distribution of oxygen, leading to the more oxidized form of nitrogen. In contrast, for eutrophic waters, the utilization of oxygen is more for decomposition processes in the deep waters and the sediment layers. This would allow the build-up of other reduced forms of nitrogen as well, released primarily from organic sources. In the temperate region, winter mixing brings up nitrogen from the deep oceans mainly as nitrate. Thus, sources of nitrogen for natural phytoplankton populations can be (a) surface water DIN, (b) DIN brought up in the surface waters from deep waters through physical processes like upwelling and vertical advection, and (c) localized inputs brought down by rivers and ground water as well as seasonal precipitation.

In locations like the central subtropical oceans, recycling of surface waters is the main source of DIN where upwelling events are not prominent due to the existence of a vertical thermocline. Moreover, due to the remoteness of these locations, localized coastal inputs are also not very common. On the other hand, in coastal oceanic waters, upwelling, vertical advection as well as localized inputs from riverine sources cumulatively contribute to nitrogen concentrations in these waters. During spring as the water stratifies, phytoplankton population increases and absorbs nitrate that eventually depletes to nanomolar levels in the surface layers. In marine environment, cyanobacterial taxa *Trichodesmium erythraeum* appear as a surface bloom, turning the ocean water red. They fix N_2 from dissolved nitrogen of ocean water. In most temperate oligotrophic and mesotrophic freshwaters, NO_3^- is present in comparatively more amount together with phosphorus. In highly eutrophic waters NH_4^+ and NO_2^- are also present in both surface and subsurface area. If oxygen is present, NO_2^- is converted to NO_3^-, whereas in anoxic waters, it is reduced to NH_4^+. In nitrate-enriched water maximum plankton diversity can be recorded. Green algal genera like *Scenedesmus* spp., *Chlorococcum* spp., *Kirchneriella* spp., *Ankistrodesmus* spp., etc., generally flourished in mesotrophic to moderately eutrophic waters. But in NO_3^-- and NH_4^+-enriched water, phytoplankton genera like *Spirulina* spp. and *Trichodesmium* spp. appear in large quantities, sometimes resulting in algal blooms.

2.2.4 Phosphorus

Phosphorus concentrations in surface waters are mainly accounted by geochemical processes that occur in a basin adjoining an aquatic habitat. Unlike N, P remains mostly in bound forms in clay minerals and different soil components (Vollenweider et al. 1998). Thus, while studying P availability, mainly two sources of P are to be considered: (1) DIP (dissolved inorganic phosphate) and DOP (dissolved organic phosphate) and (2) particulate P mainly accounted for biological availability. Unlike N, for P the more common form of the element is PO_4^{3-} that is abundant both in biomolecules and in the environment. The requirement of phosphorus by phytoplankton populations is much smaller as compared to carbon, nitrogen and silicon. Thus, phytoplankton populations flourish even under nanomolar phosphorus concentrations, as can be observed in the Eastern Mediterranean and the Aegean Sea.

DIP is mostly constituted by orthophosphate (PO_4^{3-}) with much lower concentration of monophosphates (HPO_4^{2-}) and dihydrogen phosphate (H_3PO_4). Phytoplanktons primarily utilize DIP as a phosphorus source, but under limiting conditions extracellular alkaline phosphatase (AP) promotes utilization of phosphate bound to organic substances. Thus, level of AP activity is an indication of phosphate limitation in aquatic habitats (Rengefors et al. 2003). Under P-limited conditions, if there is a sudden pulse of soluble reactive phosphorus (SRP), planktonic algal cells have the unique ability to make 'luxury consumption' and develop polyphosphate bodies, thereby creating an internal pool of phosphorus to deal with phosphate shortage.

In oligotrophic waters, DIP turnover rate is very low in winter and accounts for only 10 % of total phosphorus (TP). This results in low phytoplankton populations with low growth rates. In coastal marine waters, DIP builds up in periods of vertical mixing. Under stratification, DIP pool is depleted that results in species-wise drop in phytoplankton populations. In contrast, in eutrophic freshwater, DIP may constitute 100 % of TP due to uncontrolled discharges from point (industrial effluents, sewage) and nonpoint sources (runoffs from agricultural and urban areas) (Capone et al. 2005). The increased concentration of DIP may be surplus to algal requirements, thereby building up in water column, with a slow turnover rate. This excess phosphorus provides an opportunity for cyanobacteria populations to develop and can even reach blooming proportions.

2.2.5 Silicon

Small amounts of silicon are required by all planktons for protein synthesis. In freshwater, soluble reactive silica (SRSi) exists as monosilicic acid (H_4SiO_4) that ranges from 0.7 to 7 mg/L (25–250 µM). In oceanic environment, maximum concentrations of SRSi (~3 mg/L) present in upwelling zones. Utilization of SRSi by diatom for development leads to reduce the levels of SRSi in both freshwater and oceanic habitats. During periods of summer stratification, concentration of SRSi may reach below detectable levels (<0.1 µM) and becomes a limiting nutrient in these habitats.

Diatom cell covering or frustule is made up of polymerized silica that increases the density of diatom cell. Thus, development and abundance of diatoms is dependent on turbulent mixing conditions that render them buoyant in the euphotic zone of the water column. Polymerized silica decomposes slowly (~50 days) which is a possible hindrance in rapid recycling of silicon in the epilimnion of shallow lakes. Dead diatom cells often reach the benthos in intact forms and settle down as sediments. Dissolved silicon is available in surface waters of habitat by external inputs and turnover of the water column during mixing conditions. In oceans, as the mixing depth is much greater than lakes, silicon in diatom valves redissolves between the surface and about 1,000 m depth.

Parameters like nitrate, nitrite, ammonia, dissolved inorganic nitrogen (DIN), dissolved inorganic phosphate (DIP) and dissolved silicate (DSi) contents can be measured spectrophotometrically within 30 min of sample collection following the protocols of APHA (1998). The values thus obtained are fitted to a standard curve prepared for determination of nutrient concentrations of water samples.

The procedures are mentioned below:

Nitrate

Reagents required: Silver sulphate solution, phenol disulphonic acid, liquid ammonia.

Procedure: An aliquot of 50 mL filtered sample water is taken in a conical flask to which an equivalent amount of silver sulphate solution is added and heated slightly to precipitate any chloride content that may be present. The filtrate of sample thus obtained is evaporated to dryness in a porcelain basin. The residue obtained is dissolved in 2 mL of phenol disulphonic acid and the contents are diluted, if necessary. Subsequently, 6 mL of liquid ammonia is added to the solution to develop a yellow colour.

Absorbance is recorded at 410 nm. Concentration of nitrate was calculated from standard curve, prepared from known nitrate concentration.

Nitrite

Reagents required: EDTA solution, sulphanilic acid, α-N-napthylethylene amine, sodium acetate solution.

Procedure: Filtered sample water of known volume (50 mL) is taken in a conical flask. To it, 2 mL of each of EDTA solution, sulphanilic acid, α N–napthylethylene amine and sodium acetate solution is added in succession. The reagents are thoroughly mixed and allowed to stand for 5 min. A wine red colour developed. Absorbance is recorded at 543 nm. Concentration of nitrite is calculated from standard curve.

Ammonia

Reagents required: Nessler's reagent.

Procedure: 50 mL filtered sample water is taken in a conical flask. To it, 2 mL of Nessler's reagent is added. The reagent is thoroughly mixed and is allowed to stand for 5 min. A pale yellow colour developed. Absorbance is recorded at 640 nm. Concentration of ammonia is calculated from standard curve.

Dissolved Inorganic Phosphate (DIP)

Reagents required: Ammonium molybdate solution, stannous chloride solution.

Procedure: An aliquot of 50 mL of filtered sample water is taken in a clean conical flask and 2 mL of ammonium molybdate is added to it. This is subsequently followed by addition of five drops of stannous chloride solution. A blue colour developed. Absorbance is recorded at 690 nm after 5 min but before 12 min of the addition of the last reagent. Concentration of phosphate in the sample water is calculated from a standard curve.

Dissolved Silicate (DSi)

Reagents required: Ammonium molybdate solution, 1(N) hydrochloric acid, oxalic acid.

Procedure: 50 mL of filtered sample water is taken in a clean conical flask and 2 mL of ammonium molybdate is added to it. This is subsequently followed by addition of 0.5 mL of 1(N) hydrochloric acid. After thorough mixing, 2 mL of oxalic acid is added. A bright yellow colour developed. Absorbance is recorded at 530 nm. Concentration of dissolved silicate in the sample water is calculated from a standard curve.

2.2.6 Nutrient Uptake Model

During the growth of phytoplanktons, the minerals are consumed and several models have been proposed to establish the relations between the rate of nutrient uptake, their storage inside the cell and ultimately the growth pattern of the phytoplankton.

2.2.6.1 Michaelis–Menten Model (1913)

This model was based on the kinetics of enzyme function where ρ is considered as the nutrient transport rate (μ mole of nutrient per cell per minute). The term ρ_{max} is the maximum velocity of the nutrient transport, and S is the substrate concentration. ρ approaches to ρ_{max}, when the substrate concentration S is high and the internal store of that same nutrient (Q) is low. K_t is the half saturation constant, which equals to the values of S where $\rho = \frac{1}{2}\rho_{max}$, and unit of K_t is same as that of substrate (μ mol L^{-1}).

Therefore, according to Michaelis–Menten model, the nutrient uptake pattern is as follows:

$$\rho = \rho_{max}\left(S / K_t + S\right)$$

2.2.6.2 Droop Model

Droop (1983) proposed an equation for establishment of growth rate and internal nutrient quota as follows:

$$\mu = \mu_{max}\left(1 - Q_0\right)/Q$$

where μ is the gross growth rate or the rate of reproduction and μ_{max} is the maximum rate of reproduction. Q is the internal quota of nutrient. When Q approaches Q_0, then μ is zero.

By this model, we can understand the relationship between the growth rate and the nutrient storage within the phytoplankton cell. When phytoplankton can store nutrient at higher level, i.e. Q is more, then growth rate will also be more. When internal storage is exhausted, i.e. Q is 'zero', then growth also stops.

References

APHA. (1998). *Standard methods for the examination of water and wastewater* (20th ed.). Washington, DC: APHA-AWWA-WPCF.

Behrenfeld, M. J., & Falkowski, P. G. (1997). A consumer's guide to phytoplankton primary productivity models. *Limnology and Oceanography, 42*(7), 1479–1491.

Bergmann, S., Ihmels, J., & Barkai, N. (2004). Similarities and differences in genome-wide expression data of six organisms. *PLoS Biology, 2*, E9.

Broecker, W. S., & Henderson, G. M. (1998). The sequence of events surrounding Termination II and their implications for the cause of glacial-interglacial CO_2 changes. *Paleoceanography, 13*(4), 352.

Capone, D. G., et al. (2005). Nitrogen fixation by *Trichodesmium* spp. An important source of new nitrogen to the tropical and subtropical North Atlantic Ocean. *Global Biogeochemical Cycles, 19*(GB2024), 17. doi:10.1029/2004GB002331.

Diehl, S. (2002). Phytoplankton, light, and nutrients in a gradient of mixing depths: Theory. *Ecology, 83*, 386–398.

Droop, M. R. (1983). 25 years of algal growth kinetics – A personal view. *Botanica Marina, 26*, 99–112.

References

Edwards, M., & Richardson, A. J. (2004). Impact of climate change on marine pelagic phenology and trophic mismatch. *Nature, 430*, 881–884.

Falkowski, P. G. (2000). The global carbon cycle: A test of our knowledge of earth as a system. *Science, 290*(5490), 291–296.

Falkowski, P. G. (1995). Ironing out what controls primary production in the nutrient rich waters of the open ocean. *Global Change Biology, 1*, 161–163.

Fogg, G. E. (1991). The phytoplanktonic ways of life. *The New Phytologist, 118*, 191–232.

Goericke, R., Olson, R. J., & Shalapyouok, A. (2000). A novel niche for *Prochlorococcus* sp. in low-light suboxic environments in the Arabian Sea and the Eastern Tropical North Pacific. *Deep-Sea Research, 47*, 1183–1205.

Goldman, J. C., McCarthy, J. J., & Peavey, D. G. (1979). Growth rate influence on the chemical composition of phytoplankton in oceanic waters. *Nature, 279*, 212–215.

Hansen, P. J. (2002). Effect of high pH on the growth and survival of marine phytoplankton: Implications for species succession. *Aquatic Microbial Ecology, 28*, 279–288.

Harris, G. P., & Vollenweider, R. A. (1982). Paleolimnological evidence of early eutrophication in lake Erie. *Canadian Journal of Fisheries and Aquatic Sciences, 39*, 618–626.

Hays, G. C., Richardson, A. J., & Robinson, C. (2005). Climate change and marine plankton. *Trends in Ecology & Evolution, 20*(6), 337–344.

Hecky, R. E., Campbell, P., & Hendzel, L. L. (1993). The stoichiometry of carbon, nitrogen, and phosphorus in particulate matter of lakes and oceans. *Limnology and Oceanography, 38*(4), 709–724.

Huisman, J., & Weissing, F. J. (1995). Competition for nutrients and light in a mixed water column: A theoretical analysis. *The American Naturalist, 146*(4), 536–564.

Hutchins, D. A., Fu, F. X., Zhang, Y., Warner, M. E., Feng, Y., Portune, K., Bernhardt, P. W., & Mulholland, M. R. (2007). CO_2 control of *Trichodesmium* N_2 fixation, photosynthesis, growth rates, and elemental ratios: Implications for past, present, and future ocean biogeochemistry. *Limnology and Oceanography, 52*(4), 1293–1304.

IPCC. (2001). *Climate change 2001: The scientific basis. Contribution of Working Group I to the third assessment report of the Intergovernmental Panel on Climate Change* (J. T. Houghton, Y. Ding, D. J. Griggs, M. Noguer, P. J. van der Linden, X. Dai, K. Maskell, & C. A. Johnson, Eds., 881pp.). Cambridge/New York: Cambridge University Press.

IPCC. (2007). Summary for policymakers. In: S. Solomon, D. Qin, M. Manning, Z. Chen, M. Marquis, K. B. Averyt, M. Tignor, & H. L. Miller (Eds.), *Climate Change 2007: The physical science basis. Contribution of Working Group I to the Fourth Assessment Report of the Intergovernmental Panel on Climate Change* (996pp.). Cambridge/New York: Cambridge University Press.

Kim, J. M., Lee, K., Shin, K., Kang, J. H., Lee, H. W., Kim, M., Jang, P. G., & Jang, M. C. (2006). The effect of seawater CO_2 concentration on growth of a natural phytoplankton assemblage in a controlled mesocosm experiment. *Limnology and Oceanography, 51*(4), 1629–1636.

Lewitus, A. J., & Kana, T. M. (1994). Responses of estuarine phytoplankton to exogenous glucose: Stimulation versus inhibition of photosynthesis and respiration. *Limnology and Oceanography, 39*, 182–189.

Lomas, M. W., & Gilbert, P. M. (1999). Interactions between NH_4 and NO_3 uptake and assimilation: Comparison of diatoms and dinoflagellates at several growth temperatures. *Marine Biology, 133*, 541–551.

Michaelis, L., & Menten, M. L. (1913). Die Kinetik der Invertinwirkung. *Biochemische Zeitschrift, 49*, 333–369.

Redfield, A. C. (1958). The biological control of chemical factors in the environment. *American Scientist, 46*(3), 205–221.

Rengefors, K., Ruttenberg, K. C., Haupert, C. L., Taylor, C., Howes, B. L., & Anderson, D. M. (2003). Experimental investigation of taxon-specific response of alkaline phosphatase activity in natural freshwater phytoplankton. *Limnology and Oceanography, 48*(3), 1167–1175.

Reynolds, C. S. (2006). *The ecology of phytoplanktons*. Cambridge, UK: Cambridge University Press.

Tortell, P. D., DiTullio, G. R., Sigman, D. M., François, M., & Morel, M. (2002). CO_2 effects on taxonomic composition and nutrient utilization in an Equatorial Pacific phytoplankton assemblage. *Marine Ecology Progress Series, 236*, 37–43.

Tyrell, T. (1999). The relative influences of nitrogen and phosphorus on oceanic primary production. *Nature, 400*, 525–531.

Vollenweider, R. A., Giovanardi, F., Montanari, G., & Rinaldi, A. (1998). Characterization of the trophic conditions of marine coastal waters with special reference to the NW Adriatic Sea: Proposal for a trophic scale, turbidity and generalized water quality index. *Environmetrics, 9*, 329–357.

Wolf-Gladrow, D. A., Bijma, J., & Zeebe, R. E. (1999). Model simulation of the carbonate chemistry in the microenvironment of symbiont bearing foraminifera. *Marine Chemistry, 64*(3), 181–198.

Phytoplanktons and Primary Productivity

3

Phytoplanktons have a unique ability to sequester dissolved as well as free carbon dioxide from aquatic ecosystems and convert them as storage product. This phenomenon is regarded as the *primary productivity* of that ecosystem. As proposed by Field et al. (1998), about 50 % of global productivity through carbon sequestration is carried out in aquatic ecosystems due to the comparatively higher solubility of CO_2 than O_2 in natural water. Falkowski opined that about 98 % of productivity in oceans is accounted for phytoplankton populations. Thus, studies on productivity by phytoplankton populations in natural lotic habitat of estuaries and coasts have gained considerable attention by the plankton biologists.

The total primary productivity of a population under optimal conditions of light and temperature is regarded as the gross primary productivity (GPP). Although aquatic autotrophs are the only source of primary productivity, yet this production is consumed by the entire microbial floral and faunal populations of aquatic ecosystems. Thus, through respiration, often regarded as community respiration rate (CRR), fixed carbon is utilized as a source of energy which thereby causes a considerable decrease in the total productivity of the entire ecosystem known as net primary productivity (NPP). Thus, it can be represented as

$$NPP = GPP - CRR, or GPP = NPP + CRR$$

The relations between diversity and productivity in planktonic populations are not always cause and effect relationship. Based upon the kind of diversity and productivity, mainly three different mechanisms have been proposed:

1. Complementary use of resources (complementary effect)
2. Facilitation between species in highly diverse communities (facilitation hypothesis)
3. Higher probability that highly diverse communities include a highly productive species (sampling or selection effect)

It is a well-known fact that chlorophyll containing cells can fix CO_2 through light-assisted photosynthetic carbon fixation with the production of oxygen (O_2). Thus, dissolved oxygen content in aquatic ecosystems is actually a measure of the photosynthetic efficiency of phytoplankton populations. Thus, estimation of DO contents under different conditions of light availability can be a possible proxy for estimation of primary productivity of aquatic ecosystems. Based upon this concept, Winkler (1888) proposed a set of procedures and formulae for estimation of DO as well as productivity of aquatic habitats.

Initially, sample waters from specific habitat are collected in specially designed BOD bottles (125 mL/250 mL) below the surface of water at specific depths and are immediately stoppered below the water surface to avoid any external exchanges. The sample thus collected is immediately fixed by addition of manganese sulphate ($MnSO_4$) and alkaline potassium iodide (KI). This method of precipitation is done to stop any biological activity by planktonic organisms that may alter the actual DO content. This bottle is designated as initial bottle (IB). The precipitate on dissolution by acids is titrated using sodium thiosulphate with starch as indicator. The DO content is subsequently calculated from the following formula:

$$DO(mg/L) = \frac{x \times .025 \times 8 \times 1{,}000}{\frac{V_2(V_1 - v)}{V_1}}$$

where:
V_1 = total volume of sample taken (125 mL)
V_2 = volume taken for titration (100 mL)
v = 2 mL of reagents (1 mL $MnSO_4$ + 1 mL alkaline KI)
x = volume of $Na_2S_2O_3$ consumed for titration

These processes are repeated for two other sample bottles as well and are designated as light (LB) and dark bottles (DB) respectively. LB is incubated in natural light and DB is removed from light and kept in complete darkness for equal periods of time. After incubation, DO contents of both LB and DB are determined in the same way as has been mentioned in the previous section. The different parameters of productivity are determined from the following formulae:

$$GPP = \frac{\left[(O_2 LB) - (O_2 DB)\right] \times 1{,}000}{PQ \times t}$$

$$NPP = \frac{\left[(O_2 LB) - (O_2 IB)\right] \times 1{,}000}{PQ \times t}$$

$$CRR = \frac{\left[(O_2 IB) - (O_2 DB)\right] \times 1{,}000}{t}$$

where:
GPP = gross primary productivity
NPP = net primary productivity
CRR = community respiration rate
O_2 LB = DO content of the BOD bottle after incubation in sunlight for 3 h
O_2 IB = DO content of the BOD bottle immediately after sampling
O_2 DB = DO content of the BOD bottle after incubation in dark for 3 h
PQ = photosynthetic quotient ($\equiv 1.2$)
t = time period of incubation (light/dark) (in hours)

A combination of factors like light, CO_2 concentration, species composition and chlorophyll content play a well-orchestrated role in the overall primary productivity in aquatic ecosystems. In oligotrophic waters, due to the clear nature of the water column, light penetration is very high that enhances the epilimnion to depths of up to 60 m, as found in the Aegean Sea, Eastern Mediterranean region. In contrast to the general feeling, net primary productivity is not very high in these oligotrophic waters, although the PAR (photosynthetically active radiations) availability is quite high in the water column. This is mainly due to the nutrient status of the habitat where phosphate concentrations often reach nanomolar levels. In these limiting conditions, phytoplankton species that can acquire phosphate from organic sources by using extracellular alkaline phosphatase (AP) constitutes the phytoplankton population. Thus, the plankton population is dominated by nano- and picoplanktonic forms with significantly low chlorophyll a content and low cellular biovolumes. This accounts for the relatively low primary productivity of the Eastern Mediterranean as compared to other eutrophic habitats around the world.

An entirely different scenario is observed for eutrophic waters like the Bhagirathi–Hugli estuary. Although both light and nutrients are optimally present in these habitats, yet the high SPM (suspended particulate matter) load in the water column inhibit PAR availability at different depths of the water column. This low availability of PAR tends to reduce the photosynthetic efficiency of phytoplankton taxa that culminates in reducing the primary productivity of the aquatic ecosystem.

In eutrophic water, the nutrient-enriched habitat promotes diversification of microbial population that includes bacterial population as well. As can be expected for a food web, bacterial population acts as decomposer. For the purpose of decomposition, oxygen acts as a major source for oxidation–reduction reactions. The requirement of O_2 by microbial population for biochemical decomposition of organic matter is regarded as *biochemical oxygen demand (BOD)*. In a eutrophic habitat, the high nutrient status promotes survival of diverse life forms which subsequently add up to the organic decomposable matter load on course of their death and decay. Thus, eutrophic habitats tend to represent under saturated DO levels with high BOD values with an increase in heterotrophic population.

For determination of biochemical oxygen demand (BOD), the water samples are to be kept under optimum conditions of light and temperature without fixation with addition of 1 mL of each of K_2HPO_4, Na_2HPO_4, $7H_2O$, $MgSO_4$, anhydrous $CaCl_2$ and $FeCl_2$ and $6H_2O$. After 5 days from the day of sampling, DO contents are subsequently measured (Winkler 1888).

BOD values were determined from the following formula:

$$BOD(mg/L) = (DO_{0\ days} - DO_{5\ days})$$

Thus, a combination of different factors is indicative of the overall trophic status as well as the primary productivity of different study areas at different ecosystems around the world.

References

Field, C. B., et al. (1998). Primary production of the biosphere: Integrating terrestrial and oceanic components. *Science, 281*, 237–239.

Winkler, L. W. (1888). The determination of dissolved oxygen in water. *Berichte der Deutschen Chemischen Gesellschaft, 21*, 2843.

Community Pattern Analysis

Phytoplankton communities in aquatic ecosystems are the most important component that varies significantly on the basis of the available environmental conditions and trophic status of the habitat. Thus, analysis of the phytoplankton community is highly indicative of the condition of the habitat. Interpretation of plankton data from an ecological perspective depends upon the sampling strategy and the area of study. Thus, strategies of phytoplankton sampling may vary depending upon the ecosystem dynamics which is different for standing water (lakes and wetlands) as compared to lotic habitats (rivers and estuaries). For proper and precise data, collection sampling cannot remain restricted to a particular station or site. Several sites/stations should be sampled on the same instance to reduce uneven horizontal distribution (patchiness). Accordingly, sample collection for phytoplankton community analysis is an important aspect for correct community pattern analysis. From time to time the procedures for phytoplankton sample collection have improved significantly. Thus, in this section, we will discuss about the more commonly used phytoplankton sampling methods applicable both in estuarine and marine habitats.

4.1 Phytoplankton Sampling

In an attempt to determine the cell count of phytoplankton populations, different methods have been implemented as follows:

1. Bottle samplers
2. Plankton pumps
3. Plankton nets

4.1.1 Bottle Samplers

Sampling of water by bottle sampler is probably the simplest but well-recommended method to correctly determine the quantitative composition of the phytoplankton. A water bottle sample generally contains all but the rarest organisms in the water mass sampled and includes the whole size spectrum from the largest entities, like diatom colonies to the smallest single cells. Bottle sample method is the simplest method which is mainly used for the collection of water samples from any desired depth of shallow systems like the nearshore water, estuaries and mangroves (Fig. 4.1). The sample volume as well as the depth at which samples are to be collected can be easily controlled as per the discretion of the plankton biologists. Thus, this simple method can be utilized by both inexperienced students and experienced researchers for collection of phytoplankton samples in a water column.

4.1.1.1 Meyer's Water Sampler (Fig. 4.2)

This type of water sampler consists of ordinary glass bottles of 2 L capacities which are enclosed with a metal band. This apparatus is weighted below with a lead weight and there are two strong nylon graduated ropes: one tied to the neck of the bottle and the other to the cork that caps the open

Fig. 4.1 Simple graduated bottles for collection of phytoplankton water samples. The nozzle in front allows the observer to squeeze out the measured portions of sample for study

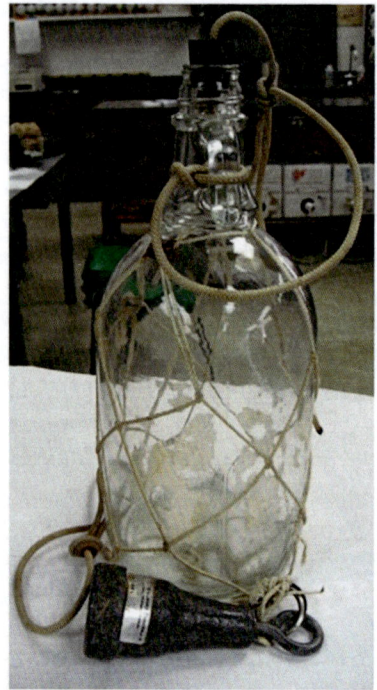

Fig. 4.2 A simple Meyer's water sampler

end of the bottle. During sample collection, the corked up apparatus (closed bottle) is brought down to the desired depth where the stopper is jerked open by a strong pull of the cork rope. Water flows into the bottles and then the cork rope is released to keep the cork closed. Afterwards, using the neck rope, the bottle containing the water sample as well as the biotic variables is taken out of the water columns. This apparatus can only be used up to the depths of 20 m.

4.1.1.2 Friedinger's Water Sampler

This water sampling apparatus is made of Plexiglas or Perspex with two hinged covers. While operation, the sampler is sent down in an open state to the desired depth and can be closed by a drop weight messenger, which falls down inside on sliding rail and closes the covers and makes the bottle water tight. By this way, the water together with the planktonic organisms of the specified column is trapped inside.

4.1.1.3 Niskin Water Sampler (Fig. 4.3)

This is a more sophisticated apparatus that is mainly used for collection of large samples in river systems as well as in oceans. It is employed for taking water samples for phytoplankton enumeration from subsurface levels to various depths. In this apparatus, several non-metallic, free-flushing bottles are used for general water sample collection. These samplers can be individually or serially attached on a hydro cable and activated by a messenger, or placed in any kind of multisampling system (like G.O., Sea Bird, Falmouth Scientific, and small multisampling system), and activated by remote or preprogrammed command.

4.1.2 Plankton Pumps

Plankton pumps are integrating samplers that pump a continuous stream of water to the surface and the phytoplankton can then be rapidly concentrated by continuous filtration. Because the pumps can collect continuously as the tube is lowered through the water column, the samples are integrated from the surface to the desired depth. This method has its disadvantages like breaking up of colonies or large *Chaetoceros* setae or long pennate cells like *Thalassiothrix* spp.

Fig. 4.3 A Go-Flo rosette Niskin water sampler

4.1.3 Plankton Nets

Phytoplankton nets are most popularly used device for sampling. Plankton nets may vary in design that range from basic tow nets (conical or with truncated neck) to more complicated device fitted apparatus for more specific sample collection (Fig. 4.4). Nets permit quantitative studies, since the mesh size will select the type of phytoplankton collected. Sampling by nets is highly selective, depending on the mesh size of the gauze, net towing speed and the species present in the water. *Chaetoceros* setae, for instance, may form a fine network inside the gauze and very small single cells, which in other cases pass through the meshes, are retained. On the other hand, nets with very fine meshes (5 or 10 µm) often filter too little water to provide an adequate diatom sample. The most useful mesh size for collecting diatoms is 25 µm. Net hauls have the advantage of a simultaneous collection and concentration of the plankton providing sufficient for species identification. A typical plankton net usable in the surface layers is conical in shape and has the following constituents: a net ring made up of stainless steel, wrapped and sealed with polythene tubing, present anteriorly. To this, a non-filtering portion made of a coarse khaki cloth is attached using button and hole system. The filtering portion

Fig. 4.4 Image of a typical phytoplankton net being hauled for phytoplankton collection (Image courtesy: www.nearhus.gr)

is made of monofilament nylon material as described earlier and is followed by again a non-filtering portion of khaki cloth. To the latter, a metal net bucket provided with a stopcock is tied with a strong twine. The determination of the volume of water filtered through any plankton net is essential for the estimation of the standing crop. The volume of water traversed by the net is determined as an approximate value by the formula $v = \pi \cdot r^2 d$, where V is the volume of the water filtered by the net, r is the radius at the mouth of the net and d is the distance through which the net is towed. The water collected through the different water samplers is either centrifuged or passed through fine mesh nylon or filter papers to separate the plankton present in it. The smaller the subsample, the fewer number of rare species will be obtained. On the other hand, there is no point in concentrating large quantities of a sample rich in one or a few species. Concentration by settling, concentration by centrifugation and concentration

Table 4.1 Showing net mesh sizes and types of planktons to be harvested

Size of aperture	Approximate open area (%)	Types of planktons
1,024	58	Largest zooplankton and ichthyoplankton
752	54	Larger zooplankton and ichthyoplankton
569	50	Large zooplankton and ichthyoplankton
366	46	Large microcrustacea
239	44	Zooplankton – microcrustacea
158	45	Zooplankton – microcrustacea and most rotifers
76	45	Net phytoplankton – macroplankton and microplankton
64	33	
53	–	
2	–	Nanophytoplankton

by filtration are the most used methods. Plankton concentration is generally used to overcome the damages caused, to certain groups of phytoplankton especially the setoid diatoms and dinoflagellates, by vacuum filtration and centrifugation. The simple plankton concentrator, which is quite gentle in its action, consists of a stiff tube (1.2 cm dia.: 10 cm height) of Perspex or PVC to the bottom of which a filter is attached. A filter paper (Whatman No. 42) or membrane filters supported by monofilament nylon netting which serves as the filter are glued at the bottom of the tube with the aid of ethylene dichloride. While using the tubes are dipped slowly into a beaker containing the phytoplankton sample. Through the filter water flows slowly upwards into the tube and is removed with a large pipette. By forcing the tube downwards, the rate of flow through the filter can be increased. On the basis of mesh size, different types of planktons can be harvested using different plankton nets (Table 4.1).

The phytoplankton net can be hauled horizontally, vertically or obliquely on the basis of sampling requirements. A vertical haul is more appropriate for collection of composite sample of the entire water column, whereas a horizontal haul remains restricted to surface water composite samples only. In case of horizontal haul, the volume of water sampled can be estimated by determining the area of the net aperture and the distance it travelled through a flow meter (Eaton et al. 2003). The basic drawback for this hauling method is that it will accommodate both phytoplanktons and zooplanktons. Thus, phytoplankton biologists often use a zooplankton sampler within a phytoplankton sampler to reduce the number of zooplanktons in the sample, although there remains a risk that larger-sized phytoplankton may get arrested in the zooplankton net. Thus, on completion of the hauling process, samples are collected by unscrewing the end fitting and subsequent collection in sample tubes/containers.

4.1.3.1 Preservation and Fixation of Plankton Samples

Generally 2–5 % formalin is used for preservation and fixation of phytoplankton samples. Commercially available formaldehyde is suitably diluted to desired conditions for fixation. Marine samples are mostly preserved in 5 % neutralized formalin in seawater. Excess seawater is generally removed by filtration for 1–2 days to reduce precipitation of salts from seawater. Eighty percent of pure methyl alcohol was also used as an effective preservative although it often produces shrinkage and discolouration. Formalin–acetic acid–alcohol (FAA) is a good preserving and killing agent for cytological studies of planktonic populations except dinoflagellates. In case of dinoflagellates, it causes loosening of thecal plates that often causes improper fixation of samples. In recent times, Lugol's iodine solution has popularized as a preserving agent for phytoplankton especially of small sizes. The iodine component fixes, preserves and colours the plankton, whereas the other component, acetic acid, preserves the flagella and cilia. This acts as an excellent preservative if the samples are stored in the dark.

4.2 Biomass Estimation

Quantification of phytoplankton biomass is an important aspect as it works as a possible proxy for primary productivity in aquatic ecosystems and gives a measure of the amount of organic

4.2 Biomass Estimation

material available for zooplankton consumption. Thus, phytoplankton population can be quantified under two different heads:

Total Biomass

In this measurement, the entire biomass is measured. Chlorophyll *a* estimation has remained as the most preferred method for this estimation as chlorophyll *a* acts as the main light-harvesting pigment in all groups of phytoplankton taxa recorded as in case of higher plants as well.

Species and Group Biomass

Here, indirect estimates of populations are made on the basis of cellular counts and biovolumes. In these cases, determination of group-specific photopigment contents can be an ideal proxy. The importance of such estimates lies not only in assessing the productivity but can also be a possible reference for the determination of the diversity of the study area. For phytoplankton community pattern analysis, different parameters have to be considered to determine the contribution of individual phytoplankton taxa to the entire phytoplankton population. Some of the commonly used parameters are as follows:

4.2.1 Cell Counts

Cell count is an important parameter that is to be investigated to determine the diversity of phytoplankton populations. Earlier plankton biologists implemented this method as it provided data on the abundance as well as density of individual taxa in a population. Furthermore, cell count data of single taxa from a mixed population provide us with the information about the proportion of the population contributed by those taxa. This allows the scientific community to apply different biotic indices to make an assessment of the diversity of the population in question. Thus, such calculations are indicative about not only the species composition but also of the ecosystem functioning. The drawback for this method is that it does not take into consideration the shape and volume of the cell which may significantly affect the photosynthetic efficiency of the taxa. As an example, spherical cells of pico- and microplankton may have similar cell counts, but due to their dimensions, microplankton will be more productive as compared to picoplankton in a natural ecosystem. In case of cyanobacteria, the S/V (surface area–volume) value is an important parameter due to the buoyant nature and hydrographic properties. Thus, although the cell counts may be the same for two different cyanobacterial taxa, the sinking rate in a water column can be significantly different as evident from Stoke's law. Moreover, the pigment composition can also be a determining factor in the overall productivity of the phytoplankton population as it is the primary regulator of photosynthetic efficiency of algal cells with regard to carbon sequestration from aquatic habitats. So a mere calculation of cell counts with inadequate taxonomic identification of phytoplankton taxa may not be a correct representation of the phytoplankton community composition at the species level. Thus, cell counts provide us with data on abundance and density of specific taxa in a population of phytoplankton.

4.2.2 Utermohl Sedimentation Method for Cell Counts

Collected phytoplankton samples are preserved in Lugol's solution so as to make them heavier for easier sedimentation. In this method, a glass tubing of specific length is selected and one end is sealed with a large cover slip using waterproof adhesive (Fig. 4.5). A specific volume of the preserved samples is poured in these chambers and allowed to settle overnight. Once the phytoplankton settles on the floor of the tube, i.e. the cover slip, it is immediately observed under an inverted microscope for enumeration of phytoplankton taxa. This is done by employing an ocular micrometer by standardization with stage micrometer. This measurement technique would allow the estimation of phytoplankton cell counts for a specific area of the sedimentation chamber cover slip. Precautions are to be taken so as to count every representative field of view.

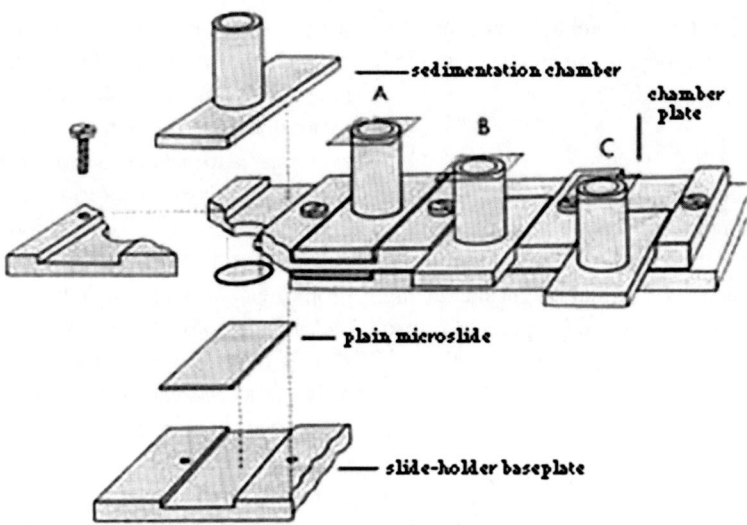

Fig. 4.5 A typical Utermohl sedimentation tube (Image courtesy: www.aquaticresearch.com)

4.2.3 Biovolume

Cellular biovolumes of individual algal taxa as well as phylogenetic group have long been used as an important parameter in determining the species composition of phytoplankton populations at an individual species level. Microalgal biovolume is commonly calculated to assess the relative abundance (as biomass or carbon) of co-occurring algae varying in shape and/or size. This is mainly due to the highly diverse shapes and sizes of algal cells that range from the picoplanktonic prochlorophytes to that of diatoms which measure more than 1 mm in diameter (Reynolds 1984). In natural mixed phytoplankton populations, high numbers of small-sized species might actually contribute only a minor fraction of the overall biomass. Other larger-sized species that are much less abundant in numbers might dominate the overall biomass. Thus, determinations of cell counts are often inadequate as a measure of relative algal biomass (Smayda 1978; Wetzel and Likens 1991). Biovolumes are also calculated for conversion of cell counts to carbon equivalents so as to estimate the fluxes of organic carbon in aquatic communities. Phytoplankton carbon calculation from biovolume eliminates the error due to detrital particulate matter contained in particulate organic carbon (Montagnes et al. 1994).

Although plankton biologists around the world understood the need to estimate biovolumes, yet no standardized formula or calculations were proposed. Afterwards, the formulae were mostly developed at the discretion of the scientists that were working on different populations (Rott 1981). The problem was especially more pronounced for complex-shaped genera like dinoflagellates, diatoms and desmids. Scientists like Kovala and Larrance (1966) used an accurate but complex approach, whereas Edler (1979) applied fairly simple methods that might not have accurately represented cell shape (Hillebrand et al. 1999).

In recent times, more advanced methods like electronic particle counting (Boyd and Johnson 1995), flow cytometry (Steen 1990), microscopic image analysis (Krambeck et al. 1981; Estep et al. 1986) and combined systems (Sieracki et al. 1998) have been implemented to measure phytoplankton cellular biovolumes. However, different drawbacks and expensive equipments do not allow us to recommend a single most accurate technique for the measurement of cellular biovolumes. The flow cytometry method is limited to the level of easily discernible groups (e.g. algae of different classes or pigment composition). This method often gives erroneous results for benthic samples that are often cohesive and contaminated with sediment

4.2 Biomass Estimation

Table 4.2 Geometric shapes and equations for the calculation of biovolume (Hillebrand et al. 1999)

Shape	Equation
1. Sphere	$V = 4/3\, \pi \cdot r^3 = \pi/6 \cdot d^3$
2. Prolate spheroid	$V = \pi/6 \cdot d^2 \cdot h$
3. Ellipsoid	$V = \pi/6\, a \cdot b \cdot h$
4. Cylinder	$V = \pi \cdot r^2 \cdot h = \pi/4 \cdot d^2 \cdot h$
5. Cylinder + 2 half spheres	$V = \pi \cdot r^2 \cdot h + 4/3 \cdot \pi \cdot r^2 = \pi \cdot d^2 \cdot (h/4 + d/6)$
6. Cylinder + 2 cones	$V = \pi/4 \cdot d^2 \cdot h + 2 \cdot \pi/12 \cdot d^2 \cdot z = \pi/4 \cdot d^2 \cdot (h + z/2)$
7. Cones	$V = 1/3 \cdot \pi \cdot r^2 \cdot z = \pi/12 \cdot d^2 \cdot z$
8. Double cone	$V = 2/3 \cdot \pi \cdot r^2 \cdot z = \pi/6 \cdot d^2 \cdot z$
9. Cone + half sphere	$V = 1/3 \cdot \pi \cdot r^2 \cdot z + \frac{1}{2} \cdot 4/3\, \pi \cdot r^3 = \pi/12 \cdot d^2 \cdot (z + d)$
10. Rectangular box	$V = a \cdot b \cdot c$
11. Prism on elliptic base	$V = \pi/4\, a \cdot b \cdot c$
12. Elliptic base with transapical constrictions	Same as above, where means of c are considered
13. Prism on parallelogram base	$V = \frac{1}{2}\, a \cdot b \cdot c$
14. Half-elliptic prism	$V = \frac{1}{2} \cdot \frac{1}{4} \cdot \pi \cdot a \cdot 2b \cdot c = \pi/4\, a \cdot b \cdot c$
15. Sickle-shaped prism	$V = \frac{1}{4} \cdot \pi \cdot c \cdot (a \cdot b - a_2 \cdot b_2)$
16. Monoraphidioid	$V = d^2/4 \cdot \left\{ \dfrac{(2b - d + a) \cdot \pi^2}{12} + \dfrac{(2b - d + a)}{2} \right\}$
17. Cymbelloid	$V = 4/6\, \pi \cdot b^2 \cdot a \cdot \beta/360$
18. Prism on triangle base	$V = \frac{1}{2} \cdot l \cdot m \cdot h$
19. Pyramid	$V = 1/3 \cdot l_1 \cdot l_2 \cdot h$
20. Elliptic prism with transapical inflations	$V = \pi/4 \cdot c \cdot (a \cdot b + i^2)$

where V volume, r radius, d diameter, h height, a apical axis (length), b transapical axis (width), c pervalvar axis (height), z height of cone, l length of one side (l_1 and l_2, if sides are unequal), m height of a triangle, β angle between the two transapical sides and i diameter of inflation

particles. Computer-mediated image analysis technique is more applicable for bacterial systems (Psenner 1993) where taxonomic resolution is inadequate. Moreover, for proper measurements, these methods are time consuming and are related to direct microscopic measurements (cf. Krambeck et al. 1981). More recently, Sieracki et al. (1998) proposed a new flow-through analysing system for plankton samples, but this system sacrifices taxonomic information as well.

Hillebrand et al. (1999) worked out and proposed a more conclusive method for the determination of biovolumes on the basis of geometric shapes of algal cells. A standard set of 20 geometric shapes was developed on the basis of the morphology of different microalgal cells. This method proposed that cell biovolumes should be calculated on an individual cell basis even for coenobial, colonial and filamentous forms. In this work, a comprehensive list of individual algal taxa belonging to different phylogenetic groups was prepared and each taxa was given a specific shape. Based upon that, for each shape, specific formulae for measuring the biovolume (V) were recommended. In recent times, this set of formulae has been accepted as the more authenticated method for measurements of cell biovolumes around the world (Table 4.2).

4.2.4 Chlorophyll and Photopigments

Phytoplankton is perhaps the most important component of pelagic ecosystem since it traps almost all the energy used by the ecosystem. Consequently, phytoplankton biomass estimates with respect to algal carbon content are highly important. Unfortunately such estimations are often extremely difficult due to the size variations of phytoplankton taxa. Accordingly, such estimates are made from other parameters, which require many calculations and/or the use of imprecise conversion factors (Geider et al. 1997). Measurements of photopigment concentrations are widely used to estimate algal biomass (Smayda 1978). Chlorophyll a is common to all photosynthetic organisms. Furthermore, it is the most abundant photosynthetic pigment and it is relatively easy and rapid to quantify. Consequently, its concentration is used extensively for estimating phytoplankton biomass. A variety of techniques are at present available, offering varying degrees of accuracy. However, the ratio of chlorophyll a to cell carbon depends on external and internal factors, such as phytoplankton taxonomic composition, cellular physiological conditions, temperature, nutrient concentrations and light intensity (Reynolds 1984).

The relationship between chlorophyll a and phytoplankton biovolume has been widely studied (Kalchev et al. 1996), where both linear and allometric relationships have been found between these parameters (Tolstoy 1977; Desortová 1981). The spatiotemporal variations among chlorophyll and biovolume mainly depend upon the taxonomic composition of phytoplankton populations, like the life form of the predominant group and their average cell size. Influence of environmental factors like changes in available light intensities, nutrient load or species predominance is some of the other parameters that can influence the 'chlorophyll maxima' in natural aquatic ecosystems.

Earlier, the fluorometric method was mainly used for the quantitative analysis of chlorophyll a and phaeopigments. However, the presence of chlorophyll b and/or chlorophyll c produced erroneous results. Chlorophyll b generally not abundant in surface water can be as high as 0.5 times of chlorophyll a concentration in the region of deep chlorophyll maxima, causing underestimations of the chlorophyll a concentration and overestimations of the phaeopigment concentrations. Divinyl chlorophyll also accounted for such incorrect estimations.

Thus, natural water samples are collected, filtered through specific filters and extracted using different solvents, and absorbance is read at wavelengths from 450 to 700 nm. Different formulae have been developed to estimate different fractions of chlorophyll from natural phytoplankton samples. Some of the more well-known methods are given below.

4.2.4.1 Formulae of Chlorophyll Estimation (90 % Acetone Extract)

Chlorophyll a

$$\text{Chl } a \left[\text{mg m}^{-3} \right] = \left(11.85\, D_{663-665} - 1.54\, D_{647} - 0.08\, D_{630} \right) v\, l^{-1}\, V^{-1}$$

D = absorbance at wavelength indicated by subscript, after correction by the cell-to-cell blank and subtraction of the cell-to-cell blank corrected absorbance at 750 nm
v = volume of acetone
l = cell (cuvette) length
V = volume of filtered water

Chlorophyll b

$$\text{Chl } b \left[\text{mg m}^{-3} \right] = \left(-5.43\, D_{663-665} + 21.03\, D_{647} - 2.66\, D_{630} \right) v\, l^{-1}\, V^{-1}$$

D = absorbance at wavelength indicated by subscript, after correction by the cell-to-cell blank and subtraction of the cell-to-cell blank corrected absorbance at 750 nm
v = volume of acetone
l = cell (cuvette) length
V = volume of filtered water

Chlorophyll c

$$\text{Chl } c \left[\text{mg m}^{-3} \right] = \left(-1.67\, D_{663-665} - 7.6\, D_{647} + 24.52\, D_{630} \right) v\, l^{-1}\, V^{-1}$$

D = absorbance at wavelength indicated by subscript, after correction by the cell-to-cell blank and subtraction of the cell-to-cell blank corrected absorbance at 750 nm
v = volume of acetone
l = cell (cuvette) length
V = volume of filtered water

Zeaxanthin (cyanobacteria and a 'bit' in Chlorophytes)

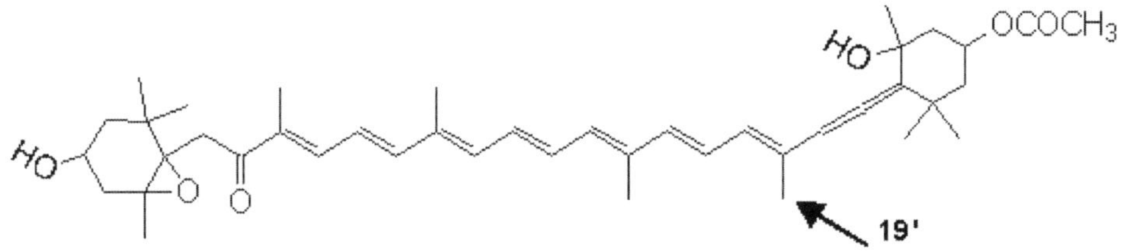

Fucoxanthin (Chrysophytes)
19'-butanoyloxy- and 19'-hexanoyloxy-fucoxanthins (prymnesiophytes)

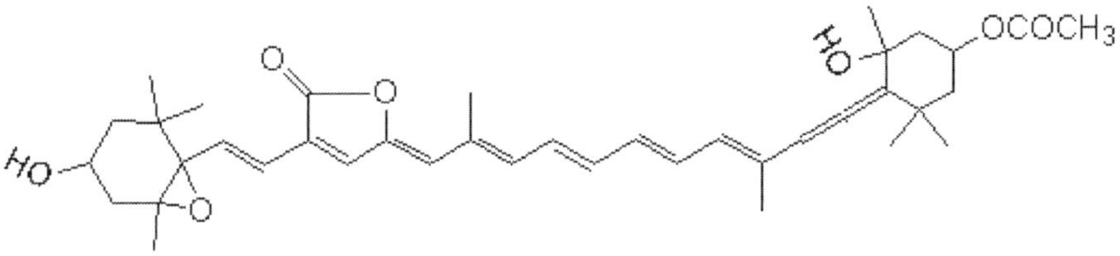

Peridinin (Pyrrhophyta, Dinoflagellates)

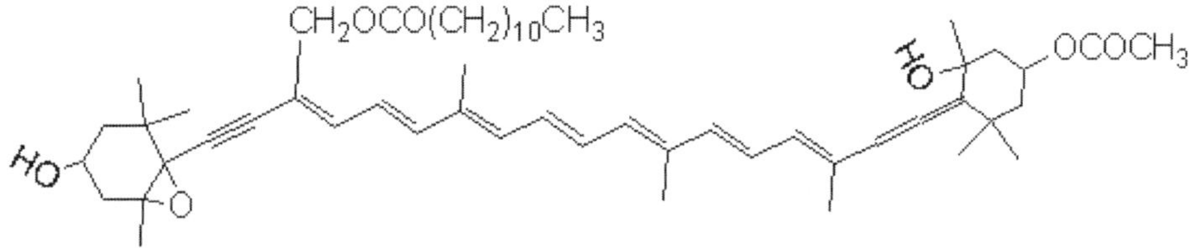

Gyroxanthin diester (Florida Red Tide, Karenia brevis: Pyrrhophyta)

Fig. 4.6 Selected chemotaxonomic biomarker carotenoids

As mentioned before, the interference of phaeopigments and other fractions of chlorophyll often produced incorrect estimates. Thus, in recent times a different approach is implemented where after estimation of chlorophyll a, the entire fraction is acidified to convert all chlorophyll to phaeopigments. Subsequent fluorometric readings are taken for phaeopigments and both values are taken for calculations by applying a correction factor for phaeopigment interference (Holm-Hansen and Riemann 1978; Herbland et al. 1985).

In recent times, pigment analyses have not only remained restricted to chlorophyll but other pigments have also been analysed as well as a proxy for biomass estimates with respect to species composition. This method is regarded as 'chemotaxonomy' where class-specific photosynthetic accessory pigments (PAPs) are estimated through HPLC, e.g. chlorophyll b (chlorophytes), fucoxanthin plus chlorophyll c (chrysophytes, diatoms and relatives), gyroxanthin-diester (Florida red tide, *Karenia brevis*), peridinin (dinoflagellates) and the divinyl chlorophylls a/b (prochlorophytes). Additionally, there are many taxon-specific (or abundant, 'zea') photoprotectorant pigments (PPPs), such as zeaxanthin ('zea', cyanobacteria), myxoxanthophyll (cyanobacteria), keto-carotenoids (echinenone, canthaxanthin, cyanobacteria), lutein (chlorophytes) and alloxanthin (cryptophytes), that are estimated as well (Fig. 4.6).

Pigment-based chemotaxonomy has gained increasing favour for rapid spatiotemporal investigations of microalgal communities, such as phytoplankton distributions in lakes and oceans. It is possible to assign a numerical relationship between the marker pigment and chlorophyll *a*, the biomass marker. The percent composition of the community is calculated by the relative abundance of the taxon-specific chlorophyll *a*.

Pigment-based chemotaxonomy can be extremely advantageous and cost-effective in large-scale ecosystem research and monitoring programs. Recently, uses of CHEMTAX software are in practice, where HPLC data for class-specific photopigments helps in determining the quantitative species composition of the phytoplankton population. Although it is a well-established process, it cannot possibly be a replacement for taxonomic identification and subsequent biovolume or cell count estimates. The drawback of the chemotaxonomy method is that it deals only a class or division level and cannot be applied at a taxon-specific level. Thus, diversity assessment by application of biotic indices cannot be done in this type of community composition study.

4.3 Species Diversity Index

Species diversity index is a statistical measure that indicates the abundance of species in a particular population. This clearly suggests that the population with more number of taxa will show a higher diversity index as compared to another population where the number of individuals for each species may be same but the total number of taxa or species is less as compared to the previous population of similar organisms. The commonly used diversity indices are simple transformations of the effective number of species or taxa, but each diversity index can also be interpreted in its own right as a measure corresponding to some real phenomenon.

The Shannon–Weiner's Index is one of the more commonly used diversity indices in ecological literature, where it is also known as the Shannon diversity index, the Shannon–Wiener index, the Shannon–Weaver index, the Shannon entropy, etc. The measure was originally proposed by Claude Shannon to quantify the entropy (uncertainty or information content) in strings of text. The idea is that the more different letters there are, and the more equal their proportional abundances in the string of interest, the more difficult it is to correctly predict which letter will be the next one in the string. The Shannon entropy quantifies the uncertainty (entropy or degree of surprise) associated with this prediction. It is calculated as follows:

$$H' = -\sum_{i=1}^{R} p_i \log p_i$$

Here, p_i can be represented as n_i/N – where n_i is the number of individuals of ith species in a population and N is the total number of individuals of all species recorded in the population. In ecology, p_i is often the proportion of individuals belonging to the ith species in the data set of interest. Then the Shannon entropy quantifies the uncertainty in predicting the species identity of an individual that is taken at random from the data set.

The base of the logarithm used when calculating the Shannon entropy can be chosen freely. Shannon himself discussed logarithm bases 2, 10 and e each of which corresponds to a different measurement units, which have been called binary digits (bits), decimal digits (decits) and natural digits (nats) for the bases 2, 10 and e, respectively.

Simpson index is used to measure the degree of concentration when individuals are classified into types. The measure equals the probability that two entities taken at random from the data set of interest represent the same type:

$$\lambda = \sum_{i=1}^{R} p_i^2$$

This also equals the weighted arithmetic mean of the proportional abundances p_i of the types of interest, with the proportional abundances themselves being used as the weights. Proportional abundances are by definition constrained to values between zero and unity, but their weighted arithmetic mean, and hence λ, can never be smaller than $1/S$, which is reached when all types are equally abundant.

4.3 Species Diversity Index

Since mean proportional abundance of the types increases with decreasing number of types and increasing abundance of the most abundant type, λ obtains small values in data sets of high diversity and large values in data sets of low diversity. The other popular indices have been the inverse Simpson index (1/λ) and the Gini–Simpson index (1−λ). Both of these have also been called the Simpson index in the ecological literature, so care is needed to avoid accidentally comparing the different indices as if they were the same.

4.3.1 Species Evenness

Species evenness refers to how close in numbers each species in an environment is. Mathematically it is defined as a diversity index, a measure of biodiversity which quantifies how equal the community is numerically. The evenness of a community can be represented by Pielou's evenness index:

$$J' = \frac{H'}{H'_{max}}$$

where H' is the number derived from the Shannon diversity index and H'_{max} is the maximum value of H', equal to

$$H_{max} = -\sum_{i=1}^{S} \frac{1}{S} \ln \frac{1}{S} = \ln S$$

J' is constrained between 0 and 1. The lesser the variation in communities between the species, the higher J' is. Thus, species evenness provides us with an opportunity to determine the contribution of individual taxon to the total population. High species evenness suggests that every taxon in the population has almost a similar number of individual representatives in the population. This also emphasizes that the habitat allows diversification of different populations with different requirements and life strategies. On the contrary, low species evenness is indicative of the fact that the contribution of individual taxon to the total population is variable, thereby ascertaining that the habitat is more suitable for proliferation of selected taxa that have favourable environmental conditions. Thus, under blooming conditions, species evenness shows minimum value as in that period a single species dominates with almost negligible representatives of other taxon.

4.3.2 Species Richness

Species richness is a measure of the number of different species represented in a set or collection of individuals in a natural population. Species richness is simply a count of species, and it does not take into account the abundance of the species or their relative abundance and distributions. The purpose of such estimation can be different on the basis of the quantifying individuals that are taken into consideration. Thus, for correct estimation of species richness, identification of individuals is important. Habitat heterogeneity is another determining factor in calculation of species richness. If samples are collected from different habitats, the build-up in the number of new species in each habitat will be higher as compared to samples collected from the same habitats. Thus, species diversity and richness are a closely knit phenomenon where although they are mutually interdependent, species diversity index is a more authenticated approach as it takes into consideration the number of individuals of each species as well.

Thus, here, the species diversity index (H') is − 2.091923, species evenness (J') is 0.908, and species richness is 10 calculated on the basis of Table 4.3.

In recent times, plankton studies have not remained restricted to morphometric analysis only, but molecular phylogenetic analyses have gained considerable impetus as well. Most of the present works have focussed on picoeukaryotes as they are not very easily detectable under the light microscope. Furthermore, it has been opined by different groups of scientists that due to their low surface–volume ratio, they have the ability to remain

Table 4.3 Calculation for biotic indices

Species	No. of individuals (n_i)	p_i (n_i/N)	$p_i \ln p_i$
A	123	0.134	−0.26933
B	46	0.049	−0.14778
C	72	0.078	−0.198982
D	22	0.024	−0.089513
E	89	0.097	−0.226305
F	182	0.198	−0.32066
G	111	0.121	−0.255547
H	56	0.061	−0.170609
I	19	0.021	−0.081127
J	201	0.218	−0.33207
Total	921	1.001	−2.091923

Species evenness: $H'/\text{Ln}S = -2.091923/\text{Ln}(10) = -2.091923/2.302585 = 0.908$

buoyant, thereby showing greater photosynthetic efficiency. Furthermore, results from different parts of the world have shown that in oligotrophic waters, the majority of planktonic population is accounted by picoeukaryotes. Thus, community composition study on the basis of molecular studies is a well-practiced method. Many of the works available have focussed on 18S rRNA gene for community analysis due to their highly conserved nature through evolutionary timescale with works from the equatorial Pacific Ocean (Staay et al. 2001), the Antarctic Polar Front (Lo'pez-Garcı'a et al. 2001), the Mediterranean and Scotia Sea as well as the North Atlantic Ocean (Dı'ez et al. 2001). The use of target gene sequences for molecular analysis has not remained restricted to 18S rRNA genes only, but other sequences like ITS and rbcL have been exploited as well.

4.4 Multivariate Analysis

Several studies over the years have conclusively established that planktonic populations in natural habitat are not dependent on a single parameter, but a combination of biotic and abiotic variables regulate the spatiotemporal dynamics of phytoplanktons. Thus, a multivariate statistical approach should be taken for this, where interrelationships among biotic and abiotic variables can be well represented in a single graphical representation. James and McCulloch (1990) also opined that 'It is no longer possible to gain a full understanding of Ecology and Systematics without some knowledge of multivariate analysis'. Thus, here we make an attempt to discuss some of the more commonly used multivariate procedures that are presently being employed for phytoplankton study, which includes data preparation as follows:

4.4.1 Preparation of Data Sets

The initial multivariate data set consists of a table of objects in rows and measured variables for those objects in columns. It is essential to correctly identify as to what are the objects and variables respectively. This distinction among objects and variables is essential to correctly implement multivariate analysis because procedures that analyse relationships among objects or among variables are different. It is assumed that objects are independent, whereas variables are interconnected or interrelated among each other during representation in a multivariate analysis.

4.4.2 Data Transformations

In multivariate data tables, measured variables can be binary, quantitative, qualitative, rank ordered, classes, frequencies or even a mixture of those types. If variables are measured in different ranges, then units of measurements (e.g. environmental parameters) of the variables have to be transformed in an appropriate format before performing further analyses. These transformations will help in developing 'dummy' numerical values for original values of qualitative variables that are subsequently utilized in multivariate procedures. Data transformations can be categorized into two main types:
- *Standardization* is a method mainly used to minimize the effects of magnitude difference

4.4 Multivariate Analysis

with respect to scales or units. This type of transformation is mainly applied for environmental parameters where range and scale of measurement are often different from each other. A common procedure is to apply the z-score transformation to the values of each variable. For each variable, it consists of:

1. Computing the difference between the original value and the mean of the variable (i.e. centring)
2. Dividing this difference by the standard deviation of the variable

- *Normalization* transformations are mainly implemented to correct the distribution shapes of certain variables, which depart from normal distribution. This would help to obtain more homogenous variances for variables that would allow better application of multivariate procedures. Different mathematical transformations can be used to normalize the x values of a variable like:

The arcsin ($\sqrt{x}$) transformation can be applied to percentages or proportions.

Log($x+c$) to variables departing strongly from a normal distribution.

$\sqrt{(x+c)}$ where c is a constant generally added to avoid mathematically undefined computations. The c constant is generally chosen so that the smallest nonzero value is obtained. The constant should also be of the same order of magnitude as the variable (Legendre and Legendre 1998).

To make community composition (either presence–absence or abundance) data containing many zeros suitable for analysis by linear methods, Hellinger transformation is the preferred method to yield good results of multivariate analysis (Legendre and Gallagher 2001). These transformations are important where samplings are done more randomly with no specific pattern of sample collection and data mining.

4.4.3 Exploratory Analysis

Multivariate exploratory methods are implemented to understand the specific patterns in data sets, and the possible explanation for those patterns in the data set depends solely on the expertise and discretion of the researcher who is implementing these methods. Thus, care should be taken in regard to data transformation methods as well as the selection of the specific multivariate procedure that the researcher wishes to implement for desired results.

4.4.3.1 Cluster Analysis and Association Coefficients

The basic purpose of cluster analysis is to group the objects on the basis of dissimilarities in a group of variables. In other words, cluster analysis maximizes between group variations and minimizes within group variations, so as to reduce the dimensionality of the data sets only to a few groups or rows (James and McCulloch 1990; Legendre and Legendre 1998). Cluster analysis is mainly used for microbial diversity study or to determine the differences between DNA or amino acid sequences in different group of samples.

Cluster analysis of a data table is mainly carried out in two parts. Firstly, a specific association coefficient is to be found on the basis of which similarity or dissimilarity matrix is to be developed. Secondly, the given data set is analysed accordingly and the calculated matrix is represented either as a horizontal tree (hierarchical clustering) or as a distinct group of objects (k-means clustering). The choice of appropriate and ecologically meaningful association coefficients is particularly important because it directly affects the values that are subsequently used for the categorization of objects.

4.4.3.2 Principal Component Analysis (PCA)

PCA has been applied to numerous phenotypic and genotypic (e.g. fingerprinting patterns) data sets, and it is one of the most popular exploratory analyses. The PCA procedure basically calculates new synthetic variables (principal components), which are linear combinations of the original variables. The aim is to represent the objects (rows) and variables (columns) of the data set in a new system of coordinates (generally on two or three axes or dimensions) where the maximum amount of variation from the original data set can be depicted. PCA plots can be developed either on the basis of variance–covariance matrix or on a correlation matrix. The first approach is followed when the same units or data types are used (e.g. abundance of different species). The aim is then to preserve and to represent the relative positions of the objects and the magnitude of variation between variables in the reduced space. PCA on a correlation matrix is rather used when variables are measured in different units or scales (e.g. different environmental parameters). The two approaches lead to different principal components and different distances between projected objects in the ordination; hence, the interpretation of the relationships must be made with care. Indeed, for correlation matrices, variables are first standardized (i.e. they become independent of their original scales), and so distances between objects are also independent from the scales of the original variables. All variables thus contribute to the same extent to the ordination of objects, regardless of their original variance. PCA results are generally displayed as a biplot (Jolicoeur and Mosimann 1960), where the axes correspond to the new system of coordinates, and both samples (dots) and taxa (arrows) are represented. The direction of a species arrow indicates the greatest change in abundance, whereas its length may be related to a rate of change. Depending on whether a distance or a correlation biplot is chosen, different interpretations can be made from the ordination diagram. The interpretation of the relationships between samples and species differs and is directly affected by the scaling chosen.

PCA is successful when most of the variance is accounted for by the largest (generally the first two or three) components. The amount of variance accounted for by each principal component is given by its 'eigenvalue'. The cumulative percentage of variance accounted for by the largest components indicates how much proportion of the total variance is depicted by the actual ordination. High absolute correlation values between the synthetic variables (principal components) and the original variables are useful to identify which variables mainly contribute to the variation in the data set, and this is referred to as the loading of the variables on a given axis.

4.4.3.3 Correspondence Analysis (CA)

CA has generally been used in microbial ecology to determine whether patterns in microbial OTU distribution could reflect differentiation in community composition as a function of seasons, geographical origin or habitat structure (Olapade et al. 2005; Edwards et al. 2006; Kent et al. 2007). The overall aim of the method is to compare the correspondence between samples and species from a table of counted data (or any dimensionally homogenous table) and to represent it in a reduced ordination space (Hill 1974). Noticeably, instead of maximizing the amount of variance explained by the ordination, CA maximizes the correspondence between species scores and sample scores. The technique is popular among ecologists because CA is particularly recommended when species display unimodal relationships with environmental gradients.

4.4.3.4 Nonmetric Multidimensional Scaling (NMDS)

NMDS is generally efficient at identifying underlying gradients and at representing relationships based on various types of distance measures. The NMDS algorithm ranks distances between objects and uses these ranks to map the objects nonlinearly onto a simplified, two-dimensional ordination space, so as to preserve their ranked differences, and not the original distances (Shepard 1966). In NMDS ordination, the proximity between objects corresponds to their similarity, but the ordination distances do not correspond to the original distances among objects. Because NMDS preserves the order of objects, NMDS ordination axes can be freely rescaled, rotated or inverted, as needed for a better visualization or interpretation. NMDS is more computer intensive than eigenanalyses such as PCoA, PCA or CA.

References

Boyd, C. M., & Johnson, G. W. (1995). Precision of size determination of resistive electronic particle counters. *Journal of Plankton Research, 17*, 41–58.

Desortova, B. (1981). Relationships between chlorophyll-a concentration and phytoplankton biomass in several reservoirs in Czechoslovakia. *Internationale Revue der gesamten Hydrobiologie und Hydrographie, 66*, 153–169.

Diez, B., Pedros-Alio, C., & Massana, R. (2001). Genetic diversity of eukaryotic picoplankton in different oceanic regions by small-subunit rRNA gene cloning and sequencing. *Applied and Environmental Microbiology, 67*, 2932–2941.

Eaton, D. R., Brown, J., Addison, J. T., Milligan, S. P., & Fernand, L. J. (2003). Edible crab (Cancer pagurus) larvae surveys off the east coast of England; implications for stock structure. *Fisheries Research, 65*, 191–199.

Edler, L. (Ed.). (1979). *Phytoplankton and chlorophyll: Recommendations on methods for marine biological studies in the Baltic Sea* (Baltic Marine Biologists Publication No. 5, p. 38). Uppsala, Sweden.

Edwards, I. P., Burgmann, H., Miniaci, C., & Zeyer, J. (2006). Variation in microbial community composition and culturability in the rhizosphere of *Leucanthemopsis alpina* (L) heywood and adjacent bare soil along an alpine chronosequence. *Microbial Ecology, 52*, 679–692.

Estep, K. W., MacIntyre, F., Hjorleifsson, E., & Sieburth, J. M. (1986). MacImage: A user friendly image-analysis system for the accurate mensuration of marine organisms. *Marine Ecology Progress Series, 33*, 243–253.

Geider, R. J., MacIntyre, H. L., & Kana, T. M. (1997). Dynamic model of phytoplankton growth and acclimation: Responses of the balanced growth rate and the chlorophyll a: Carbon ratio to light, nutrient-limitation and temperature. *Marine Ecology Progress Series, 148*, 187–200.

Herbland, A., Bouteiller, A. L., & Raimbault, P. (1985). Size structure of phytoplankton biomass in the equatorial Atlantic Ocean. *Deep Sea Research Part A. Oceanographic Research Papers, 32*(7), 819–836.

Hill, M. O. (1974). Correspondence analysis: A neglected multivariate method. *Applied Statistics, 23*, 340–354.

Hillebrand, H., Dürselen, C. D., Kirschtel, D., Pollingher, U., & Zohary, T. (1999). Biovolume calculation for pelagic and benthic microalgae. *Journal of Phycology, 35*(2), 403–424.

Holm-Hansen, O., & Riemann, B. (1978). Chlorophyll a determination: Improvements in methodology. *Oikos, 30*(3), 438–447.

James, F. C., & McCulloch, C. E. (1990). Multivariate analysis in ecology and systematics: Panacea or pandora's box? *Annual Review of Ecology and Systematics, 21*, 129–166.

Jolicoeur, P., & Mosimann, J. E. (1960). Size and shape variation in the painted turtle. *Growth, 24*, 339–354.

Kalchev, R. K., Beshkova, M. B., Boumbarova, C. S., Tsvetkova, R. L., & Sais, D. (1996). Some allometric and non-allometric relationships between chlorophyll-a and abundance variables of phytoplankton. *Hydrobiologia, 341*, 235–245.

Kent, A. D., Yannarell, A. C., Rusak, J. A., Triplett, E. W., & McMahon, K. D. (2007). Synchrony in aquatic microbial community dynamics. *ISME Journal, 1*, 38–47.

Kovala, P. E., & Larrance, J. D. (1966). *Computation of phytoplankton cell numbers, cell volume, cell surface and plasma volume per liter from microscopical counts* (Special Report 38; 21 +Appendix). Seattle: Department of Oceanography, University of Washington.

Krambeck, C., Krambeck, H. J., & Overbeck, J. (1981). Microcomputer-assisted biomass determination of plankton bacteria on scanning electron micrographs. *Applied and Environmental Microbiology, 42*, 142–149.

Legendre, P., & Gallagher, E. D. (2001). Ecologically meaningful transformations for ordination of species data. *Oecologia, 129*, 271–280.

Legendre, P., & Legendre, L. F. J. (1998). *Numerical ecology* (pp. 1–870). Amsterdam: Elsevier.

Lopez-Garcia, P., Moreira, D., & Rodriguez-Valera, F. (2001). Diversity of free-living prokaryotes from a deep-sea site at the Antarctic Polar Front. *FEMS Microbiology Ecology, 36*, 193–202.

Montagnes, D. J. S., Berges, J. A., Harrison, P. J., & Taylor, F. J. R. (1994). Estimating carbon, nitrogen, protein, and chlorophyll a from volume in marine phytoplankton. *Limnology and Oceanography, 39*, 1044–1060.

Olapade, O. A., Gao, X., & Leff, L. G. (2005). Abundance of three bacterial populations in selected streams. *Microbial Ecology, 49*, 461–467.

Psenner, R. (1993). Determination of size and morphology of aquatic bacteria by automated image analysis. In P. F. Kemp, B. F. Sherr, E. B. Sherr, & J. J. Cole (Eds.), *Handbook of methods in aquatic microbial ecology* (pp. 339–345). Boca Raton: Lewis Publishers.

Reynolds, C. S. (1984). *The ecology of freshwater phytoplankton* (pp. 1–396). Cambridge: Cambridge University Press.

Rott, E. (1981). Some results from phytoplankton counting intercalibrations. *Schweizerische Zeitschrift für Hydrologie, 43*, 34–62.

Shepard, R. N. (1966). Metric structures in ordinal data. *Journal of Mathematical Psychology, 3*, 287–315.

Sieracki, C. K., Sieracki, M. E., & Yentsch, C. M. (1998). An imaging in-flow system for automated analysis for marine microplankton. *Marine Ecology Progress Series, 168*, 285–296.

Smayda, T. J. (1978). From phytoplankton to biomass. In A. Sournia (Ed.), *Phytoplankton manual* (Monographs on oceanographic methodology 6; pp. 273–279). Paris: UNESCO.

Staay, S. Y. M., Wacher, R. D., & Vault, D. (2001). 18srDNA sequences from picoplankton reveal unsuspected eukaryotic diversity. *Nature, 409*, 607–609.

Steen, H. B. (1990). Characters of flow cytometers. In M. R. Melamed, T. Lindmo, & M. L. Mendelsohn (Eds.), *Flow cytometry and sorting* (2nd ed., pp. 11–25). New York: Wiley-Liss.

Tolstoy, A. (1977). Chlorophyll-a as a measure of phytoplankton biomass. *Acta Universitatis Uppsaliensis, 416*, 1–30.

Wetzel, R. G., & Likens, G. E. (1991). *Limnological analyses* (2nd ed., 391 pp). New York: Springer.

Case Study 5

5.1 Studies from Eastern Mediterranean Region: A Review

Several reports are available from different ecological niche. An important example is the Mediterranean region where most of the water bodies represent oligotrophic condition. Many works have concentrated on the Eastern Mediterranean Sea, an extreme oligotrophic environment (Krom et al. 2003) at the far end of a prominent west–east with increasing oligotrophy gradient (Turley et al. 2000). This ultra-oligotrophic condition is testified by high light penetrance (Berman et al. 1984a; Ignatiades 1998); low nutrient concentrations; very low values for phytoplankton; primary productivity and cell abundance (Sournia 1973; Berman et al. 1984a, b; Dowidar 1984; Azov 1986; Bonin et al. 1989; Psarra et al. 2000; Christaki et al. 2001), with a dominance of small-size phytoplankton (Li et al. 1993; Yacobi et al. 1995; Ignatiades 1998; Ignatiades et al. 2002); and outstandingly low bacterial abundance and production (Robarts et al. 1996). Several groups from the Mediterranean countries have worked on the phytoplankton species composition and productivity from the Eastern Mediterranean region. Assessments of the trophic status of habitat by application of empirical indices, statistical analysis and other analytical methods have drawn significant attention from different groups of plankton biologists from this region. Such characterization have been carried out in the Adriatic region using the OECD statistical methodology by Vollenweider and Kerekes (1982) as well as by Vollenweider et al. (1998). Similar assessments were also performed from selected areas of the Aegean Sea (Saronikos Gulf, Island of Rhodes, Mytilini Island) with the use of nutrient and/or phytoplankton species data and the application of statistical analyses (Ignatiades et al. 1992; Stefanou et al. 2000), ecological indices (Karydis and Tsirtsis 1996) and simulation modelling (Tsirtis 1995).

A detailed study was carried out on the chlorophyll a and primary productivity in the Aegean Sea from the northern to the southern open sea environment (Ignatiades et al. 2002) and from inshore to offshore waters (Ignatiades 2005). Although this area primarily represents oligotrophy, due to localized nutrient enrichment, there are mesotrophic as well as eutrophic conditions also developed. Thus, experimental work was carried out from northern and southern Aegean Sea along with inshore and offshore waters of Saronikos Gulf. Samples were collected from depths of 1–120 m (Ignatiades et al. 2002) in the Aegean Sea, whereas samples were collected from 1 to 60 m depth at the Saronikos Gulf. Abiotic variables like salinity, temperature, chlorophyll a and primary productivity were measured in situ for each sampling depth. Samples thus collected for primary productivity estimates were incubated with ^{14}C–$NaHCO_3$, and ^{14}C incorporation were measured in a liquid scintillation counter. Statistical methods were used to establish the relation between Chl *a* and primary productivity.

Results showed that a distinct gradient was evident from the northern to the southern open waters of the Aegean Sea and from inshore to offshore coastal waters of the Saronikos Gulf. The open water was cooler than the inshore–offshore waters whereas salinity levels in the northern Aegean were lower than in the southern Aegean. As expected, the open and offshore waters were nutrient poor as compared to the inshore waters. Based on the nutrient scaling criteria as proposed by Ignatiades et al. (1992), the nutrient levels of the open waters (northern and southern Aegean) were oligotrophic, whereas those of the offshore (Saronikos) waters were mesotrophic and inshore were eutrophic. Depending upon optical classification of seawater by Jerlov (1997), the habitat waters of the sampling stations were also categorized as oligotrophic (the northern and southern Aegean Sea), mesotrophic (the offshore waters of the Saronikos Gulf) and eutrophic (the inshore waters of the Saronikos Gulf). Analysis of the results of Chl a and primary productivity clearly suggested that the concentration levels of both parameters are related to the origin of the water samples.

In another work, carbon flux of planktonic food web was studied along with oligotrophic gradient (Siakou-Frangou et al. 2002). It has long been proposed that carbon flux in ocean is regulated by the magnitude of primary production and biogeochemical processes within the photic zone. Moreover, the complexity of pelagic food webs and the interactions between its different components are mainly responsible for the regulation of carbon flow. In oligotrophic waters like the Mediterranean region, different studies have established that both carbon and nutrients are remineralized and recycled within a complex microbial community dominated by minute producers and consumers (Caron et al. 1999; Azam et al. 1983; Sherr and Sherr 1988; Roman et al. 1995). Thus, simultaneous estimates of the production and biomass of phytoplankton, bacteria, heterotrophic nano- and microplankton and mesozooplankton are important parameters for the assessment of the carbon flux (Nielsen et al. 1993; Nielsen and Hansen 1995; Richardson et al. 1998; Bradford-Grieve et al. 1999).

Thus, like the previous part, this work was also carried out in the Aegean Sea, an area located between the Black Sea and the other seas of the eastern basin of the Mediterranean region (Ionian and Levantine Seas). Due to the involvement of so many different marine systems with variable salinity and nutrient status, significant variability is observed in regard to the oligotrophic status of the Aegean Sea at different sections. Thus, as a part of the EU project, this work was taken up to assess the organic carbon partitioning between autotrophic and heterotrophic plankters of different sizes and to investigate the carbon flow in the photic zone among the different areas of the Aegean Sea.

Here again, southern and northern Aegean Sea were taken into consideration for sampling that represented different degree of oligotrophy. Samplings were done on board R/V AEGAEO in two contrasting seasons of March and September, when the water column was well mixed and stratified. Total seven stations were selected from North Aegean Sea (N1–N7) and 4/5 stations were selected from South Aegean Sea (S1–S3 and S6, S7). Hydrographic measurements (temperature, pressure, conductivity) were carried out by CTD sampler from Niskin bottles, and inorganic nutrients, phytoplankton biomass and production, bacteria biomass and production, heterotrophic nanoflagellates (HNAN) and ciliates were collected at 2–100 m depths. Inorganic nutrients were measured on board spectrophotometrically. Water samples were size fractionated and chlorophyll concentrations were measured as a proxy for carbon biomass using the conversion factor of Malone et al. (1993). Photosynthetic productivity is also measured in situ by ^{14}C method. Incubation experiments with ^{14}C–$NaHCO_3$ was done and hourly measurements were recorded using a liquid scintillation counter. Epifluorescence microscopy was used to determine heterotrophic bacteria and heterotrophic nanoflagellate population as described by Christaki et al. (1999). Bacterial abundance data were converted into biomass (Lee and Fuhrman 1987), and biovolume–carbon conversion was done for flagellates (Børsheim and Bratbak 1987). Bacterial production (BP) was estimated by the ^3H-leucine method

(Kirchman 1993; Christaki et al. 1999). Similar counts and subsequent biovolume–carbon conversions and productivity measurements were done for ciliates as well. Larger mesozooplankton populations were hauled. Since calanoid copepods dominated the population, grazing of autotrophs were measured from gut fluorescence and gut evacuation rates (Dam and Peterson 1988).

Results from hydrographic studies showed that stratification in the North Aegean was much pronounced as compared to South Aegean Sea. Nutrient concentrations represented a highly contrasting status as compared to the eutrophic status, as observed by the present authors during their study in coastal West Bengal, India. Nitrate concentrations ranged from 0.05 to 2.5 µM and phosphate concentrations ranged from 0.02 to 0.08 µM in the entire area, starting from North to South Aegean Sea, with low seasonal variations. Phytoplankton counts were comparatively higher in March than in September, yet seasonal variations were not very pronounced.

Autotrophic biomass was relatively high (1,488–2,568 mg C m^2) with no significant difference between the areas. At all the sampling areas, picoplanktons (<3 µm) dominated the autotrophic component, although there were differences in the population of nanophytoplanktons at the two sampling regions. Whereas in North Aegean Sea they were <10 µm, at South Aegean Sea they were mostly larger than 10 µm. Moreover, the abundance of coccolithophorids was higher in North Aegean as compared to South Aegean Sea.

Among the heterotrophic populations, distinct patterns were observed. Although bacterial component accounted as the largest contributor to the carbon content, their biomass did not vary significantly between regions. On the other hand, although ciliates accounted for a very small proportion of the biomass, the abundance was significantly higher in the south as compared to the northern region. In contrast, mesozooplankton population decreased significantly from northern to southern region.

The carbon partitioning picture in this area did not quite replicate the typical picture, where autotrophic component outnumbers the heterotrophic components. The study area showed a gradual decrease in the autotrophic component with a subsequent increase in the heterotrophic component from northern to southern regions which were probably due to the abundance microheterotrophs. Thus, this work further established the oligotrophic condition for the Aegean Sea as was found for other regions of Eastern Mediterranean as well (Berman et al. 1984a; Krom et al. 1993; Robarts et al. 1996; Mazzocchi et al. 1997) in regard to low nutrient, plankton biomass and productivity. It further establishes an oligotrophic gradient from north to south in the Aegean Sea. Although nutrient concentrations were similar in the entire sampling area, there was a gradual differentiation in plankton community from Northeast to South Aegean. An inverse correlation between nutrient concentrations and autotrophic biomass in the North Aegean can be characterized as 'an anomalous interrelationship' that has been recorded previously in the Gulf of California (Hernandez-Becerril 1987). Experimental studies have shown that there is nutrient inflow from the Black Sea that caused localized enrichment, but due to the high rate of assimilation, nutrient concentrations are often depleted. Abundance patterns of bacteria, ciliates, heterotrophs and microheterotrophs further establish microheterotrophs as an essential component in regulating the spatial and temporal of carbon partitioning in the Aegean Sea.

Thus, from the studies it can be said that lotic aquatic ecosystems around the world represent significant variations with regard to the habitat. It can range from ultra-oligotrophic to highly eutrophic conditions. Moreover, light penetration can be significantly altered due to suspended matter load in the water column. This may in turn result in reduction in the photic zone ratio in the water column. Thus, although the incident radiation may be high, the net reproductivity as well as the dissolved oxygen content in the habitat may not complement the incident radiation. Planktonic populations are highly responsive to such alteration in the habitat including light availability. Thus, sudden shift in abiotic variables like light and temperature can account for major oscillations in the water column properties which in turn cascades in regulating the phytoplankton population of the study area.

5.2 Case Study I: Phytoplankton Diversity of East Calcutta Wetland: A Ramsar Site

The phytoplankton study in Indian subcontinent extensively started in the mid-twentieth century and was primarily focused on the diversity and taxonomic study. The value of phytoplanktons and other algae as direct or indirect feed for fishes and their usefulness as indicator of water quality has long been well recognized. With the progress of inland fisheries in India, studies on productivity and diversity of phytoplanktons in Inland waters have gained considerable importance. Many aquatic ecosystems especially the sewage-fed ponds are generally affected by eutrophication, which indicates the inorganic nutrient supplies exceeding the phytoplankton growth demands (Fischer et al. 1988; McComb et al. 1995). Bioassay tests were conducted by several authors to determine the relation between nutrient characteristics and phytoplankton abundance (Redfield 1958; Ryther and Dunstan 1971; Pearl et al. 1990; Siep 1994).

The practice of fish culture in shallow waste ponds is quite popular in Indian subcontinent. Different aspects of wastewater ecology and productivity of this type of ecosystem have been studied by a number of authors (Sen 1941). According to Ray Chaudhuri et al. (2008), shallow fish-producing water bodies in West Bengal, India (called Bheris) have distinct architecture, resulting in extensive purification of waste. Such freshwater fish ponds of East Kolkata Wetland Complex (22°27′N 88°27′E) are also declared as a 'Ramsar site', by Ramsar Convention in 19 August 2002. (Ramsar Bureau List was established under the Article 8 of Ramsar Convention.) The Government of India declared this wetland as 'Wetland of International Importance'.

Thus, East Calcutta Wetland can be cited as best example of integrated resource recovery. The utility of Kolkata municipal waste on life and growth of fish was reported by Nayar (1944) and Bose (1944). Roy et al. (1981) further reported the use of Kolkata municipal waste for Bidyadhari–Kulti Fishery complex. The general ecology and biodiversity of Fauna of many ponds have also been recorded by many authors (Mukherjee 1996; Chakraborty 1988; Jana 1998; Mukherjee et al. 2002). Mukherjee et al. (2010) also reviewed that a Bheri is a biological complex system both at quantitative levels as compared to rain water as well as waste water-fed ponds. Pradhan et al. (2008) suggested that phytoplankton growth could be an important factor responsible for greater fish production and could also act as biomonitor for water quality assessment in the Bheris. However, over growth of plankton results in bloom which could be a problem for pond management. Fishes also play a crucial role by maintaining a proper balance of phytoplankton growth, and the planktons convert the nutrients available from waste into consumable form as food for fish (Ghosh 1999). The water and effluent generated form Bheris are used for cultivation of vegetables, which were found to show no harmful accumulation (Ray Chaudhuri et al. 2007). In the present study phytoplankton diversity of waste-fed fish pond and their taxonomic documentation have been done in detail.

5.2.1 Study Area

One such freshwater eutrophic fish pond of the Ramsar site was our study pond – the Captain Bheri. It is situated at eastern region of Kolkata between 88°27′ east latitude and 22°27′ north latitude, at the south of Salt Lake City on Eastern bypass, covering an area of 450 m². This pond serves the dual purpose of recycling sewage water of Kolkata metropolitan city and for cultivating fishes extensively. The sewage water includes municipal waste and small-scale industrial effluents of Eastern Kolkata's urban and semiurban areas.

The variation in physicochemical factors recorded throughout the year was also recorded. The temperature of water varied from 15.3 to 31 °C, pH from 7.5 to 8.62. This suggests the alkaline nature of the habitat water. Different nutrients like nitrate, nitrite, phosphate, ammonium nitrogen, etc., were also measured. Nitrate was found to vary between 0.0775 and 0.286 mg/ml. Nitrite was found to range from 0.885 to 4.4 mg/L. Phosphate ranged from 0.057 to 0.277 mg/L. Ammonium nitrogen concentration varied between 0.45 and 0.187 mg/L.

5.2.2 Results

A total of 55 taxa were recorded during the study period which were found to belong to the groups of Cyanobacteria, Chlorophyta, Bacillariophyta and Euglenophyta. Chlorophyta population was found to comprise of 30 species. On the other hand 9 taxa of Cyanophyta, 8 taxa of Euglenophyta and 8 taxa of Bacillariophyta were also recorded (Table 5.1). From the pie chart of the population, it is evident that chlorophytes were found to be maximum comprising of 55 % of the total population, followed by cyanobacteria with 16 % of the total population, and then euglenophytes and bacillariophytes (15 % and 14 %, respectively) (Fig. 5.1).

Description of few taxa are given below which were restricted to this area only; the rest are given in case studies II and III

Division – Cyanophyta
Class – Myxophyceae
Order – Chroococcales
Family – Chroococcaceae
Chroococcus dispersus (Keissl.) Lemmermann (Plate 5.1, Fig. 1)
Lemmermann 1904 p. 102; Prescott 1982, Pl. 100, Fig. 7

Table 5.1 Showing name of the phytoplankton taxa recorded

S. No.	Name of taxa
	Cyanophyta
1.	*Chroococcus dispersus*
2.	*Coelosphaerium dubium*
3.	*Merismopedia glauca*
4.	*Merismopedia minima*
5.	*Merismopedia trolleri*
6.	*Synechococcus elongatus*
7.	*Planktolyngbya contorta*
8.	*Arthrospira platensis*
9.	*Spirulina subsalsa*

(continued)

Table 5.1 (continued)

S. No.	Name of taxa
	Chlorophyta
10.	*Chlorococcum humicola*
11.	*Pediastrum duplex*
12.	*Pediastrum duplex* var. *clathratum*
13.	*Pediastrum tetras* var. *tetras*
14.	*Pediastrum tetras* var. *tetraodon*
15.	*Coelastrum microporum*
16.	*Coelastrum proboscideum*
17.	*Ankistrodesmus falcatus*
18.	*Ankistrodesmus falcatus* var. *tumidus*
19.	*Kirchneriella lunaris*
20.	*Kirchneriella contorta*
21.	*Selenastrum bibraianum*
22.	*Tetraedron minimum*
23.	*Tetraedron muticum*
24.	*Tetraedron trigonum*
25.	*Tetraedron caudatum*
26.	*Crucigenia apiculata*
27.	*Crucigenia quadrata*
28.	*Crucigenia crucifera*
29.	*Crucigenia tetrapedia*
30.	*Scenedesmus abundans*
31.	*Scenedesmus acuminatus*
32.	*Scenedesmus bicaudatus*
33.	*Scenedesmus bijuga*
34.	*Scenedesmus dimorphus*
35.	*Scenedesmus quadricauda* var. *parvus*
36.	*Scenedesmus obliquus*
37.	*Scenedesmus acutus*
38.	*Scenedesmus quadricauda*
39.	*Scenedesmus ecornis*
	Euglenophyta
40.	*Euglena gracilis*
41.	*Euglena viridis*
42.	*Phacus helikoides*
43.	*Phacus nordstedii*
44.	*Phacus tortus*
45.	*Phacus chloroplastes*
46.	*Phacus curvicauda*
47.	*Euglena proxima*
	Bacillariophyta
48.	*Pleurosigma angulatum*
49.	*Cyclotella meneghiniana*
50.	*Amphora coffeaeformis*
51.	*Navicula halophila*
52.	*Navicula microspora*
53.	*Navicula lanceolata*
54.	*Nitzschia actinastroides*
55.	*Aulacoseira granulata*

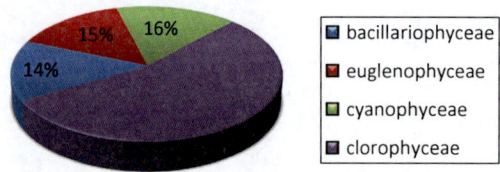

Fig. 5.1 Pie chart showing percentage abundance of phytoplankton population of the study area

Ovate or irregularly shaped colony of 4–6 spherical cells, free floating and flattened; cells are either single or arranged in small clusters, evenly distributed at some distance from one another in the mucilaginous envelop; individual cell sheaths not evident; cell contents bright blue green, cells 3–4.5 μ in diameter

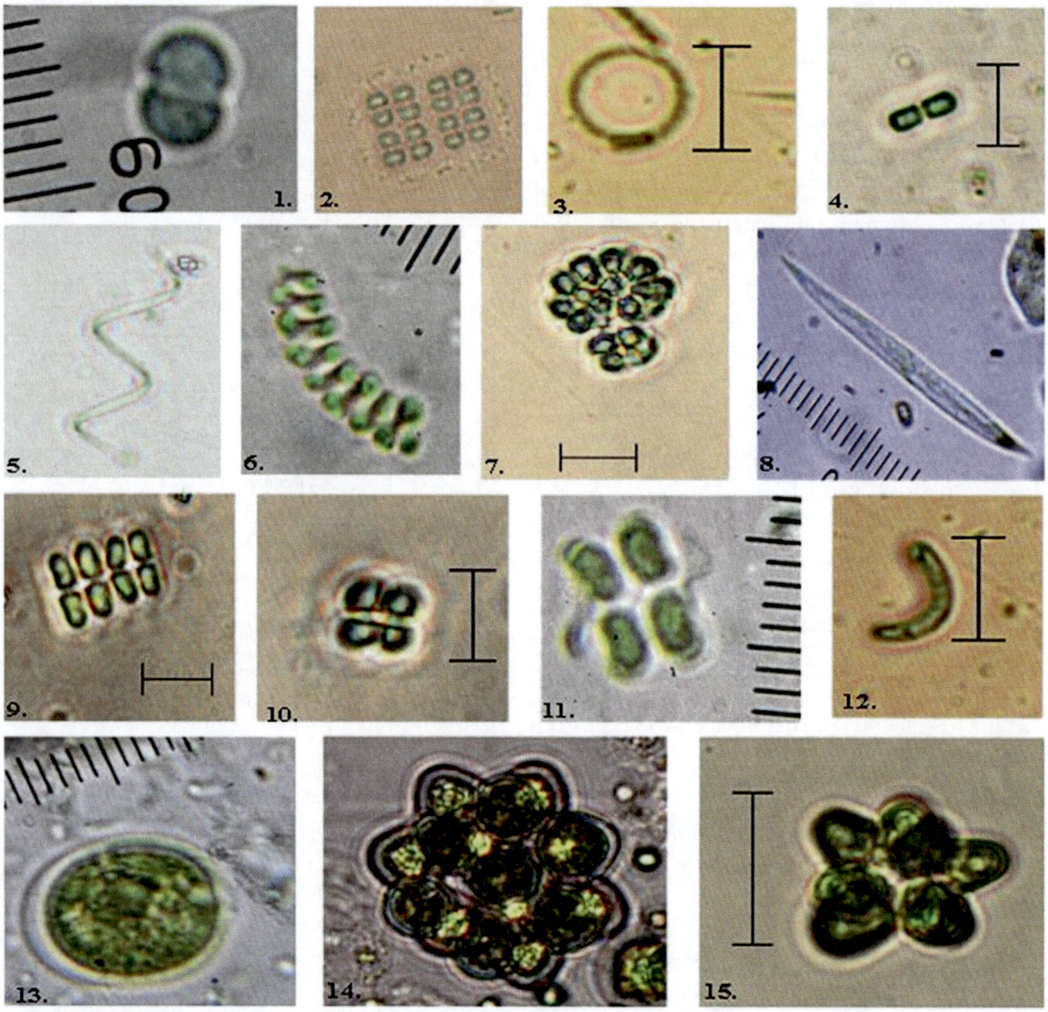

Plate 5.1 (**1**) *Chroococcus dispersus*, (**2**) *Merismopedia trolleri*, (**3**) *Planktolyngbya circumcreta*, (**4**) *Synechococcus elongatus*, (**5**) *Arthrospira platensis*, (**6**) *Spirulina subsalsa*, (**7**) *Coelosphaerium dubium*, (**8**) *Ankistrodesmus falcatus* var. *tumidus*, (**9**) *Crucigenia apiculata*, (**10**) *Crucigenia quadrata*, (**11**) *Crucigenia crucifera*, (**12**) *Kirchneriella contorta*, (**13**) *Chlorococcum humicola*, (**14**) *Coelastrum microporum*, (**15**) *Coelastrum proboscideum*

Coelosphaerium dubium Grunow in Rabenhorst (Plate 5.1, Fig. 7)

Rabenhorst 1865, p. 55; Prescott 1982, Pl. 106, Fig. 1

Spherical or sometimes irregularly shaped colony, cells densely arranged. Cells spherical, free floating, prominent colonial mucilage to form a peripheral layer, cell contents blue green, cells 5–7 μ in diameter; compound colonies as much as 300 μ in diameter

Merismopedia trolleri Bachmann (Plate 5.1, Fig. 2)

Bachmann 1920, p. 350; Prescott 1982, Pl. 101, Fig. 5

Colonial, each cell with a distinct sheath, 8–16 spherical cells evenly arranged within a transparent colonial mucilage; cell contents with pseudovacuoles, appearing brownish or purplish because of light refraction; cells 2–3.5 μ in diameter

Synechococcus elongatus Nag. (Plate 5.1, Fig. 4)

Desikachary, 1959, pl no. 25, fig 7

Cells cylindrical, 1.4–2 μ broad, 1.5–3 times as long as broad, contents homogeneous and light blue green

Order – Oscillatoriales

Family – Oscillatoriaceae

Planktolyngbya circumcreta (G.S. West) Anagnostidis et Komarek (Plate 5.1, Fig. 3)

Kommarek 2005, p. 160, fig 194

Filamentous, solitary, free floating, spirally coiled, 2–2.5 μ wide, coils 33–39 μ broad, making 2–3 turns, sheathes thin, colourless, trichomes light pale blue green

Arthrospira platensis (Nordst.) Gomont (Plate 5.1, Fig. 5)

Desikachary, 1959 pl no. 35, fig 2

Spiral thallus blue green in colour, trichomes slightly constricted at the cross walls, 6–8 μ broad, distance between spirals 8–10 μ, end cells broadly rounded

Spirulina subsalsa Oernst. ex Gomont (Plate 5.1, Fig. 6)

Desikachary, 1959 pl no. 36, figs 3, 9

Spiral thallus, trichomes 1–2 μ broad, bright blue green or yellowish green thallus, irregular spirals, spirals very close to each other, spirals 3–5 μ broad

Division – Chlorophyta
Class – Chlorophyceae
Order – Chlorococcales
Family – Chlorococcaceae

Chlorococcum humicola (Naeg.) Rabenhorst (Plate 5.1, Fig. 13)

Rabenhorst 1868, p. 58; Prescott 1982, Pl. 45, Fig. 1

Unicellular green alga, prominent cell wall, parietal chloroplast, cells spherical, solitary or in small clumps, variable in size within the same plant mass; cells 8–25 μ in diameter

Family – Hydrodictyaceae

Pediastrum duplex var. *clathratum* (A. Braun) Lagerheim (Plate 5.2, Fig. 6)

Lagerheim 1882, p. 56; Prescott 1982, Pl. 48, Figs. 6

Green algal colony, colony with larger perforations than in the typical form; walls with deep emarginations; apices of lobes of peripheral cells truncate; cells 12–20 μ in diameter

Pediastrum tetras var. *tetraedon* (Corda) Rabenhorst (Plate 5.2, Fig. 7)

Rabenhorst 1868, P. 78; Prescott 1982, pl no. 50, Fig 7

Colony 4–8 celled, outer margins of the peripheral cells with deep incisions; the lobes extended into sharp, horn-like process; cells 12–15 μ in diameter, 16–18 μ long

Pediastrum tetras (Ehrenb.) Ralfs var. *teras* (Plate 5.2, Fig. 9)

Ralfs 1844, p. 469; Prescott 1982, Pl. no 50, Figs 3, 6

Colony entire; inner cells (frequently none) with 4–6 straight sides but with one margin deeply incised; peripheral cells crenate, with a deep incision in the outer free margin, their lateral margins adjoined along two third of their length; cells 8–16 μ in diameter

Family – Coelastraceae

Coelastrum microporum Naegeli (Plate 5.1, Fig. 14)

A. Braun 1855, p. 70; Prescott 1982, Pl. 53, Fig. 3

Coenobium spherical, composed of 8–68 sheathed globose cells (sometimes ovoid, with the narrow end outwardly directed); cells interconnected by very short, scarcely discernible gelatinous processes, leaving small

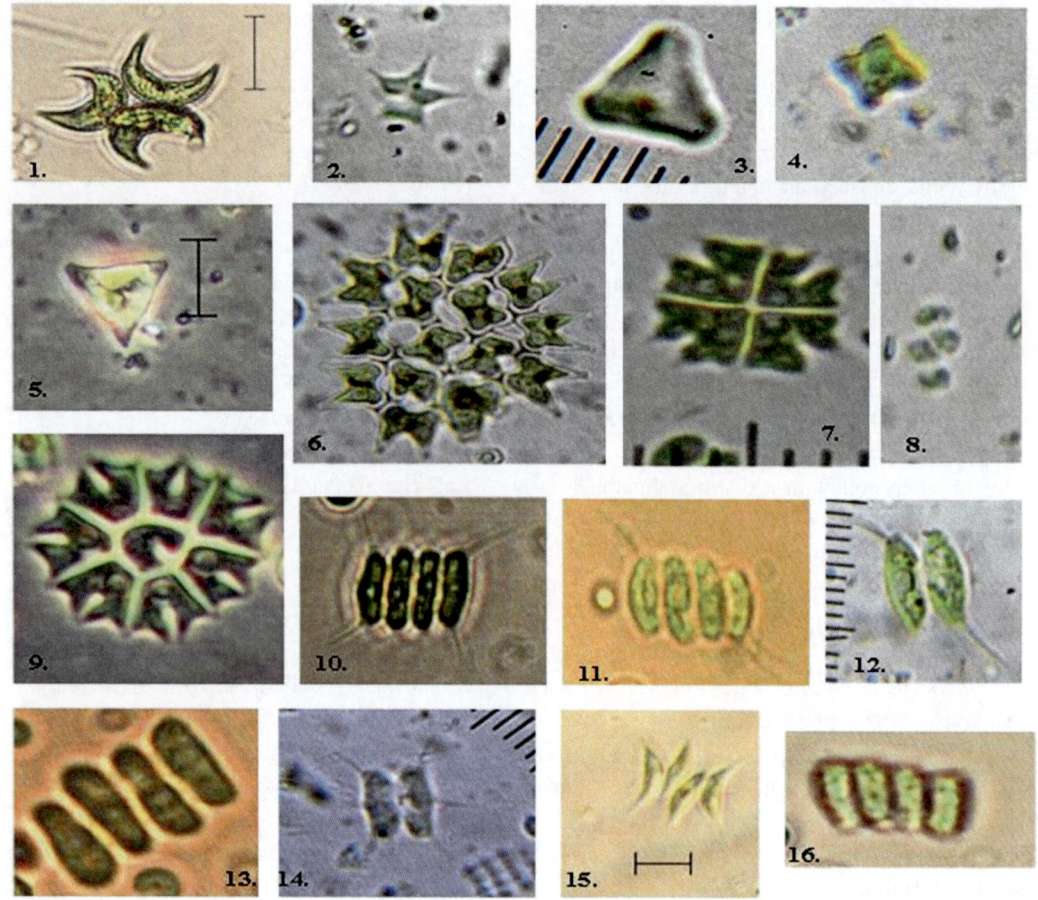

Plate 5.2 (**1**) *Selenastrum bibraianum*, (**2**) *Tetraedron caudatum*, (**3**) *Tetraedron muticum*, (**4**) *Tetraedron minimum*, (**5**) *Tetraedron trigonum*, (**6**) *Pediastrum duplex* var. *clathratum*, (**7**) *Pediastrum tetras* var. *tetraodon*, (**8**) *Scenedesmus acutus*, (**9**) *Pediastrum tetras* var. *tetras*, (**10**) *Scenedesmus quadricauda* var. *parvus*, (**11–12**) *Scenedesmus bicaudatus*, (**13**) *Scenedesmus ecornis*, (**14**) *Scenedesmus abundans*, (**15**) *Scenedesmus acuminatus*, (**16**) *Scenedesmus obliquus*

intercellular spaces; cells 8–20 μ in diameter including the sheath

Coelastrum proboscideum Bohlin (Plate 5.1, Fig. 15)

Bohlin 1897, p. 33; Prescott 1982, Pl. 53, Figs. 4, 5, 8

Coenobium pyramidal or cubical (rarely polygonal), composed of 4-8-16-32 truncate cone shaped cells with the apex of the cone directed outward, the inner or basal wall of the cell concave, the lower lateral walls of the cells adjoined about a large space in the centre of the colony; cells 8–15 μ in diameter; 4-celled colony as much as 35 μ in diameter

Family – Oocystaceae

Ankistrodesmus falcatus var. *tumidus* (West & West) G.S. West (Plate 5.1, Fig. 8)

West 1904, p. 224; Prescott 1982, Pl. 56, Fig. 9

Unicellular, cells lunate or fusiform, the ventral margin decidedly tumid in the midregion, 4.5–6.5 μ in diameter, 61–73 μ long

Kirchneriella contorta (Schmidle) Bohlin (Plate 5.1, Fig. 12)

Bohlin 1897 p. 20; Prescott 1982, Pl. 57, Figs. 7, 8.

Colonial, free floating, usually of 16 twisted, arcuate, cylindrical cells with broad, convex apices, lying irregularly scattered throughout the homogeneous, gelatinous envelope, chloroplast covering the entire wall of cells, which are 1–2 µ in diameter, 5.8–10 µ long

Selenastrum bibraianum Reinsch (Plate 5.2, Fig. 1)

Reinsch 1867, p. 64; Prescott 1982 pl no. 57, fig. 9

Colony ovate in outline, composed of 4–16 lunate or sickle shaped cells with sharp apices and arranged so that the convex surfaces are apposed and directed towards the centre of the colony; cells 5–8 µ in diameter, 20–38 µ long; distance between apices 16–42 µ

Tetraedron minimum (A. Braun) Hansgirg (Plate 5.2, Fig. 4)

Hansgirg 1888, p. 131; Prescott 1982, pl no. 60, figs 12–15

Unicellular, cells small, flat, tetragonal, the angles rounded and without spines or processes, lobes sometimes cruciately arranged; margins of the cell concave with one frequently incised; cells 6–20 µ in diameter

Tetraedron muticum (A. Braun) Hansgirg (Plate 5.2, Fig. 3)

Hansgirg 1888, p. 131; Prescott 1982, pl no. 60, figs. 16, 17.

Unicellular; cells small, flat, triangular; the angles without spines or furcations; sides of the cell emarginated or slightly convex; cells 6–18 µ in diameter

Tetraedron trigonum (Naeg.) Hansgirg (Plate 5.2, Fig. 5)

Hansgirg 1888, p. 130; Prescott 1982, pl no. 61, figs. 11, 12.

Cells flat, 3-angled, the angles tapering to sharply rounded, spine-tipped apices; margins convex; sides of the cell body concave or straight; cells are 19–29.8 µ in diameter

Tetraedron caudatum (Corda) Hansgirg (Plate 5.2, Fig. 2)

Hansgirg 1888, p. 131; Prescott 1982, pl no. 59, figs. 17, 24, 25.

Cells flat, 5-sided, the angles rounded and tipped with a short sharp spine; the sides between the angles concave, but with one margin narrowly and deeply incised; cells in their longest dimension 8-15-(22)

Family – Scenedesmaceae

Crucigenia apiculata (Lemn.) Schmidle (Plate 5.1, Fig. 9)

Schmidle 1901, p. 234; Prescott 1982, Pl. 65, Fig. 3

Colonial, 4 ovate rhomboidal or somewhat triangular cells arranged cruciately, cells 3–7 µ in diameter, 5–10 µ long; colony 6–12.5 µ wide, 9–8 µ long

Crucigenia quadrata Morren (Plate 5.1, Fig. 10)

Morren 1830, pp. 415, 426; Prescott 1982, Pl. 65, Fig. 10

Colony free-floating, consisting of a circular plate of 4 triangular cells, arranged about a small central place, chloroplasts parietal discs, as many as 4 in a cell; pyrenoids not always present; cells 2.5–6 µ in diameter, 3.7 µ long; multiple quadrate colonies formed by the close arrangement of component quardets

Crucigenia crucifera (Wolle) Collins (Plate 5.1, Fig. 11)

Collins 1909, p. 170; Prescott 1982, Pl. 65, Fig. 4

Colony consisting of 4-sided cells arranged about a central square opening, the outer free walls longer and concave, the outer free angles of the cells rounded, the lateral adjoin other cells, the inner walls colonies resulting from the adherence of quartets of cells by persisting mother cell walls; cells 3.5–5 µ in diameter, 5–7 µ long; colony 9–11 µ wide, 14–16 µ long

Scenedesmus abundans (Kirch.) Chodat (Plate 5.2, Fig. 14)

Chodat 1913, p. 77; Prescott 1982, pl. 62, fig 21

Cells oblong or ovate, in a linear series of 4, the terminal cells with 1 or 2 polar spines and 2 spines on the lateral wall, the inner cells with the spines at each pole, cells 4–7 µ in diameter, 7–12 µ long

Scenedesmus acuminatus (Lag.) Chodat (Plate 5.2, Fig. 15)

Chodat 1902, p. 211, G.W. Prescott 1982, pl no. 62, Fig no. 16

Cells arranged in a curved series of 4 (rarely 8) cells strongly lunate, with sharply pointed apices, the convex walls adjoined inwardly, the concave faces directed outwards; cells 3–7 µ in diameter, 30–40 µ long

Scenedesmus bicaudatus Deuds (Plate 5.2, Figs. 11 and 12)

Jaiswal & Tiwari, p. 94 pl. 13, fig 8

Coenobium 2–8 celled, with linear or slightly alternate in arrangement, cells elongated, outer cells with a long curved spine at alternate poles; inner cells without spines, oval to cylindrical, length 10–12 μ, width 4–6 μ

Scenedesmus obliquus (Turp.) Kuetzing (Plate 5.2, Fig. 16)

Kuetzing 1833, p. 609; Prescott 1982, pl no. 63, Fig 17

Colony composed of 2–8 (usually 4 or 8) fusiform cells arranged in a single series; apices of the cells apiculate; wall smooth; cells 4.2–9 μ in diameter, 14–21 μ long

Scenedesmus quadricauda var. *parvus* G.M. Smith (Plate 5.2, Fig. 10)

Smith 1916a, p. 480; Prescott 1982, pl no 64, Fig 6

Colony composed of 2–16 cylindrical–ovate cells arranged in a single series; outer cells with a long spine at each pole; inner cells with spineless walls; cells 4–6.5 μ in diameter, 12–17 μ long

Scenedesmus ecornis (Ehrenb.) Chodat (Plate 5.2, Fig. 13)

Colonies of 4 cells attached side by side along 3/4 of their side, arranged in a zigzag way; cell body elliptical; cell wall smooth, inner spaces absent, without spiny or dented projections, 7–20 μ long, 4–10 μ in diameter

Scenedesmus acutus Meyen (Plate 5.2, Fig. 8)

Jaiswal & Tiwari, p. 95 pl. 13, fig 12

Colony 4 celled, cells linear or slightly alternate in arrangement, terminal cells with concave outerface, length 20–24 μ and breadth 5–7 μ

Division – Euglenophyta
Class – Euglenophyceae
Order – Euglenales
Family – Euglenaceae

Euglena gracilis Klebs (Plate 5.3, Figs. 1 and 2)

Klebs 1883, p. 303; Prescott 1982, pl no. 85, fig. 17

Unicellular, comparatively larger cell, grass green in colour, cells short fusiform to ovoid; chloroplast many evenly distributed throughout the cell, cell 8–20 μ diameter, 35–50 μ long

Euglena proxima Dangeard (Plate 5.3, Fig. 3)

Dangeard 1902, p. 154; Prescott 1982, pl no. 85, fig. 25

Larger fusiform cells, narrowed posteriorly to a blunt tip, periplast spirally striated, chloroplast numerous, cells 14–21 μ in diameter, 50–95 μ long

Euglena viridis (O.F. Müller) Ehrenberg (Plate 5.3, Fig. 4)

Large cell, 40–65 μm long, 14–20 μm wide; anterior end rounded, posterior end pointed; fusiform during locomotion; highly plastic when stationary; chloroplasts more or less bandform, radially arranged; nucleus posterior

Phacus chloroplastes Prescott (Plate 5.3, Fig. 8)

Prescott 1982, pl no. 87, figs. 15, 16.

Cells broad, pyriform; at posterior end a straight or vey slightly deflected caudus is formed. broadly rounded anterior end with a median papilla; periplast longitudinally striated; margin of cell entire, chloroplasts in parietal bands, cells 20–22 μ in diameter, 29–31 μ long

Phacus curvicauda Swirenko (Plate 5.3, Fig. 7)

Swirenko 1915, p. 333 G.W. Prescott 1982, pl no. 87, figs. 14

Cells broadly ovoid to sub orbicular in outline, slightly spiral in the posterior part, which is extended into a caudus that curves obliquely, chloroplasts numerous, 24–26 μ in diameter, 28–30 μ long

Phacus nordstedtii Lemmermann (Plate 5.3, Fig. 5)

Lemmermann 1904, p. 125; Prescott 1982, pl no. 88, figs. 1

Cells napiform, nearly spherical and with a long straight sharply pointed caudus, broadly rounded anteriorly, periplast spirally striated, cells 18–19 μ in diameter, 35–40 μ long

Phacus tortus (Lemm.) Skvortzow (Plate 5.3, Fig. 10)

Skvortzow 1928, p. 110; Prescott 1982, pl no. 88, figs. 20

Cells broadly fusiform, broadest at the anterior end of the cell, tapering and spirally twisted in the posterior position to form a long straight caudus, cells 38–50 μ in diameter, 85–95 μ long

Phacus helikoides Pochmann (Plate 5.3, Fig. 6)

Pochmann 1942, p. 212; Prescott 1982, pl no. 87, fig. 9

Cells elongate fusiform, twisted throughout the entire length, briefly narrowed anteriorly and bilobed, tapering posteriorly to a spirally twisted long straight caudus, margin entire with two or three bulges, cells 30–50 μ in diameter, 70–120 μ long

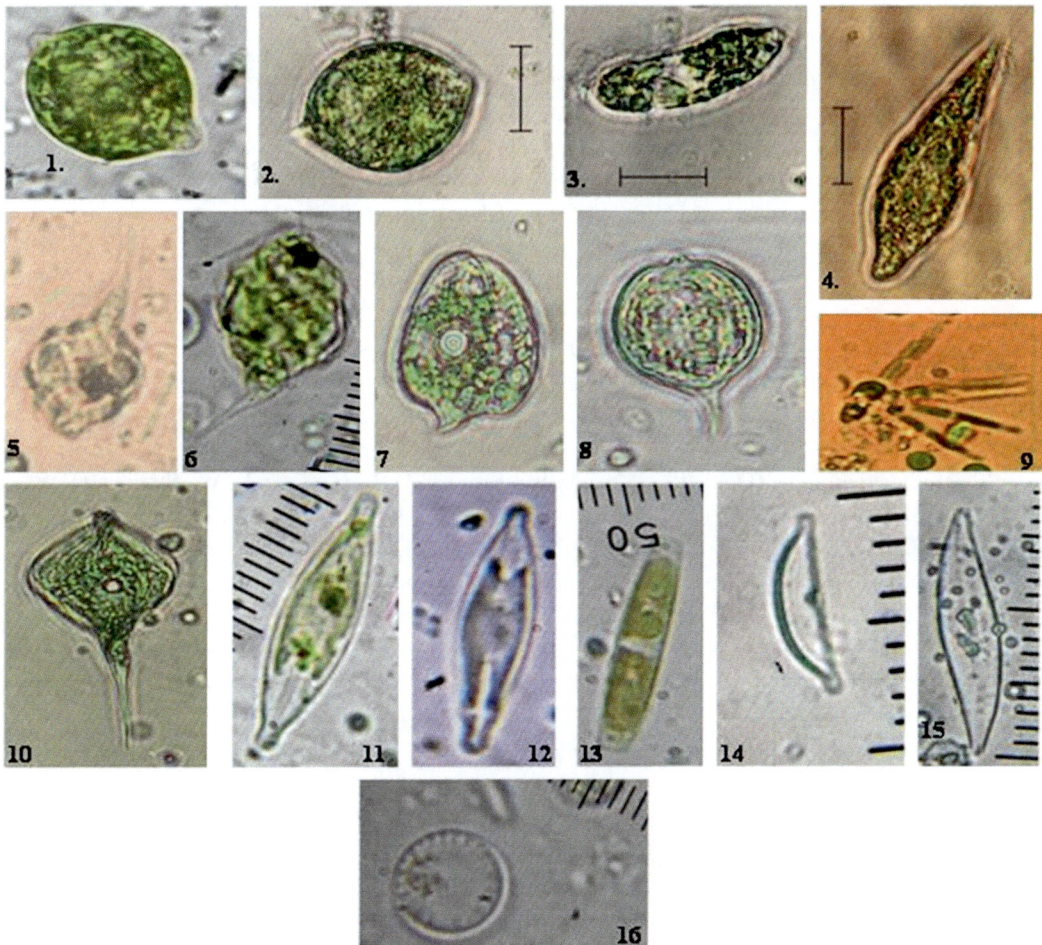

Plate 5.3 (1–2) *Euglena gracilis*, (3) *Euglena proxima*, (4) *Euglena viridis*, (5) *Phacus nordstedtii*, (6) *Phacus helikoides*, (7) *Phacus curvicauda*, (8) *Phacus choloroplastes*, (9) *Nitzshia actinastroides*, (10) *Phacus tortus*, (11) *Navicula halophila*, (12) *Navicula micropora*, (13) *Navicula lanceolata* (*side view*), (14) *Amphora coffeaeformis*, (15) *Pleurosigma angulatum*, (16) *Cyclotella meneghiniana*

Division – Bacillariophyta
Class – Bacillariophyceae
Order – Thalassiophysales
Family – Catenulaceae
Amphora coffeaeformis Agardh (Plate 5.3, Fig. 14)
Frustules in girdle view very prominent elliptic lanceolate, truncate. Valves arcuate on the dorsal margin and straight or slightly concave on the ventral margin. Striae delicate
Order – Thalassiosirales
Family – Stephanodiscaceae
Cyclotella meneghiniana Kutz. (Plate 5.3, Fig. 16)
Venkataraman, 1939, p. 303, fig 11

Frustules discoid in valve view, rectangular and undulated in girdle view; margin well defined, coarsely striated and striae wedge shaped. Diameter 11–30 μ
Order – Naviculales
Family – Naviculaceae
Navicula microspora Kant & Gupta (Plate 5.3, Fig. 12)
Kant and Gupta 1998, p. 27, pl. 127, fig. 12.
Frustules elliptical to lanceolate, rostrate apices, pseudoraphae at the centre, axial area broad, striation not visible in fresh material, longer than broad, 42–54 μ long and 10–14.5 μ broad

Navicula lanceolata (Agardh) Ehrenberg (Plate 5.3, Fig. 13)

Valves linear–lanceolate to lanceolate, with very slightly protracted, bluntly rounded apices. Striae radiate at the centre, becoming convergent at the apices, with a broadly rounded to squarish central area. Length 28–70 µ, width (8)-9-12 µ

Navicula halophila (Grun.) Cleve (Plate 5.3, Fig. 11)

Valves lanceolate with slightly produced and capitate ends. Axial area narrow, linear, central area slightly widened in the middle. Striations parallel and slightly convergent at the ends

Family – Pleurosigmataceae

Pleurosigma angulatum (Quekett) W. Smith (Plate 5.3, Fig. 15)

Valves lanceolate, slightly sigmoid, ends subacute, 116 µ long, 16.5 µ broad; raphae more sigmoid than valve, excentric near the ends

Order – Bacilllariales

Family – Bacillariaceae

Nitzschia actinastroides Van Goor (Plate 5.3, Fig. 9) Huber – Pestalozii 1942, p. 472, pl. CXXXVIII, fig. 560.

Rustules straight, liner, much longer than broad, 60–90 µ long and 2–4 µ broad, striation not clearly visible in fresh material, 4–5 frustules joined at one end

References

Agardh, C. A. (1827). *Aufzählung einiger in den östreichischen Ländern gefundenen neuen Gattungen und Arten von Algen, nebst ihrer Diagnostik and beigefügten*. Bemerkungen.

Bachmann, H. (1920). *Merismopedia trolleri* nov. spec. *Zeitschrift für Hydrologie, 1*(3–4), 350.

Barinova, S., et al. (2010). The influence of the monsoon climate on phytoplankton in the Shibpukur pool of Shiva temple in Burdwan, West Bengal, India. *Limnological Review, 12*(2), 47–63.

Bohlin, K. (1897). *Die Algender ersten Regnell'schen Expedition. I*. Protococcoideen. Bih.

Bose, B. C. (1994). Calcutta sewage-fisheries culture. *Proceedings of the National Institute of Sciences of India, 10*, 443–459.

Braun, A. (1855). *Algarum unicellularum genera nova velminus cognita* (p. 70). Pls. 1–6.

Chakraborty, S. (1988). East Kolkata Wetlands its significance, preservation protection and developments. *Meenbrata (Special issue on Wetland)*, 58–60.

Chodat, R. (1902). *Algues vertes de la Suisse. 1* (No. 3, pp. 1–373).

Chodat, R. (1913). *Monograp;hie dialuges en culture pure*. Ibd., 4 (Fasc. 2, pp. 1–226).

Collins, F. S. (1909). *The green algae of North America* (Tufts College Studies, pp. 79–170). Scientific series 2.

Deshikachary, T. V. (1959). *Cyanophyta* (pp. 1–686). New Delhi: Indian Council of Agricultural Research.

Fischer, T. R., Harding, L. W., Stanley, D. W., & Ward, L. G. (1988). Phytoplankton, nutrients, and turbidity in the Chesapeake, Delaware and Hudson estuaries. *Estuarine and Coastal Shelf Science, 27*, 61–93.

Ghosh, D. (1999). *Participatory management in wastewater treatment and reuse in West Bengal*.

Hansgirg, A. (1888). Ueber die Susswasser—gattungen Trochiscia Ktz [Astericium Corda, Polyedrium Nag., Cerasterias Reinsch]. *Hedwigia 27*, 126–132.

Huber-Pestalozii, G. (1942). *Das phytoplankton des Süßwassers. 2* (Teil, 2. Hälfte, 545 p). Stuttgart: Schweizerbart'she Verlagsbuchhandlung.

Hustedt, F. (1952). Neue und weing bekannte Diatomeen, IV. *Botaniska Notiser Hf4*, 366–410. 134 Figs, *Flora 10*(40), 625–640.

Jadhav, S. B. (2008). *Studies on quality of Lentic water resources of Wadi Ratnagiri* (162 p). M.Phil. thesis, Shivaji University Kolhapur.

Jaiswal, K. K., & Tiwari, G. L. (2003). *Chlorococcales (green algae)*. Allahabad: Bioved Research Society.

Jana, B. B. (1998). Sewage fed aquaculture: The Kolkatas model. *Ecological Engineering, 11*, 73–85.

Kant, S., & Gupta, P. (1998). *Algal flora of Ladakh* (341 p). Jodhpur: Scientific Publication.

Klebs, G. (1883). Über die Organisation einiger Flagellatengruppen und ihre Beziehungen zu Algen und Infusorien. *Untersuchungen aus dem Botanischen Institut zu Tübingen, 1*, 233–362.

Komarek, J., & Anagnostidis, K. (2008). *Cyanoprokaryota*. Heidelberg: Spectrum.

Krishna Kumari, L., et al. (2002). Primary productivity in Mandovi-Zuari estuaries in Goa. *Journal of the Marine Biological Association of India, 44*(1&2), 1–13.

Kützing, F. T. (1834 [1833]). Synopsis diatomearum oder Versuch einer systematischen Zusammenstellung der Diatomeen. *Linnaea 8*, 529–620, Plates XIII–XIX [79 figs].

Lagerheim, G. (1882). *Bidrag till kannedomen om Stockholmstraktens Pediasteer, Protococcaceer och Palmellaceer*. Oefv. Kongl. Sv, Vet.-Akad, Forhandl., 39 (No. 2, pp. 47–81; Pls. 2, 3).

Lemmermann, E. (1904). Das Plankton schwedischer Gewasser. *Arkiv för Botanik, 2*(2), 1–209. Pls. 1, 2.

McComb, A. J., & Lukatelich, R. J. (1995). In A. J. McComb (Ed.), *The Peel-Harvey estuarine system, Western Australia*.

Morren, C. (1830). Memoire sur un vegetalmicroscopique dun nouveau genre, propose sous le nomme Microsoter, ou conservateau des perites choses. *Annales des Sciences Naturelles – Biblioteca, 20*(Ser. 1), 404–426.

Mukherjee, M. (1996). Pisciculture and environment: An economic evaluation of sewage-fed fisheries in East Kolkata. *Science, Technology and Development, 14*(2), 73–79.

Mukherjee, M., Nath, U., Kashem, A., & Chattopadhyay, M. (2002). Peri-urban aquaculture and its environmental status: Kolkata experience. *Fishing Chimes, 22*(i), 9–15.

Mukherjee, I., Bhaumik, P., Mishra, M., Ranjan Thakur, A., & Ray Chaudhuri, S. (2010). Bheri-a unique example of biological complex system. *Journal of Biological Sciences, 10*(1), 1–10.

Nayar, K. K. (1994). The effect of Calcutta sewage of fish life. *Proceedings of the National Institute of Sciences of India, 10*, 147–156.

Pearl, H. W., Rudek, J., & Mallin, M. A. (1990). Stimulation of phytoplankton production in coastal waters by natural rainfall inputs: Nutritional and trophic implications. *Marine Biology, 107*, 247–254.

Pradhan, A., Bhaumik, P., Das, S., Mishra, M., Khanam, S., et al. (2008). Phytoplankton diversity as indicator of water quality for fish cultivation. *American Journal of Environmental Sciences, 4*, 406–411.

Prescott, G. W. (1982). *Algae of the Great Western Lakes area* (pp. 1–997). Dubuque: W.C. Brown Co.

Rabenhorst, L. (1864–1868). *Florae Europaea Algaum Aque dulcis et submarinae* (3 Vols.). Leipzig.

Ralfs, J. (1844). On the British Desmidieae. *Annals and Magazine of Natural History, 14*(Ser. 1), 469.

Ray Chaudhuri, S., Salodkar, S., Sudarsan, M., & Thakur, A. R. (2007). Integrated resource recovery at East Calcutta Wetland: How safe ISB this? *American Journal of Agricultural and Biological Sciences, 2*, 75–80.

Ray Chaudhuri, S., Mishra, M., Nandy, P., & Thakur, A. R. (2008). Waste management: A case study of ongoing traditional practices at east Calcutta wetland. *American Journal of Agricultural and Biological Science, 3*, 315–320.

Redfield, A. C. (1958). The biological control of chemical factors in the environment. *American Scientist, 46*, 205–222.

Reinsch, P. F. (1867). Die Algenflora des mittleren Theiles von Franken, enthaltend die vom Autor bis jetzt in diesen Gebieten beobachteten Susswasseralgenetc. *Abh Naturh Ges Nurnberg, 3*, 64.

Roy, P., Saha, S. B., & Banerjee, K. K. (1981, December 11–13). *A case study of Calcutta municipal wastes for fish culture in Bidyadhari-Kulti complex*. W.B. international symposium of water resources conservation, pollution and abatement, University of Roorkee.

Ryther, J. H., & Dunstan, W. M. (1971). Nitrogen, phosphorus and eutrophication in the coastal marine environment. *Science, 171*, 1008–1013.

Schmidle, W. (1901). *Algologische Notizen. XV.* Allg Bot Ges, 19, p. 234.

Sen, S. (1941). *Fisheries of West Bengal*. Alipore: West Bengal Government Press.

Sewell, R. B. S. (1935). A study of fauna of Salt Lake, Calcutta. *Records of the Indian Museum, 36*, 45–49, 58–60.

Shafi, N., et al. (2013). Phytoplankton dynamics of Nigeen Lake in Kashmir Himalaya. *International Journal of Environment and Bioenergy, 6*(1), 13–27.

Siep, K. L. (1994). Phosphorous and nitrogen limitation of algal biomass across trophic gradients. *Aquatic Sciences 56*, 16–28.

Smith, G. M. (1916). A monograph of the algal genus *Scenedesmus* based upon pure culture studies. *Transactions of the Wisconsin Academy of Sciences Arts and Letters, 18*, 422–530.

Sterner, R. W. (1994). Seasonal and spatial patterns in macro and micronutrient limitation in Joe Pool Lake, Texas. *Limnology and Oceanography, 39*, 535–50.

Thajuddin, N., & Subramanian, G. (1992). *Survey of Cyanobacterial Flora of the Southern East Coast of India, Botanica Marina* (Vol. 35, pp. 305–314).

Venkataraman, G. (1939). A systematic account of some south Indian diatoms. In *Proceedings of the Indian Academy of Sciences* (Vol. X, No. 6, Sec. B).

West, W., & West, G. S. (1904). *Fresh water algae from the Orkneys and Shetlands* (p. 224) Edinburgh: Transactions of the Botanical Society.

5.3 Case Study II: Phytoplankton Diversity of Indian Sunderbans

The Sunderbans mangrove forest is one of the largest mangrove forest in the world (140,000 ha). It is situated in the southeast part of Asia or the southeastern coast of India and Bangladesh, on the delta of the Ganges, Brahmaputra and Meghna rivers on the mouth of Bay of Bengal. The Indian site lies between 21°31′ to 22°53′N and 88°37′ to 89°09′E. Due to its beauty and richness of wildlife, it was declared a world natural heritage site by UNESCO on 1974. The site is intersected by a complex network of small-forested islands and mudflats and some interconnected tidal rivers, creeks and canals, therefore forming a complex network of tidal waterways.

The land is constantly moulded and altered by tidal action, with erosion along estuaries and deposition of silts from seawater. The Sunderbans consist of three wildlife sanctuaries (Sundarbans West, East and South) including the man eater Royal Bengal Tiger.

Sunderbans' beauty lies in its unique natural surroundings. Thousands of meandering streams, creeks, rivers and estuaries have enhanced its charm. Forest areas are dominated by typical mangrove plants, like Sundri and Gewu and patches of Nypa palm. Sunderban is the natural habitat of the world famous Royal Bengal Tiger, spotted deer, crocodiles, jungle fowl, wild boar, lizards, monkey and an innumerable variety of beautiful birds. The entire area is flooded with brackish water during high tides which mix with freshwater from inland rivers. The monsoon rains, flooding, delta formation, and tidal influence combine in the Sundarbans to form a dynamic landscape that is constantly changing. The ecology and the biodiversity of the Sunderban algae have been studied by a very few authors (Prain 1903; Naskar and Mandal 1999; Sen and Naskar 2003). Several authors studied the mangrove ecosystem with reference to human habitation and settlement, development of agricultural fields and brackish water fisheries. The algal flora of Suderbans has extensively studied by the present group from taxonomic point of view and as a potential source for biodiesel (Mukhopadhyay and Pal 2002; Choudhury and Pal 2008; Satpati et al. 2013)

Phytoplankton population of Sunderbans rivers varied to a greater extent with salinity gradient. In present study, phytoplanktons were collected from the different islands of Sunderbans like Basanti, Patharpratima, Bakkhali and Lothian Islands and surroundings river systems like Vidya river, Kholnar khal, Muriganga river, Saptamukhi river, Matla river and Bishalakshi khal. Mid river samplings were done for phytoplankton collection. Sampling section represents from freshwater zone to marine zone. Salinity varies from 0 to 22 psu, temperature profiles were recorded as 22–34 °C and an average pH was measured as 7.98.

5.3 Case Study II: Phytoplankton Diversity of Indian Sunderbans

Phytoplankton community was dominated by diatoms (Biacillariophyceae) followed by Chlorophyceae, and other algal groups recorded were Cyanophyceae, Euglenophyceae and Dinophyceae. A total of 34 taxa belonging to 5 groups were recorded. Species diversity was found to be maximum in summer (March) and minimum in winter season (November) in all the sample stations.

5.3.1 List of Phytoplankton Genera Recorded from Sunderbans

Cyanobacteria	1. *Merismopedia glauca*
	2. *Merismopedia minima*
	3. *Spirulina platensis*
	4. *Gloeocapsa punctata*
Chlorophyceae	5. *Chlorococcum infusionum*
	6. *Pediastrum tetras*
	7. *Crucigenia tetrapedia*
	8. *Scenedesmus quadricauda*
	9. *S. bijuga*
	10. *S. dimorphus*
	11. *Closterium tumidium*
Euglenophyceae	12. *Euglena gracilis*
	13. *E. polymorpha*
	14. *Phacus longicauda*
	15. *P. segretii*
Bacillariophyceae	16. *Achnanthes hauckiana*
	17. *Pleurosigma angulatum*
	18. *P. normanii*
	19. *Navicula halophila*
	20. *N. cincta*
	21. *N. minima*
	22. *N. ignorata*
	23. *N. peregrina*
	24. *Coscinodiscus excentricus*
	25. *C. excentricus*
	26. *Cyclotella meneghiniana*
	27. *Cyclotella striata*
	28. *Nitzschia obtuse*
	29. *N. amphibian*
	30. *Melosira nummuloides*
	31. *Fragilaria intermedia*
	32. *F. oceanic*
Dinophyceae	33. *Ceratium hirundinella*
	34. *Noctiluca scintillans*

5.3.2 Taxonomic Account of a Few Phytoplankton Taxa Recorded from Sunderbans (Rest Are in Case Studies I and III)

Division – Euglenophyta
Class – Euglenophyceae
Order – Euglenales
Family – Euglenaceae
Euglena polymorpha Dangeard (Plate 5.4, Fig. A)
   [Dangeard, 1902, p. 175. Pl. 85, Figs. 21, 22; Prescott, 1982, p. 393.]
   Cells ovoid to pyriform to subcylindric, narrowed gradually posteriorly to a short blunt tip; periplast with spiral striations; chloroplasts many and disc-like with laciniate margins, with one pyrenoid; cells 20–26 µ in diameter, 80–90 µ long
Phacus segretii var. *ovum* Prescott (Plate 5.4, Fig. B)
   [Prescott, 1982, p. 369, pl. 88, fig. 23.]
   Cells larger than in the typical form, broadly ovoid, 29 µ in diameter and 40 µ long.
   Occurrence, Jharkhali pond (N 22°01.142′, E 088°41.168′), fresh water
Division – Heterokontophyta
Class – Bacillariophyceae
Order – Pennalales
Suborder – Araphidineae
Family – Fragilariaceae
Subfamily – Fragilarioideae
Fragilaria brevistriata Grun. forma *elongata* f. nov (Plate 5.4, Fig. F)
   [Venkataraman 1939, p. 305, fig. 25, 26]
   Valves linear lanceolate with rounded ends. Striae short and marginal. Pseudoraphae broad. Length 31.20 µ, breadth 4.31 µ, striae 13 in 10 µ
Fragilaria oceanica Cleve. (Plate 5.4, Fig. H)
   [Subrahmanyan 1946, p. 165, figs. 336–339]
   Frustules in girdle view linear–rectangular, forming a very compact ribbon-like chain. Valves broadly lanceolate with rounded ends, 11.5 µ long, and 6.5 µ broad
Suborder – Monoraphidineae
Family – Achnanthaceae
Subfamily – Achnanthoideae
Achnanthes hauckiana var. *rostrata* (Plate 5.4, Fig. I)
   [Schulz 1926, p 191, fig. 40]

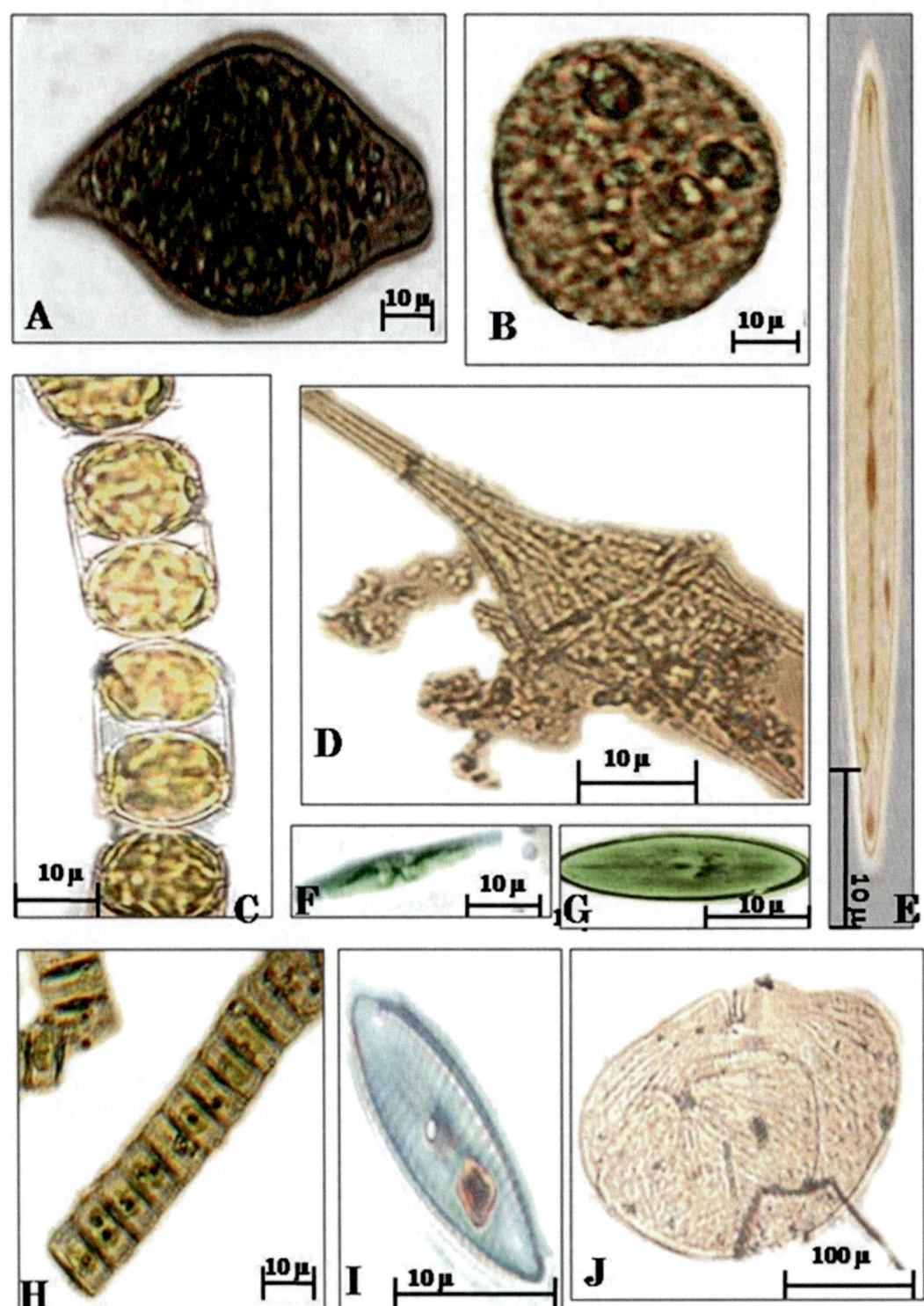

Plate 5.4 (**A**) *Euglena polymorpha*, (**B**) *Phacus segretii*, (**C**) *Melosira numuloides*, (**D**) *Ceratium hirudinella*, (**E**) *Navicula ignorata*, (**F**) *Fragilaria brevistriata*, (**G**) *Navicula cincta*, (**H**) *Fragilaria oceanica*, (**I**) *Acnanthes hauckiana*, (**J**) *Noctiluca scintillans*

Valves elliptic lanceolate with slightly truncate ends. Raphe thin, thread-like, axial area narrow, central areas somewhat broadened. The length is 20.41 µ and breadth 6.806 µ. Number of striae 20 in 10 µ

Suborder – Biraphidineae
Family – Naviculaceae
Subfamily – Naviculoideae
Navicula cincta (Ehr.) Kutz., var. Heufleri Grun. (Plate 5.4, Fig. G)
[G. Venkataraman 1939, p. 326, Fig 89]
Valves linear lanceolate with obtuse ends. Axial area narrow. Central area small broadened. Striations at the end convergent, 26.56 µ long and 5.81 µ broad, number of striations 10 in 10 µ

Navicula ignorata Krasske (Plate 5.4, Fig. E)
[Hustedt, 1952, p. 442, fig. 819; Foged 1979, p. 87, pl. XLIII, fig. 5]
Valves linear–lanceolate, with somewhat bent ends. The length is 57.44 µ and the breadth is 3.59 µ

Class – Coscinodiscophyceae
Order – Melosirales
Family – Melosiraceae
Melosira nummuloides var. *lesdiana* Pantocsek (Plate 5.4, Fig. C)
[Agardh, 1827, p. 307, fig -312.]
Cells cylindrical to subspherical, forming chains. Valve face flat or domed, covered with small spines or granules; a more or less well-developed corona

Division – Dinophyta
Class – Dinophyceae
Order – Noctilucales
Family – Noctilucaceae
Noctiluca scintillans (Macartney) (Plate 5.4, Fig. J)
Large, round to kidney-shaped cells, with a striated tentacle, one flagellum and a eukaryotic nucleus; cytoplasm may contain photosynthetic symbionts and gametes are gymnodinoid, phagotrophic with food vacuoles containing prey; chloroplasts absent; cells are 200–250 µ in diameter

Ceratium hirudinella (O.F. Muell.) Dujardin (Plate 5.4, Fig. D)
[Prescott, 1982, p. 437, pl. 92, figs. 4, 5]

Cells with polygonal plates, epicone usually larger than hypocone; cells are triangular with horns with prominent transverse furrows; cells are 45 µ long and 15 µ broad

References

Agardh, C. A. (1827). *Aufzählung einiger in den östreichischen Ländern gefundenen neuen Gattungen und Arten von Algen, nebst ihrer Diagnostik and beigefügten*. Bemerkungen.

Choudhury, A., & Pal, R. (2008). Diversity of planktonic diatoms from West Bengal coast with special reference to taxonomic accounts. *Phytomorphology, ISPM, 58*(1 and 2), 29–42–41.

Hustedt, F. (1952). Neue und weing bekannte Diatomeen, IV. *Botaniska Notiser Hf4*, 366–410. 134 Figs, *Flora 10*(40), 625–640.

Mukhopadhyay, A., & Pal, R. (2002). A report on biodiversity of algae from coastal West Bengal (South & North 24-parganas) and their cultural behavior in relation to mass cultivation programme. *Indian Hydrobiology, 5*(2), 97–107.

Naskar, K. R. & Mandal, R. N. (1999). *Ecology and biodiversity of Indian mangrove* (Vols. 1 & 2). New Delhi: Daya Publishing House.

Prain, D. (1903). Flora of sundarbans. *Records of the Botanical Survey of India, 2*, 231–390.

Prescott, G. W. (1982). *Algae of the Western Great Lakes area* (2nd ed.). Dubuque: WM Brown & Co.

Satpati, G. G., Barman, N., & Pal, R. (2013). A study on green algal flora of Indian sundarbans mangrove forest with special reference to morphotaxonomy. *Journal of Algal Biomass Utilization, 4*(1), 26.

Schulz, P. (1926). Die Kieselalgen der Danziger Bucht mit Einschluss derjenigen aus glazialen und postglacialen Sedimenten. *Bot Arch Königsberg, 13*(3/4), 149–327.

Sen, N., & Naskar, K. (2003). *Algal flora of sundarbans Mangal*. New Delhi: Daya Publishing House.

Subrahmanyan, R. (1946). A systematic account of the marine plankton diatoms of the Madras Coast. *Proceedings of the Indian Academy of Science, 24B*, 85–197.

5.4 Case Study III: Phytoplankton Dynamics of Eastern Indian Coast

5.4.1 Study Area

The Bhagirathi–Hugli estuary, situated at the eastern coast of India, is a deltaic offshoot of the River Ganges and lies approximately between 21°31′–23°20′N and 87°45′–88°45′E. It is a tropical coastal estuary and is associated with the Sunderbans Mangrove Biosphere Reserve which is part of world's largest delta, the Ganges–Brahmaputra delta. The estuary is funnel shaped with the breadth and cross-sectional area at the mouth being 25 km and 156,250 m², which decreases to 6 km and 36,799 m² at the head end. Climatic condition is dominated by NE and SW monsoon where the annual rainfall varied between 188 and 245 cm, with 75–85 % of the total annual rainfall occurring in the monsoon months. The pronounced influence of south west monsoon winds resulted in heavy seasonal precipitation which may be as high as 500 mm in a month. This being a tropical coastline showed a prolonged summer (April–May–June) with an average temperature of 35 ± 5 °C and a long monsoon season (July–August–September) with an average temperature of 35 ± 2 °C. Winter (November–December–January) is comparatively mild in this region of the world with the mean temperature of 12 ± 5 °C. Thus, this region shows a progressive increase in temperature from January to May followed by a decline with the onset of southwest monsoon from June onwards and reaches minimum during the end of the year. From the estuarine and coastal area, we selected four main stations for our study which was mainly based on the salinity gradient of each sampling station (Diamond Harbour, Kakdweep, Junput and Digha). The sampling stations were designated as freshwater zone (Diamond Harbour), estuarine zone (Kakdweep) and marine zone (Junput and Digha).

Diamond Harbour (22°8.78′N and 88°9.0′E) is located at a distance of 56 km from Kolkata, the state capital of West Bengal, India (Fig. 5.2). Just upstream of this coastal station, Rupnarayan River merges with the Hugli River which is further joined by the Haldi River at a distance of 20 km downstream. As this coastal station is located at the upstream region of the Bhagirathi–Hugli estuary, accordingly, salinity was relatively low (0–4.5 psu) and reached as low as 0 psu in the monsoon months. Thus, this station was designated as the freshwater station (Fig. 5.2). Our second sampling station Kakdwip (21°53′0″N, 88°11′0″E) is located at a distance of 36 km from Diamond Harbour and was considered as an estuarine location (salinity, 4–17 psu). At this station, the mixing conditions were more prevalent between the freshwater of Bhagirathi–Hugli and marine waters of the Bay of Bengal, which resulted in the mesohaline conditions of this region (Fig. 5.2).

For our marine samples along a relatively high salinity gradient (26–36 psu), two coastal stations were chosen (Junput, 21°43′N, 87°49′E, and Digha, 21°37′N, 87°31′E) from the confluence of the Hugli River and Bay of Bengal and were together designated as the coastal marine region (Fig. 5.2).

5.4.2 Nutrient and Phytoplankton Dynamics of Coastal West Bengal

Phytoplankton population at coastal West Bengal is in a dynamic state and brought about primarily by interplay of several physicochemical and biological factors. Environmental variables and nutrient concentrations fluctuated significantly both on a monthly as well as on a seasonal basis. Such variability in the nutrient status and physicochemical environment of the habitat brought about fluctuations in the phytoplankton population as well as in the biotic indices. Accordingly, the findings have been represented in different sections as follows:

(a) Phytoplankton diversity of the study area
(b) Analysis of physicochemical parameters of the study area

5.4 Case Study III: Phytoplankton Dynamics of Eastern Indian Coast

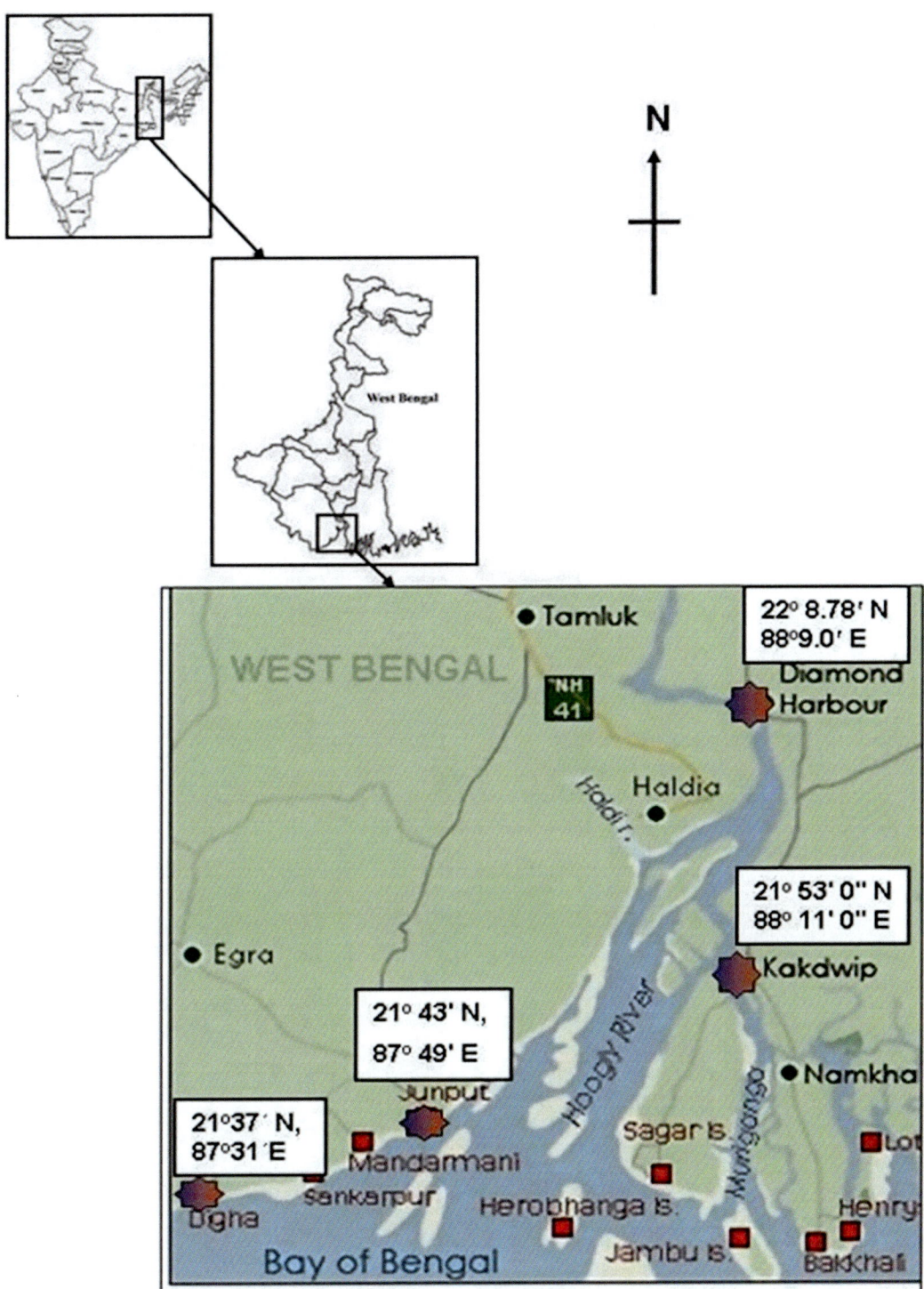

Fig. 5.2 Map of the study area at Coastal West Bengal

(c) Phytoplankton population study (cell count) by application of biotic indices
(d) Estimation of Primary productivity (GPP, NPP and CRR) and Oxygen Concentrations (DO and BOD)
(e) Implementation of Multivariate Procedures for community pattern analysis of the study area

5.4.2.1 Phytoplankton Diversity of the Study Area

The phytoplankton population of the study area was represented by 75 taxa belonging to three different algal divisions like Cyanobacteria, Chlorophyta and Bacillariophyta. Among the taxa recorded, 8 species were from Cyanobacteria, 17 species belonged to Chlorophyta and 51 species from Bacillariophyta (Table 5.2). As evident, diatoms

Table 5.2 List of phytoplankton genera recorded from the entire study area (viz. freshwater station, estuarine station and coastal marine region)

S. No.	Name of taxa
1.	*Gloeocaspapunctata*
2.	*Microcystis aeruginosa*
3.	*Merismopedia glauca*
4.	*Merismopedia minima*
5.	*Merismopedia punctata*
6.	*Gloeothece rupestris*
7.	*Spirulina meneghiniana*
8.	*Spirulina platensis*
	Chlorophyta
9.	*Pediastrum duplex* var. *rotundatum*
10.	*Pediastrum simplex*
11.	*Pediastrum simplex* var. *duodenarium*
12.	*Pediastrum tetras*
13.	*Ankistrodesmus falcatus*
14.	*Ankistrodesmus falcatus* var. *stipitatus*
15.	*Dictyosphaerium pulchellum*
16.	*Shroederia judayi*
17.	*Kirchneriella lunaris*
18.	*Scenedesmus bijuga*
19.	*Scenedesmus dimorphus*
20.	*Scenedesmus quadricauda*
21.	*Crucigenia rectangularis*
22.	*Crucigenia tetrapedia*
23.	*Tetrastrum staurogeniaeforme*
24.	*Rhizoclonium riparium*
	Bacillariophyta
25.	*Aulacoseira granulata* var. *angustissima*
26.	*Skeletonema costatum*

(continued)

Table 5.2 (continued)

S. No.	Name of taxa
27.	*Thalassiosira decipiens*
28.	*Cyclotella meneghiniana*
29.	*Cyclotella striata*
30.	*Coscinodiscus granii*
31.	*Coscinodiscus excentricus*
32.	*Coscinodiscus excentricus* var. *fasciculata*
33.	*Actinocyclus normanii* f. *subsala*
34.	*Leptocylindrus danicus*
35.	*Bacteriastrum delicatulum*
36.	*Bacteriastrum varians*
37.	*Chaetoceros curvisetus*
38.	*Chaetoceros diversus*
39.	*Chaetoceros messanensis*
40.	*Chaetoceros wighami*
41.	*Eucapmia zoodiacus*
42.	*Ditylum brightwellii*
43.	*Biddulphia alternans*
44.	*Odontella aurita*
45.	*Biddulphia dubia*
46.	*Biddulphia heteroceros*
47.	*Biddulphia mobiliensis*
48.	*Odontella rhombus*
49.	*Thalassionema nitzschoides*
50.	*Thalassiothrix frauenfeldii*
51.	*Asterionella japonica*
52.	*Cocconeis dirupta*
53.	*Gyrosigma beaufortianum*
54.	*Gyrosigma acuminatum*
55.	*Gyrosigma obtusatum*
56.	*Pleurosigma normanii*
57.	*Pleurosigma salinarum*
58.	*Diploneis weissflogii*
59.	*Diploneis interrupta*
60.	*Navicula minima* Grunow
61.	*Navicula mutica* fo. *cohni*
62.	*Navicula peregrina*
63.	*Navicula quadripartita*
64.	*Amphiprora gigantea*
65.	*Cymbella naviculiformis*
66.	*Bacillaria paradoxa*
67.	*Nitzschia amphibia*
68.	*Nitzschia bilobata* var. *minor*
69.	*Nitzschia ignorata*
70.	*Nitzschia obtusa*
71.	*Nitzschia punctata*
72.	*Nitzschia sigmoidea*
73.	*Nitzschia delicatissima*
74.	*Nitzschia pacifica*
75.	*Surirella fastuosa* var. *recedens*

5.4 Case Study III: Phytoplankton Dynamics of Eastern Indian Coast

accounted for the bulk of the phytoplankton population which represented about 69 % of the total population. Although diatom taxa were recorded from all the three sampling areas of freshwater zone, estuarine zone and marine zone, cyanobacterial population was restricted to the freshwater zone, i.e. Diamond Harbour only. On the other hand, green algal genera were recorded from the freshwater and estuarine locations, but none were recorded from the marine coastal region. Among the different members of Bacillariophyta, centric diatoms were more abundant at the estuarine to marine region, whereas pennate diatoms were recorded from freshwater to estuarine region. On an individual site basis, the phytoplankton population at Diamond Harbour was most diverse, in comparison to other two zones, and was represented by 54 taxa. The population at marine coastal region (Digha and Junput) was represented by 41 diatom genera only. The population at the estuarine location (Kakdweep) was least diverse as it was represented by 25 algal taxa altogether. Each site exhibited distinct seasonal phytoplankton assemblages which varied spatiotemporally in response to different environmental conditions.

The phytoplankton population at freshwater zone (Diamond Harbour) was represented by 54 phytoplankton taxa of which 8 species were blue green, 16 taxa were green algae and 30 genera were of diatoms (Table 5.3). The phytoplankton population showed distinct patterns where green algal population was mostly observed during the warmer summer and monsoon months whereas diatoms were more available in the cooler months of

Table 5.3 Floristic list of phytoplankton genera recorded during the study period from the freshwater station (Diamond Harbour)

Division	Name of genera	Period of availability	Abundance
Cyanobacteria	*Merismopedia minima* (MM)	July to Nov	+
	Merismopedia punctata (MP)	August to Nov	+
	Merismopedia glauca (MG)	June, July	+
	Spirulina platensis (SP)	Apr to Sept	+
	Spirulina meneghiniana (SM)	July, Aug	+
	Gloeocapsa punctata (Glcsp)	Apr to July	+
	Gloeothece rupestris (Glthc)	July	+
	Microcystis aeruginosa (MAero)	Oct to Dec	+ +
Chlorophyta	*Ankistrodesmus falcatus* (AF)	Apr to Sept	+
	Ankistrodesmus falcatus var. *stipitatus*	Apr to June	+
	Kirchneriella lunaris (KL)	Apr to Nov	+
	Shroederia judayi (SJ)	Mar to Aug	+
	Crucigenia rectangularis (CR)	May, June	+ +
	Crucigenia tetrapedia (CT)	May, June	+
	Crucigenia quadrata (CQ)	Rare	+
	Scenedesmus quadricauda (SQ)	Mar to June	+ +
	Scenedesmus bijuga (Sbi)	Mar to June	+ +
	Scenedesmus dimorphus (Sdi)	Mar to June	+
	Pediastrum simplex	June, July	+ +
	Pediastrum tetras (Ptet)	Mar to June	+
	Pediastrum duplex var. *Rotundatum* (Pdro)	Mar to June	+ +
	Pediastrum simplex var. *duodenarium* (Psduo)	Mar to June	+
	Dictyosphaerium pulchellum (Dictyo)	Mar, Apr	+ +
	Rhizoclonium riparium (RR)	Mar to Sept (irregular)	+

(continued)

Table 5.3 (continued)

Division	Name of genera	Period of availability	Abundance
Bacillariophyta	*Odontella aurita* (OAu)	Nov to Feb	+
	Biddulphia dubia (Bdu)	Nov to Feb	+
	Gyrosigma obtusatum (GOb)	irregular	+
	Gyrosigma acuminatum (GAcu)	Aug to Oct	+ +
	Gyrosigma beaufortianum (Gbeau)	Oct to Apr	+
	Pleurosigma salinarum (Psal)	Oct to Mar	+ +
	Cyclotella meneghiniana (CM)	Sept to Oct	+ +
	Cyclotella striata (CS)	Mar and Sept	+
	Coscinodiscus excentricus (CE)	Mar and Sept	+ +
	Actinocyclus normanii f. *subsala* (AN)	Oct to Apr	+
	Stephanodiscus hantzschii (SH)	Nov to Feb	+
	Ditylum brightwellii (DB)	Mar, Apr	+
	Bacteriastrum varians (BV)	Sept, Oct	+
	Bacteriastrum delicatulum (Bdeli)	Aug to Oct	+
	Aulacoseira granulata (Aug)	Aug to Oct	+
	Thalasiothrix frauenfeldii (TF)	Nov to Feb	+ +
	Thalassionema nitzschioides (TN)	Nov to Mar	+ +
	Bacillaria paxillifer (Bpax)	Nov to Feb	+
	Amphiprora gigantea (AG)	Nov to Jan	+
	Leptocylindrus danicus (LD)	Feb to Sept	+ + +
	Cymbella naviculiformis (CN)	Nov to Feb	+
	Navicula peregrina (Nper)	Feb to June	+
	Navicula quadripartita (Nquad)	Oct to Mar	+
	Nitzschia ignorata (Nig)	Nov, Dec	+
	Nitzschia amphibia (Namp)	Rare	+
	Nitzschia obtusa (Nob)	Oct to Dec	+
	Nitzschia punctata (Npu)	Nov to Feb	+ +
	Nitzschia delicatissima (Ndeli)	Nov to Feb	+ +
	Navicula minima (Nmi)	Oct to Feb	+ +
	Navicula mutica (Nmu)	Rare	+ +

$+ = \leq 1 \times 10^4$ cells/L
$+ + = \geq 1 \times 10^4$ to $\leq 5 \times 10^4$ cells/L
$+ + + = \geq 5 \times 10^4$ to $\leq 1 \times 10^5$ cells/L
$+ + + + = \geq 1 \times 10^5$ to 20×10^5 cells/L

post-monsoon and winter. Appearance and abundance of cyanobacterial population was restricted mainly to the monsoon and post-monsoon periods. The most dominant taxon recorded during the entire sampling period was *Leptocylindrus danicus* (Table 5.4). Other abundant phytoplankton genera were *Thalassiothrix frauenfeldii*, *Thalassionema nitzschoides*, *Pediastrum tetras*, *P. duplex* var. *rotundatum*, *P. simplex* var. *duodenarium* and *Scenedesmus* spp. In an analysis of the seasonality of phytoplankton species composition throughout the study period, following trends were noted in the freshwater zone.

Table 5.4 Showing percentage abundance of dominant phytoplankton genera at the freshwater zone (Diamond Harbour)

Season	Dominant species	Maximum contribution to total population (%)
Summer	*Leptocylindrus danicus*	9.53
Monsoon	*Leptocylindrus danicus*	18.54
Post-monsoon	*Microcystis aeruginosa*	21
Winter	*Nitzschia delicatissima*	10.07

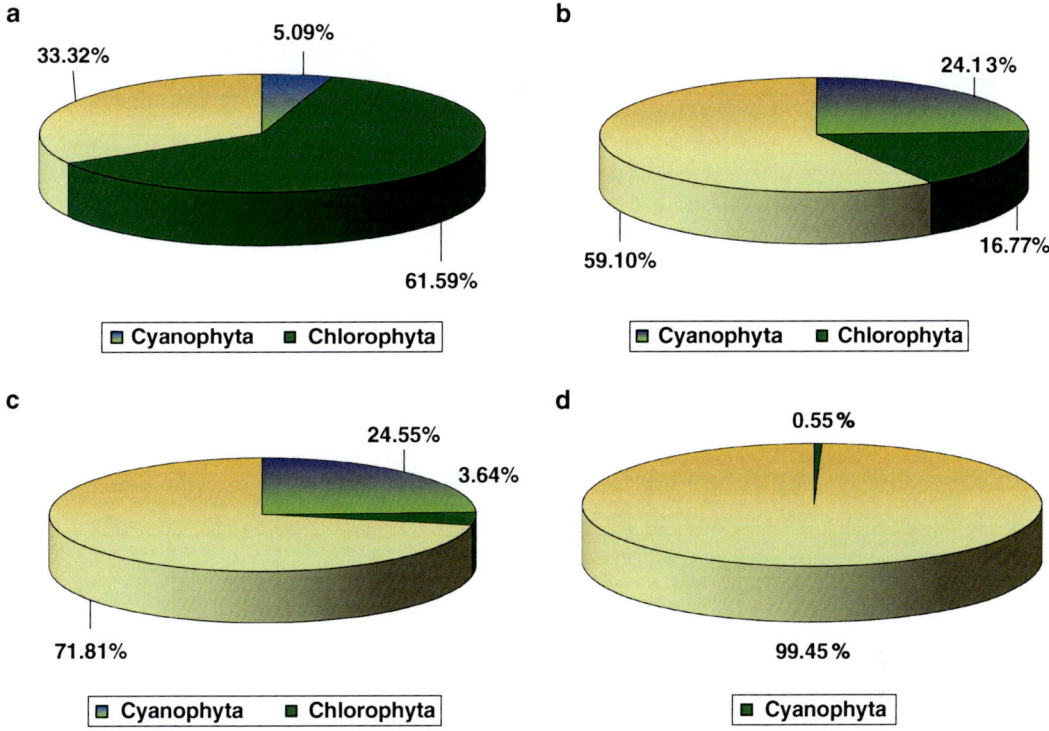

Fig. 5.3 Percentage seasonal contributions of different algal divisions to the total phytoplankton population: (**a**) summer, (**b**) monsoon, (**c**) post-monsoon and (**d**) winter at the freshwater station (Diamond harbour)

In summer, green algal population was high and made up to 61.59 % of the total phytoplankton population whereas diatom and blue-green algae accounted for 33.32 % and 5.09 % of the population, respectively (Fig. 5.3a). Interestingly, it was observed that the most abundant taxon was the diatom genus *Leptocylindrus danicus* that made up 9.53 % of total population, followed by green algal genera like *Pediastrum duplex* var. *rotundatum*, *Ankistrodesmus falcatus* and *Pediastrum simplex* var. *duodenarium* contributing to 8.94 %, 8.82 %, and 7.32 % of the total phytoplankton population, respectively.

In monsoon period, dominance of diatoms persisted (59.1 % of the total population), although there was a rise in cyanobacterial population (24.13 % of total population) together with green algae, contributing to 16.77 % of the total population (Fig. 5.3b). On a generic level, dominance of the diatom genus *Leptocylindrus danicus* persisted contributing to 18.54 % of the total population. Among the green algal genera, the most abundant taxon was *Kirchneriella lunaris* that made up 10.54 % of the total phytoplankton population. The cyanobacterial population was composed of taxa like *Spirulina* spp., *Merismopedia* spp., *Microcystis aeruginosa*, etc.

In post-monsoon period, as seasonal precipitation reduced, a further shift in phytoplankton species composition was observed (Fig. 5.3c). The green algal availability decreased further and contributed to only 3.64 % of the population. On the other hand, diatom population continued to increase and made up to 71.81 % of the total population in this period. Cyanobacterial population also flourished in this period and *Microcystis aeruginosa* appeared as the most abundant taxon, contributing 24.55 % of the total population. Diatom population was represented by several genera, among which the more abundant taxa were *Thalassionema nitzschoides*, *Pleurosigma salinarum*, *Nitzschia amphibia* and *Gyrosigma obtusatum* each of which made up more than 5 % of the total population.

As winter appeared, almost the entire population was represented by diatoms (99.45 %), rest

Table 5.5 Floristic list of phytoplankton genera recorded during the study period from the estuarine station

Division	Name of genera	Period of availability	Abundance
Chlorophyta	*Kirchneriella lunaris* (KL)	Mar to Sept	+ +
	Ankistrodesmus falcatus (AF)	Mar to Oct	+ +
	Scenedesmus bijuga (Sbi)	Mar to June	+ +
	Scenedesmus dimorphus (Sdi)	Mar to June	+ +
	Scenedesmus quadricauda (Squa)	Apr to May	+
	Crucigenia tetrapedia(CT)	Apr to Sept (irregular)	+
	Rhizoclonium riparium (RR)		+ +
Bacillariophyta	*Thalassiothrix frauenfeldii* (TF)	June to Feb	+ +
	Cyclotella meneghiniana (CM)	May to Nov	+ +
	Cyclotella striata (CS)	May to Nov	+ +
	Ditylum brightwellii (DB)	May to Nov	+ +
	Gyrosigma obtusatum (GOb)	May to Nov	+ +
	Gyrosigma acuminatum (GAcu)	May to Nov	+ +
	Bacteriastrum varians (BV)	Sept to Oct	+
	Bacteriastrum delicatulum (Bdeli)	Aug and Sept	+
	Stephanodiscus hantzschii (SH)	irregular	+
	Thalassionema nitzschoides (TN)	Oct to Apr	+
	Aulacoseira granulata (AuG)	Oct to Apr	+
	Pleurosigma salinarum (PS)	Oct to Apr	+
	Amphiprora gigantea (AG)	Oct to Mar	+
	Coscinodiscus excentricus (CEx)	Oct to Mar	+
	Nitzschia delicatissima (Ndeli)	Oct to Apr	+
	Odontella aurita (Oau)	Nov to Mar	+
	Biddulphia dubia (Bdu)	Nov to Mar	+

$+ = \leq 1 \times 10^4$ cells/L
$+ + = \geq 1 \times 10^4$ to $\leq 5 \times 10^4$ cells/L
$+ + + = \geq 5 \times 10^4$ to $\leq 1 \times 10^5$ cells/L
$+ + + + = \geq 1 \times 10^5$ to 20×10^5 cells/L

were green algae without any blue green algal members (Fig. 5.3d). Total phytoplankton cell count was significantly higher in this period as compared to other seasons. This was due to increase in cell count of individual taxa. The most abundant taxon was *Nitzschia delicatissima* that contributed to 10.07 % of the population (Table 5.4). Pennate diatom availability tended to be higher and was represented by taxa like *Thalassionema nitzschoides*, *Pleurosigma salinarum*, *Bacillaria paxillifer* (syn. *Nitzschia paradoxa*, www.itis.gov) *Thalassiothrix frauenfeldii*, etc. Each of the taxon accounted for more than 5 % of the population. Centric diatom taxa like *Biddulphia dubia* and *Coscinodiscus excentricus* also made significant contributions to the phytoplankton population.

At Kakdweep, the estuarine coastal station, minimum phytoplankton diversity was recorded as compared to other sampling stations. The population was represented by green algae and diatoms only with no record of cyanobacterial taxa. The phytoplankton population at the estuarine zone was represented by 25 taxa, out of which 8 species were green algae and 17 genera were diatoms (Table 5.5). Here also, a distinct pattern of phytoplankton population was recorded, where green algae flourished mainly in the summer and monsoon months, but diatoms were more abundant in the post-monsoon and winter periods. Here, unlike other stations, a single genus could not be ascertained as most abundant taxon.

In summer, green algal population was dominant, contributing to 69.46 % of the total phytoplankton population along with diatoms that accounted for 30.54 % of the population (Fig. 5.4a).

5.4 Case Study III: Phytoplankton Dynamics of Eastern Indian Coast

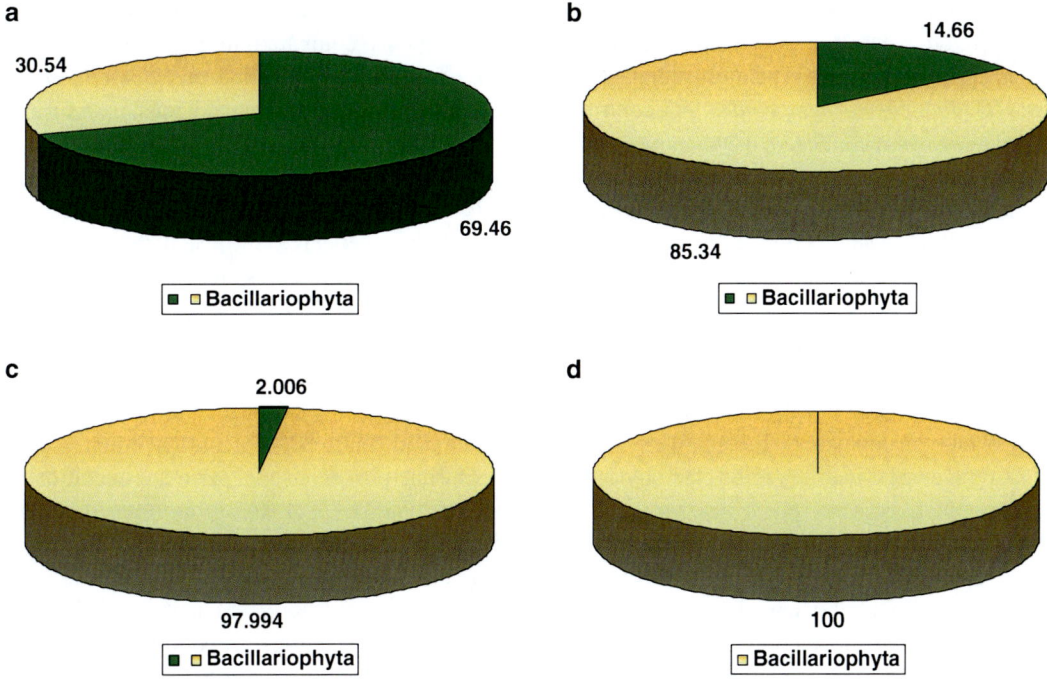

Fig. 5.4 Percentage seasonal contributions of different algal groups to the total phytoplankton population: (**a**) summer, (**b**) monsoon, (**c**) post-monsoon and (**d**) winter at the estuarine station (Kakdweep)

The most abundant taxon was *Kirchneriella lunaris* that made up to 13.81 % of total population. Other dominant taxa were *Scenedesmus brevicauda* and *S. dimorphus* that contributed to 12.53 % and 11.43 % of the population, respectively (Table 5.6).

In monsoon, a shift in phytoplankton species composition was recorded with a drop in green algal population (14.66 %) and a rise in diatom population (85.34 %) (Fig. 5.4b). On a generic level, the most abundant taxon was *Thalassiothrix frauenfeldii* with a contribution of 13.73 % to the total population (Table 5.6). Other major genera that flourished in this period were *Ditylum brightwellii* and *Stephanodiscus hantzschoides*, each of which accounted for more than 10 % of the population. Green algal population was represented by genera like *Rhizoclonium riparium*, *Kirchneriella lunaris*, *Ankistrodesmus falcatus* and *Crucigenia tetrapedia* although their total counts were much lower as compared to diatoms.

In the post-monsoon period, as seasonal precipitation reduced, phytoplankton species composition changed. The green algal availability decreased further and contributed to only 2.006 %

Table 5.6 Showing percentage abundance of dominant phytoplankton genera at the estuarine zone (Kakdweep)

Season	Dominant species	Maximum contribution to total population
Summer	*Kirchneriella lunaris*	13.81
Monsoon	*Thalassiothrix frauenfeldii*	13.73
Post-monsoon	*Thalassionema nitzschoides*	15.54
Winter	*Thalassionema nitzschoides*	20.49

of the total phytoplankton population. On the other hand, diatom population continued to increase and made up to 98 % of the total population in this period (Fig. 5.4c). Diatom population was represented by several genera, among which the most abundant taxa was *Thalassionema nitzschoides* contributing more than 15 % of the total population. The other dominant genera were *Paralia sulcata*, *Nitzschia delicatissima*, *Aulacoseira granulata* and *Thalassiothrix frauenfeldii*.

As winter appeared, almost the entire population was represented by diatoms (Fig. 5.4d). Total cell count was significantly higher in this period as compared to other seasons with an increase in cell count of individual taxa. The most abundant taxon was *Thalassionema nitzschoides* that contributed to 20.49 % of the population. Centric diatom abundance tended to be higher as compared to pennate taxa, represented by genera like *Aulacoseira granulata*, *Pleurosigma salinarum*, *Coscinodiscus excentricus* and *Biddulphia dubia*. Each genus accounted for more than 5 % of the population. Pennate diatom taxa like *Thalassiothrix frauenfeldii*, *Nitzschia delicatissima* and *Amphiprora gigantea* also made significant contributions to the phytoplankton population (Table 5.6).

At the coastal marine region, the phytoplankton population was represented by 41 taxa belonging to 26 genera. The phytoplankton flora was represented mainly by diatoms. Only two genera of Dinophyta, namely, *Dinophysis* and *Gymnodinium*, were recorded during the study period (Table 5.7). A few genera like *Bacillaria paxillifer*, *Nitzschia pacifica*, *Amphiprora gigantea*, *Biddulphia heteroceros*, *Aulacoseira granulata*, *Coscinodiscus granii*, *Surirella fastuosa*, *Chaetoceros curvisetus*, *Eucampia zoodiacus* and *Aulacodiscus johnsonii* occurred rarely in the study area. From the present study a distinct seasonal trend in phytoplankton population of this coastal station could be elucidated as follows.

In summer, a total of 12 taxa represented the phytoplankton population. Centric diatom genera (82.69 %) were more abundant as compared to pennate genera (17.31 %) (Fig. 5.5a). The most abundant taxa were *Odontella rhombus* and *Asterionella japonica*, with contributions of 24.06 % and 11.59 % to the total population, respectively (Table 5.8). The population of *Odontella rhombus* gradually increased in the summer months with cell count of > 700 cells/mL in June 2005. *Cocconeis dirupta* appeared in the month of May 2005 that accounted for 25 % of the phytoplankton population. Rest of the diatom genera contributed 5–25 % of the total phytoplankton population and appeared seasonally. The sudden appearance of *Skeletonema costatum* with a high cell count was another major taxon recorded in the month of June 2006 that contributed to 50.31 % of the total plankton population.

In monsoon also, the centric diatom population was high as compared to pennate population (Fig. 5.5b). The population of *Odontella rhombus* continued as dominant genus in monsoon months, accounting for 13.48 % of total population. Other relatively abundant taxa were *Odontella mobiliensis* and *Ditylum brightwellii*, each with cell counts of > 100 cells/mL. Members of *Chaetoceros* spp. appeared only during monsoon period at this zone. Taxa like *C. diversus* (7.64 %) and *C. messanensis* (7.73 %) were quite abundant at this time. During this period, there was a steady rise in abundance of centric diatoms which made up to 80.48 % of total count.

During post-monsoon period, population of centric diatoms (62.4 %) began to recede with a rise in pennate population (37.6 %). The most abundant genus was *Biddulphia alternans* contributing to 16.05 % of population (Fig. 5.5c). Other dominant centric members were *Biddulphia dubia* and *Odontella aurita*. Among pennate members, taxa like *Thalassiothrix frauenfeldii*, *Thalassionema nitzschoides*, *Diploneis weissflogii* and *Diploneis interrupta* were important members of the total phytoplankton population.

As winter approached, there was a significant drop in freshwater input to the coastal marine waters along with a gradual decrease in temperature as well. This further resulted in shifting of species composition of the population. In this period, there was a sharp rise in pennate diatom population primarily due to a high growth rate of the taxon *Asterionella japonica*. Pennate genera contributed for >80 % of total population with *A. japonica* contributing for 77.35 % of the population (Fig. 5.5d). Centric diatom population was represented by several taxa like *Coscinodiscus* spp., *Biddulphia* spp., *Paralia sulcata*, *Ditylum brightwellii*, *Stephanodiscus hantzschii*, etc.

5.4.2.2 Analysis of Physicochemical Parameters of the Study Area

The eastern coast of India is rich in nutrients and light which is sufficient to support phytoplankton population. Therefore, light and nutrients never acted as limiting factors in the study area although the phytoplankton production was hampered due to the silt contents along with other anthropogenic

5.4 Case Study III: Phytoplankton Dynamics of Eastern Indian Coast

Table 5.7 Floristic list of phytoplankton genera recorded during the study period from coastal marine region

Division	Name of genera	Period of availability	Abundance
Bacillariophyta	*Asterionella japonica* (AJ)	Almost throughout the year	+ + + +
	Odontella rhombus (OR)	Apr to Oct	+ + +
	Odontella mobilensis (OM)	Mar to Oct	+ +
	Odontella aurita (OAu)	Oct to Mar	+ +
	Biddulphia alternans (Balt)	Aug to Jan	+ +
	Biddulphia dubia (Bdu)	Oct to Mar	+ +
	Biddulphia heteroceros (BH)	Oct, Nov	+
	Gyrosigma obtusatum (GOb)	Mar to July	+
	Gyrosigma acuminatum (GAcu)	Feb to Apr	+
	Pleurosigma normanii (PN)	Dec to Mar	+
	Cyclotella meneghiniana (CM)	Almost throughout the year	+ +
	Coscinodiscus perforatus (CP)	Almost throughout the year	+ +
	Coscinodiscus centralis (CC)	Nov to Apr	+ +
	Coscinodiscus excentricus (CE)	Apr to July	+ +
	Actinocyclus normanii f. *subsala* (AN)	June to Oct	+ +
	Coscinodiscus granii (CG)	Oct, Nov	+
	Nitzschia delicatissima (ND)	Apr to July	+ +
	Nitzschia sigmoidea (NS)	Apr	+
	Cyclotella striata (CS)	May to Sept	+
	Stephanodiscus hantzschoides (SH)	Nov (2006)	+ +
	Thalassiosira decipiens (TD)	Mar (2007)	+
	Cocconeis dirupta	Aug to Mar	Rare
	Ditylum brightwellii (DB)	May (2005)	+ +
	Chaetoceros curvisetus	Almost throughout the year	+ +
	Surirella fastuosa (SF)	June, July (2005)	Rare
	Bacteriastrum varians (BV)	June, Oct	+
	Chaetoceros diversus (CD)	Mar	+
	Diploneis interrupta (DI)	July to Sept	+ +
	Diploneis weissflogii (DW)	July to Nov	+
	Aulacoseira granulata (AuG)	Oct, Nov	+
	Chaetoceros wighami (CW)	July to Nov	+
	Chaetoceros messanensis (ChM)	July to Oct	+
	Thalassiothrix frauenfeldii (TF)	July to Oct	+ +
	Thalassionema nitzschioides (TN)	Aug to Oct	+ +
	Bacillaria paxillifer (BPax)	Aug to Feb	+ +
	Nitzschia pacifica n. sp. (Npac)	Nov to Jan	+ +
	Amphiprora gigantea (AG)	Mar, Oct	+
	Eucampia zoodiacus	Oct, Nov	+
		Oct, Nov	Rare
	Skeletonema costatum	June (2006)	Rare
		March (2006)	Rare
Dinophyta	*Gymnodinium* sp.	May (2005)	
	Dinophysis sp.	May (2005)	

$+ = \leq 1 \times 10^4$ cells/L
$+ + = \geq 1 \times 10^4$ to $\leq 5 \times 10^4$ cells/L
$+ + + = \geq 5 \times 10^4$ to $\leq 1 \times 10^5$ cells/L
$+ + + + = \geq 1 \times 10^5$ to 20×10^5 cells/L

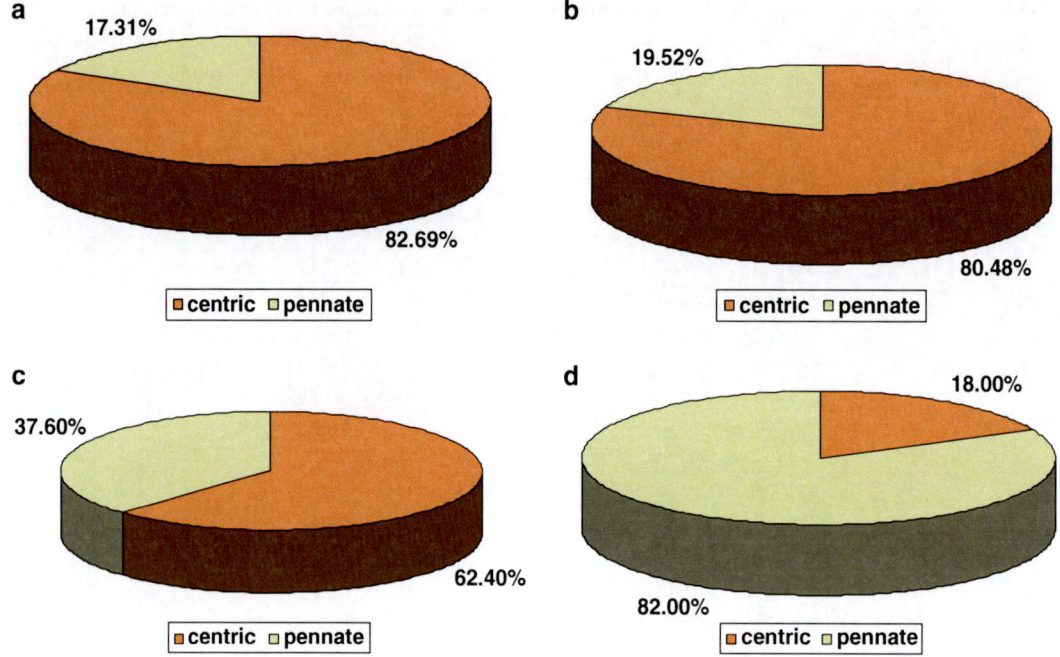

Fig. 5.5 Percentage seasonal contributions of centric and pinnate diatoms to the total phytoplankton population: (**a**) summer, (**b**) monsoon, (**c**) post-monsoon and (**d**) winter at the coastal marine region (Junput and Digha)

Table 5.8 Showing percentage abundance of dominant phytoplankton genera at the coastal marine zone (Junput and Digha)

Season	Dominant species	Maximum contribution to total population (%)
Summer	*Skeletonema costatum*	26.63
Monsoon	*Odontella rhombus*	13.48
Post-monsoon	*Biddulphia alternans*	16.05
Winter	*Asterionella japonica*	77.35

factors. Salinity variations were pronounced that varied from 0 to 36 psu. Seasonal precipitation and heavy riverine inflow especially in the monsoon months played an important role in lowering the salinity levels at all stations. This being a tropical coastline, there is a prolonged summer (April–May–June) with an average temperature of 35±5 °C and a long monsoon season (July–August–September) with an average temperature of 35±2 °C. Winter (November–December–January) is comparatively mild with a mean temperature of 12±5 °C. The pH level varied from 7 to 8.3 in the entire study area. The nutrient content especially with respect to dissolved inorganic nitrogen (DIN) and dissolved inorganic phosphate (DIP) were comparatively higher in the freshwater station and estuarine station than at the marine region. On the contrary, dissolved silicate (DSi) contents showed an opposite trend with higher values at the marine region in comparison to the other stations.

Results showed that the habitat water of the freshwater station was weakly alkaline with the pH ranging from 7.0 to 8.3 with the mean pH of 7.57 (Fig. 5.6a). During the drier months of summer and winter, the pH was relatively high, ranging from 7.5 to 8.3 showing pH maxima in summer for both 2005 and 2006. As expected, water temperature was also maximum in summer and minimum in winter. Salinity was relatively low with a mean of 2.2 psu which reached as low as 0 psu in monsoon period due to high influx of freshwater from perennial rivers, along with heavy seasonal precipitation (Fig. 5.6c). Maximum salinity at this hypo saline station was recorded in winter, reaching as high as 5 psu in January 2006, when seasonal precipitation as

5.4 Case Study III: Phytoplankton Dynamics of Eastern Indian Coast

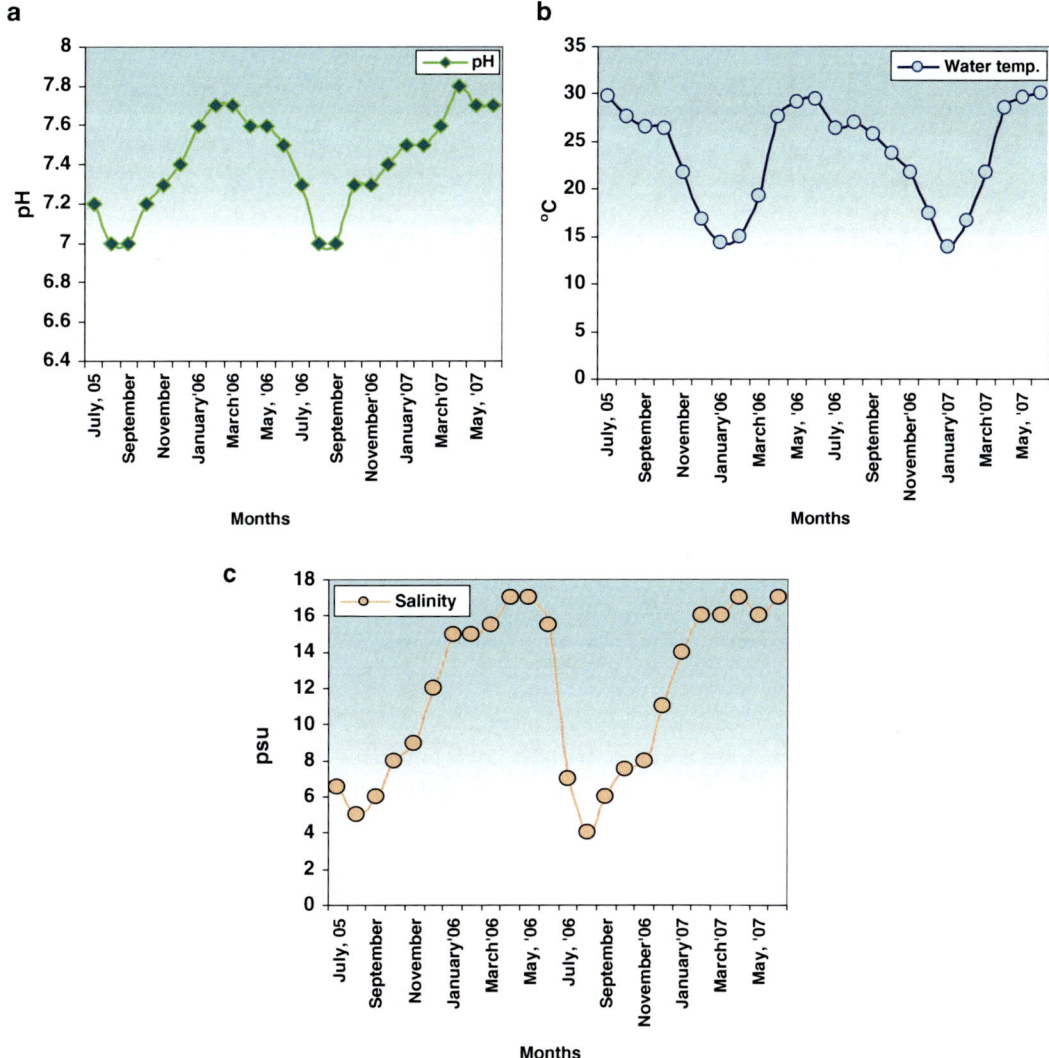

Fig. 5.6 Monthly variations in (**a**) pH, (**b**) water temperature and (**c**) salinity at the freshwater station (Diamond Harbour)

well as riverine inflow of freshwater was minimal. In monsoon, minimum values for both salinity and pH were recorded.

DIN contents of the habitat water were high along the entire study period, which ranged from 50.12 μM (February 2007) to 104.36 μM (September 2006) (Fig. 5.7a). The concentrations of nitrite (1.086–11.263 μM) and ammonia (0.294–4.41 μM) (Fig. 5.7a, b) nitrogen were low in comparison to nitrate (43.66–95.42 μM) nitrogen. Nitrate nitrogen accounted for 85–90 % of DIN (Fig. 5.7a). Dissolved inorganic phosphate (DIP) content of the habitat water varied from 2.23 to 9.44 μM, showing maximum value in September 2005 (9.44 μM) and minimum in December 2006 (2.23 μM) (Fig. 5.7c). Dissolved silicate (DSi) contents in the habitat waters were comparatively higher – ranging from 42.61 (September 2005) to 90.77 μM (November 2005) (Fig. 5.7d). Seasonally, both DIN and DIP contents of the habitat waters were maximum in monsoon (DIN, 90.22 μM; DIP, 7.45 μM in 2005), intermediate in post-monsoon (DIN, 75.31 μM; DIP, 5.17 μM in 2005) and minimum

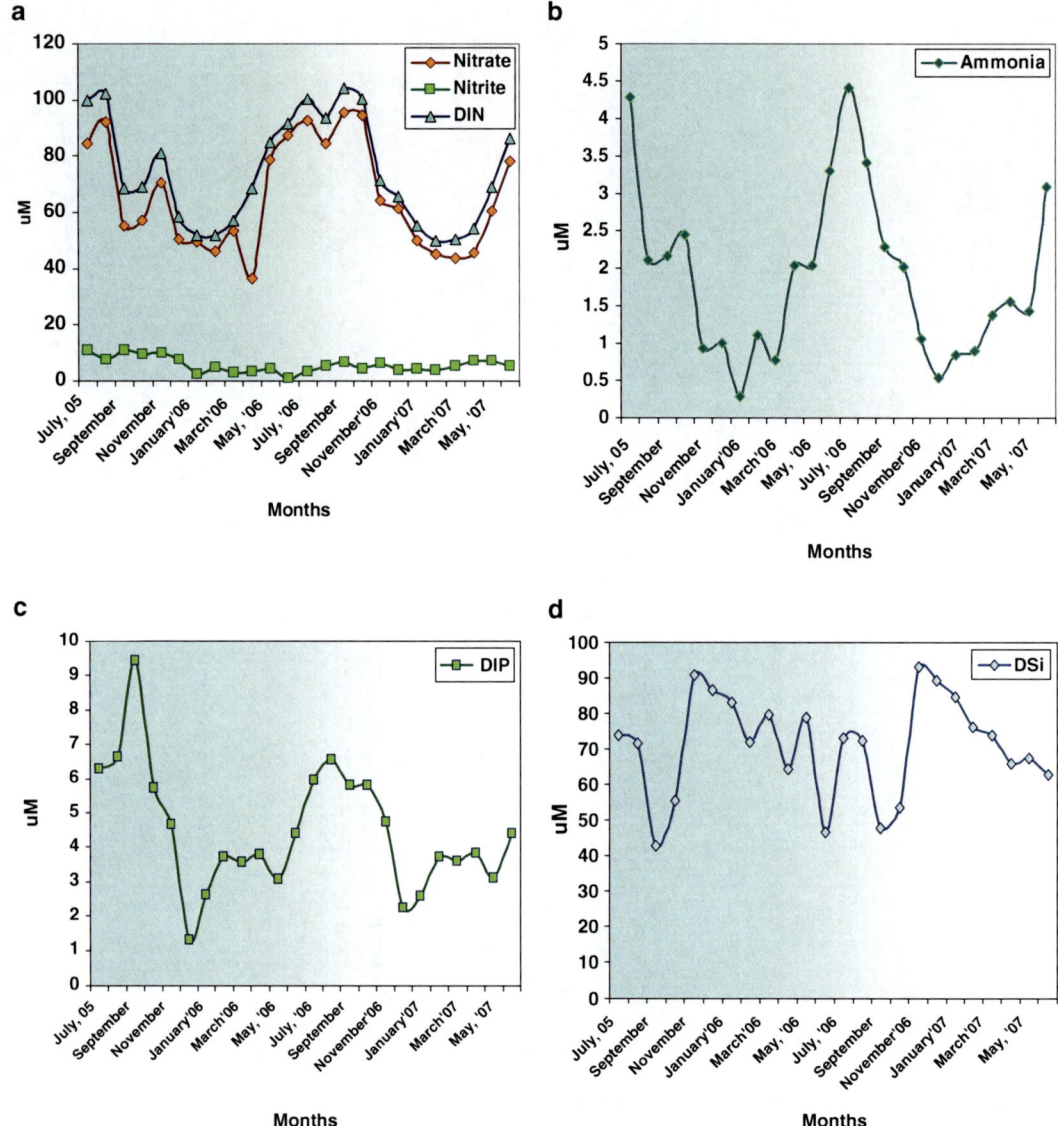

Fig. 5.7 Monthly variations in (**a**) DIN (dissolved inorganic nitrogen), (**b**) ammonia, (**c**) DIP (dissolved inorganic phosphate) and (**d**) DSi (dissolved phosphate) at the freshwater station (Diamond Harbour)

in winter (DIN, 54.24 μM; DIP, 3.21 μM in 2005–2006). This trend persisted for the entire study period with insignificant inter-annual variation. DSi represented an opposite pattern where lower values were recorded in the monsoon months and higher values in winter months.

Significant negative correlation was established between DIN and phytoplankton cell counts ($R^2=0.25$, $r=-0.49$ at $p<0.05$) (Fig. 5.8a). On the other hand, diatom population showed significant positive correlation with DSi ($R^2=0.26$, $r=0.51$ at $p<0.05$) (Fig. 5.8b) but negative correlation with seasonal temperature variations ($R^2=0.36$, $r=-0.6$ at $p<0.05$) (Fig. 5.8c).

In the habitat water of the estuarine station (Kakdweep), the pH level ranged from 7 to 7.8 (Fig. 5.9a) and the water temperature from 13.9 °C (January 2007) to 30 °C (June 2007) (Fig. 5.9b). Maximum salinity level was recorded in summer (April and May 2006, April and June

5.4 Case Study III: Phytoplankton Dynamics of Eastern Indian Coast

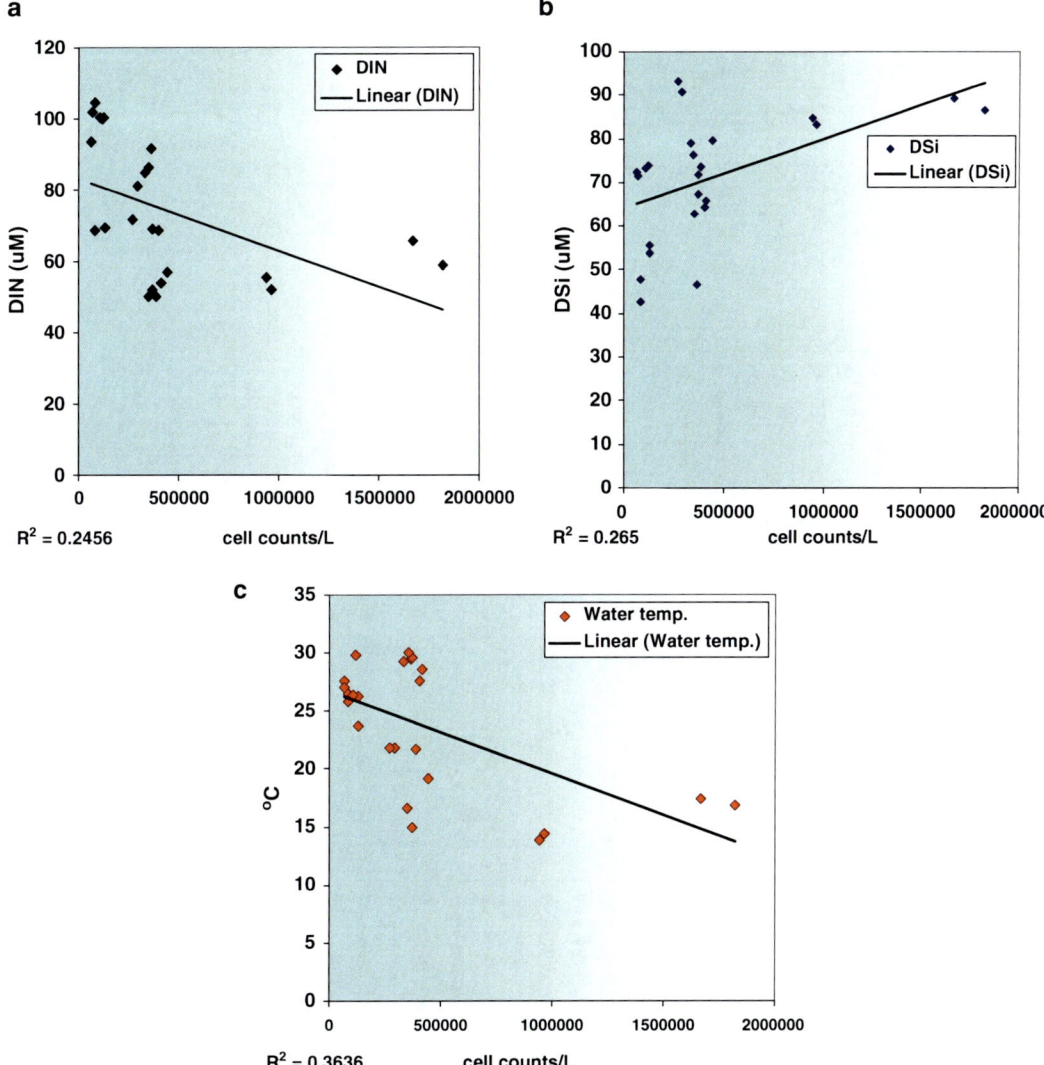

Fig. 5.8 Graphical representation of significant correlation between (**a**) cell count and DIN, (**b**) cell count and DSi and (**c**) cell count and water temperature at the freshwater station (Diamond Harbour)

2007, 17 psu), which dropped in monsoon period (August 2006, 4 psu) (Fig. 5.9c).

DIN (dissolved inorganic nitrogen) contents fluctuated seasonally as well as temporally (Fig. 5.10a). Nitrate content was maximum in August 2006 (87.25 µM) with minimum value in June 2007 (45.32 µM), which suggested that comparatively low levels of nitrate in summer and high amount during monsoon months. Unlike nitrate, ammonia content was significantly lower with the mean concentration being only 0.69 µM (Fig. 5.10b). Nitrite concentration was intermediate that ranged from 40.22 µM (August 2005) to 14.39 µM (November 2005). Phosphate concentrations did not represent a very regular and recurrent pattern of variation in the data set which ranged from 2.15 to 9.37 µM (Fig. 5.10c). Silicate concentration was quite high in comparison to phosphate levels throughout the year, ranging from 49.93 µM to 129.82 µM with highest value in August 2005 and lowest in December 2006 (Fig. 5.10d). Therefore, in the study area

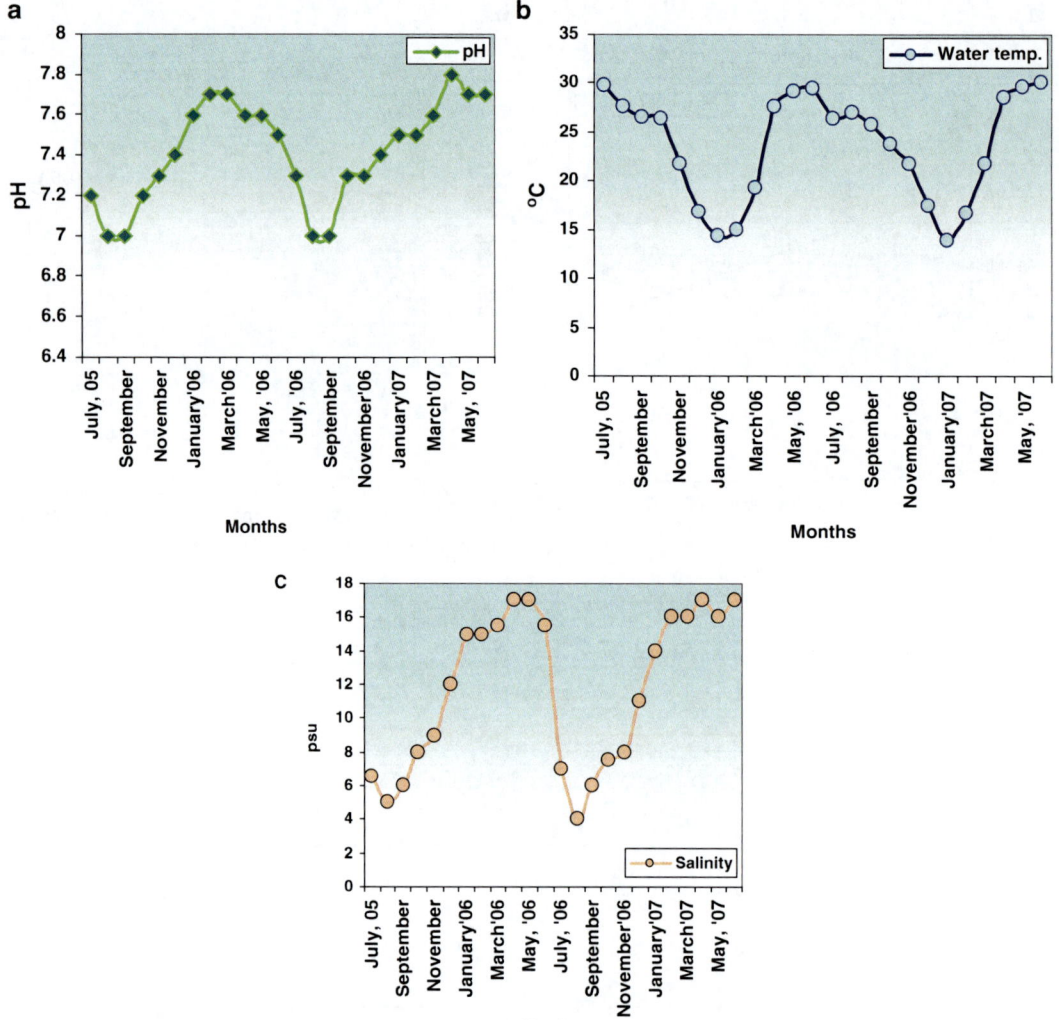

Fig. 5.9 Monthly variations in (**a**) pH, (**b**) water temperature and (**c**) salinity at the estuarine station (Kakdweep)

nutrient concentration was maximum in monsoon period and minimum in the winter and summer months.

Correlation matrix and 2-D scatter plots were performed based on Pearsonian 'r' values which showed that cell count had negative significant correlation with nitrite ($R^2 = 0.27$, $r = -0.52$, $p < 0.05$) (Fig. 5.11a), ammonia ($R^2 = 0.21$, $r = -0.45$, $p < 0.05$) (Fig. 5.11b), DIN ($R^2 = 0.2$, $r = -0.45$, $p < 0.05$) (Fig. 5.12a) and DSi ($R^2 = 0.51$, $r = -0.71$, $p < 0.05$) (Fig. 5.12b) with no significant correlation with nitrate and DIP (Table 5.7). Among the physical parameters significant negative correlation was established between cell count and temperature only ($R^2 = 0.31$, $r = -0.55$, $p < 0.05$) (Fig. 5.12c).

The marine coastal region showed different physicochemical statuses in comparison to the other two stations. The temperature and pH value of the coastal water of the study area ranged from 15–30 °C to 7–7.6, respectively (Fig. 5.13a, b). Minimum temperature was observed in winter and minimum pH value in monsoon. Maximum salinity level was recorded in winter (36 psu), which dropped in monsoon period (26 psu) (Fig. 5.13c). As evident from Fig. 5.16, DIN con-

5.4 Case Study III: Phytoplankton Dynamics of Eastern Indian Coast

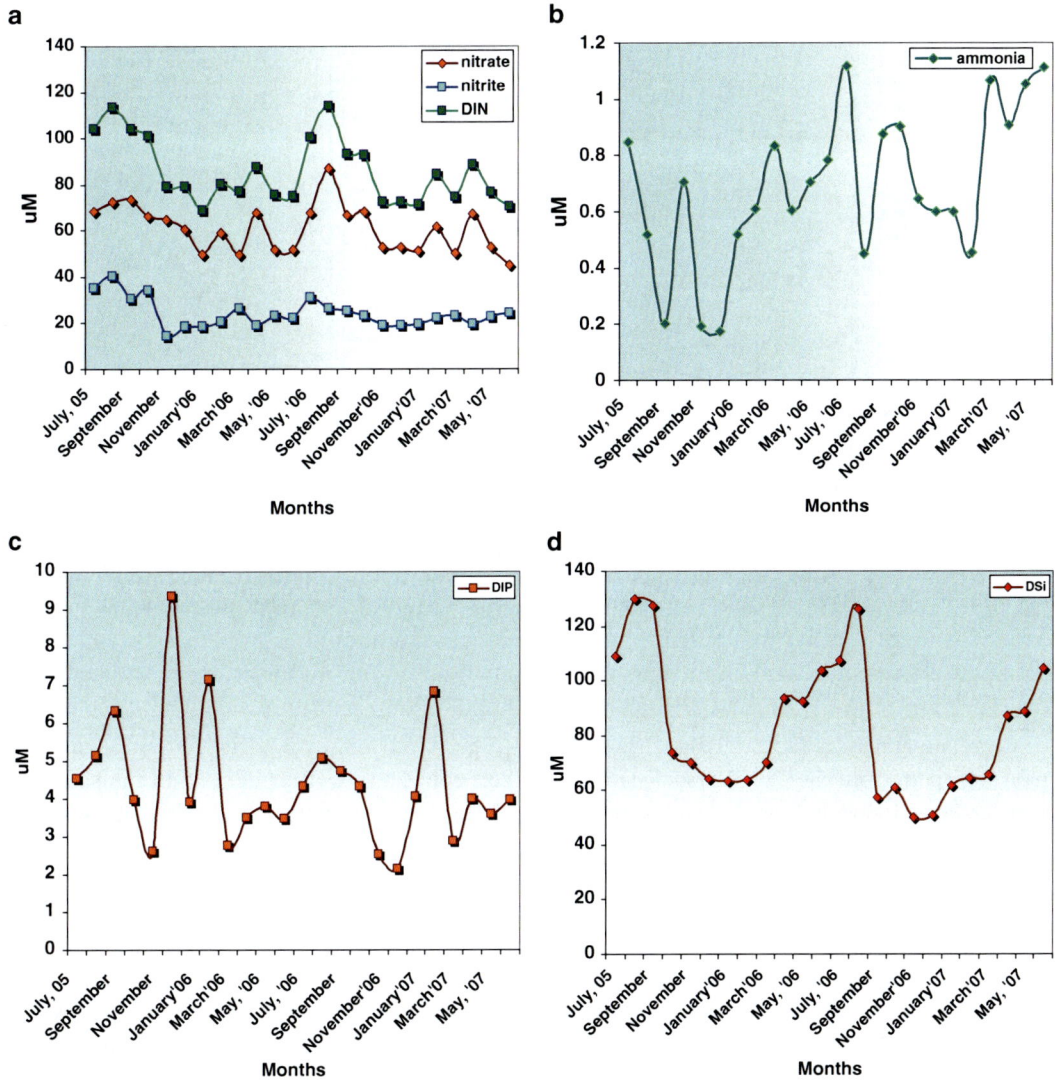

Fig. 5.10 Monthly variations in (**a**) DIN (dissolved inorganic nitrogen), (**b**) ammonia, (**c**) DIP (dissolved inorganic phosphate) and (**d**) DSi (dissolved phosphate) at the estuarine station (Kakdweep)

tent was maximum in October 2006 (28.48 μM) with minimum value in December 2005 (14.32 μM) (Fig. 5.14a). Like the freshwater and estuarine stations, nitrate concentrations were primarily responsible for the variations of DIN contents of the habitat waters. Nitrite concentrations were intermediate which was maximum in September 2005 (3.67 μM) and minimum in March 2007 (1.45 μM). Ammonia concentrations remained low and seldom reached above 1 μM levels in the habitat waters (Fig. 5.14b). Maximum phosphate concentration was estimated in August 2005 (9.41 μM) and minimum in December 2005 and January 2006 (2.23 μM; Fig. 5.14c). Therefore, in the study area nutrient concentration was maximum in monsoon period. Silicate concentration was quite high in comparison to nitrate and phosphate levels throughout the year, ranging from 19.97 to 127.32 μM with highest value in August 2005 and lowest in January 2006 (Fig. 5.14d) favouring the diatom growth.

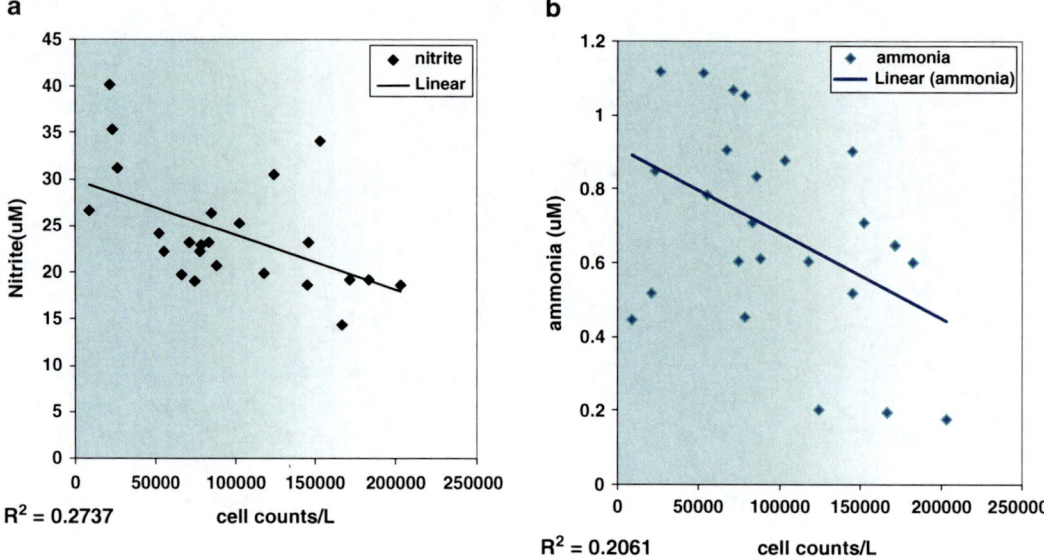

Fig. 5.11 Graphical representation of significant correlation between (**a**) cell count and nitrite and (**b**) cell count and ammonia at the estuarine station (Kakdweep)

Correlation matrix (Table 5.9) and 2-D scatter plots were performed based on Pearsonian r values, which showed that cell count had significant negative correlation with DIN ($R^2=0.28$, $r=-0.52$, $p<0.05$) (Fig. 5.15a) and dissolved silicate ($R^2=0.23$, $r=-0.48$, $p<0.05$) (Fig. 5.15b) and nonsignificant with phosphate and temperature but positive significant correlation with salinity ($R^2=0.3$, $r=0.54$, $p<0.05$) (Fig. 5.15c) and pH ($R^2=0.22$, $r=0.57$, $p<0.05$) (Fig. 5.15d) at $n=24$ (Table 5.8).

5.4.2.3 Phytoplankton Population in Terms of Cell Count and Their Biotic Indices

The findings of our study showed that along coastal West Bengal, a significant variation of phytoplankton population was related to seasonal variations and physicochemical parameters of habitat waters. On an average, among all the sampling stations, the total cell count was maximum at marine coastal zone (mean=986 cells/mL) and minimum at the estuarine location (mean=97 cells/mL) while the freshwater zone occupied an intermediate position (mean=439 cells/mL) (Fig. 5.16). Seasonal variation with respect to cell count was well evident at all the sampling stations, where winter season was the most productive and monsoon was the least. Gradual decreases in cell count were recorded from the summer to monsoon period whereas gradual increases in cell count from the post-monsoon to the winter period were observed. Cell count of individual species may have played a significant role in determining the diversity of this coastal region. At the freshwater region, total cell counts ranged from 65 cells/mL (August 2006) to 1,820 cells/mL (December 2005). At the estuarine location, it ranged from 9 cells/mL (August 2006) to 203 cells/mL (December 2005). Likewise, cell count ranged from 345 cells/mL (April 2006) to 2,020 cells/mL (January 2006) at the marine coastal region.

Diversity of phytoplankton population at coastal West Bengal showed distinct variations on a monthly as well as on a seasonal basis in terms of different biotic indices. Diversity, measured on the basis of Shannon–Wiener Index (SWI), was maximum at the freshwater station (Mean $H'=2.67$), intermediate at the estuarine station

5.4 Case Study III: Phytoplankton Dynamics of Eastern Indian Coast

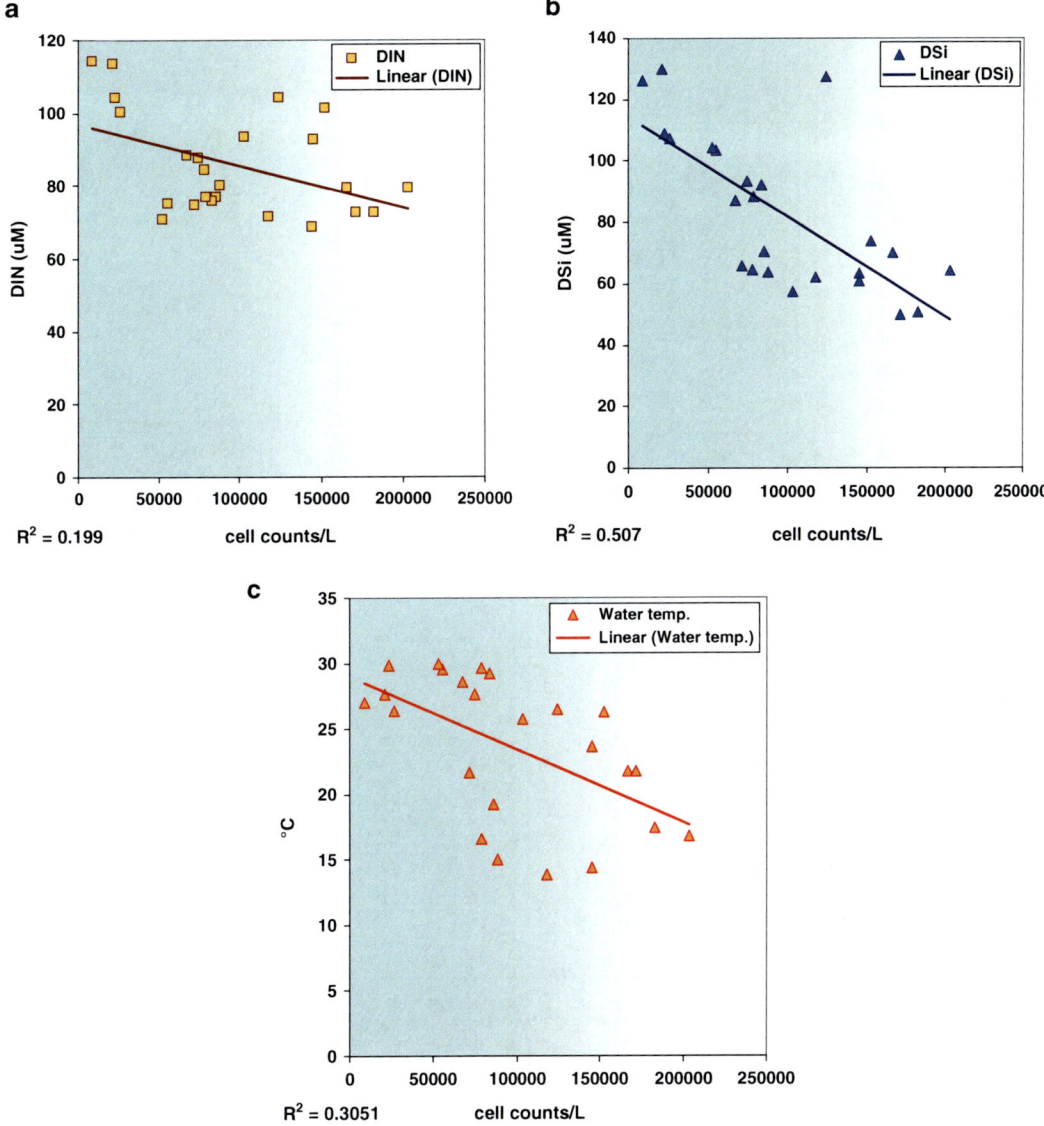

Fig. 5.12 Graphical representation of significant correlation between (**a**) cell count and DIN, (**b**) cell count and DSi and (**c**) cell count and water temperature at the estuarine station (Kakdweep)

(Mean $H' = 2.278$) and minimum at the marine region (Mean $H' = 1.90$) (Fig. 5.17). Higher values for SWI were recorded in the transition period (March) between winter and summer months at the freshwater station, whereas it was maximum during post-monsoon period at the estuarine and the coastal stations. On the other hand, minimum values for SWI were recorded in winter at marine and estuarine region, whereas it was minimum in monsoon at the freshwater station.

Species evenness is a measure of the contributions of individual taxa to the phytoplankton population. It was found that seasonal variation in species evenness was more pronounced at the marine region as compared to the other stations.

At the freshwater station (Diamond Harbour), SWI varied from 2.13 to 3.23. On a monthly basis, maximum value was recorded in November 2006 ($H' = 3.23$) whereas minimum value was in October 2005 ($H' = 2.13$) (Fig. 5.18). Maximum

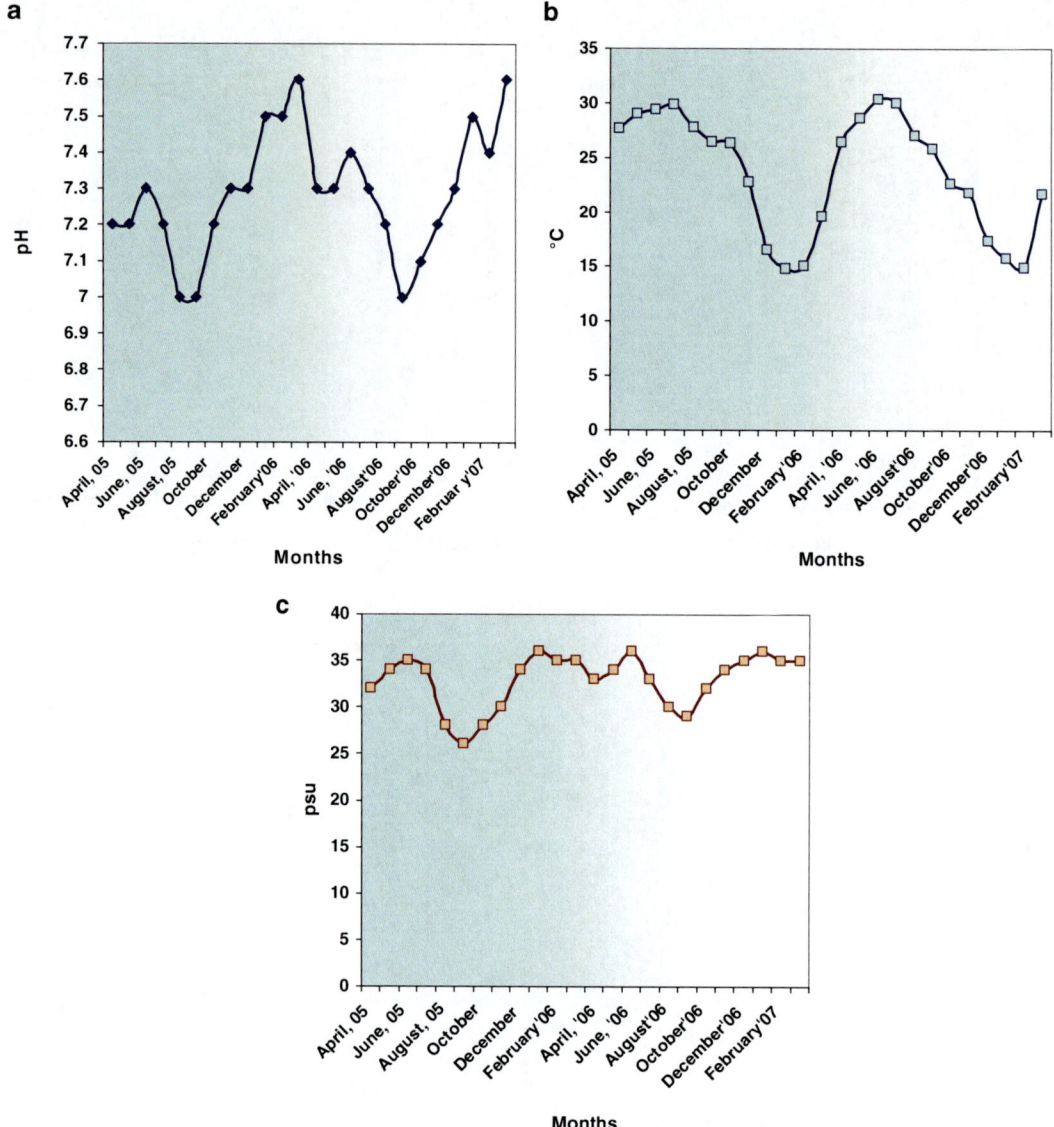

Fig. 5.13 Monthly variations in (**a**) pH, (**b**) water temperature and (**c**) salinity at the coastal marine region (Digha and Junput)

SWI was recorded in the transition period between the summer and winter months (March). The subsequent summer months (April–May–June) also showed a comparatively high SWI values. SWI values were intermediate in the post-monsoon period although there was significant interannual variations. On the contrary, minimum SWI values were obtained in the monsoon months (Fig. 5.19). The mean species evenness (e) was relatively high as compared to the other coastal stations ($e = 2.13$). Variation in species evenness was not very pronounced in this station although seasonal fluctuations were observed. Seasonally, it was maximum during summer ($e = 2.17$) and minimum in winter ($e = 2.09$) closely followed by the post-monsoon period ($e = 2.1$) (Fig. 5.20).

At the estuarine coastal station (Kakdweep), SWI varied from 1.88 to 2.53. Highest value of SWI was recorded in the post-monsoon period (October–November) whereas lowest value of the same was recorded in winter (December–

5.4 Case Study III: Phytoplankton Dynamics of Eastern Indian Coast

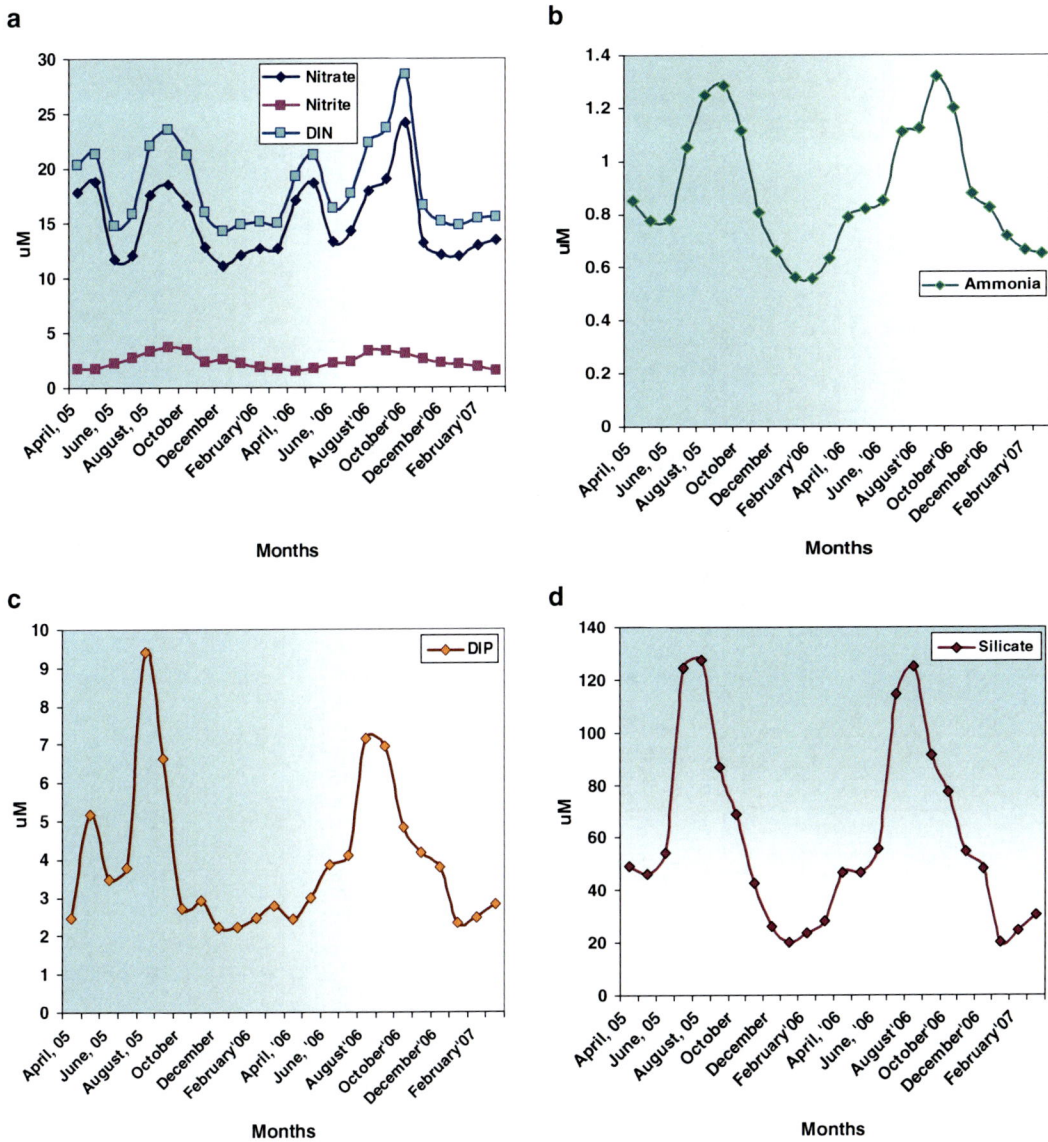

Fig. 5.14 Monthly variations in (**a**) DIN (dissolved inorganic nitrogen), (**b**) ammonia, (**c**) DIP (dissolved inorganic phosphate) and (**d**) DSi (dissolved phosphate) at the coastal marine region (Digha and Junput)

January–February) (Figs. 5.20 and 5.21). The subsequent summer months also showed a comparatively high SWI values which dropped slightly in the monsoon periods. The mean species evenness (e) was higher as compared to the freshwater station ($e = 2.17$). Variation in species evenness was not very pronounced in this station although seasonal fluctuations were observed. Seasonally, it was maximum during monsoon ($e = 2.21$) followed by the transition period (March) and minimum in post-monsoon ($e = 2.14$) closely followed by summer ($e = 2.15$) (Fig. 5.21).

At the marine coastal region, SWI varied from 0.33 to 2.76. Highest value of SWI was recorded in the post-monsoon period (October–November), whereas lowest value of the same was recorded in winter (Fig. 5.22). A gradual increase in SWI was

Table 5.9 Correlation matrix plot between cell count, productivity and environmental variables at the freshwater station

	Cell count	DIN	DIP	DSi	Saln	pH	Water temp.	BOD	DO	GPP	NPP	CRR
Cell count		−0.50	−0.75	0.51	0.61	0.21	−0.60	−0.64	0.72	0.89	0.91	0.05
DIN	−0.50		0.58	−0.36	−0.80	−0.42	0.65	0.64	−0.64	−0.60	−0.49	−0.40
DIP	−0.75	0.58		−0.59	−0.78	−0.63	0.52	0.71	−0.73	−0.74	−0.64	−0.35
DSi	0.51	−0.36	−0.59		0.59	0.22	−0.56	−0.21	0.44	0.64	0.55	0.30
Saln	0.61	−0.80	−0.78	0.59		0.46	−0.80	−0.71	0.79	0.76	0.63	0.49
pH	0.21	−0.42	−0.63	0.22	0.46		0.06	−0.49	0.24	0.23	0.06	0.47
Water temp.	−0.60	0.65	0.52	−0.56	−0.80	0.06		0.45	−0.70	−0.71	−0.70	−0.20
BOD	−0.64	0.64	0.71	−0.21	−0.71	−0.49	0.45		−0.80	−0.68	−0.59	−0.33
DO	0.72	−0.64	−0.73	0.44	0.79	0.24	−0.70	−0.80		0.76	0.72	0.19
GPP	0.89	−0.60	−0.74	0.64	0.76	0.23	−0.71	−0.68	0.76		0.96	0.17
NPP	0.91	−0.49	−0.64	0.55	0.63	0.06	−0.70	−0.59	0.72	0.96		−0.09
CRR	0.05	−0.40	−0.35	0.30	0.49	0.47	−0.20	−0.33	0.19	0.17	−0.09	

Correlations (marked correlations are significant at $p < .05000$ $N = 24$)

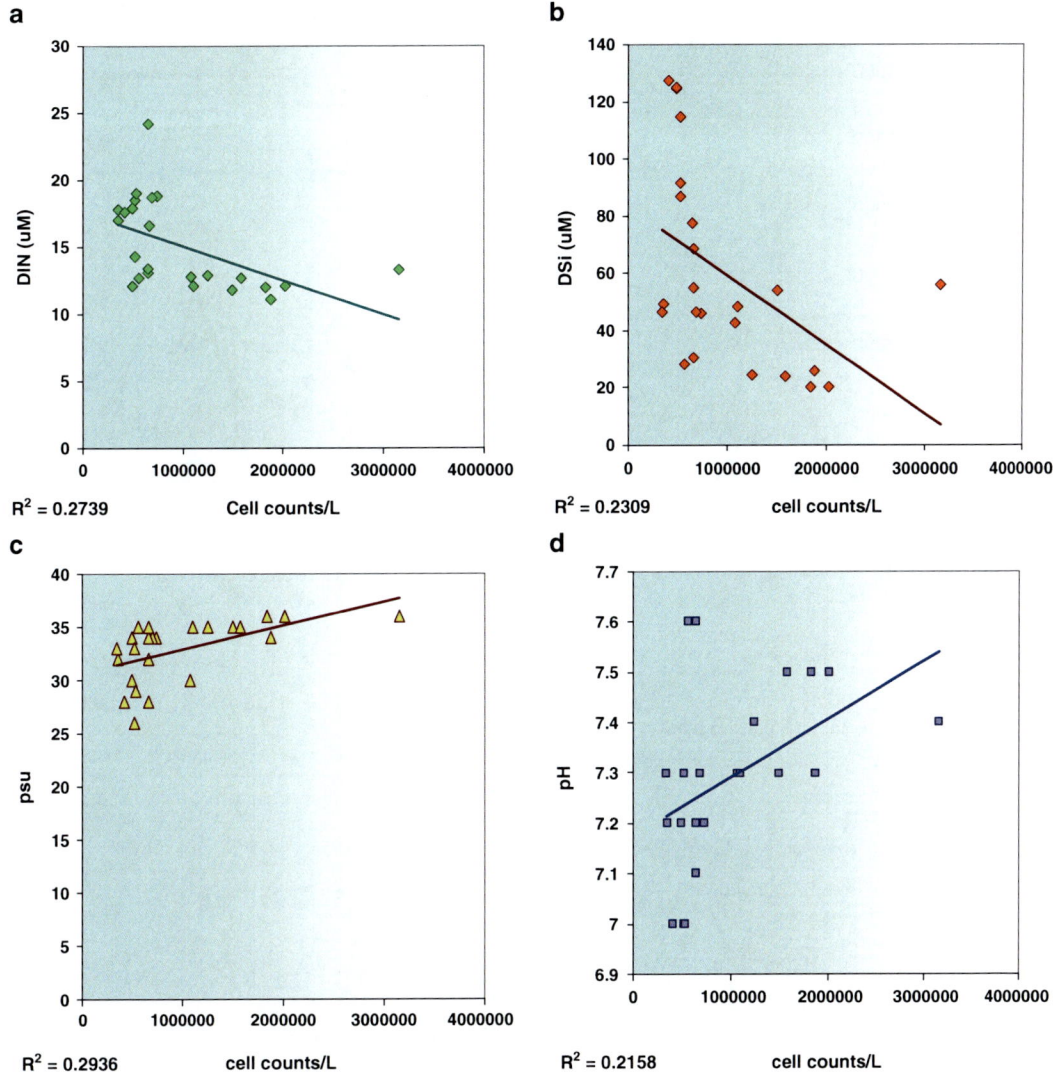

Fig. 5.15 Graphical representation of significant correlation between (**a**) cell count and DIN, (**b**) cell count and DSi and (**c**) cell count and salinity and (**d**) cell count and pH at the coastal marine region (Digha and Junput)

recorded from July onwards reaching maximum in October (2.76) (Fig. 5.22). In winter months, SW Index dropped drastically when single species abundance of *Asterionella japonica* was very high and contributed to more than 90 % of the total phytoplankton population (Fig. 5.23). Similarly, SWI values were relatively low in summer months as well, when *Odontella rhombus* population contributed to >47 % of the total population.

Seasonally, the coastal marine region showed a similar pattern in diversity as was observed in the freshwater and estuarine stations. Highest value of SWI was recorded in the post-monsoon period (October–November) whereas lowest value of the same was recorded in winter (December–January–February) (Fig. 5.24). The subsequent summer months also showed a comparatively low SWI values which gradually increased in the monsoon and post-monsoon

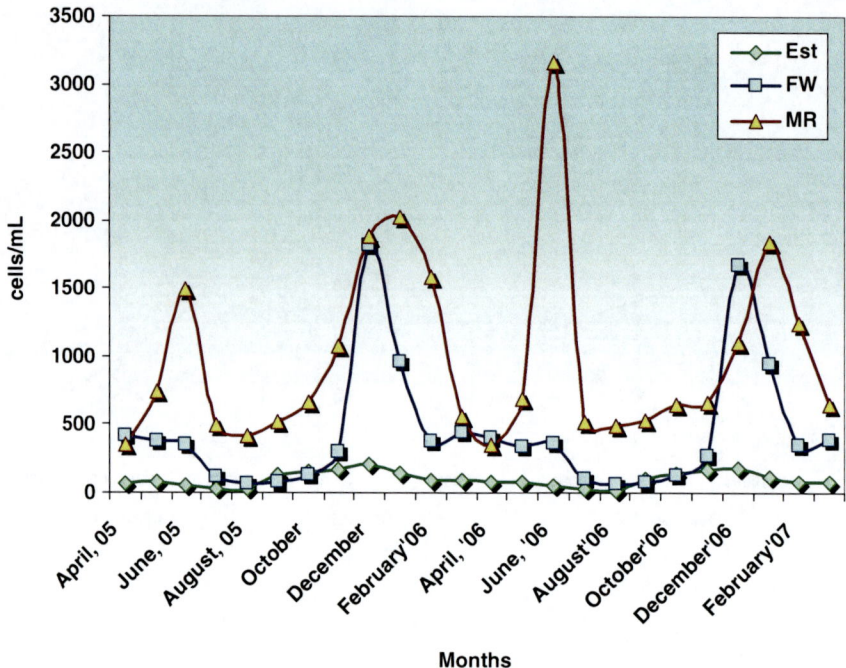

Fig. 5.16 Monthly variations in total cell count (cells/mL) at the three sampling stations (*FW* freshwater station, *Est* estuarine station, and *MR* coastal marine region)

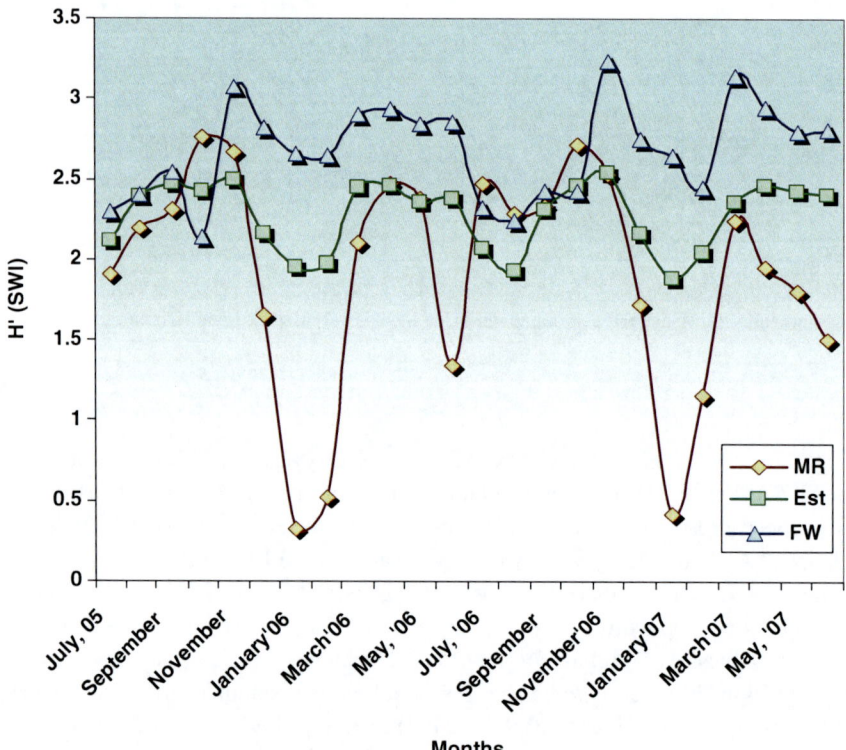

Fig. 5.17 Monthly variations in Shannon–Wiener Index at the three sampling regions (*FW* freshwater station, *Est* estuarine station, and *MR* coastal marine region)

5.4 Case Study III: Phytoplankton Dynamics of Eastern Indian Coast

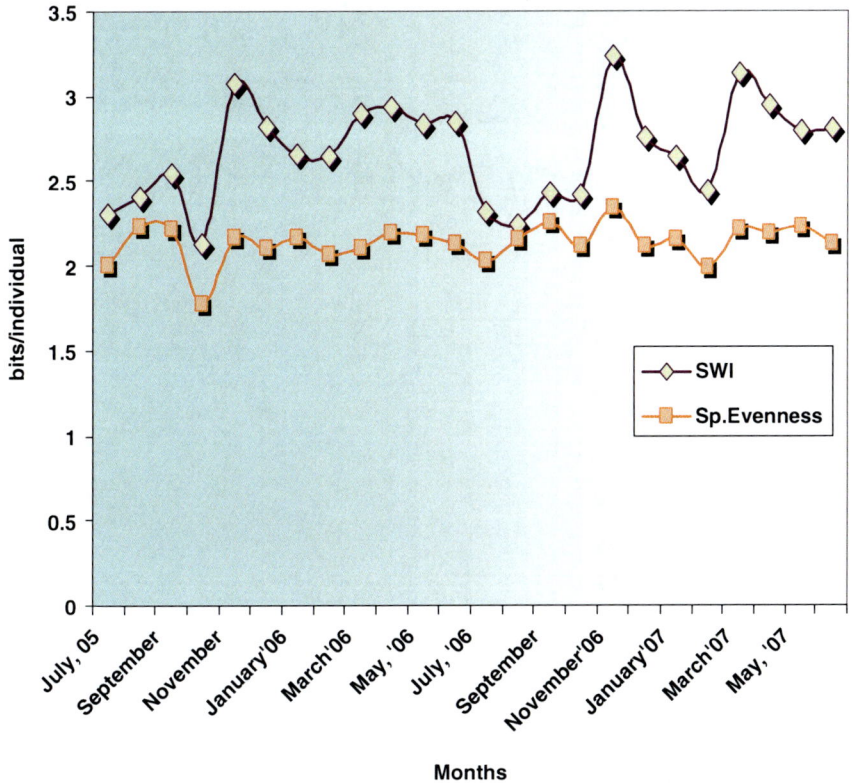

Fig. 5.18 Monthly variations in Shannon–Wiener Index (*SWI*) and species evenness at the freshwater station (Diamond Harbour)

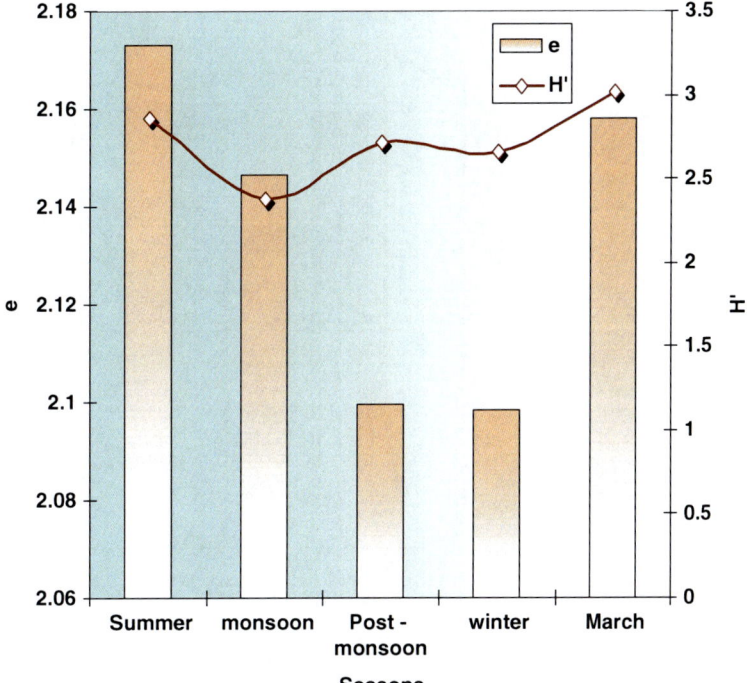

Fig. 5.19 Seasonal variations in Shannon–Wiener Index (*SWI*) and species evenness at the freshwater station (Diamond Harbour)

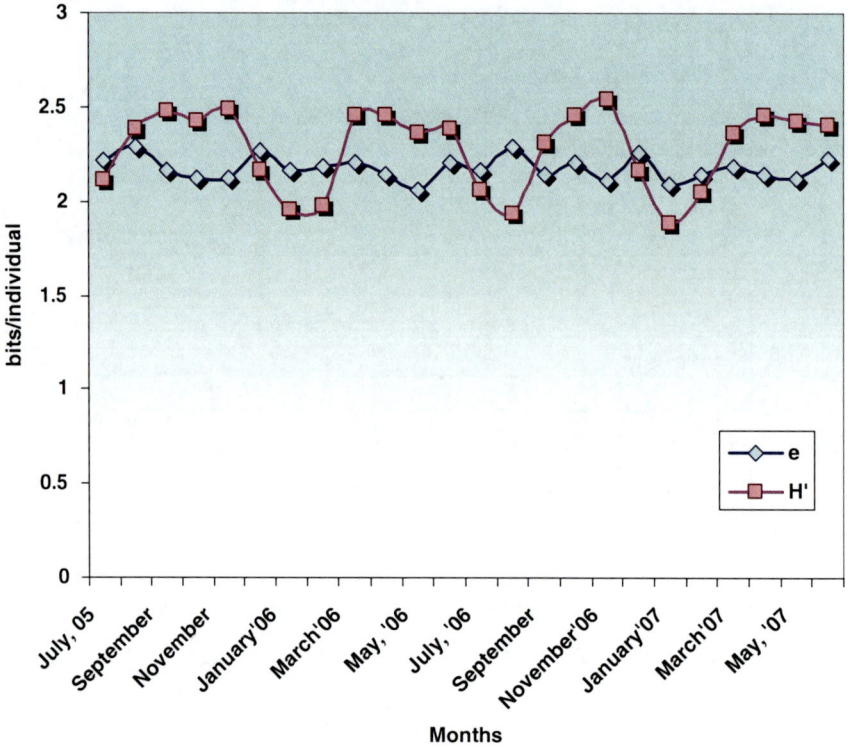

Fig. 5.20 Monthly variations in Shannon–Wiener Index (*SWI*) and species evenness at the estuarine station (Kakdweep)

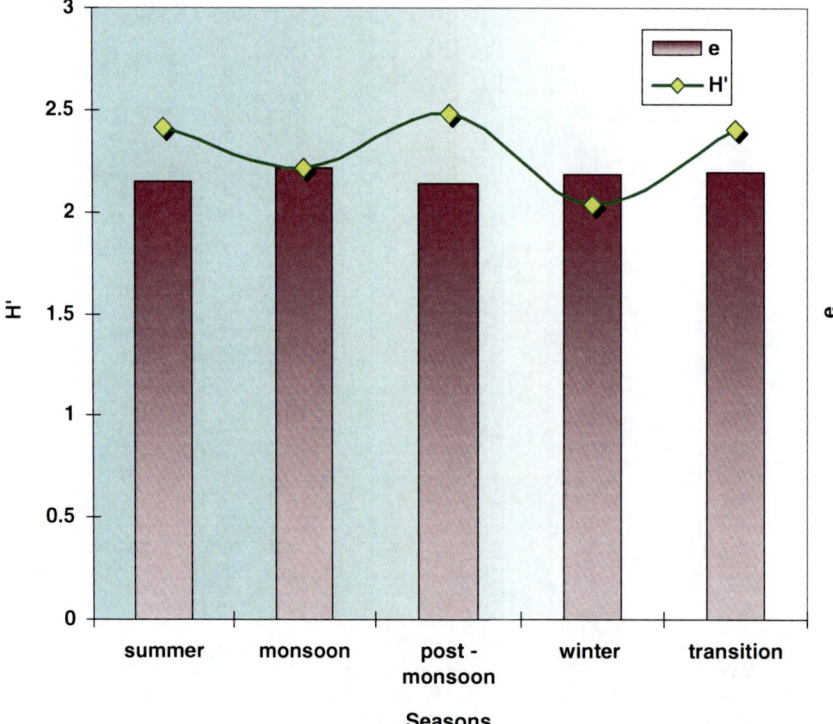

Fig. 5.21 Seasonal variations in Shannon–Wiener Index (*SWI*) and species evenness at the estuarine station (Kakdweep)

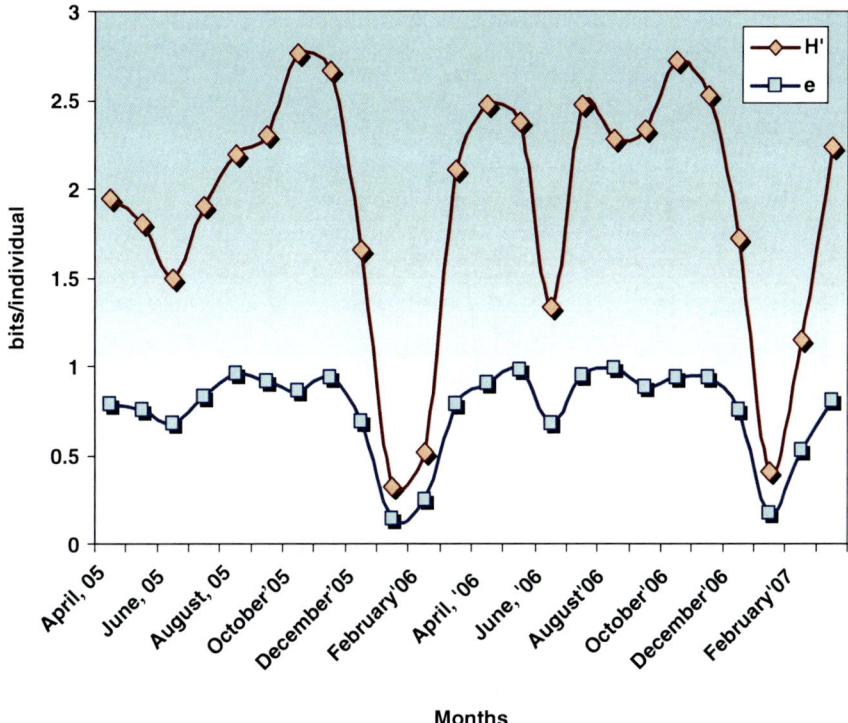

Fig. 5.22 Monthly variations in Shannon–Wiener Index (*SWI*) and species evenness at the coastal marine region (Junput and Digha)

periods. The mean species evenness (*e*) was much lower as compared to the other stations (*e* = 0.75). Variation in species evenness was pronounced in this station both on a monthly as well as on a seasonal basis. Seasonally, it was maximum during monsoon (*e* = 0.92) followed closely by the post-monsoon period (*e* = 0.92) and minimum in winter (*e* = 0.42) (Fig. 5.24). The pattern of variations for both SWI and species evenness was almost similar with significant decrease in winter and summer.

5.4.2.4 Primary Productivity (GPP and NPP) and Oxygen Concentrations

Results show that coastal West Bengal promoted phytoplankton production although there were significant seasonal variations. The study area was highly suitable for the development and diversification of phytoplankton populations especially for diatoms. On an individual site basis, phytoplankton productivity was maximum at the coastal marine region and minimum at the estuarine region with the freshwater station occupying an intermediate position. CRR values also represented a similar pattern as was observed for primary productivity. Seasonally, winter months were most productive and monsoon periods were least productive. DO values being a reflection of the photosynthetic efficiency of phytoplankton population showed a similar seasonal pattern as was observed for primary productivity. In contrast, BOD values were minimum in winter and maximum in monsoon at all the sampling stations.

At the freshwater station, winter months were most productive with respect to carbon equivalents as was evident from GPP values. Typically, monsoon periods were least productive both in respect to phytoplankton productivity (GPP) as

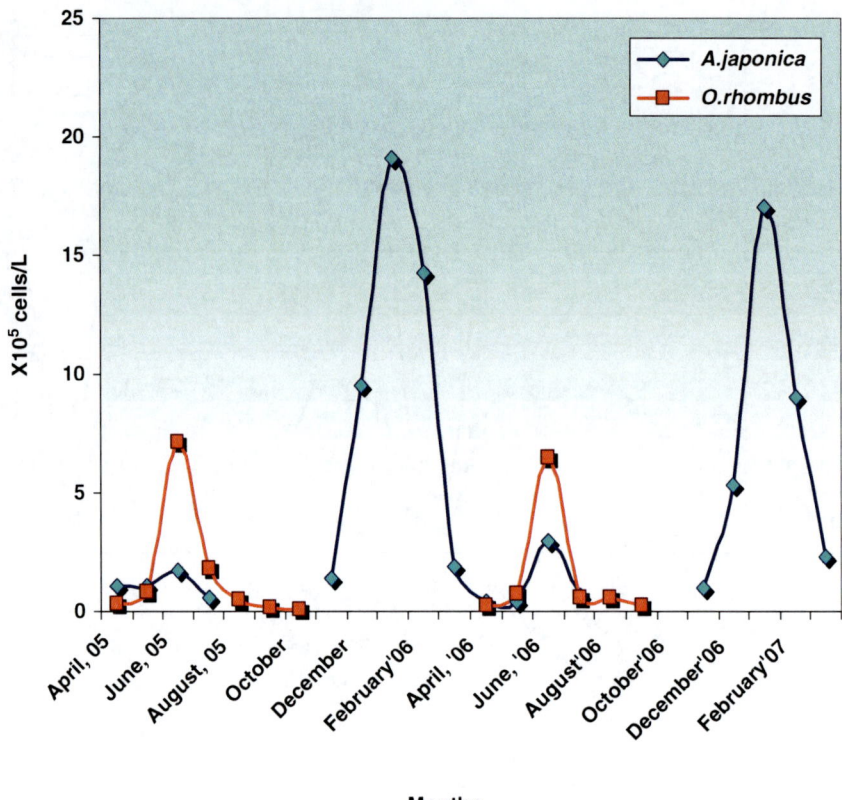

Fig. 5.23 Monthly variations of total cell counts of the two dominant species (*Asterionella japonica* and *Odontella rhombus*) at the coastal marine region

Fig. 5.24 Seasonal variations in SW Index and species evenness at coastal marine region

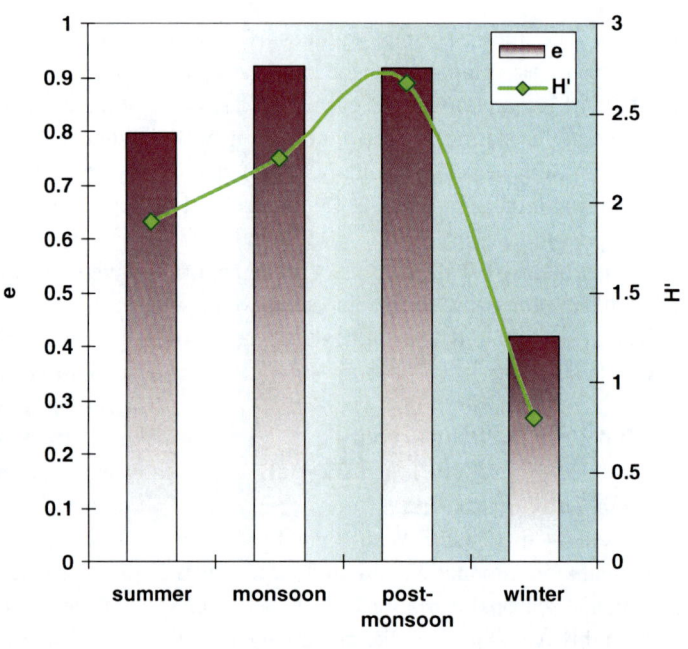

5.4 Case Study III: Phytoplankton Dynamics of Eastern Indian Coast

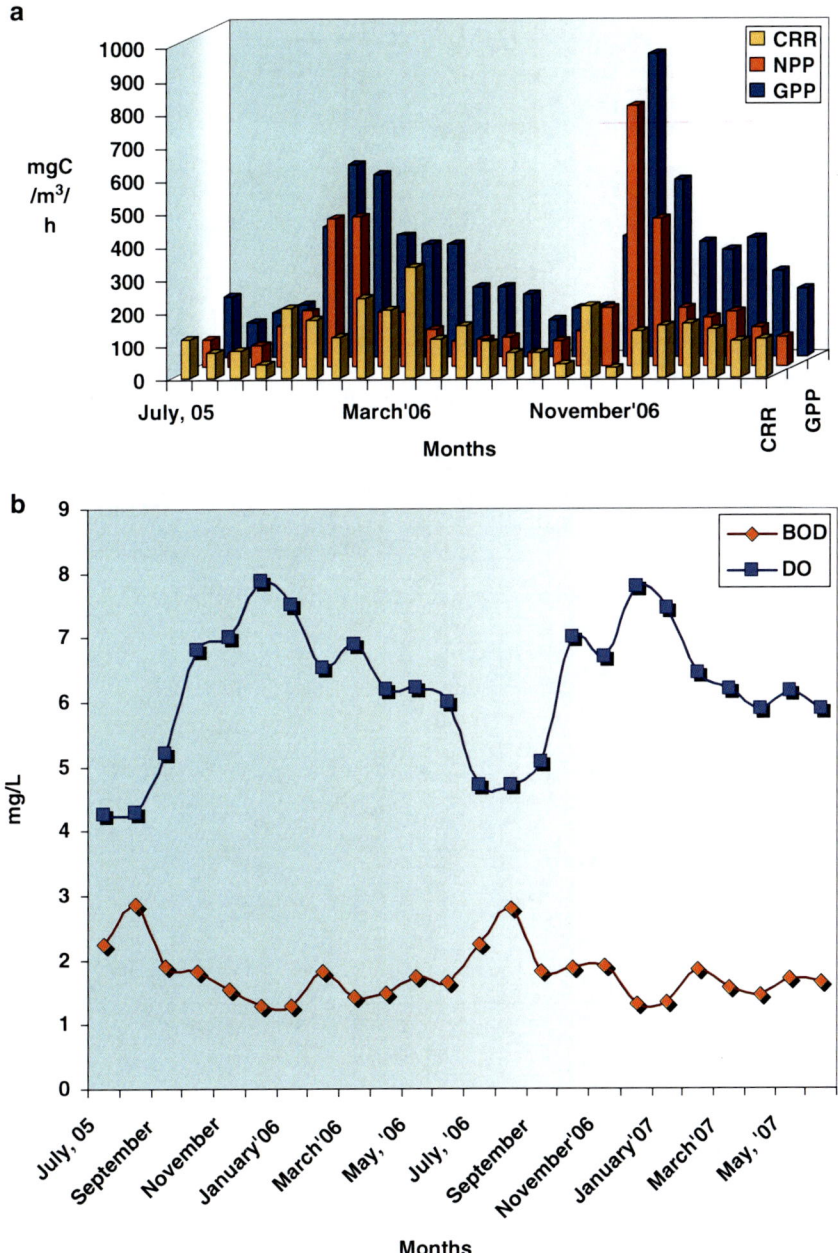

Fig. 5.25 Monthly variation in (**a**) productivity (NPP, GPP and CRR) and (**b**) oxygen concentrations (DO and BOD) at the freshwater station (Diamond Harbour)

well as community productivity (NPP) with insignificant interannual variations. As can be expected, CRR varied significantly as well, where maximum CRR was in winter (178.67 mgC/m^3/h) and minimum in monsoon (86.37 mgC/m^3/h) (Fig. 5.25a). A highly positive significant correlation between GPP and cell count ($R^2=0.8$, $r=0.9$, $p \leq 0.05$) was established (Fig. 5.26a).

Dissolved oxygen (DO) was present in optimal quantity in the habitat waters that ranged from 4.25 mg/L (July 2005) to 7.882 mg/L (December 2005) (Fig. 5.25b). Seasonally, DO

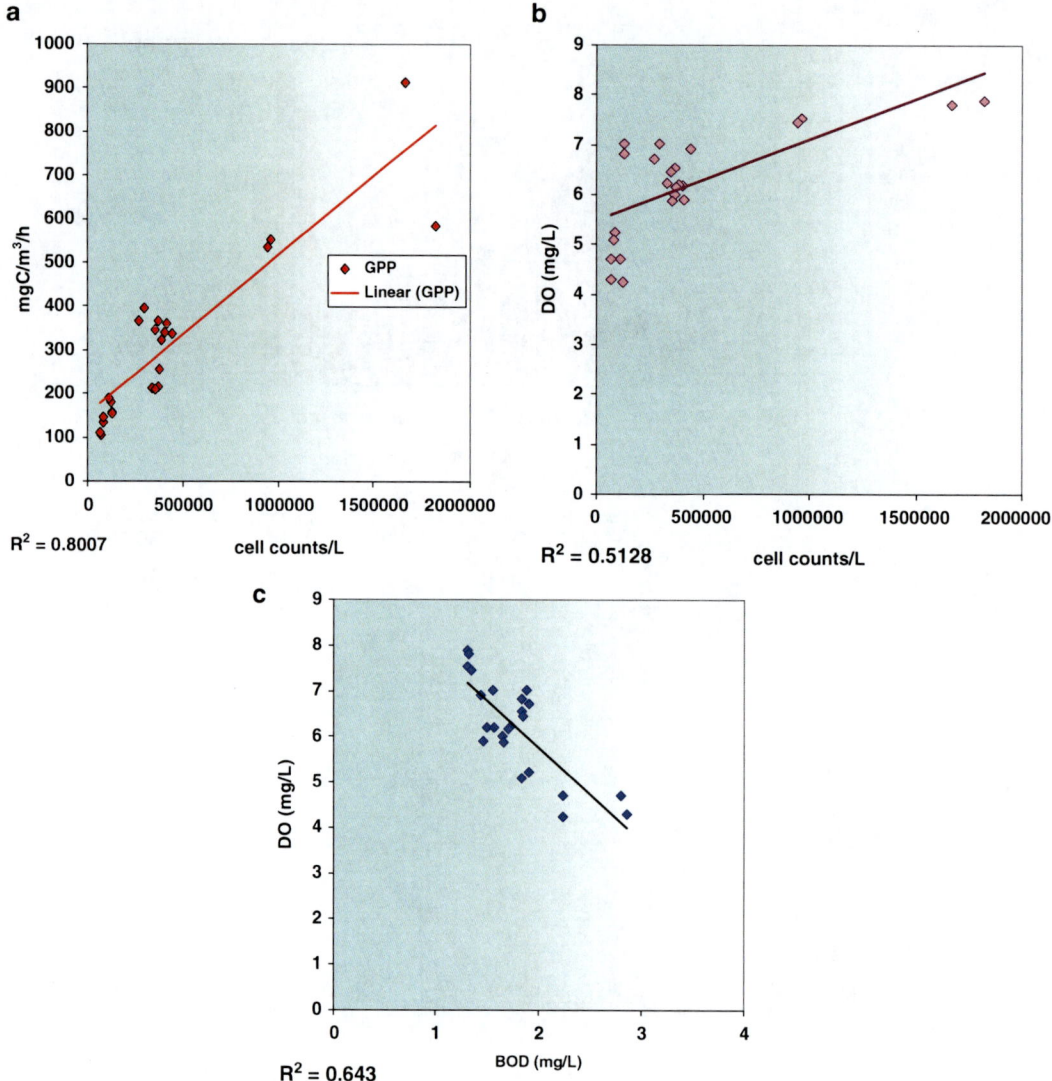

Fig. 5.26 Graphical representation of significant correlation between (**a**) cell count and GPP, (**b**) cell count and DO and (**c**) BOD and DO at the freshwater station

showed a similar trend with that of productivity and cell count that was maximum in winter (7.32 mg/L) and minimum in monsoon (4.59 mg/L). DO contents in the habitat waters had positive significant correlation with cell count ($R^2=0.51$, $r=0.68$, $p \leq 0.05$, $n=24$) (Fig. 5.26b) whereas BOD, which represented the heterotrophic oxygen requirements, showed an opposite pattern with maximum values in monsoon and minimum values in winter. Thus, a negative correlation was observed between DO and BOD ($R^2=0.64$, $r=0.8$, $p \leq 0.05$, $n=24$) (Fig. 5.26c).

Phytoplankton productivity [Gross Primary productivity (GPP)] at the estuarine station was maximum in December 2006 (227.77 mgC/m^3/h) when total phytoplankton cell count was maximum as well (203 cells/mL) and minimum in August 2006 (58.95 mgC/m^3/h) (Fig. 5.27a). Seasonally, maximum productivity with respect to carbon equivalents was recorded in winter when cell counts showed highest values as well

5.4 Case Study III: Phytoplankton Dynamics of Eastern Indian Coast

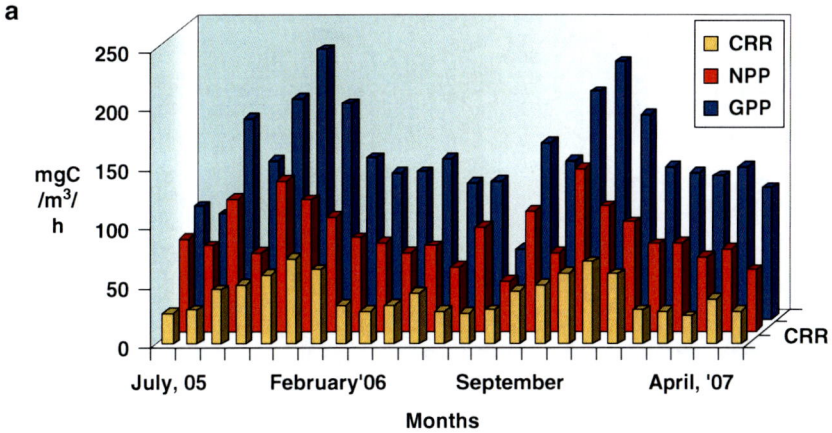

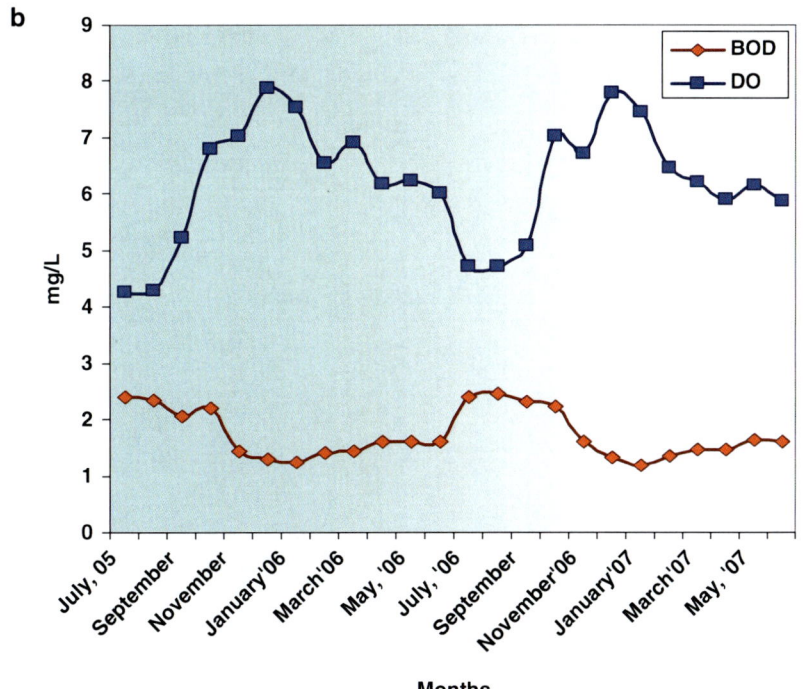

Fig. 5.27 Monthly variation in (**a**) productivity (NPP, GPP and CRR) and (**b**) oxygen concentrations (DO and BOD) at the estuarine station (Kakdweep)

(136 cells/mL). On the contrary, minimum productivity with respect to both carbon equivalents as well as total phytoplankton cell counts was recorded in the monsoon months (cells/mL) (Fig. 5.27a). Results of 2-D scatter plot shows highly significant positive correlation between total phytoplankton cell count and GPP as well ($R^2 = 0.84$, $r = 0.92$, $p < 0.05$, $n = 24$) (Fig. 5.28a). Almost similar patterns were observed for net primary productivity (NPP), which was maximum in November 2006 (137.78 mgC/m^3/h) and minimum in August 2006 (42.72 mgC/m^3/h). An intermediate significant positive correlation was established between phytoplankton cell count and NPP ($R^2 = 0.48$, $r = 0.7$, $p < 0.05$, $n = 24$) (Fig. 5.28b). Community respiration rate (CRR) being a measure of catabolic loss of carbon equivalents due to respiration was minimum in April 2007

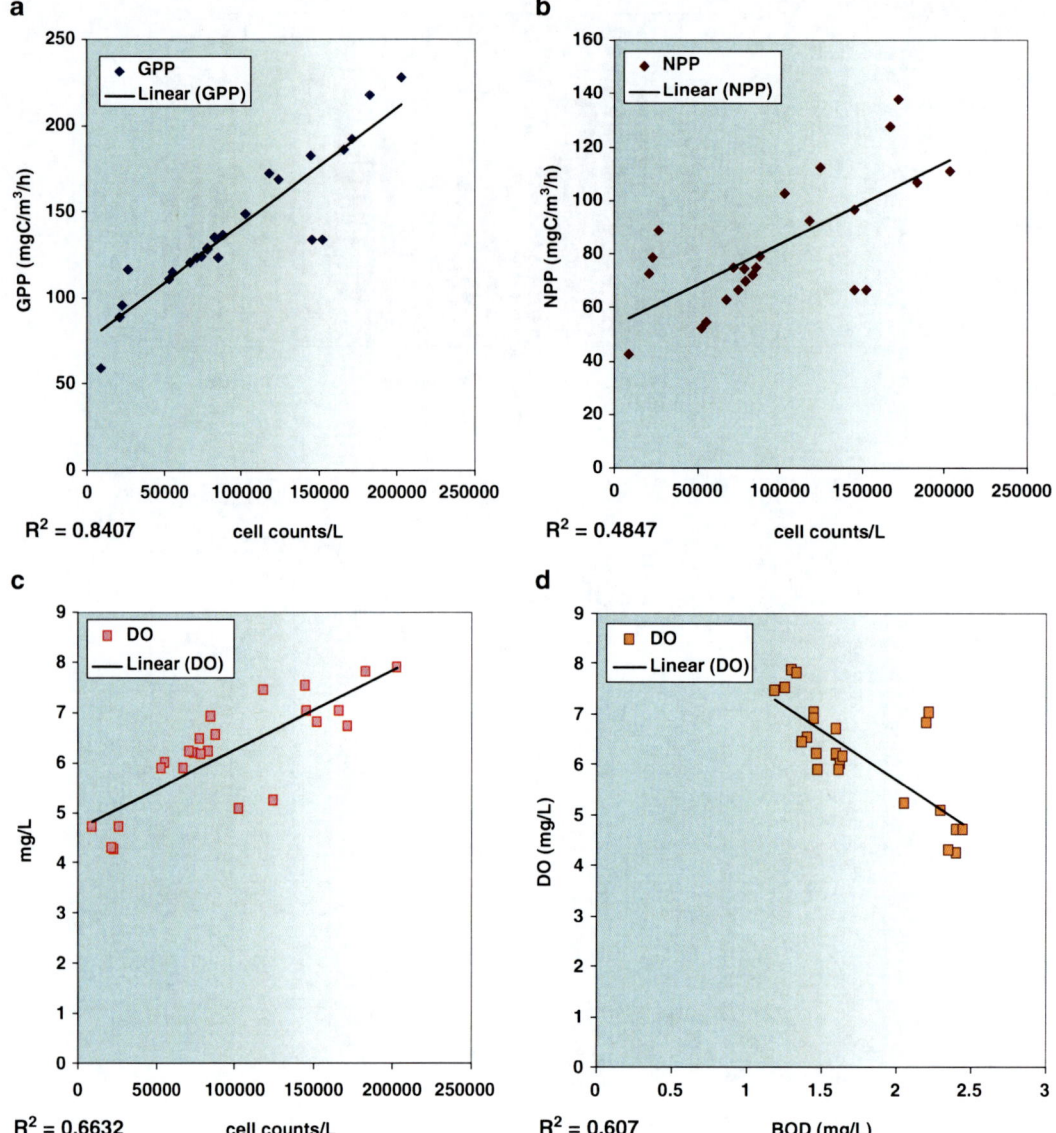

Fig. 5.28 Graphical representation of significant correlation between (**a**) cell count and GPP, (**b**) cell count and NPP, (**c**) cell count and DO and (**d**) BOD and DO at the estuarine station (Kakdweep)

(24. 158 mgC/m^3/h) whereas it was maximum in December 2006 (72.2 mgC/m^3/h) (Fig. 5.27a).

Dissolved oxygen content is a reflection of the photosynthetic activity of the phytoplankton biomass. Accordingly DO values were higher in those months where plankton count was high, with a maximum of 7.88 mg/L in December 2005 and minimum of 4.25 mg/L in July 2006 (Fig. 5.27b). BOD value ranged from 1.18 to 2.45 mg/L, with highest value in monsoon (August 2006) and lowest in winter (January 2007) when DO content was high (Fig. 5.27b). From the correlation matrix plot, it was observed that there was a positive correlation between cell count and dissolved oxygen content ($R^2=0.66$, $r=0.81$, $p<0.05$, $n=24$) (Fig. 5.28c) and negative correlation with BOD ($R^2=0.61$, $r=-0.46$, $p<0.05$, $n=24$) (Fig. 5.28d).

5.4 Case Study III: Phytoplankton Dynamics of Eastern Indian Coast

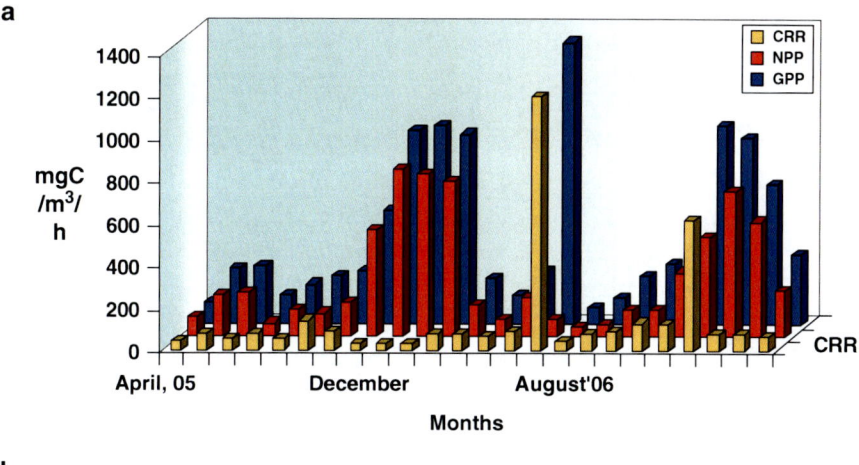

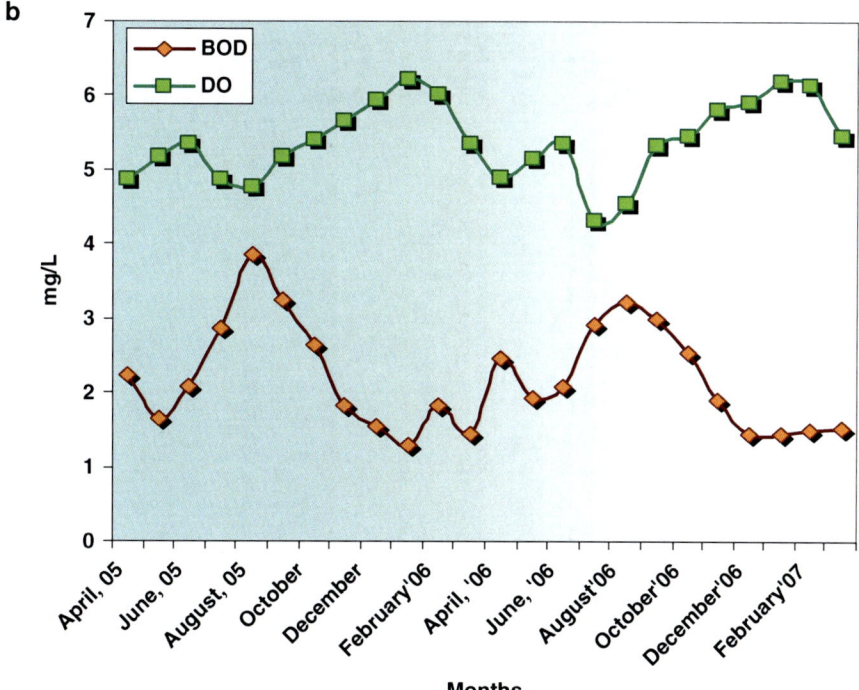

Fig. 5.29 Monthly variation in (**a**) productivity (NPP, GPP and CRR) and (**b**) oxygen concentrations (DO and BOD) at the coastal marine region (Digha and Junput)

At the coastal marine region, maximum productivity [GPP] was recorded in June 2006 (1,330 mgC/m^3/h) and minimum productivity was recorded in July 2006 (77.78 mgC/m^3/h) (Fig. 5.29a). Highest NPP value was recorded in December 2005 (788.88 mgC/m^3/h) when GPP (913.34 mgC/m^3/h) also was relatively high whereas it was minimum in July 2006 (44.44 mgC/m3/h). An abruptly maximum CRR value was recorded in June 2006 (1,191.67 mgC/m^3/h) when GPP was high as well although NPP was much low (83.33 mgC/m^3/h) (Fig. 5.29a).

On a seasonal basis, highest productivity was recorded in winter, followed by summer with the lowest productivity in monsoon. Regarding productivity, there was interannual variation where productivity in post-monsoon period was higher than summer in 2005, unlike in 2006.

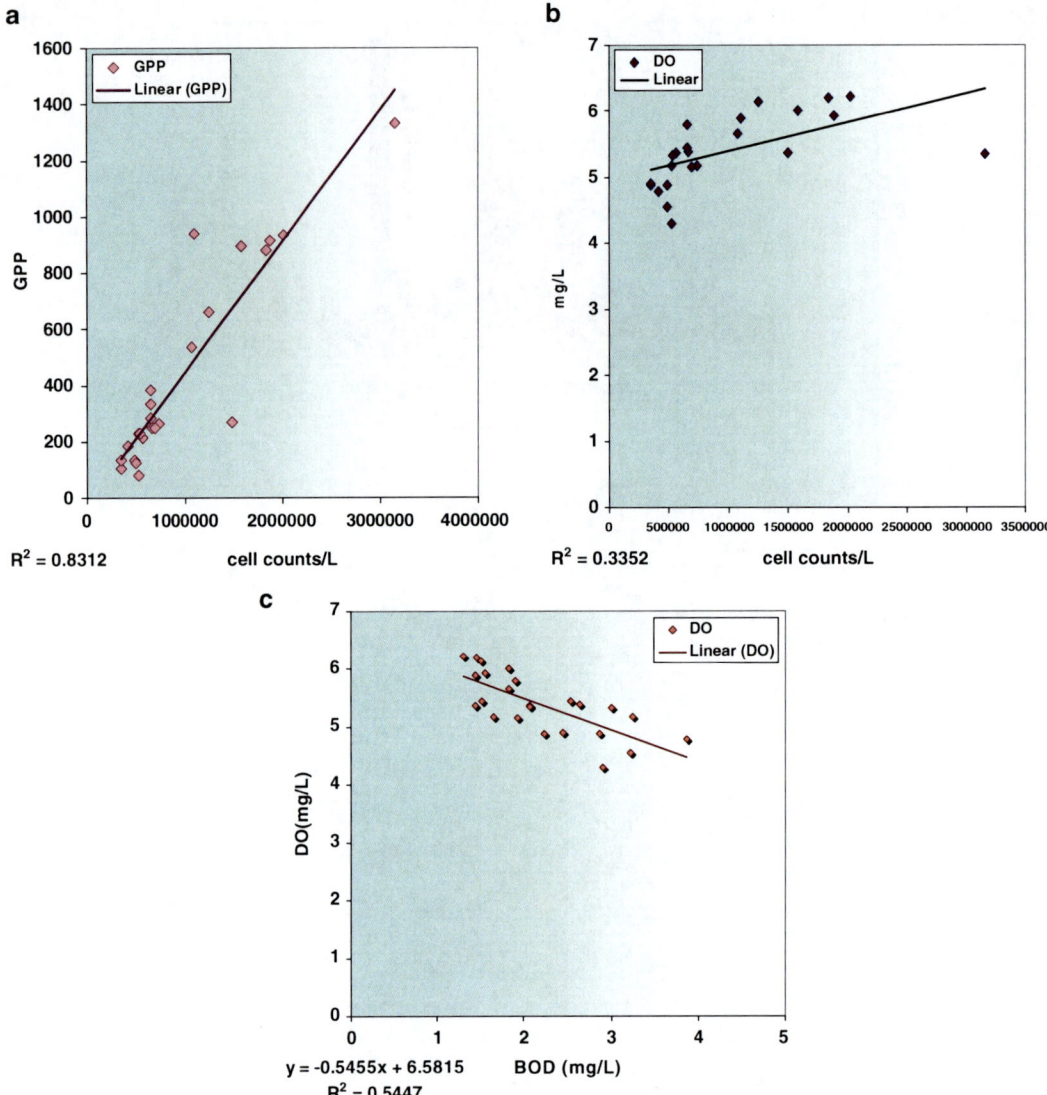

Fig. 5.30 Graphical representation of significant correlation between (**a**) cell count and GPP, (**b**) cell count and DO and (**c**) BOD and DO at the coastal marine region

Gross productivity in 2006–2007 (440.863 mgC/m³/h) was higher as compared to 2005–2006 (383.361 mgC/m³/h). Net primary productivity showed similar pattern as was with GPP with highest values in winter, intermediate values in summer and post-monsoon with lowest in monsoon. On the contrary, CRR values were relatively higher in monsoon as compared to the summer and post-monsoon periods

DO values were relatively low, with a maximum of 6.22 mg/L in January 2006 and minimum of 4.3 mg/L in July 2006 (Fig. 5.29b). DO content was maximum in winter and minimum in monsoon. BOD value ranged from 1 to 4 mg/l, with highest value in monsoon (August 2005) and lowest in winter (January 2006) when DO content was high. From the correlation matrix plot, there was a positive correlation between cell count and

5.4 Case Study III: Phytoplankton Dynamics of Eastern Indian Coast

Table 5.10 Correlation matrix plot between cell count, productivity and environmental variables at the estuarine station

	Cell count	DIN	DIP	DSi	Saln.	BOD	DO	Water temp.	pH	CRR	NPP	GPP
Cell count		-0.45	0.02	-0.71	-0.01	-0.46	0.81	-0.55	0.00	0.91	0.70	0.92
DIN	-0.45		0.30	0.64	-0.75	0.87	-0.74	0.45	-0.74	-0.36	-0.24	-0.54
DIP	0.02	0.30		0.13	-0.12	0.07	-0.07	-0.23	-0.19	0.03	0.00	0.06
DSi	-0.71	0.64	0.13		-0.28	0.58	-0.78	0.70	-0.33	-0.58	-0.45	-0.65
Saln	-0.01	-0.75	-0.12	-0.28		-0.78	0.45	-0.17	0.94	-0.15	-0.29	0.07
BOD	-0.46	0.87	0.07	0.58	-0.78		-0.78	0.61	-0.75	-0.35	-0.24	-0.57
DO	0.81	-0.74	-0.07	-0.78	0.45	-0.78		-0.70	0.47	0.71	0.34	0.75
Water temp.	-0.55	0.45	-0.23	0.70	-0.17	0.61	-0.70		-0.20	-0.54	-0.45	-0.60
pH	0.00	-0.74	-0.19	-0.33	0.94	-0.75	0.47	-0.20		-0.15	-0.26	0.07
CRR	0.91	-0.36	0.03	-0.58	-0.15	-0.35	0.71	-0.54	-0.15		0.70	0.89
NPP	0.70	-0.24	0.00	-0.45	-0.29	-0.24	0.34	-0.45	-0.26	0.70		0.82
GPP	0.92	-0.54	0.06	-0.65	0.07	-0.57	0.75	-0.60	0.07	0.89	0.82	

Correlations (marked correlations are significant at $p < .05000 \, N = 24$)

Table 5.11 Correlation matrix plot between cell count, productivity and environmental variables at the coastal marine region

	Cell count	DIN	DIP	DSi	Saln	BOD	DO	Water temp.	pH	GPP	NPP	CRR
Cell count		−0.52	−0.36	−0.48	0.54	−0.49	0.58	−0.34	0.46	0.91	0.56	0.59
DIN	−0.52		0.62	0.54	−0.69	0.66	−0.48	0.47	−0.75	−0.53	−0.58	−0.12
DIP	−0.36	0.62		0.77	−0.65	0.78	−0.50	0.44	−0.73	−0.37	−0.48	0.02
DSi	−0.48	0.54	0.77		−0.61	0.91	−0.77	0.67	−0.71	−0.56	−0.67	−0.03
Saln	0.54	−0.69	−0.65	−0.61		−0.78	0.45	−0.39	0.77	0.52	0.42	0.25
BOD	−0.49	0.66	0.78	0.91	−0.78		−0.74	0.64	−0.77	−0.58	−0.65	−0.10
DO	0.58	−0.48	−0.50	−0.77	0.45	−0.74		−0.85	0.46	0.73	0.86	0.08
Water temp.	−0.34	0.47	0.44	0.67	−0.39	0.64	−0.85		−0.53	−0.58	−0.87	0.14
pH	0.46	−0.75	−0.73	−0.71	0.77	−0.77	0.46	−0.53		0.47	0.48	0.09
GPP	0.91	−0.53	−0.37	−0.56	0.52	−0.58	0.73	−0.58	0.47		0.71	0.60
NPP	0.56	−0.58	−0.48	−0.67	0.42	−0.65	0.86	−0.87	0.48	0.71		−0.12
CRR	0.59	−0.12	0.02	−0.03	0.25	−0.10	0.08	0.14	0.09	0.60	−0.12	

Correlations (marked correlations are significant at $p < .05000 \, N = 24$)

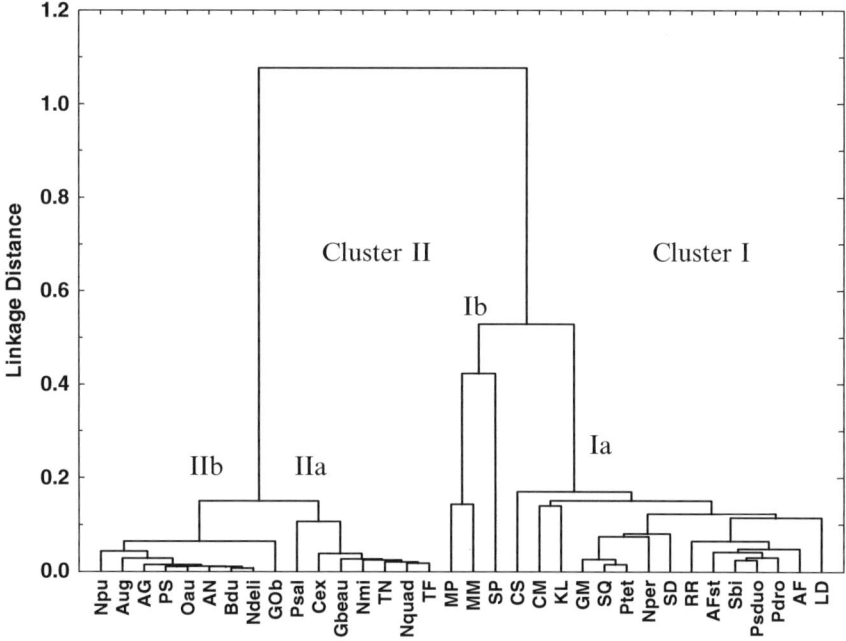

Fig. 5.31 UPGA cluster diagram based on seasonal abundance of different phytoplankton genera recorded from the freshwater station

dissolved oxygen content ($R^2 = 0.34$, $r = 0.58$, $p \leq 0.05$, $n = 24$) (Fig. 5.30b). Similarly, a significant negative correlation was established between DO and BOD ($R^2 = 0.54$, $r = 0.74$, $p \leq 0.05$, $n = 24$) (Fig. 5.30c) (Tables 5.10 and 5.11).

5.4.2.5 Multivariate Procedures

In the present work, cluster analysis (CA) was used to enumerate the diversity patterns in phytoplankton populations along coastal West Bengal. Moreover, we also tried to find out the probable role of the measured environmental variables in analysing the diversity patterns with the application of principal component analysis (PCA).

Figure 5.31 depicts the results obtained by using the 34 most abundant species during the entire study period at the freshwater station (Diamond Harbour). Two distinct groups of phytoplankton association were observed that are designated as I and II. The ultimate outcome of the 34 taxa through cluster analysis reveals the presence of two major clusters among the investigated taxa, one comprising of 16 species and other 18 species. These two major clusters are mainly separated from each other in accordance to their preferences for different temperature gradients, i.e. Cluster I consists of taxa that had a preference for the warmer periods whereas Cluster II consists of taxa that were mainly abundant in the cooler months.

Cluster I comprised of late spring (transition period), summer and monsoon population. This group primarily represented the populations of the relatively warmer periods when the mean temperature was 28 ± 5 °C. In this group of phytoplankton population, green and blue-green algal members were more abundant in comparison to diatoms. Cluster Ia represented the population of late spring and summer with predominance of green algal genera like *Pediastrum* spp. (*P. tetras*, *P. simplex* var. *duodenarium* and *P. duplex* var. *rotundatum*) and *Ankistrodesmus* spp. (*A. falcatus*, *A. falcatus* var. *stipitatus*) with only two species of diatoms which were *Leptocylindrus danicus* (LD) and *Nitzschia peregrina* (Nper). On a closer observation in respect to linkage distances, the measured distances among members of *Pediastrum* spp. and *Ankistrodesmus* spp.

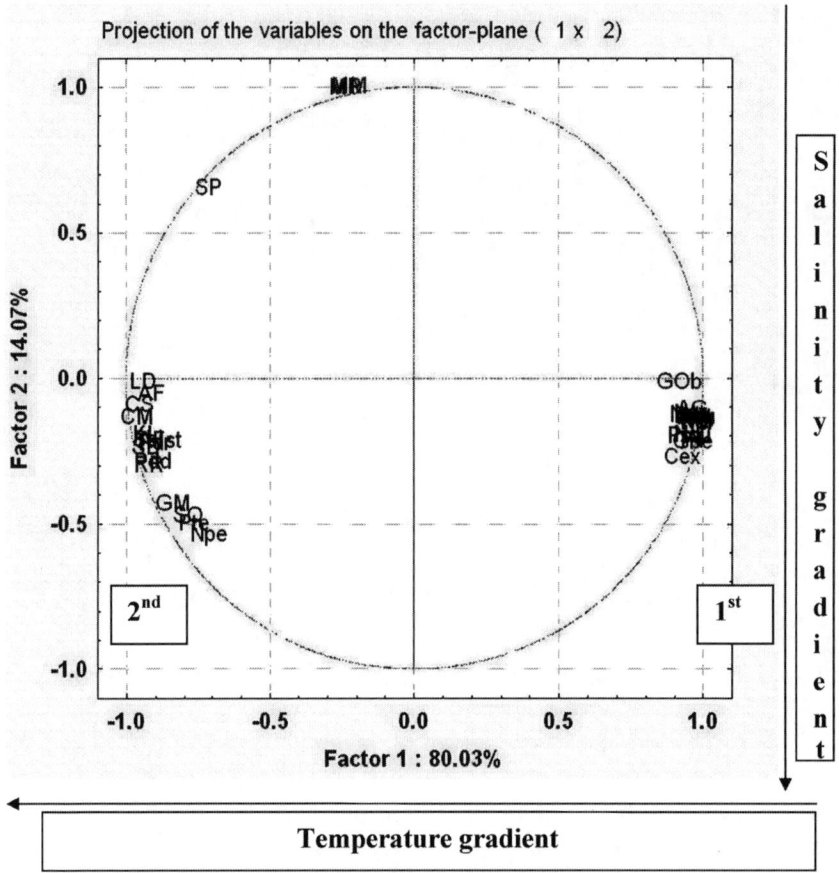

Fig. 5.32 PCA plot of Factor 1 vs. Factor 2 showing the pattern of species orientation based on environmental variables and magnitude of abundance at the freshwater station

ranged from 0.02 to 0.1 which further testifies the very close association among these members in the community. Cluster Ib clustered species that were abundant only during monsoon that included only the cyanobacterial members like *Spirulina platensis* (SP), *Merismopedia minima* (MM) and *Merismopedia punctata* (MP).

Cluster II represented the population of the autumn and winter periods with predominance of diatoms. Cluster IIa accounted for the population of autumn period which was mainly represented by pennate diatom taxa like *Navicula minima* (Nmi), *Nitzschia quadripartita* (Nquad), *Pleurosigma salinarum* (Psal) and *Gyrosigma beaufortianum* (Gbeau). Cluster IIb represented the population of winter months where along with the pennate population of autumn period, some centric diatoms appeared like *Biddulphia dubia* (Bdu), *Odontella aurita* (Oau), *Aulacoseira granulata* (Aug).

The juxtaposition of taxa was further observed in principal component analysis (PCA) of investigated taxa (Fig. 5.32). The results of PCA confirmed the results of the Cluster diagram. The highly clustered appearance of species scores in the PCA plot can be attributed to the low linkage distances between species in each subgroup of the cluster diagram. The principal component analysis of data matrix (based on correlation coefficient) resulted into extraction of 17 principal components among the taxa. In PCA, principal components exhibiting maximum amount of variations and having eigenvalues above 1.00 are generally considered. In the present investigation,

5.4 Case Study III: Phytoplankton Dynamics of Eastern Indian Coast

the first six principal components are represented by eigenvalues greater than 1.0; however only the first two components (PC 1 and PC 2) explain maximum amount of variations As such, the first two components have been taken into consideration for preparation of PCA plot. PC 1 and PC 2 together explain 80.03 % and 14.07 % of the variance, respectively, that cumulatively accounts for 94.1 % of variance among the data.

The two-dimensional diagram showed the distribution of taxa in all the four quadrants (Fig. 5.32). The distribution of taxa in the PCA plot follows the cluster formation of phenogram. PCA also reveals the environmental variables responsible for separation of taxa along the different dimensions.

Based upon factor versus variables scores along each component, PC 1 can be designated as a negative temperature gradient from right to left (Fig. 5.32), whereas PC 2 can be designated as a positive salinity gradient from top to bottom. As evident from the PCA plot, mainly two groups of phytoplankton populations could be observed. The 1st group consists of those species (e.g. *Gyrosigma obtusatum*, *G. beaufortianum*, *Coscinodiscus excentricus*, *Amphiprora gigantea*, *Odontella aurita*, *Actinocyclus normanii*, *Biddulphia dubia*, *Nitzschia delicatissima*, *Pleurosigma salinarum*, etc.) that have a high positive loading along PC 1 with intermediate negative loadings along PC 2. On closer observation, it can be further observed that the differences in loadings among the different members in this group are low, thereby suggesting their similar patterns of occurrence with almost similar affinities to temperature and salinity. On the other hand, the 2nd group consists of those members (e.g. *Leptocylindrus danicus*, *Ankistrodesmus falcatus*, *A. falcatus* var. *stipitatus*, *Cyclotella meneghiniana*, *C. striata*, *Rhizoclonium riparium*, *Nitzschia peregrina*) that have high negative loadings along PC 1 with intermediate negative loadings along PC 2. Within this group variation in loading values of different species suggests greater variations as compared to the members of group 1. This suggests that although the population have similar preferences for temperature and salinity, their abundance and contributions to the entire population may not be similar during the entire period of study when temperature was relatively high. A 3rd group of species was observed which was represented by few species with intermediate negative loadings along PC 1 and very high positive loadings along PC 2. This suggests that genera like *Merismopedia minima* and *M. punctata* mainly flourished under conditions of intermediate temperature and very low salinity, being as low as 0 psu.

From the estuarine station (Kakdweep), similar cluster analysis and principal component analysis were performed with the 25 most abundant taxa recorded. Here again, two distinct groups of phytoplankton population were recorded which were designated as Cluster I and Cluster II, respectively. Cluster I comprised of 8 taxa whereas Cluster II comprised of 17 taxa. As was observed for the freshwater, the grouping of phytoplankton population in the cluster diagram at the estuarine station was also based upon temperature gradients, i.e. Cluster I consists of taxa that had a preference for the warmer periods whereas Cluster II consists of taxa that were mainly abundant in the cooler months (Fig. 5.33).

Cluster I comprised of late spring (transition period), summer and monsoon population. This cluster primarily represented the populations of the relatively warmer periods when the mean temperature was 28 ± 5 °C. This cluster of phytoplankton was composed of green algae and diatoms with greater abundance of diatom genera. Cluster Ia represented the population of late spring and summer with 7 green algal taxa like *Scenedesmus* spp. (*S. dimorphus*, *S. brevicauda*, *S. bijuga* and *S. quadricauda*), *Ankistrodesmus falcatus*, *Kirchneriella lunaris* and *Crucigenia tetrapedia*. On the contrary, Cluster Ib was represented by 10 phytoplankton taxa of which all except *Rhizoclonium riparium* were diatom. This cluster primarily represented the population that dominated the habitat waters during the late monsoon and post-monsoon periods. The more abundant taxa in this cluster were *Cyclotella meneghiniana*, *Thalassiothrix frauenfeldii*, *Gyrosigma obtusatum*, *G. acuminatum*, etc. The comparatively high linkage distances between the different taxa within the same cluster suggest that

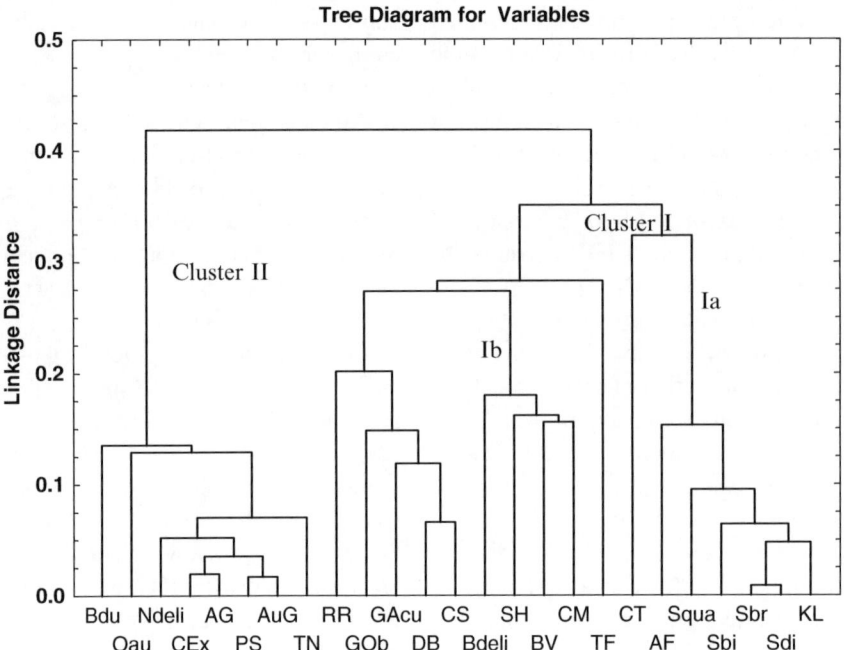

Fig. 5.33 UPGA cluster diagram based on seasonal abundance of different phytoplankton genera recorded from the estuarine station

although their availability were recorded under similar environmental conditions, their cell counts varied significantly during the sampling periods.

Cluster II represented the population of the late post-monsoon and winter periods when diatoms were the only representative taxa with no availability of any green algae. Cluster II was mainly represented by both centric and pennate diatom taxa. The pennate diatom population was represented by taxa like *Nitzschia delicatissima*, *Thalassionema nitzschoides* and *Amphiprora gigantea* whereas the centric diatom population was represented by *Odontella aurita*, *Biddulphia dubia*, *Coscinodiscus excentricus*, *Paralia sulcata* and *Aulacoseira granulata*.

In an attempt to better understand the phytoplankton species composition of the estuarine region as responses to environmental variables, PCA was implemented by taking into account the availability data of the 21 most abundant taxa. Based on eigenvalues, the PCA plot was done taking in consideration PC 1 and PC 2 that together explained 71.35 % of the variance within the data set.

The two-dimensional diagram showed the distribution of taxa in all the four quadrants (Fig. 5.34). The distribution of taxa in the PCA plot follows the cluster formation of phenogram. PCA also reveals the environmental variables responsible for separation of taxa along the different dimensions.

Here again, like our freshwater station, based on the factor loading values of each environmental variable along PC 1 and PC 2, PC 1 could be designated as a negative temperature gradient from right to left (Fig. 5.34) whereas PC 2 can be designated as a positive salinity gradient from top to bottom. As evident from the PCA plot, mainly three groups of phytoplankton populations could be observed. The 1st group consists of those species (e.g. *Biddulphia dubia*, *Odontella aurita*, *Thalassionema nitzschoides*, *Amphiprora gigantea*, *Aulacoseira granulata*) that have a high positive loading along PC 1 with low negative loadings along PC 2. On closer observation, it can be further observed that the differences in loadings among the different members in this group are low, thereby suggesting their similar

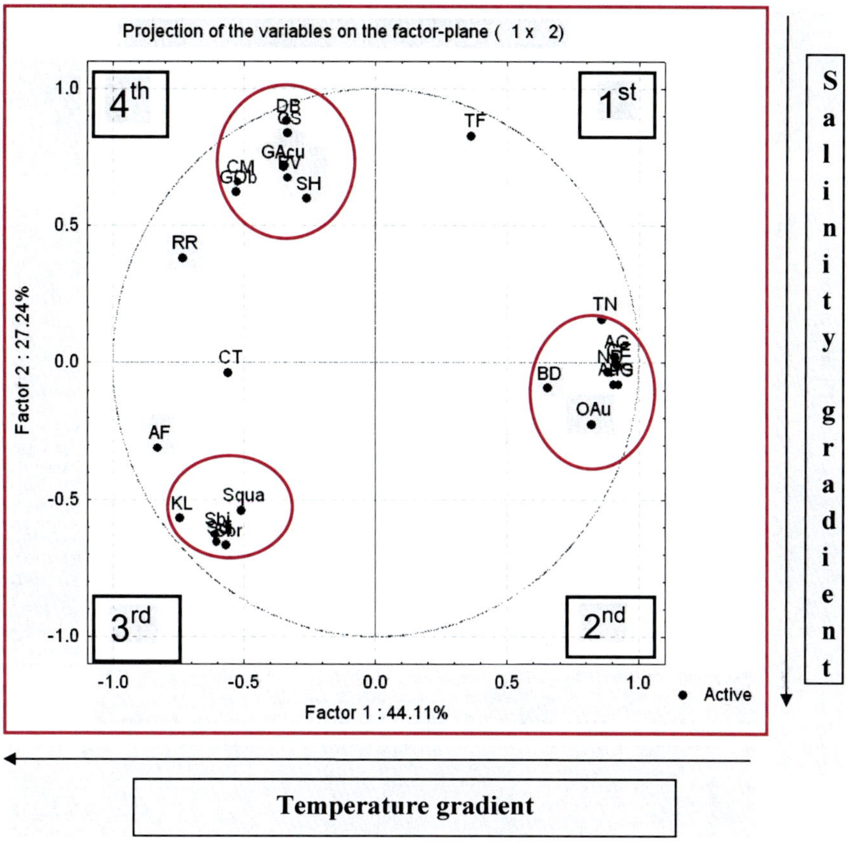

5.4 Case Study III: Phytoplankton Dynamics of Eastern Indian Coast

Fig. 5.34 PCA plot of Factor 1 vs. Factor 2 showing the pattern of species orientation based on environmental variables and magnitude of abundance at the estuarine station

patterns of occurrence with almost similar affinities to temperature and salinity. On the other hand, the 2nd group consists of those members (e.g. *S. brevicauda, S. bijuga, S. quadricauda, Ankistrodesmus falcatus*) that have intermediate to high negative loadings along PC 1 with intermediate negative loadings along PC 2. This suggests that the population have similar preferences for temperature and salinity, with almost similar patterns of abundance. A 3rd group of species was observed which was represented by species with intermediate negative loadings along PC 1 and very high positive loadings along PC 2 (*Ditylum brightwellii, Cyclotella meneghiniana, C. striata, Gyrosigma obtusatum, G. acuminatum, Bacteriastrum varians* and *Stephanodiscus hantzschii*). This shows that these genera mainly flourished under relatively oligohaline conditions when temperature was comparatively low.

As evident in the spatial orientation of the different phytoplankton genera recorded from the estuarine habitat, the plot was more spatially distributed as compared to the PCA from our freshwater station. Accordingly, in an attempt to understand whether if any seasonal succession pattern was operative at this station, multidimensional scaling (MDS) was performed. The applicability of this procedure was that the orientation in MDS plot was done on two dimensional spaces where nonlinear relationships between species scores were taken into consideration as well.

In the MDS plot (Fig. 5.35) of the phytoplankton population at the estuarine station, 3 distinct groups were configured. Starting from an anticlockwise direction (Group 1), members *Scenedesmus* spp. appeared along with *Kirchneriella lunaris* (KL), *Ankistrodesmus falcatus* (AF) and *Crucigenia tetrapedia* that

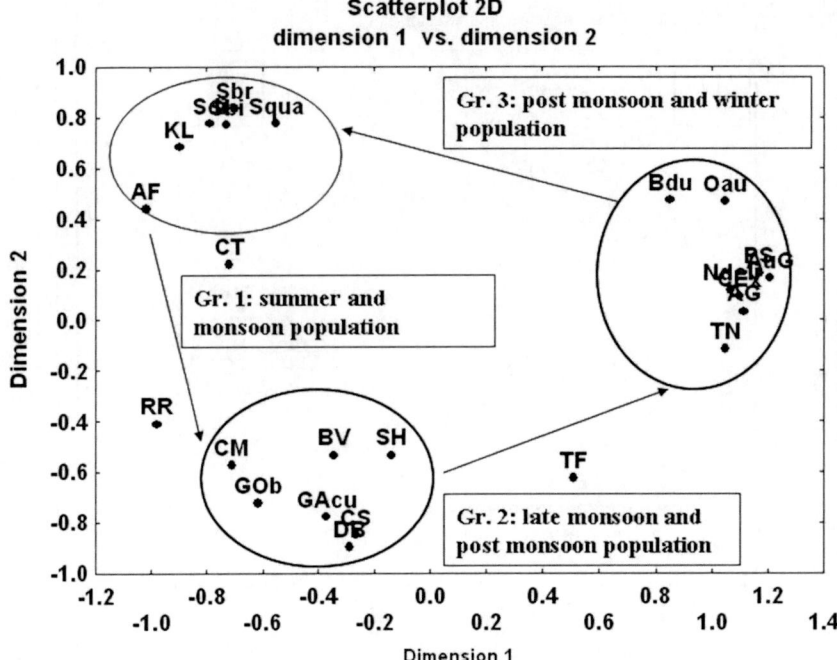

Fig. 5.35 Multidimensional scaling (*MDS*) of different species considering Dimension 1 and Dimension 2 at the estuarine station

represented the population of the summer and early monsoon months. *Rhizoclonium riparium* (RR) began to flourish in late summer months and continued to increase in population to the monsoon period. In the post-monsoon period, as seasonal precipitation receded, there were significant alterations of the available nutrient concentration and accordingly the population fluctuated with the abundance of diatom taxa (Group 2) like *Cyclotella meneghiniana* (CM), *C. striata* (CS), *Gyrosigma obtusatum*, *G. acuminatum*, *Ditylum brightwellii*, *Bacteriastrum varians* and *Stephanodiscus hantzschii* which were available only in the months of October and November with intermittent presence in the late monsoon months. In winter season a different population flourished which was represented by genera like *Biddulphia dubia* (Bdu), *O. aurita* (OAu), *N. delicatissima* (Ndeli), *Thalassionema nitzschoides* (TN), *Amphiprora gigantea* (AG), *Aulacoseira granulata* (AuG), *Pleurosigma salinarum* (PS) and *Coscinodiscus excentricus* (CEx) (Group 3). These genera appeared mostly in the late post-monsoon period of November and flourished in the winter months and gradually diminished with the approach of the summer months.

Similar cluster analysis and principal component analysis were performed with the 36 most abundant taxa recorded from the coastal marine region. Here, three distinct groups of phytoplankton population were recorded which were designated as Cluster I, Cluster II and Cluster. Cluster I comprised of 15 taxa, Cluster II comprised of 18 taxa and Cluster III was represented only by 3 taxa. Unlike the freshwater and estuarine stations, the linkage distances were comparatively higher which suggest that the within cluster association between species was less specific with respect to the period and magnitude of abundance, as was observed for the other stations. Furthermore, each cluster was further divided into subgroups which represented different associations of phytoplankton species assemblages (Fig. 5.36).

Cluster I was represented by 3 taxa, namely, *Nitzschia sigmoidea* (NS), *Gyrosigma acuminatum* (Gacu) and *Asterionella japonica* (AJ).

5.4 Case Study III: Phytoplankton Dynamics of Eastern Indian Coast

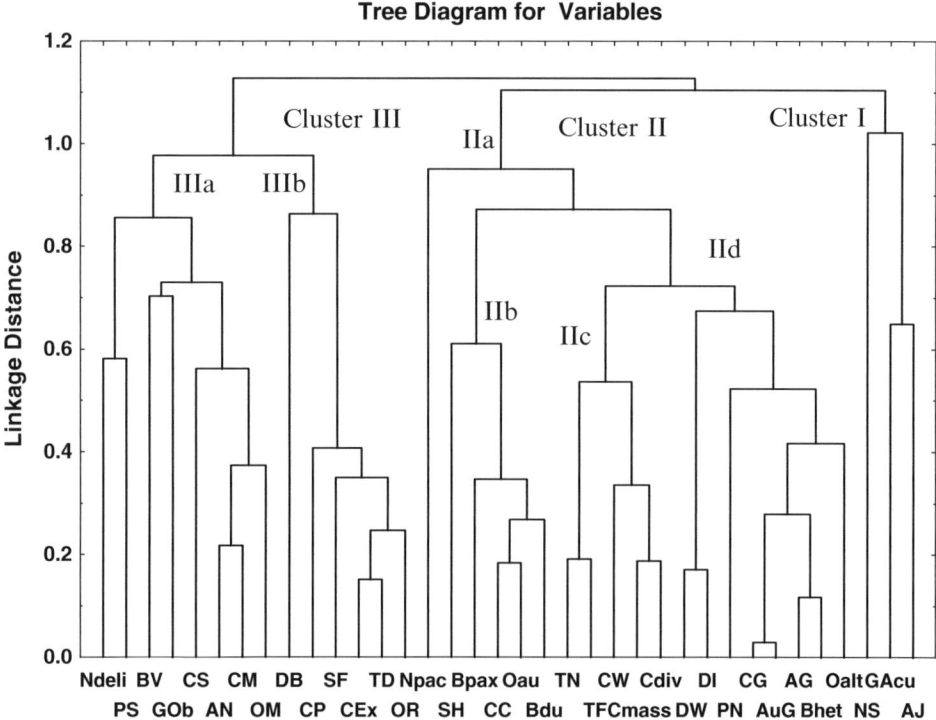

Fig. 5.36 UPGA cluster diagram based on seasonal abundance of different phytoplankton genera recorded from the coastal marine region

All the taxa showed different periods of availability. Among them, A. japonica showed a very high abundance in winter months and contributed maximally to the phytoplankton population of winter.

Cluster II consisted of four subgroups of which IIa was represented by the single species *Nitzschia pacifica* (Npac) that had a very restricted appearance only in the post-monsoon period with low cell counts. Subgroup IIb was composed of five species which were *Stephanodiscus hantzschii* (SH), *Bacillaria paxillifer* (BPax), *Coscinodiscus centralis* (CC), *Odontella aurita* (Oau) and *Biddulphia dubia* (Bdu). These taxa appeared as a component of the phytoplankton population in the late monsoon to post-monsoon periods and continued to flourish till the early summer months (March and April) that accounted for a significant proportion of the phytoplankton population. Subgroups IIc and IId consisted of the taxa appeared mainly in the late monsoon and post-monsoon periods but did not flourish significantly.

Cluster III comprised of two subgroups that are represented as IIIa and IIIb. Subgroup IIIa consisted of 8 species which are *Nitzschia delicatissima* (Ndeli), *Paralia sulcata* (PS), *Bacteriastrum varians* (BV), *Gyrosigma obtusatum* (GOb), *Cyclotella striata* (CS), *Actinocyclus normanii* (AN), *Cyclotella meneghiniana* (CM) and *Odontella mobiliensis* (OM). These taxa primarily appeared in the early summer period (March and April) and continued to flourish in the early monsoon months (July to September) with relatively low cell counts. Subgroup IIIb was represented by 6 species like *Odontella rhombus* (OR), *Coscinodiscus perforatus* (CP), *Coscinodiscus excentricus* (CEx), *Thalassiosira decipiens* (TD), *Ditylum brightwellii* (DB) and *Surirella fastuosa* (SF). These taxa appeared in the summer months (April–June) and continued to flourish in the monsoon to post-monsoon

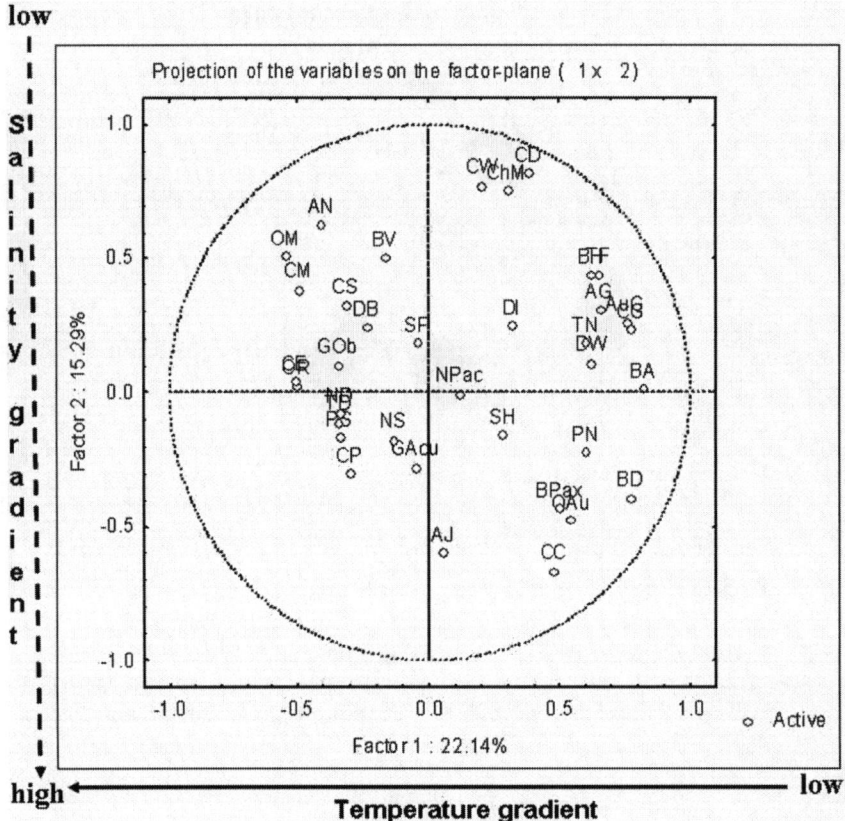

Fig. 5.37 PCA plot of PC1 vs. PC2 showing the pattern of species orientation based on environmental variables and magnitude of abundance at the coastal marine region

periods with comparatively higher cell counts than in the subgroup IIIa.

As evident, the cluster analysis did not show well-demarcated species associations, as was observed for the freshwater station. Accordingly, to further understand the species-specific responses of phytoplankton populations to environmental variables, principal component analysis was carried out as well.

In our principal component analysis (PCA) study of the species composition at the marine coastal region, the factor loading matrix indicates the correlation between each principal component to each of the species. In our study, species that appeared only once in the total sampling period were not considered. Out of 36 species considered, 11 had their strongest correlation with highest factor loading with the first two principal components (PC 1 and PC 2). Accordingly the PCA plot was done considering PC 1 and PC 2. These two components explained 37.43 % of the variation within the species data. The covariates in the PCA plot were grouped together based on their factor loading values along PC 1/PC 2. Covariability between the species plotted along PC 1 and PC 2 was relatively high because the explained variance by the first two principal components is about 6.5 times higher than it would have been if the time series of the 36 algal species were not correlated at all. Thus, more than one-third of the variation in the 36 algal species during the study period is accounted for by the 1st and 2nd principal components in our plot. On plotting PC 1 against PC 2, the variability in individual species occurrence becomes evident, where the distance between the plotted species points provide a relative measure of the degree of similarity/dissimilarity between species with respect to both their seasonal occurrence and magnitude of abundance.

5.4 Case Study III: Phytoplankton Dynamics of Eastern Indian Coast

The temporal pattern of occurrences of each individual species was also accounted by plotting PC 1 vs. PC 2. As evident from our PCA plot, (Fig. 5.37) a negative temperature gradient was established along PC 1. Accordingly, genera with a high positive factor loading along PC 1 flourished in the cooler months with comparatively low water temperature. Thus, genera with positive factor loadings on PC 1 are more abundant in the post-monsoon and winter months when the average temperature is about 12–18 °C. On the contrary, genera with negative factor loading along PC 1 had a preference to flourish in the summer months when the average water temperature is comparatively higher (30–36 °C). PC 2 axis represents the relative degree of variation in temporal occurrence with low to high gradient of species availability during an annual cycle. As evident from the plot PC 2 also represent a salinity gradient from top to bottom in which the species align as per their salinity requirement. Species having a high positive correlation with this vector generally exhibited a specific seasonal occurrence where salinity requirement was low (as the gradient is from low to high).

Accordingly, from the PCA plot (1st quadrant) it was evident that *Biddulphia alternans* (BA) along with other genera like *Thalassionema nitzschoides* (TN), *Diploneis interrupta* (DI) and *Diploneis weissflogii* (DW) are more abundant in the post-monsoon and winter months. Taxa like *C. granii* (CG), *A. gigantea* (AG) and *Aulacoseira granulata* (AuG) have intermediate positive factor loading on PC 1 which are the representative of only the particular post-monsoon period with low or no availability in other seasons. Genera like *Chaetoceros wighami* (CW), *C. messanensis* (ChM) and *C. diversus* (CD) had a very high positive loading along PC 2 with intermediate factor loading along PC 1 which can be clearly explained by the fact that their availability was restricted to the monsoon season (water temperature 28–32 °C) and they never flourished in any other season. Genera with a positive factor loading on PC 1 but negative loading on PC 2 (2nd quadrant) were mostly available in the late post-monsoon to winter months with a high abundance. This is because during this period average water temperature is low, but with a decrease in both seasonal and riverine precipitation there is a rise in salinity. Accordingly, genera like *Biddulphia dubia* (BD), *Odontella aurita* (OAu) and *Coscinodiscus centralis* (CC) attained their peak growth in this period with high abundance which culminated with a very high cell count of *A. japonica*. Genera in the 3rd quadrant had intermediate negative factor loading for both PC 1 and PC 2. Hence, these genera flourished in the relatively warmer months when salinity was about 32–35 psu but with a low cell count. *Nitzschia sigmoidea* (NS) and *Gyrosigma acuminatum* (GAcu) were available only in the months of March and April. *Nitzschia delicatissima* (ND) and *Paralia sulcata* (PS) were exclusively available only in the months of April to July during the entire study period. Thus, the findings clearly suggested that these genera represented the population of the transitory period from winter to summer months. Finally, the 4th quadrant comprised of those genera which were abundant in the warmer months but were significantly less in the winter months of December and January with a relatively high cell count of individual species. Along with genera like *O. rhombus* (OR) and *C. excentricus* (CE), genera like *Odontella mobiliensis* (OM), *Cyclotella meneghiniana* (CM) and *Actinocyclus normanii* f. *subsala* (AN) had negative loading on PC 1 but intermediate positive loading along PC 2, suggesting their relatively even abundance with high density in the summer and monsoon periods.

For further confirmation of this seasonal pattern of species based upon their abundance data and preference for temperature and salinity, multidimensional scaling (MDS) was performed (Fig. 5.38). The MDS configuration plot of all genera (Dimension 1 vs. Dimension 2) clearly demarcates distinct groups based upon the seasonal preference of the individual species.

In the MDS plot 5 distinct groups were configured. Starting from an anticlockwise direction from summer months (Group 1), the genera *O. rhombus* (OR), *Coscinodiscus excentricus* (CE) and *O. mobiliensis* (OM) appeared together as the dominant genera of the summer months along with *C. meneghiniana* (CM) that was available in

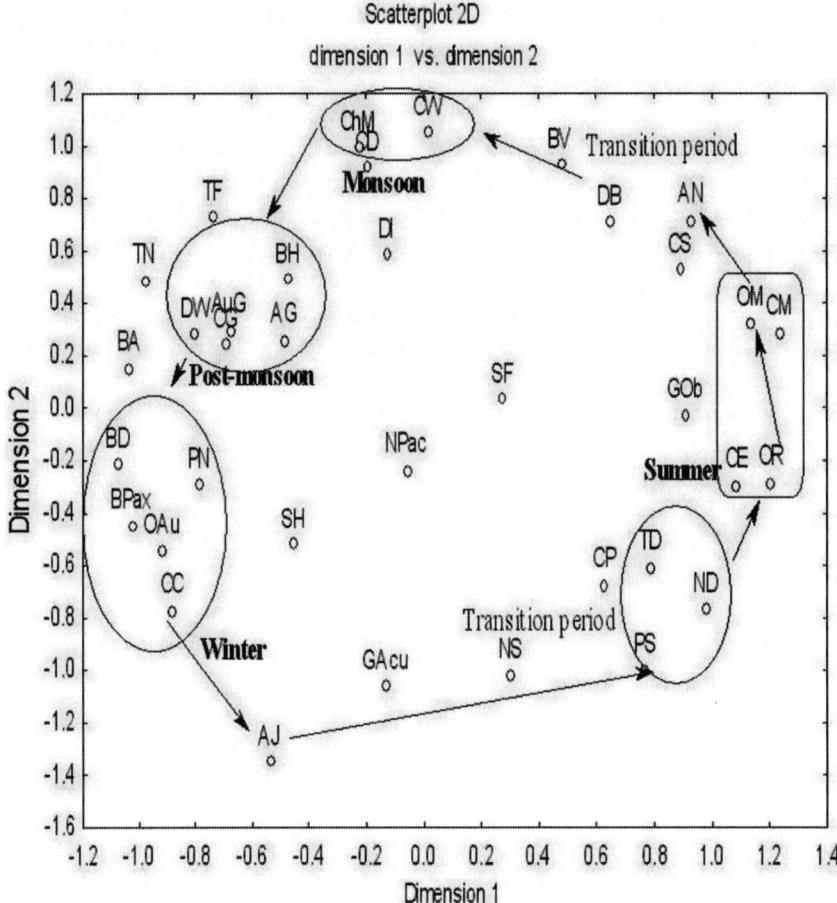

Fig. 5.38 Multidimensional scaling of different species considering Dimension 1 and Dimension 2 at the coastal marine region

other seasons as well but had a preference for the summer months. Genera like *Ditylum brightwellii* (DB), *Cyclotella striata* (CS) and *A. normanii* f. *subsala* (AN) began to flourish in the summer months and continued to increase in population to the monsoon period and accordingly they are representative of the transition from a summer to a monsoon season. With the advent of the monsoon months there is significant alteration of the available nutrient concentration and accordingly the population fluctuates with the abundance of representative genera (Group 2) like *Chaetoceros wighami* (CW), *C. messanensis* (ChM) and *C. diversus* (CD). As monsoon is prolonged there is significant enhancement of nutrient input but reduction in mean salinity. Accordingly, phytoplankton population is highly specific which is evident from the fact that the different species of *Chaetoceros* do not flourish at any other season of our sampling period except monsoon. In the post-monsoon period phytoplankton population changes with the appearance of representative genera like *B. heteroceros* (BH), *A. granulata* (AuG), *A. gigantea* (AG), *C. granii* (CG) and *D. weissflogii* (DW) (Group 3) which were available only in the months of October and November with intermittent presence in the late monsoon months. In winter season a different population flourished which was represented by genera like *Biddulphia dubia* (BD), *O. aurita* (OAu) *Pleurosigma normanii* (PN), *B. paxillifer* (BPax) and *Coscinodiscus centralis* (CC) (Group 4). These genera appeared mostly in the late post-monsoon period of November and flourished in the winter months and gradually diminished with the approach of the summer months. This winter population finally culminated with the abrupt rise in *A. japonica* (AJ) population where only a very few individuals of other species developed. Genera like *G. acuminatum* (GAcu) appeared in the late winter months of February and flourished in March till the early summer months of April. Likewise *Nitzschia sigmoidea* (NS) and *N. delicatissima* (ND) (Group 5) appeared in the early summer month of April and accordingly they are considered as representatives of the transitory population between a winter and summer phytoplankton population. *Surirella fustuosa* (SF) and *N. pacifica* (NPac) appeared irregularly with very low cell count. On the contrary, *Stephanodiscus hantzschoides* (SH) appeared all throughout the year and did not show any seasonal preference. Accordingly, these genera appeared in the middle of the MDS configuration in a scattered manner and did not belong to any particular group.

Phytoplankton population of coastal area is controlled by various physical, chemical and biological factors which ultimately regulates the phytoplankton dynamics of particular ecosystem. Results showed that coastal West Bengal represented a diverse phytoplankton population which is primarily composed of members of Cyanophyta, Chlorophyta and Bacillariophyta. Members of Bacillariophyta or diatoms made up the bulk of the population, followed by green and blue-green algae. This was evident from the findings which showed that out of the 75 phytoplankton taxa recorded, 51 were diatoms. In the freshwater and estuarine region, the phytoplankton flora was mainly represented by members of Cyanophyta, Chlorophyta and diatoms. On the other hand, the marine coastal region was represented by diatoms only. The diatom population primarily comprised of the members belonging to the tribes Coscinodisceae, Chaetocerae, Biddulphieae (Centrales) and Naviculeae (Pennales) (Cupp 1943). A high availability of diatoms as a major component of coastal and estuarine plankton population has been reported from other parts of the Indian subcontinent as well. A detailed account of the phytoplankton population along Madras coast was prepared by Venkataraman (1939) where he reported 98 taxa belonging to 33 genera. In another significant work on Madras coast, 134 species belonging to 64 genera were reported by Subrahmanyan (1946). Later, a total of 57 species of diatoms belonging to 36 genera were reported from the Gulf of Kutch, western coast of India (Gopalakrishnan 1972). In a 1-year study from the Mandovi–Zuari estuary a total of 48 phytoplankton species were identified – of which 37 species were diatoms and only 2 taxa were blue-green algae (Krishna Kumari et al. 2002). In a more recent work from the Zuari estuary, Goa, 66 planktonic diatom species belonging to 29 genera were recorded, of them 36 species were Pennales, whereas 30 species were of Centrales (Redekar and Wagh 2000). The dominance of diatoms as a component from marine coastal regions has been reported from other stations along the Bay of Bengal as well (Madhupratap et al. 2003). Thus, abundance of diatoms has been reported from all the Indian estuaries and coastal waters – where members of Chaetoceraceae, Coscinodiscaceae and Naviculaceae are more abundant (Venkataraman and Wafar 2005). Hence, our findings in the present study along coastal West Bengal were well in agreement with other records from different estuaries of the Indian subcontinent.

From the results it is evident that species composition pattern did not show significant interannual variations. Here, almost the same genera appeared in the same pattern in successive years. Enormous freshwater run-offs from several major rivers may result in strong vertical stratification that inhibits vertical mixing (Gopalakrishnan and Sastry 1985). According to them, lack of intense upwelling of this coast is due to the equator ward flow of the freshwater plume, which resulted in overwhelming of the offshore Ekman transport in the coast of Bay of Bengal. These conditions accounted for the similar seasonal plankton dynamics with minimal interannual variability. This lack of upwelling events can be further attributed to the fact that the Bay of Bengal, especially the northern parts, represents a very narrow shelf (Qasim 1977; SenGupta et al. 1977; Radhakrishna et al. 1978).

Among diatoms on a generic level, *Coscinodiscus*, *Leptocylindrus*, *Chaetoceros*, *Biddulphia*, *Odontella*, *Gyrosigma*, *Ditylum*, *Thalassionema*, *Thalassiothrix*, *Navicula* and *Nitzschia* have been recorded as dominant taxa from this region. Along with the diatom genera, green algal genera like *Kirchneriella*, *Ankistrodesmus*, *Pediastrum*, *Scenedesmus* and the blue green algal genera like *Merismopedia* and *Microcystis* were available in this region. Abundance of *Chaetoceros* spp. and *Bacteriastrum* spp. were observed only during the monsoon period mainly at the estuarine and marine region along coastal West Bengal. In other reports, dominance of pennate members like *Nitzschia* spp. have been reported from the freshwater regions while genera like *Coscinodiscus* spp. were more abundant in the high-salinity zones of both Zuari River and Mandovi River estuary (Matondkar et al. 2006). Dominance of members of *Chaetoceros* spp. from June to September (monsoon months) has also been reported from the neritic zone of the Urdaibai estuary at northern Spain (Trigueros and Orive 2001).

From our study along coastal West Bengal, it was established that each station had different dominant representative taxa such as *Leptocylindrus danicus* at freshwater station, *Thalassiothrix frauenfeldii* at estuarine station and *Asterionella japonica* at marine region. A similar high abundance of *Asterionella japonica* in winter months has been reported from the adjacent Orissa coast where it made up almost 99 % of the total population (Sasamal et al. 2005). In the estuarine and marine region, abundance of euryhaline taxa like *Skeletonema costatum*, *Thalassiosira decipiens*, *Thalassionema nitzschoides* was recorded. High abundance of chain forming diatom genera like *Leptocylindrus danicus*, *Thalassiosira decipiens* and *Skeletonema costatum* have been reported from the Alborán Sea, a continuation of southwest Mediterranean Sea (Mercado et al. 2005). These taxa are common in other estuaries around the world as well which are subjected to large salinity variations due to precipitation and riverine discharge, including Apalachee Bay (Curl 1959), Perdido Bay, (Livingston 2001), Indian River Lagoon (Badylak and Phlips 2004), Florida and in many estuaries of North Carolina (Mallin 1994). The rise in larger-sized pennate diatoms along with colonial forms like *Asterionella japonica*, *Thalassionema nitzschoides*, *Thalassiothrix frauenfeldii*, *Gyrosigma beaufortianum*, *Pleurosigma salinarum*, *Bacillaria paxillifer*, *Nitzschia sigmoidea*, etc. during the cooler post-monsoon and winter months can be attributed to the strong northeastern winds (De Jonge and van Beusekom 1995). The availability of these taxa at highly diverse habitats at different parts of the world further suggests that these euryhaline phytoplankton genera are cosmopolitan in distribution and can develop under diverse physicochemical conditions.

Diversity indices are calculated on the basis of total biomass data obtained from cell count method and the number of individuals, which indicate the community pattern of the particular ecosystem. Results showed that diversity was highest at freshwater station and minimum at the estuarine station, with marine coastal region occupying an intermediate position. This was in agreement with earlier reports from Santa Catalina basin, off the coast of California and Norwegian Sea, where low diversity has been reported due to physical variability in the estuarine and the sandy coast (Jumars 1976). Our study showed that, in the month of July or in monsoon season, nutrient concentration was high with well mixing due to seasonal precipitation and riverine influx of nutrient rich water. This resulted in the development of a suitable condition for phytoplankton diversification that accounted for maximum SW Index in the post-monsoon period with stabilized nutrient pool. Thus, in the present study, post-monsoon period (October–November) appeared to be most conducive for phytoplankton diversity showing maximum diversity index throughout the study area, although total phytoplankton cell counts were maximum in winter. Observation of maximum diversity in post-monsoon period agrees well with the earlier reports for Hugli estuary, north east coast of India (De et al. 1994) and Vellar estuary (Hangovan 1987).

High species evenness with negligible variations at the freshwater and estuarine stations signifies that the percentage contributions of individual taxon to the total phytoplankton population were almost similar during the entire sampling period. This was in contrast to the marine region where it was maximum in monsoon months and minimum in winter. Decreasing tendency of SW Index and species evenness indicate shifting of phytoplankton community from high species richness to bloom formation, as reported by many authors in eutrophic and hypertrophic lakes and reservoirs (Jacobsen and Simonsen 1993; Padisak 1993). Thus, the drop in biotic indices in winter at the marine station was primarily due to very high abundance of *Asterionella japonica* population that accounted for more than 90 % of the phytoplankton population. A similar finding was recorded for the summer months at the coastal marine station as well with the development of a different population (*Odontella rhombus*). In the monsoon period, the water column was highly disturbed due to seasonal precipitation, riverine inflows and alternating air currents due to cyclonic weather. As a result, phytoplankton count was low with intermediate value for species evenness in the monsoon period.

Results of cluster analysis and multidimensional scaling further testified the specific seasonal patterns of phytoplankton availability. Cluster analysis maximizes between group associations and minimizes within group association (Ramette 2007). Thus, the very narrow distances between the subgroups within each cluster suggest the close association between the different algal taxa. On the other hand, the high linkage distances between each cluster further suggest that each population is highly demarcated from each other. MDS plots help in orienting variables in a two-dimensional space where nonlinear relationships are also taken into consideration. Thus, while understanding the seasonal patterns among phytoplankton population from both these multivariate procedures, it is clearly evident that seasonal succession at the freshwater station was not very pronounced. On the other hand, seasonal patterns were more evident at the estuarine station. At the coastal marine region, seasonal succession was clearly evident where each season was clearly represented by distinct species assemblages.

Results showed that both productivity and cell counts were maximum at the coastal marine region and minimum at the estuarine region. Highly significant correlations between phytoplankton cell counts and GPP clearly established that GPP was actually the total photosynthetically fixed carbon by the phytoplankton population. Seasonally, as can be expected from cell count data, productivity was maximum in winter and minimum during the monsoon period along West Bengal coast. Planktonic photosynthetic productivity is one of the major contributors to the overall productivity of open aquatic ecosystems. The efficiency of energy transfer from phytoplankton to consumers and ultimate production at upper trophic levels vary with algal species composition. It is well known that diatom-dominated marine upwelling systems sustain more fish biomass per unit of phytoplankton biomass than cyanobacteria-dominated lakes, which was also evident in coastal West Bengal. Unlike, the coastal marine and freshwater regions, the low productivity at the estuarine station can be attributed to the excessive fishing activity and transportation of vessels from Kakdweep. Such commercial exploitation could have resulted in highly disturbed water column that may have accounted for an unstable stratification, resulting in decrease in productivity. NPP is a measure of available photosynthetically fixed carbon after eliminating the catabolic loss of organic matter due to respiration (CRR). Accordingly, a drop in NPP and rise in CRR clearly shows that carbon utilization was comparatively high in comparison to carbon assimilation. Such an observation can be attributed to the fact that a rise in nutrient-rich freshwater influx in the habitat waters under warm conditions led to an increase in the heterotrophic population which accounted for the enhanced carbon utilizations. A similar result was also obtained from the Mandovi River estuary (Verlencar and Qasim 1985) having highest productivity in the post-monsoon season with intermediate values in the pre-monsoon period and the lowest productivity in monsoon. A drop in primary productivity in the monsoon months and a rise of the same in the post-monsoon period was also reported from Lake Tana in Ethiopia (Wondie et al. 2007).

Variations in DO contents showed a similar pattern of variation as was observed for cell counts and productivity in both freshwater and estuarine stations. Thus, DO levels at the freshwater station were comparatively higher than at the estuarine station. Although the freshwater and estuarine stations represented optimum DO levels, DO levels at the marine coastal region represented an undersaturated condition although cell counts were maximum among all the three stations. During the monsoon months, heavy seasonal precipitation promotes an enhanced freshwater inflow with high suspended matter content, resulting in a very turbid water column at the marine region. This turbid nature of the water column may result in a decrease in the photic zone. Thus, although the incident irradiance was high, the average irradiance in the water column was relatively low which may have accounted for the drop in DO levels in the habitat waters at the marine coastal region with low photosynthetic activity. Another significant observation was the rise in BOD levels in the monsoon months and a drop in the same during the winter periods. The significant negative correlations between DO and BOD levels at all the sampling stations indicate the high heterotrophic growth during the monsoon months which resulted in the decreased DO levels.

It is evident from the findings that physicochemical properties of habitats play a determining role in the development and proliferation of phytoplankton populations at coastal West Bengal. In this tropical coastal area, optimal light and nutrient availability in association with other physical parameters like pH, temperature and salinity allowed diverse populations to develop with high cell counts.

Temperature in association with salinity played a significant role in determining the phytoplankton species composition. The restrictive appearance of cyanobacterial taxa in the freshwater region during summer and early monsoon period is a testimonial of the fact that cyanobacteria had a distinct preference for hypo-saline conditions with high temperature. Studies on cyanobacterial populations from other parts of the world further established this correlation between temperature and blue-green algal abundance as was observed at the Great Barrier Reef in Australia (Ayukai 1992), the Mediterranean (Caroppo 2000) and Indiana River Lagoon in Florida (Badylak and Phlips 2004). Chlorophycean members were mainly restricted from the freshwater to estuarine stations which suggest that green algal population developed well under oligohaline to mesohaline conditions when temperature was relatively high. In contrast to green and blue-green algal taxa, diatoms flourished mainly under mesohaline to hyper-saline conditions with low temperature. Thus, diatoms were available from freshwater to marine region although there was a shift in the population with increase in centric diatom population as the river approached the sea.

Results of PCA further underline the role of temperature and salinity in determining the species composition at each coastal station. In PCA, new synthetic ordinates are developed, based upon the linear relationships between variables. Accordingly, PCA plots of species abundance in relation to environmental variables clearly suggest that temperature and salinity are the most important factors that determine the species composition at all the stations along coastal West Bengal. An increase in Bacillariophycean members under conditions of low temperature in winter (fall) and spring has also been observed from Nakdong River, South Korea (Ha et al. 1998), where almost similar conditions of temperature and salinity were observed as was along coastal West Bengal.

The pH level varied from 7 to 8.3 at the entire sampling area which indicates the neutral to alkaline nature of the habitat waters. Several reports from different coastal regions of the world represent similar neutral to alkaline pH gradients (Skirrow 1975; Pegler and Kempe 1988; Brussaard et al. 1996; Hinga 2002). Salinity seems to play a significant role in pH changes as it influences the various equilibrium constants of temperature and alkalinity.

Phytoplankton development and succession is dependent not only on physical parameters but also on chemical factors like dissolved oxygen and nutrient availability (Reynolds 1984). Nitrogen and phosphorus are the most essential components for development of biological populations with no exception for phytoplankton as well. In an aquatic environment nitrogen, sources are mainly available as nitrate and nitrite, whereas phosphorus source is available in the form of phosphate. It has long been proposed that marine and estuarine phytoplankton abundance is N limited, whereas freshwater phytoplankton is P limited. Relatively high levels of DIN and DIP were maintained during the entire study period at all the stations which suggest that nutrient levels were well above the optimum levels and did not act as limiting factors. Cultural eutrophication was also an important factor for the relatively high DIN and DIP levels in this region. At the freshwater station DIN and DIP contents were comparatively higher than DSi. On the other hand, DIN and DSi contents were higher in the estuarine station and in the marine coastal region. Moreover, during the monsoon months, as seasonal precipitation increased, it resulted in a rise in inflow of nutrient-rich freshwater from perennial sources and from other tributaries. This accounted for an increase in nutrients especially DIN and dilution of phytoplankton population. Thus, the DIN-rich habitat waters allowed the development of cyanobacterial taxa like *Gloeocapsa* sp., *Gleothece* sp., *Merismopedia* spp., *Microcystis aeruginosa*, *Spirulina* spp., which are efficient nitrogen fixers. Diatoms accounted for a major proportion of the phytoplankton population especially during the post-monsoon and winter months. Diatom production can be limited by the availability of dissolved silicate (DSi) and DSi depletion relative to other major nutrients (Kilham 1971; Schelske and Stoermer 1971; Malone et al. 1980). DSi levels were maintained in optimum conditions throughout the study area. The comparatively low DSi concentrations in winter months were primarily due to excessive growth of planktonic diatom population which can be inferred from correlation studies as well. Thus, although salinity and temperature were the primary regulatory factors in the species composition and succession patterns at coastal West Bengal, nutrient availability also played an important role in phytoplankton dynamics.

This region represented a highly diverse phytoplankton population where seasonal succession was relatively pronounced. Temperature and salinity gradients were the principle determining factors for the seasonal variability in species composition, although nutrient concentrations also played an important role. The freshwater station was more affected due to anthropogenic nutrient loadings, whereas the marine coastal region was least. The estuarine station was a disturbed ecosystem that resulted in less diversification of phytoplankton population. The marine region was comparatively more stable from an ecological point of view where different aspects of population dynamics were more evident as compared to the other stations along coastal West Bengal.

5.4.2.6 Taxonomic Account of Common Phytoplankton Taxa Recorded from Indian Coast

Division – Cyanophyta
Class – Cyanophyceae
Order – Chroococcales
Family – Chroococcaceae

1. *Gloeocapsa punctata* Naegeli (Plate 5.5, Fig. A)

   [Desikachary, 1959, p. 115, pl. 23, Fig. 2]

   Plant mass blue green, floating, consisting of small aggregated of 8–16 individuals which are spherical and inclosed by thick sheaths; cells 2 μ in diameter; contents blue–green, homogenous

2. *Microcystis aeruginosa* Kuetz. emend. Elekin (Plate 5.5, Fig. B)

   [Desikachary, 1959, p. 93, pl. 17, Figs. 1, 2 & 6, pl. 18, Fig. 10]

   An irregularly lobed, saccate and clathrate colony of numerous spherical cells which are much crowded within a gelatinous matrix (several colonies invested by a common tegument), colonial mucilage hyaline and homogenous; cell contents blue green, highly granular; cells 3.5 μ in diameter

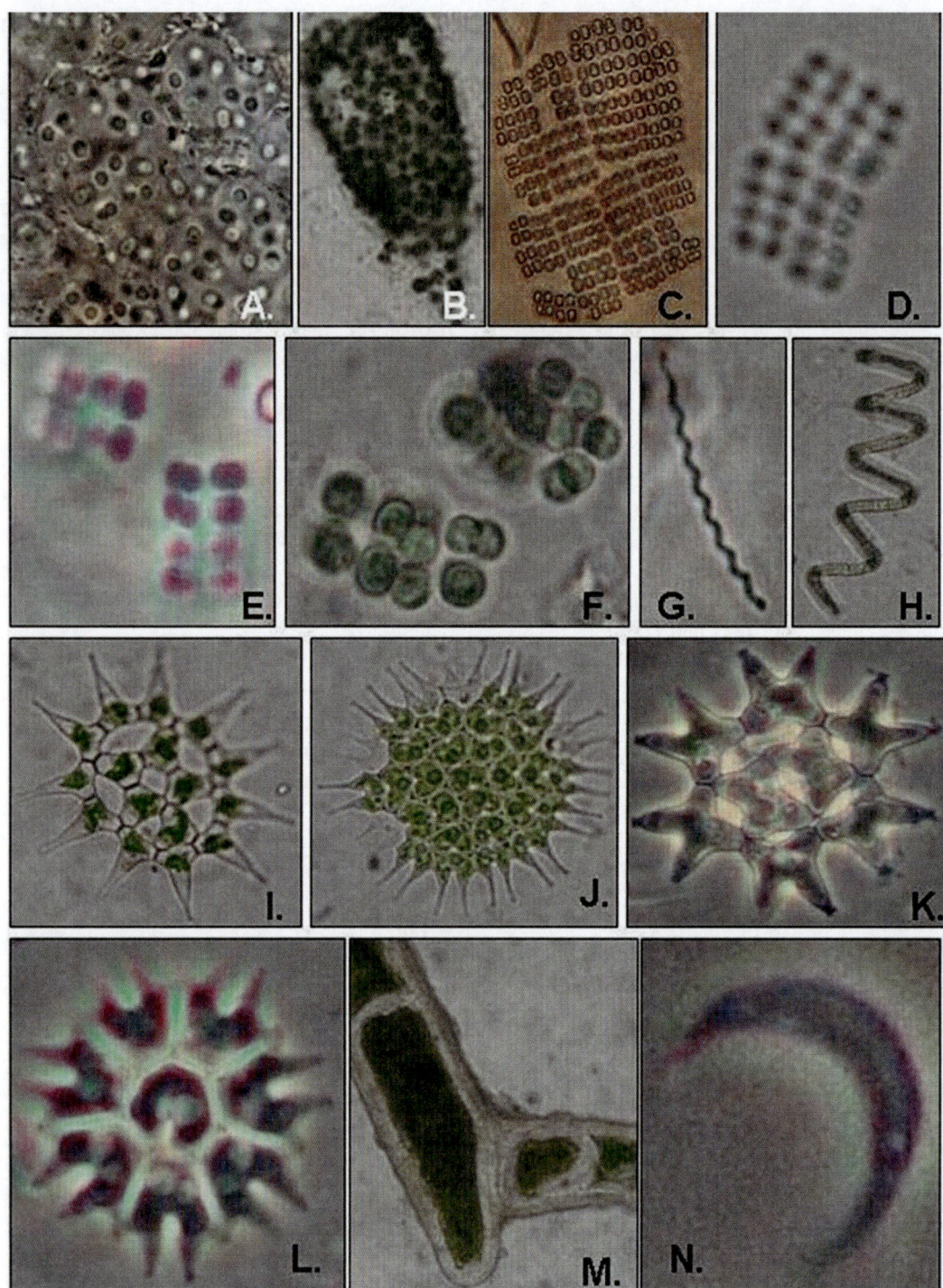

Plate 5.5 (**A**) *Gloeocapsa punctata* Naegeli, (**B**) *Microcystis aeruginosa* Kuetz. *emend.* Elekin, (**C**) *Merismopedia glauca* (Ehrenb.) Naegeli, (**D**) *Merismopedia minima* Beck, (**E**) *Merismopedia punctata* Meyen, (**F**) *Gloeothece rupestris* (Lyngb.) Bornet *in* Wittrock & Nordstedt, (**G**) *Spirulina meneghiniana* Zanard. *ex* Gomont, (**H**) *Spirulina platensis* (Nordst.) Geitler, (**I**) *Pediastrum simplex* (Meyen) Lemmermann, (**J**) *Pediastrum simplex* var. *duodenarium* (Bailey) Rabenhorst, (**K**) *Pediastrum duplex* var. *rotundatum* Lucks, (**L**) *Pediastrum tetras* (Ehrenb.) Ralfs, (**M**) *Rhizoclonium riparium* (Roth) Harvey, (**N**) *Ankistrodesmus falcatus* (Corda) Ralfs

3. *Merismopedia glauca* (Ehrenb.) Naegeli (Plate 5.5, Fig. C)
[Desikachary, 1959, p. 155, pl. 29, Fig. 5]
Colony of 16–64 ovate cells, very regularly arranged to form quadrangular colonies; 5 μ in diameter; 30-celled colony 30 μ wide; cell contents bright blue green, homogenous

4. *Merismopedia minima* Beck (Plate 5.5, Fig. D)
[Desikachary, 1959, p. 154, Pl. 29, Fig 11]
Cells four to many in small colonies, 0.498–0.664 μ broad

5. *Merismopedia punctata* Meyen (Plate 5.5, Fig. E)
[Desikachary, 1959, p. 155, pl. 23, Fig. 5 & pl. 29, Fig. 6]
A rectangular plate of 32–128 ovate cells, usually loosely arranged, sometimes in compact groups of 4–8 individuals, the groups widely separated within a broad, gelatinous envelope; cells 3 μ in diameter; cells contents homogenous, blue green

6. *Gloeothece rupestris* (Lyngb.) Bornet in Wittrock and Nordstedt (Plate 5.5, Fig. F)
[Prescott, 1982, p. 462, Pl. 103, Figs. 2, 3]
Cells ovate, irregularly arranged throughout a copious colourless or brownish gelatinous matrix, in 2's and 4's in small families surrounded by a definite sheath; plant mass free floating, cells 4–6–(9) μ in diameter, 15 μ long

Order – Oscillatoriales
Family – Oscillatoriaceae

7. *Spirulina platensis* (Nordst.) Geitler (Plate 5.5, Fig. H)
Thallus blue green; 6 μ broad, not attenuated at the ends, more or less regularly spirally coiled; spirals 30 μ broad, distances between the spirals 45 μ; cells nearly as long as broad, 6 μ long, cross-walls granulated; end cells broadly rounded

8. *Spirulina meneghiniana* Zanard. ex Gomont (Plate 5.5, Fig. G)
[Desikachary, 1959, p. 195, Pl. 36, Fig 8]
Trichome 1.99 μ broad, flexible, irregularly spirally coiled, forming a thick blue-green thallus, spiral 4.684 μ broad

Family – Hydrodictyaceae

9. *Pediastrum duplex* var. *rotundatum* Lucks (Plate 5.5, Fig. K)
[Prescott, 1982, p. 224, Pl. 48, Fig. 8]
Marginal cells with stout lobes which have convex rather than parallel margins; apices of lobes closer together than in the typical plant

10. *Pediastrum simplex* (Meyen) Lemmermann (Plate 5.5, Fig. I)
[Prescott, 1982, p. 227, Pl. 50, Fig. 2]
Colony entire, composed of 16–32–64 smooth-walled cells, inner cells 5- or 6-sided; peripheral cells with the outer free wall extended to form a single tapering, horn-like process with concave margins; cells 12–18 μ in diameter

11. *Pediastrum simplex* var. *duodenarium* (Bailey) Rabenhorst (Plate 5.5, Fig. J)
[Prescott, 1982, p. 227, Pl. 50, Figs. 4, 5]
Colony perforate, composed of 36 cells with their inner margins concave, the outer margins of inner cells forming a long process, peripheral cells forming a stout process; cells 12 μ in diameter, 27 μ long; 36-celled colony 135 μ in diameter

12. *Pediastrum tetras* (Ehrenb.) Ralfs (Plate 5.5, Fig. L)
[Ralfs, 1844, p. 469]
Colony entire; inner cells (frequently none) with 4–6 straight sides but with one margin deeply incised; peripheral cells crenate with a deep incision in the outer margin, their lateral margins adjoined along $2/3$ of their length; cells 8–12–(16) μ in diameter

Family – Oocystaceae

13. *Ankistrodesmus falcatus* (Corda) Ralfs (Plate 5.5, Fig. N)
[Prescott, 1982, p. 253, Pl. 56, Figs. 5, 6]
Cells somewhat spindle shaped, solitary, not enclosed in a colonial sheath; chloroplast one, a parietal plate without pyrenoids; cells 4 μ in diameter; 65 μ long, sometimes longer

14. *A. falcatus* var. *stipitatus* (Chod) Lemmermann (Plate 5.6, Fig. Q)
[Prescott, 1982, p. 254, Pl. 56, Figs. 14, 15]
Cells lunate (rarely almost straight) attached at one pole to filamentous algae or other submerged aquatics; usually gregarious, forming clusters of 2–8; cells 3 μ in diameter; 20 μ long

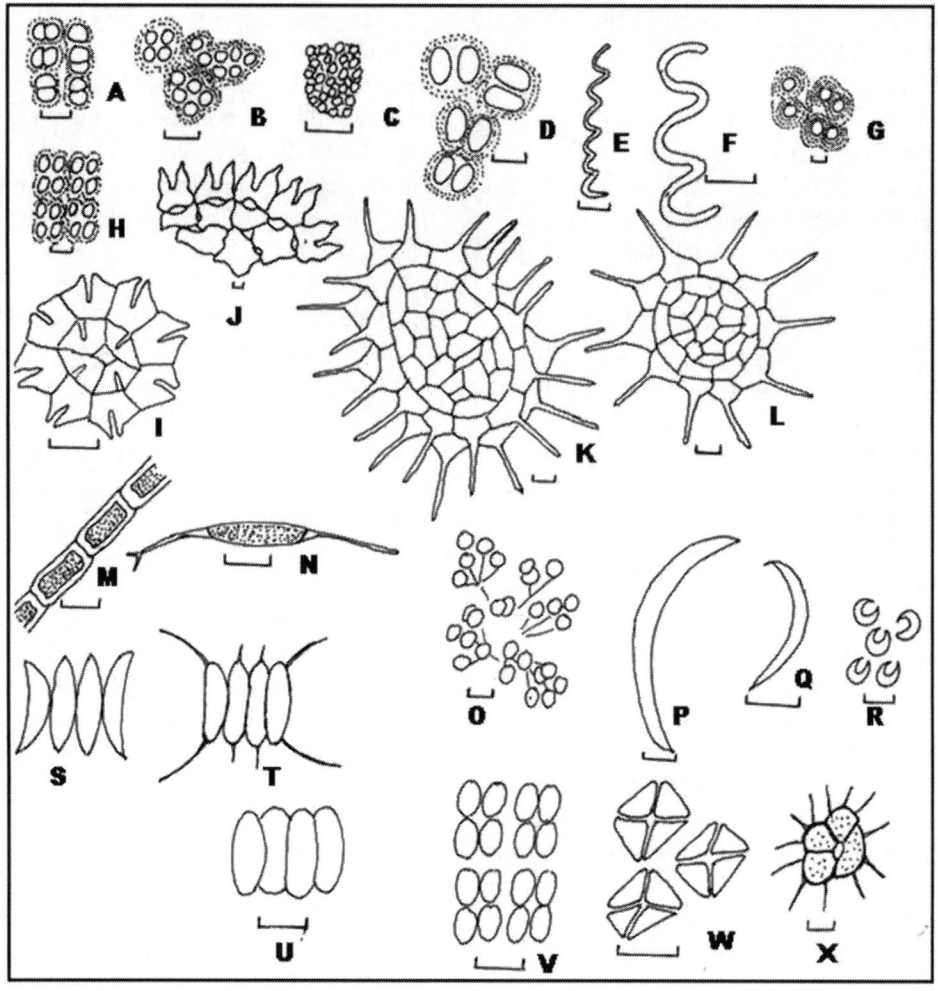

Plate 5.6 Explanation of plates (*Camera lucida* drawings of phytoplankton genera recorded) (**A**) *Gloeocapsa punctata* Naegeli, (**B**) *Merismopedia glauca* (Ehrenb.) Naegeli, (**C**) *Microcystis aeruginosa* Kuetz. emend. Elekin, (**D**) *Merismopedia minima* Beck, (**E**) *Spirulina meneghiniana* Zanard. *ex* Gomont, (**F**) *Spirulina platensis* (Nordst.) Geitler, (**G**) *Gloeothece rupestris* (Lyngb.) Bornet *in* Wittrock & Nordstedt, (**H**) *Merismopedia punctata* Meyen, (**I**) *Pediastrum tetras* (Ehrenb.) Ralfs, (**J**) *Pediastrum duplex* var. *rotundatum* Lucks, (**K**) *Pediastrum simplex* var. *duodenarium* (Bailey) Rabenhorst, (**L**) *Pediastrum simplex* (Meyen) Lemmermann, (**M**) *Rhizoclonium riparium* (Roth) Harvey, (**N**) *Schroederia judayi* G. M. Smith, (**O**) *Dictyosphaerium pulchellum* Wood, (**P**) *Ankistrodesmus falcatus* (Corda) Ralfs, (**Q**) *Ankistrodesmus falcatus* var. *stipitatus* (Chod) Lemmermann, (**R**) *Kirchneriella lunaris* (Kirch.) Moebius, (**S**) *Scenedesmus dimorphus* (Turp.) Kuetzing, (**T**) *Scenedesmus quadricauda* (Turp.) de Brébisson *in* de Brébisson & Godey, (**U**) *Scenedesmus bijuga* (Turp.) Langerheim, (**V**) *Crucigenia rectangularis* (A. Braun) Gay, (**W**) *Crucigenia tetrapedia* (Kirch.) West & West, (**X**) *Tetrastrum staurogeniaeforme* (Schröder) Lemmermann

15. *Dictyosphaerium pulchellum* Wood (Plate 5.6, Fig. O)
[Prescott, 1982, p. 238, Pl. 51, Figs. 5–7]
Colony spherical composed of as many as 32 spherical cells arranged in series of four dichotomously branched threads, inclosed in mucilage; cells 3–10 µ in diameter

16. *Schroederia judayi* Smith (Plate 5.6, Fig. N)
[Prescott, 1982, p. 256, Pl. 57, Figs. 5, 6]

Cells fusiform, straight, the poles narrowed and extended into long setae, one of which terminates into short bifurcations; one chloroplast with single pyrenoid; cells 4 µ in diameter, 55 µ long including the setae which are 13 µ long

17. *Kirchneriella lunaris* (Kirch.) Moebius (Plate 5.6, Fig. R)

[Prescott, 1982, p. 258, Pl. 58, Fig. 2]

Colony composed of numerous cells arranged in groups of 4–16 within a closed, gelatinous envelope; cells flat, strongly curved crescents with rather obtuse points; chloroplast covering the convex wall; cells 5 µ in diameter, 10 µ long; colonies 100–250 µ in diameter

Family – Scenedesmaceae

18. *Scenedesmus bijuga* (Turp.) Langerheim (Plate 5.6, Fig. U)

[Prescott, 1982, p. 276, Pl. 63, Figs. 2, 7]

Colony composed of 2–8 cells in a single (rarely alternate) flat series; cells ovate, without teeth or spines; cells 6 µ in diameter, 12 µ long

19. *Scenedesmus dimorphus* (Turp.) Kuetzing (Plate 5.6, Fig. S)

[Prescott, 1982, p. 277, Pl. 63, Figs. 8, 9]

Colony composed of 4 or 8 fusiform cells arranged in a single series; the inner walls with straight, sharp apices; the outer cells lunate, strongly curved with acute apices; cells 4 µ in diameter; 19 µ long

20. *Scenedesmus quadricauda* (Turp.) de Brébisson in de Brébisson and Godey (Plate 5.6, Fig. T)

[Prescott, 1982, p. 277, Pl. 63, Figs. 8, 9]

Colony consisting of 2-4-8 oblong, cylindrical cells usually in one series (sometimes in two alternate series); outer cells with a long curved spine at each pole; inner cells without spines or with mere papillae at the apices; cells variable in size, 12 µ in diameter, 20 µ long

21. *Crucigenia rectangularis* (A. Braun) Gay (Plate 5.6, Fig. V)

[Prescott, 1982, p. 285, Pl. 65, Figs. 7, 8]

Colony free floating consisting of ovate cells, very regularly arranged about a rectangular central space in two pairs, with the apices adjoining; cells 5.5 µ in diameter, 8 µ long

22. *Crucigenia tetrapedia* (Kirch.) West & West (Plate 5.6, Fig. W)

[Prescott, 1982, p. 285, Pl. 65, Fig. 9; Pl. 66, Fig. 1]

Colony free floating consisting of four triangular cells cruciately arranged about a minute central space; outer free wall and lateral walls straight, the angles acutely rounded; cells 6 µ in diameter; frequently forming a rectangular plate of 16 cells (four quartets)

23. *Tetrastrum staurogeniaeforme* (Schröder) Lemmermann (Plate 5.6, Fig. X)

[Pl. 66, Fig 3, Pg 286, Prescott]

A colony of 4 triangular cells, cruciately arranged about a small rectangular space; lateral margins of the cell straight, the outer free walls convex and furnished with hairlike setae; cells 3.81–4.15 µ in diameter, colony 10.74 µ width, setae 4.15–4.98 µ long

Order – Cladophorales

Family – Cladophoraceae

24. *Rhizoclonium riparium* (Roth) Harvey (Plate 5.6, Fig. M)

[Krishnamurthy, 2000, pl. 5, fig. 2]

It forms a membranaceous layer; the algae is soft and woolly to touch and intricately woven into loosely lying dense mats. The colour of the algae is bright grass green. The diameter of the filament is 60.35 µ. The length of the cell varies from 89.81 to 97.96 µ.

Division – Bacillariophyceae
Section – Centricae
Subfamily – Discoidae
Tribe – Coscinodisceae
Subtribe – Melosirinae

25. *Aulacoseira granulata* var. *angustissima* (Plate 5.7, Fig. Z)

[*Melosira granulata* (Ehr.) Ralfs var. *angustissima* Müll, Hustedt, 1930a, Bd VII, Teil 1, pp. 250, Fig. 104 d; Venkataraman, 1939, p. 297, Fig. 2]

Filaments long with narrow and long cells, the height of the cells being several times the diameter, diameter 5 µ, height of half cell 12 µ; number of punctae in the lower cell is 10 in 10 µ

Subtribe – Skeletoneminae

26. *Skeletonema costatum* (Greville) Cleve (Plate 5.7, Fig. X)

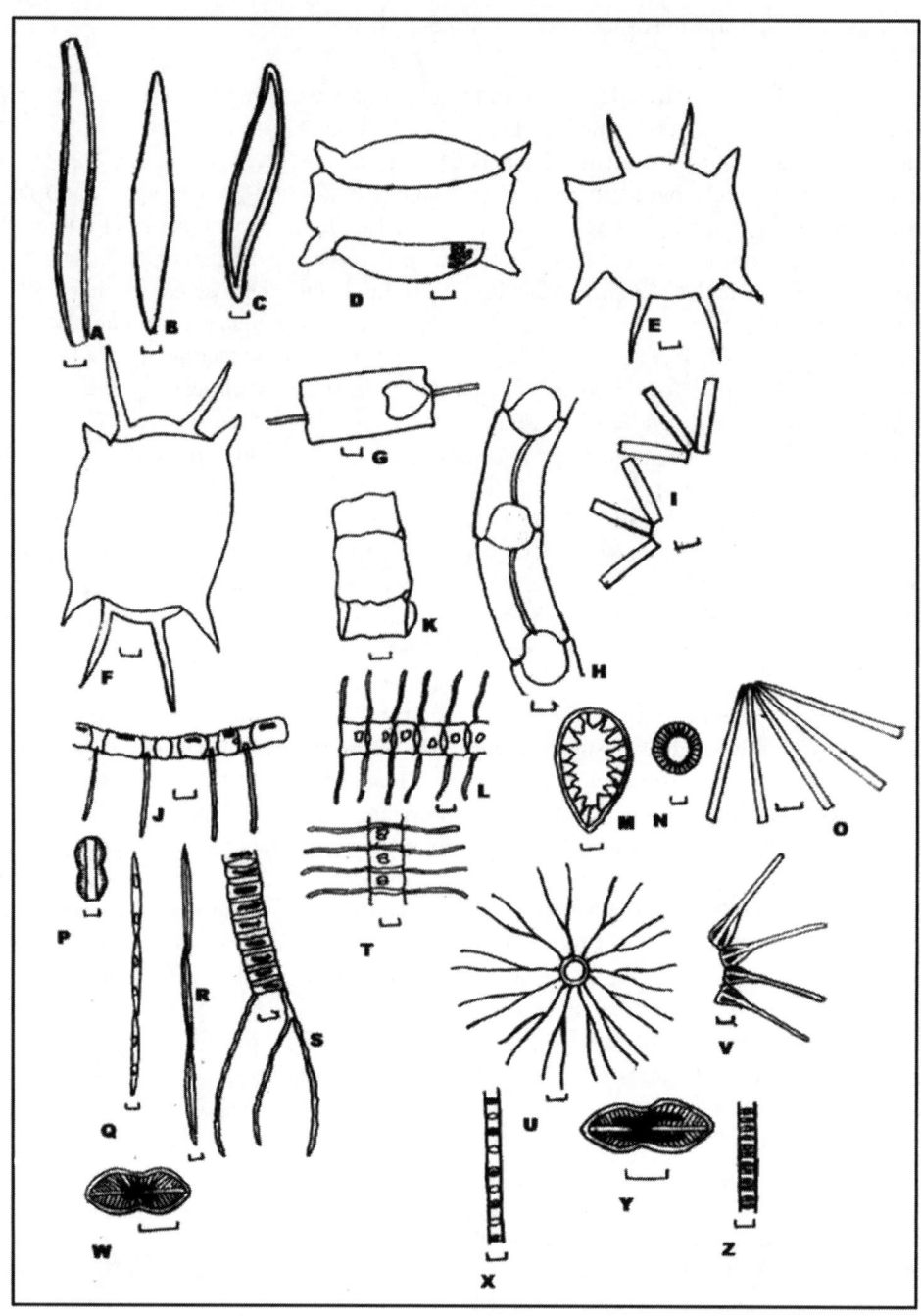

Plate 5.7 (**A**) *Nitzschia sigmoidea* Ehrenberg, (**B**) *Pleurosigma normanii* Ralfs, (**C**) *Gyrosigma obtusatum* (Sull. & Wormley) Boyer, (**D**) *Odontella rhombus* (Ehrenb.) Kützing, (**E**) *Odontella mobiliensis* (J. W. Bailey) Grunow, (**F**) *Biddulphia heteroceros* Grunow, (**G**) *Ditylum brightwellii* (West) Grunow, (**H**) *Eucapmia zoodiacus* Ehrenberg, (**I**) *Thalassionema nitzschoides* Grunow, (**J**) *Chaetoceros curvisetus* Cleve, (**K**) *Biddulphia alternans* (Bailey) van Heurck, (**L**) *Chaetoceros messanensis* Castracane, (**M**) *Surirella fastuosa* var. *recedens* (A. Schmidt) Cleve, (**N**) *Cyclotella meneghiniana* Kutzing, (**O**) *Thalassiothrix frauenfeldii* Grunow, (**P**) *Nitzschia bilobata* var. *minor* Grunow, (**Q**) *Nitzschia pacifica* n. sp., (**R**) *Nitzschia delicatissima* Cleve, (**S**) *Chaetoceros diversus* Cleve, (**T**) *Chaetoceros wighami* Brightwell, (**U**) *Bacteriastrum varians* Lauder, (**V**) *Asterionella japonica* Cleve, (**W**) *Diploneis weissflogii* (A. Schmidt) Cleve, (**X**) *Skeletonema costatum* (Greville) Cleve, (**Y**) *Diploneis interrupta* (Kutzing) Cleve, (**Z**) *Aulacoseira granulata* var. *angustissima*

[Hustedt, Bd. VII, Teil 1, p. 311, Fig. 149; Subrahmanyan, 1946, p. 89, Fig. 7, 8 and 9]

Frustules weakly silcified, lens shaped with rounded ends forming long slender straight chains with the aid of marginal spines which run parallel to the axis of the chain, space between the cells longer than the cells, chromatophores two plates which at times dissected, no visible structures on the valve, diameter of cells 10 µ

27. *Thalassiosira decipiens* (Grunow) Jørgensen (Plate 5.8, Fig. C)

[Subrahmanyan, 1946, p. 89, Fig. 19]

Cells disc shaped, valves flat with minute spines along the border, valve aerolated, areolae in three or more systems, their size becoming smaller towards the border, in the centre, about 12 in 10 µ and towards border 15 in 10 µ; diameter 60 µ

Subtribe – Coscinodiscinae

28. *Cyclotella meneghiniana* Kutzing (Plate 5.7, Fig. N)

[Venkataraman, 1939, pp. 299, Figs. 11 and 14; Subrahmanyan, 1946, p. 92, Figs. 25, 26 and 27]

Frustules discoid in valve view, margin well defined, coarsely striated and the striae wedge shaped, striae 8 in 10 µ, diameter 15 µ

29. *Cyclotella striata* (Kutzing) Grunow (Plate 5.8, Fig. F)

[Subrahmanyan, 1946, p. 92, Fig. 31]

Cells disc shaped, 20 µ in diameter, valves with more or less broad evenly striated border, striae 10–12 in 10 µ. Central portion with plexes and coarsely punctuate

30. *Coscinodiscus granii* Gough (Plate 5.8, Fig. B)

[Hustedt, 1930b, Bd. VII, Teil 1, pp. 436, Fig. 237; Venkataraman, 1939, pp. 300, Figs. 16 and 17; Cupp, 1943, pp. 56, Fig. 21; Subrahmanyan, 1946, p. 96, Figs. 33, 35 and 39]

Cells with excentric arched valves, central areolae in definite rosette. Eight aerolae in 10 µ near centre, 10 midway to margin and 11 near margin; on edge of valve mantle 13 in 10 µ, chamber openings small, dot-like, diameter 105 µ.

31. *Coscinodiscus excentricus* Ehrenberg (Plate 5.8, Fig. A)

[Hustedt 1930b, Bd. VII, Teil 1, pp. 388, Fig. 201; Cupp, 1943, pp. 52, Fig. 14; Subrahmanyan, 1946, p. 93, Figs. 29 and 30]

Cells disc shaped., valves almost flat, narrow margin, areolae hexagonal, arranged in slightly curved, nearly parallel rows, based on arrangement of seven divisions, central areola with seven areolae grouped around it, areolae 7 in 10 µ at centre, 9 midway and 10 near margin, chromatophores small, numerous, diameter 65 µ

32. *Coscinodiscus excentricus* Ehrenberg var. *fasciculata* Hustedt (Plate 5.8, Fig. E)

[Hustedt, 1930b, Bd. VII, Teil 1, p. 390, Fig. 202; Subrahmanyan, 1946, p. 93, Figs. 32 and 38]

Cells disc shaped, valve aerolated, areolae in several tangential series and because of this appearing as though in radial bundles, number of aerolae in the centre 9 in 10 µ and at the border 12 in 10 µ, diameter 65 µ

33. *Actinocyclus normanii* f. *subsala* (Juhl.-Dannf.) Hustedt (www.itis.gov) (Plate 5.8, Fig. D)

[*Coscinodiscus radiatus* Ehrenberg Cupp, 1943, pp. 56, Fig. 20]

Cells flat, coin-shaped discs, valves flat, valve surface with coarse areolae, without rosette or central area, areolae nearly same size on whole valve, 4 in 10 µ, except at margin where they are smaller, 6 in 10 µ, inner chamber openings rather indistinct, outer membrane of the areolae apparently homogeneous, diameter 50 µ

Suborder – Solenoideae

Family – Solenieae

Subfamily – Lauderiineae

34. *Leptocylindrus danicus* Cleve (Plate 5.8, Fig. G)

[Subrahmanyan, 1946, p. 113, Figs. 109, 110]

Cells cylindrical, 10 µ in diameter and 50–80 µ in length, forming long chains. No structure visible on the valve. Chromatophores numerous and disc shaped.

Subfamily – Biddulphioideae

Tribe – Chaetocereae

35. *Bacteriastrum delicatulum* Cleve (Plate 5.8, Fig. H)

[Hustedt, 1930b, Bd. VII, Teil 1, p. 612, Fig. 353; Subrahmanyan, 1946, p. 125, Figs. 161–163]

Cells longer than broad. Setae eight, perpendicular to chain axis, basal part long. Apertures

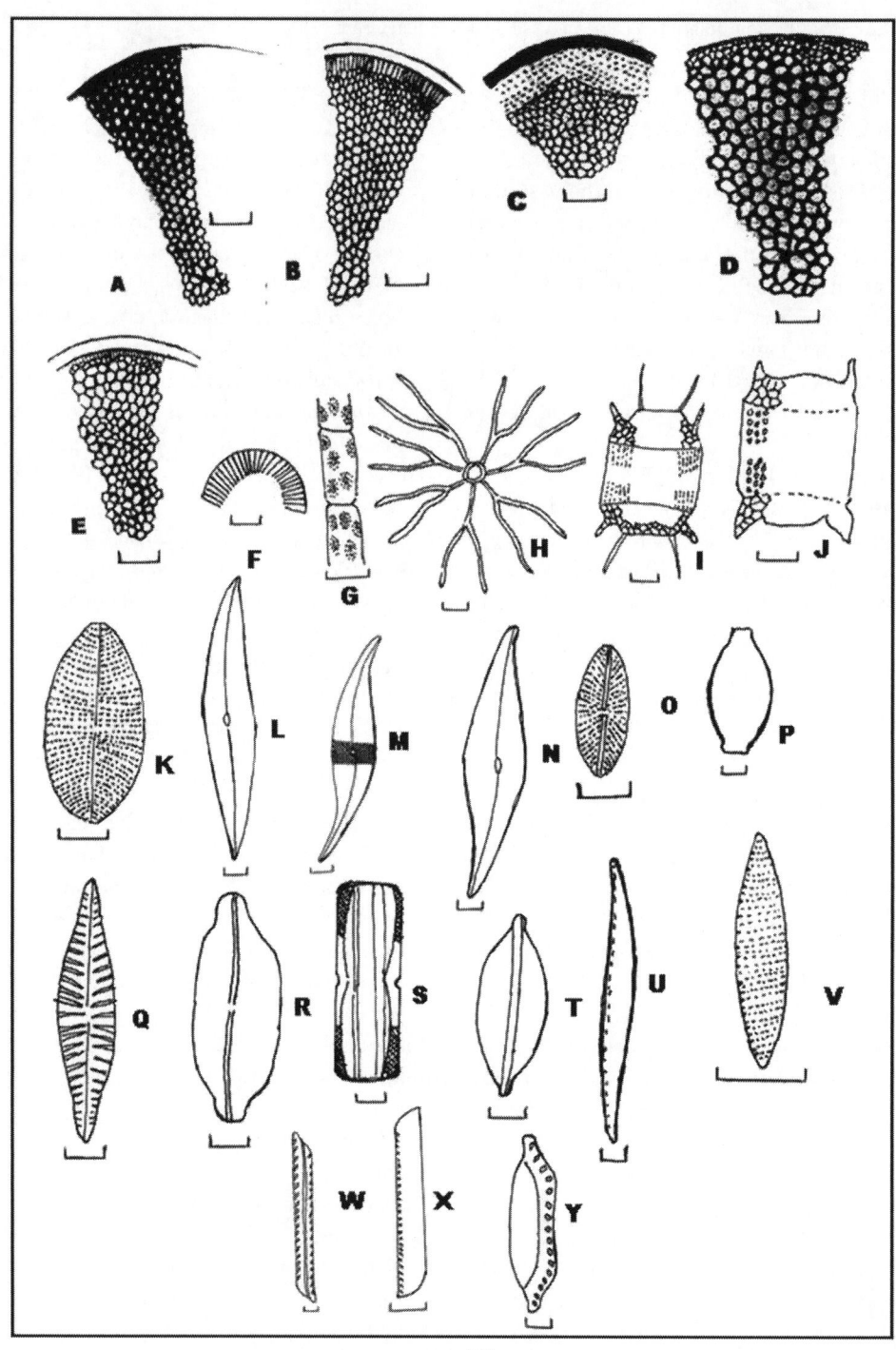

Plate 5.8 (**A**) *Coscinodiscus excentricus* Ehrenberg, (**B**) *Coscinodiscus granii* Gough, (**C**) *Thalassiosira decipiens* (Grunow) Jørgensen, (**D**) *Actinocyclus normanii* f. *subsala* (Juhl.-Dannf.) Hustedt, (**E**) *Coscinodiscus excentricus* Ehrenberg var. *fasciculata* Hustedt, (**F**) *Cyclotella striata* (Kutzing) Grunow, (**G**) *Leptocylindrus danicus* Cleve, (**H**) *Bacteriastrum delicatulum* Cleve, (**I**) *Odontella aurita*(Lyngb.) C. Agardh, (**J**) *Biddulphia dubia* (Brightwell) Cleve, (**K**) *Cocconeis dirupta* Greg., (**L**) *Gyrosigma beaufortianum* Hust., (**M**) *Gyrosigma acuminatum* (Kütz.) Rabenh., (**N**) *Pleurosigma salinarum* Grunow, (**O**) *Navicula minima* Grunow, (**P**) *Navicula mutica* Kutz. fo. *cohni* (Hilse.) Grunow, (**Q**) *Navicula peregrina* (Ehrenberg) Kutzing, (**R**) *Navicula quadripartita* Hustedt, (**S**) *Amphiprora gigantea* Grunow, (**T**) *Cymbella naviculiformis* Allerswald, (**U**) *Bacillaria paxillifer* (O. F. Müller) Hendy, (**V**) *Nitzschia amphibia* Grunow, (**W**) *Nitzschia ignorata* Kresske, (**X**) *Nitzschia obtusa* Smith, (**Y**) *Nitzschia punctata* W. Smith

large. Terminal setae bent over the chain. Diameter of cell 11 μ

36. *Bacteriastrum varians* Lauder (Plate 5.7, Fig. U)

[Subrahmanyan, 1946, p. 127, Figs. 170–172 and 175]

Cells cylindrical, setae 19, bifurcated at about middle and extended up to the tip, at right angles to the chain axis, terminal setae with fine spines arranged in spiral rows, 12 μ in diameter

37. *Chaetoceros curvisetus* Cleve (Plate 5.7, Fig. J)

[Hustedt, 1930b, Bd. VII, Teil 1, pp. 737, Fig. 426; Cupp, 1943, pp. 137, Fig. 93 (pp. 138); Subrahmanyan, 1946, p. 143, Figs. 238, 244–246]

Chains spirally curved. No distinct end cell. Apical axis of cell measuring 12 μ. Cells in broad girdle view oblong, setae starting from the corners. Aperture somewhat broadly elliptical. Setae all directed towards one side of the chain. Chromatophores a single plate with pyrenoid

38. *Chaetoceros diversus* Cleve (Plate 5.7, Fig. S)

[Hustedt, 1930b, Bd. VII, Teil 1, pp. 716, Fig. 409; Subrahmanyan, 1946, p. 142, Figs. 235, 241–243]

Cells with apical axis measuring 5 μ in length, forming straight chains which are usually short, apertures very small, setae, some hairlike; others thicker, tubular and spinous

39. *Chaetoceros messanensis* Castracane (Plate 5.7, Fig. L)

[Hustedt, 1930b, Bd. VII, Teil 1, pp. 718, Fig. 410; Subrahmanyan, 1946, p. 142, Figs. 236, 237 and 240]

Cells forming long straight chains, apical axis 12.5 μ in length, the corners of adjacent cells touching each other, apertures elliptical, bristles thin, chromatophore a single plate

40. *Chaetoceros wighami* Brightwell (Plate 5.7, Fig. T)

[Hustedt, 1930b, Bd. VII, Teil 1, pp. 724, Fig. 414; Cupp, 1943, pp. 136, Fig. 91; Subrahmanyan, 1946, p. 142, Fig. 247]

Cells somewhat tender forming chains, apical axis measuring 10 μ, cells in broad girdle view oblong with sharp corners, the corners of neighbouring cells touching each other and enclosing a narrow slitlike aperture, setae thin and fragile, inner ones perpendicular to the chain axis; end setae more or less parallel to the chain axis, chromatophore platelike

Tribe – Biddulphieae

Subtribe – Eucampiinae

41. *Eucapmia zoodiacus* Ehrenberg (Plate 5.7, Fig. H)

[Cupp, 1943, pp. 145, Fig. 103; Subrahmanyan 1946, p. 145, Figs. 248, 250–253]

Cells flattened, elliptical–linear in valve view, united in chains by two blunt processes, length of cell along apical axis 25 μ, chains spirally curved, with relatively narrow lanceolate or elliptical apertures, apertures variable in size and shape, valves distinctly sculptured, chromatophores small, numerous

Subtribe – Triceratiinae

42. *Ditylum brightwellii* (West) Grunow (Plate 5.7, Fig. G)

[Cupp, 1943, pp. 148, Fig. 107A, 107B; Subrahmanyan, 1946, p. 147, Figs. 263 and 264]

Cells prism shaped with strongly rounded angles to nearly cylindrical, usually 3–5 times as long as broad, valves triangular to circular with a central hollow spine, valve rim strengthened by small parallel ribs, girdle zone very long, chromatophores small, numerous, nucleus central, diameter 74 μ

Subtribe – Biddulphiinae

43. *Biddulphia alternans* (Bailey) van Heurck. (Plate 5.7, Fig. K)

[Cupp, 1943, p. 166, Fig. 115; Subrahmanyan, 1946, pp. 153, Figs. 277 and 282 (*Triceratium alternans*)]

Valves quadrangular, with straight or somewhat unevenly concave sides. Corners slightly elevated, rounded, separated from the central part by costae or ribs. Only a slight constriction between valve and girdle zones. Irregular ribs on both valve and girdle. Fine areolae (slime pores) on corners, 17 in 10 μ. Areolae 9 in 10 μ in centre of valve. Length along side of valve 30 μ

44. *Odontella aurita* (Lyngb.) C. Agardh (Plate 5.8, Fig. I)

[*Biddulphia aurita* (Lyngbye) Brébisson and Godey, Cupp, 1943, p. 161, Fig. 112-A (1), 112-A (2), 112-A (3)]

Valves elliptical–lanceolate, with obtuse processes inflated at the base. Centre part of valve convex, more or less flattened at the top from which long spines project. Girdle zone sharply differentiated from the valve zone by a clear depression. Cell wall strongly siliceous, areolated punctuated. Areolae 8–10 in 10 µ, on the valve in radial rows. On the girdle band in pervalvar rows, 7–10 rows in 10 µ, with 8–14 punctae

45. *Biddulphia dubia* (Brightwell) Cleve (Plate 5.8, Fig. J)

[Cupp, 1943, p. 164, Fig. 114; *Triceratium dubium* Brightwell, Subrahmanyan, 1946, p. 151, Figs. 274–276 and 278]

Valves rhombic–lanceolate. Processes obtuse. Centre part of valve convex. Valve with numerous small spines and a larger one near base of each process. Valve zone and girdle zone divided by a deep groove. Valves distinctly punctate, 9 puncta in 10 µ. Length of apical axis 54 µ

46. *Biddulphia heteroceros* Grunow (Plate 5.7, Fig. F)

[Subrahmanyan, 1946, p. 155, Figs. 288, 298 and 303]

Cells box-shaped without a sharp constriction between valve and girdle in girdle view. Horns from each pole of apical axis well developed, directed slightly away from pervalvar axis. Two strong spines on each valve a short distance from the horns. Valve between spines slightly higher that between spines and horns somewhat flat. Areolation almost the same size on valve and girdle, in regular rows, 9 in 10 µ

47. *Odontella mobiliensis* (J. W. Bailey) Grunow [www.itis.gov] (Plate 5.7, Fig. E)

[*Biddulphia mobiliensis* Bailey, Hustedt, 1930b, Bd. VII, Teil 1, p. 840, Fig. 495; Subrahmanyan, 1946, p. 155, Figs. 286, 287, 291–296 and 299, Pl. II, Figs. 1 and 2]

Cells single, length of apical axis 60 µ, valves elliptical–lanceolate, convex, with a flat or nearly flat central part, valve processes slender, directed diagonally outward, two long spines placed far apart but about equally far from the processes, directed obliquely outward, straight, cells relatively thin walled, without a sharp constriction between valve and girdle zone, sculpturing fine, reticulate, 15 areolae in 10 µ on valve and valve mantle.

48. *Odontella rhombus* (Ehrenb.) Kützing [www.itis.gov] (Plate 5.7, Fig. D)

[*Biddulphia rhombus* Ehrenberg, Hustedt, 1930b, Bd. VII, Teil 1, p. 842, Fig. 496, 497; Cupp, 1943, p. 163, Fig. 113]

Valves rhombic–elliptical, cells thick walled, strongly sculptured, processes small, short and obtuse, length of apical axis 35 µ, surface of valve convex, beset with small spines over entire surface, valve zone and girdle zone divided by deep groove, areolae on valve 9 in 10 µ, irregular at centre, then more or less regular radiating, girdle band more delicately sculptured 12 areolae in 10 µ in regular prevalvar rows

Section – Pennatae
Subsection – Araphideae
Subfamily – Fragilariodeae
Tribe – Fragilarieae
Subtribe – Fragilariinae

49. *Thalassionema nitzschoides* Grunow (Plate 5.7, Fig. I)

[Cupp, 1943, pp. 182, Fig. 133; Subrahmanyan, 1946, pp. 167, figs. 344–346]

Frustules united into zigzag chains. Cells in girdle view linear–rectangular, in valve view linear–lanceolate, poles alike, 50 µ long, 3 µ broad

50. *Thalassiothrix frauenfeldii* Grunow (Plate 5.7, Fig. O)

[Cupp, 1943, pp. 184, Fig. 135; Subrahmanyan, 1946, pp. 169, figs. 349, 351, 354–357 and 360]

Cells united into star-shaped colonies. In girdle view linear. Valves very narrow, linear, ends distinct, one-end blunt rounded, near the other end usually widened then decreased to form a wedge-shaped point. Valves 53 µ long, 7 µ wide. Valves structure less

51. *Asterionella japonica* Cleve (Plate 5.7, Fig. V)

[Venkataraman, 1939, pp. 309, Fig. 34; Cupp, 1943, pp. 188, Fig. 138; Subrahmanyan, 1946, pp. 170, figs. 361 and 371]

Cells united into starlike spiral colonies. In girdle view, very narrow linear with parallel sides, with greatly enlarged three-cornered region at the base. Cells united at the corners of enlarged region. Valves very narrow, then with a widened knob-like region at the base. Length of valve 85 µ; length of enlarged region 15 µ, width of enlarged part 6 µ. Chromatophores ½ small plates, in basal enlarged part only

Suborder – Monoraphidineae
Family – Achnanthaceae
Subfamily – Cocconeoideae

52. *Cocconeis dirupta* Greg. (Plate 5.8, Fig. K)
[Gregory, 1857, pp. 491, pl. 9, fig. 25]
Cells broadly elliptic, about 40 µ long and 30 µ broad, striae 21 in 10 µ. Raphe sigmoid. Axial area narrow, dilating into a very small central area

Subsection – Biraphideae
Subfamily – Naviculoideae
Tribe – Naviculeae

53. *Gyrosigma beaufortianum* Hust. (Plate 5.8, Fig. L)
[Foged, 1978, p. 73, 21: 8]
Valves slightly sigmoid with acute ends. Raphe central, sigmoid, central area small, rhombic, length 55.025 µ and breadth 6 µ

54. *Gyrosigma acuminatum* (Kütz.) Rabenh. (Plate 5.8, Fig. M)
[*Gyrosigma spencerii* (Quekett) Cleve, Cupp, 1943, p. 194, Fig. 144]
Valves sigmoid, lanceolate, with obtuse ends. Raphe central. Central nodule elliptical. Transverse striae 18 in 10 µ; longitudinal striae 25 in 10 µ. Length of valves 185 µ

55. *Gyrosigma obtusatum* (Sull. & Wormley) Boyer (Plate 5.7, Fig. C)
[*Gyrosigma scalproides* (Rabh.) Hustedt, 1930b, Heft 10, pp. 226, Fig. 339; Venkataraman, 1939, pp. 319, fig. 76]
Valves linear with parallel sides and obliquely rounded ends. Raphe straight, nearly central, slightly sigmoid at the ends. Longitudinal striae very faint. Length of valve along apical axis 76 µ and breadth 8 µ

56. *Pleurosigma normanii* Ralfs (Plate 5.7, Fig. B)
[Cupp, 1943, pp. 196, Fig. 148; Subrahmanyan, 1946, pp. 175, figs. 378, 379, 385 and 387]
Valves broadly lanceolate, slightly sigmoid, with subacute ends. Raphe nearly central, sigmoid. Length of valve 115 µ

57. *Pleurosigma salinarum* Grunow (Plate 5.8, Fig. N)
[Hustedt, 1930b, Heft 10, p. 228, Fig. 344; Venkataraman, 1939, p. 320, fig. 78]
Valves linear lanceolate, slightly sigmoid. Raphe central. Central nodule elongated. Length 120 µ, breadth 17 µ

58. *Diploneis weissflogii* (A. Schmidt) Cleve (Plate 5.7, Fig. W)
[Subrahmanyan, 1946, pp. 177, fig. 397]
Valves strongly constricted, with sub-elliptical ends, 28 µ long and 12.5 µ broad, at the constriction 8 µ broad. Central nodule with approximate horns.

59. *Diploneis interrupta* (Kutzing) Cleve (Plate 5.7, Fig. Y)
[Hustedt, 1930b, Heft 10, pp. 252, Fig. 400; Venkataraman, 1939, pp. 323, fig. 82]
Valves deeply constricted, the segments elliptical, rounded at the ends. Central nodule elongated, quadrate with parallel horns. Furrow linear, narrow. Costae strong usually interrupted in the middle of the valve. Length of the valve 35 µ, breadth in the middle 10 µ

60. *Navicula minima* Grunow (Plate 5.8, Fig. O)
[Hustedt 1930b, pp. 272, fig. 441]
Valves are lanceolate in shape, with tapering ends. Striations are very fine. Length 13.96 µ, breadth 3.49 µ

61. *Navicula mutica* Kutz. fo. cohni (Hilse.) Grunow (Plate 5.8, Fig. P)
[Foged 1978, pp. 93, pl. XXVIII, fig. 10]
Valves are elliptic lanceolate in shape, with rounded ends. Striations fine and delicate. Length 13.96 µ, breadth 6.98 µ

62. *Navicula peregrina* (Ehrenberg) Kutzing (Plate 5.8, Fig. Q)
[Hustedt, 1930b, Heft 10, p. 300; Venkataraman, 1939, p. 326, fig. 85]
Valves lanceolate with obtuse ends. Axial area distinct, narrow, central area broadened, elliptical, length 62.12 µ and breadth 14.9 µ

63. *Navicula quadripartita* Hustedt (Plate 5.8, Fig. R)
[Foged, 1979, pl. XXIX, Fig. 18]
Valves elliptical with broadly rounded ends. Central area widened, striations radial; the number of striae in 10 µ is 14, length 17.92 µ, breadth 5.31 µ

Subfamily – Amphiproroideae

64. *Amphiprora gigantea* Grunow (Plate 5.8, Fig. S)
[Subranmanyan, 1946, p. 184, Figs. 410 and 413]

Cells strongly constricted. Keel with hyaline margin. Junction line curved like a bow. Cells 75 µ long. Keel punctae forming obliquely decussating rows, 15 rows in 10 µ. Connecting zone with numerous longitudinal divisions

Subfamily – Gomphocymbelloideae

65. *Cymbella naviculiformis* Allerswald (Plate 5.8, Fig. T)

[Hustedt, 1930b, Heft 10, p. 356, Fig. 653; Venkataraman, 1939, p. 346, Fig. 119]

Valve elliptic lanceolate with capitate ends. Axial area narrow, linear, suddenly dialated in middle. Striations radial. Length 30.04 µ, breadth 10.12 µ, number of striations 13 in 10 µ

Subfamily – Nitzschioideae

Tribe – Nitzschieae

66. *Bacillaria paxillifer* (O. F. Müller) Hendy (Plate 5.8, Fig. U)

[*Bacillaria paradoxa* Gmelin, Subrahmanyan, 1946, p. 187, Figs. 417, 421 and 427]

Cells in girdle view linear and rectangular, united by their valves to form a mat-like colony, individual cells of which exhibit gliding movement in the living condition. Valves linear, spindle shaped in outline, 150 µ long, 8 µ broad. Keel punctae 7 in 10 µ. Transapical striae fine, 21 in 10 µ

67. *Nitzschia amphibia* Grunow (Plate 5.8, Fig. V)

[Hustedt, 1930b, Heft 10, p. 414, Fig. 793; Venkataraman, 1939, p. 353, Fig. 149]

Valves linear to linear lanceolate with the ends slightly produced and sometimes rounded. Striations coarse. Length 17.45 µ, breadth 3.49 µ, striae 15 in 10 µ

68. *Nitzschia bilobata* var. *minor* Grunow (Plate 5.7, Fig. P)

[Cupp, 1943, p. 200, Fig. 152]

Valves linear lanceolate, constricted in the middle, apiculate at the ends. Keel puncta 12 in 10 µ. Striae 25 in 10 µ. Cells broad, oblong, truncate, constricted in the middle. Length of the valve 55 µ, width of widest point 6 µ

69. *Nitzschia ignorata* Kresske (Plate 5.8, Fig. W)

[Foged, 1979, p. 215, Pl. XLII, Fig. 6]

Frustules broad, linear with ends obliquely truncate. Striations punctate, punctae very fine and linear, number of striations 10 in 10 µ. Length 44 µ, breadth 4.5 µ

70. *Nitzschia obtusa* Smith (Plate 5.8, Fig. X)

[Hustedt, 1930b, Heft 10, p. 422, Fig. 817; Venkataraman, 1939, p. 354, Fig. 147]

Frustules broad, linear with ends obliquely truncate. Striations punctate, punctae fine and linear. Length 98 µ and breadth 10 µ

71. *Nitzschia punctata* W. Smith (Plate 5.8, Fig. Y)

[Foged 1979, p. 209, pl. XL, Fig. 16]

Valves linear with slightly concave margins. Ends wedge shaped and rounded. Striae clear, 10 in 10 µ. Length 22.908 µ, breadth 5.312 µ

72. *Nitzschia sigmoidea* Ehrenberg (Plate 5.7, Fig. A)

[Hustedt, 1952, p. 147]

Cells solitary. Frustules isopolar, straight in valve view but sigmoid in girdle view; bilaterally symmetrical valves linear lanceolate, straight. The sigmoid shape of the frustule is entirely due to the form of the valve margin and girdle. Two chloroplasts per cell. Length of valve 172.5 µ

73. *Nitzschia delicatissima* Cleve (Plate 5.7, Fig. R)

[Cupp, 1943, p. 204, Fig. 158 (p. 206)]

Valve narrow, linear, acute. Cells united into stiff hairlike chains by the overlapping tips of the cells. Chains usually short, motile. Length of valve 80 µ; width 3 µ. Keel puncta 20 in 10 µ. No striae visible. Chromatophores two per cell

74. *Nitzschia pacifica* n. sp. (Plate 5.8, Fig. Q)

[Cupp, 1943, p. 204, Fig. 157]

Cells spindle shaped with more or less pointed ends. United into stiff chains by the overlapping points of the cells. Chains motile as a whole. Length of valve 94 µ, width 6 µ. Keel puncta distinct, 14 in 10 µ. Chromatophores two per cell

Subfamily – Surirelloideae

Tribe – Surirelleae

75. *Surirella fastuosa* var. *recedens* (A. Schmidt) Cleve (Plate 5.7, Fig. M)

[Cupp, 1943, p. 208, Fig. 160]
Cells wedge shaped, rounded at the angles. Valves ovate. Costae or ribs about 2.5 in 10 µ, robust, dialated at the margin. Central space lanceolate. Marginal striae distinct, 18 in 10 µ. Striae in central field 17 in 10 µ. Length of valve 60 µ, width 35 µ (Plate 5.13, Fig. N)

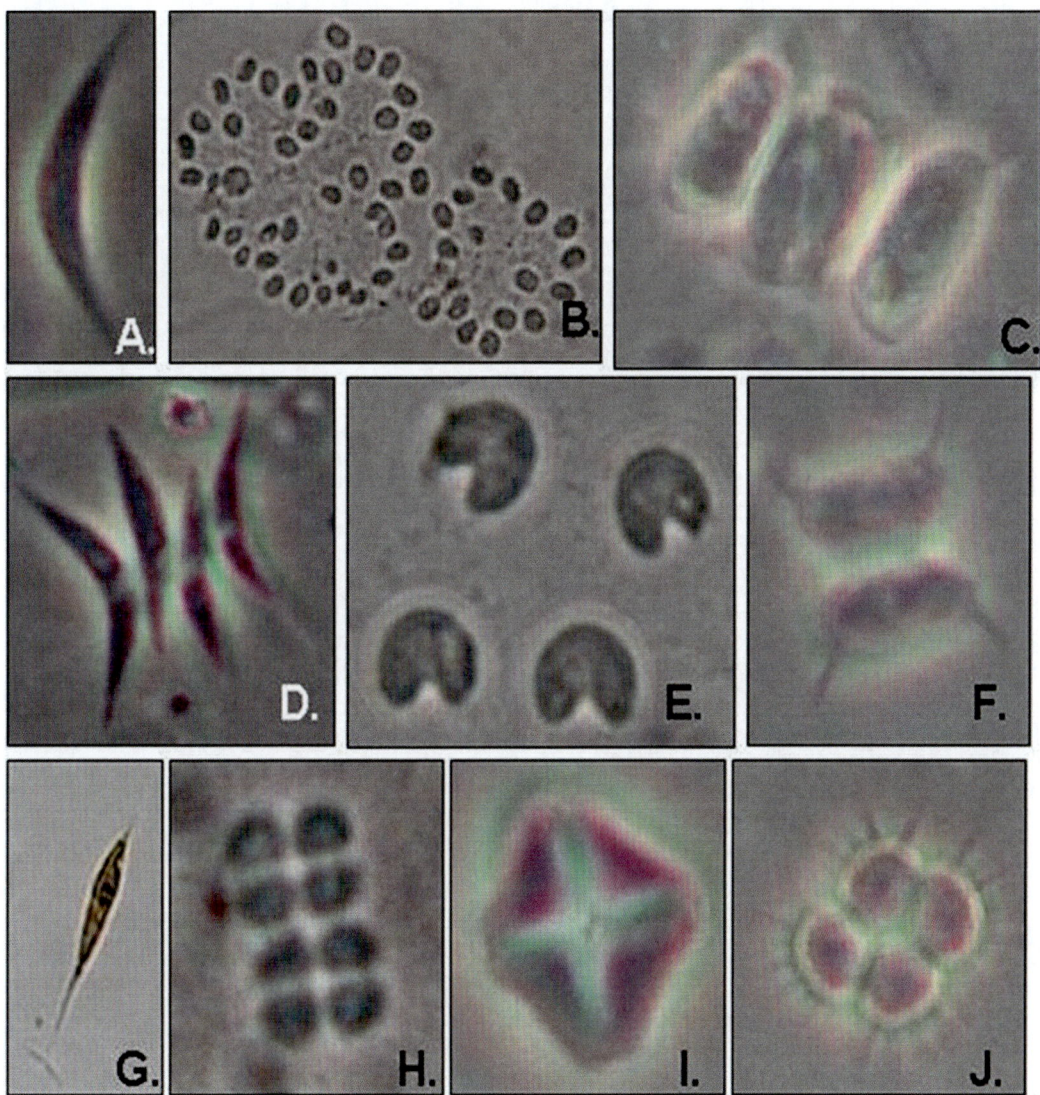

Plate 5.9 (**A**) *Ankistrodesmus falcatus* var. *stipitatus* (Chod) Lemmermann, (**B**) *Dictyosphaerium pulchellum* Wood, (**C**) *Scenedesmus bijuga* (Turp.) Langerheim, (**D**) *Scenedesmus dimorphus* (Turp.) Kuetzing, (**E**) *Kirchneriella lunaris* (Kirch.) Moebius, (**F**) *Scenedesmus quadricauda* (Turp.) de Brébisson *in* de Brébisson & Godey, (**G**) *Schroederia judayi* G. M. Smith, (**H**) *Crucigenia tetrapedia* (Kirch.) West &West, (I) *Crucigenia rectangularis* (A. Braun) Gay, (**J**) *Tetrastrum staurogeniaeforme* (Schröder) Lemmermann

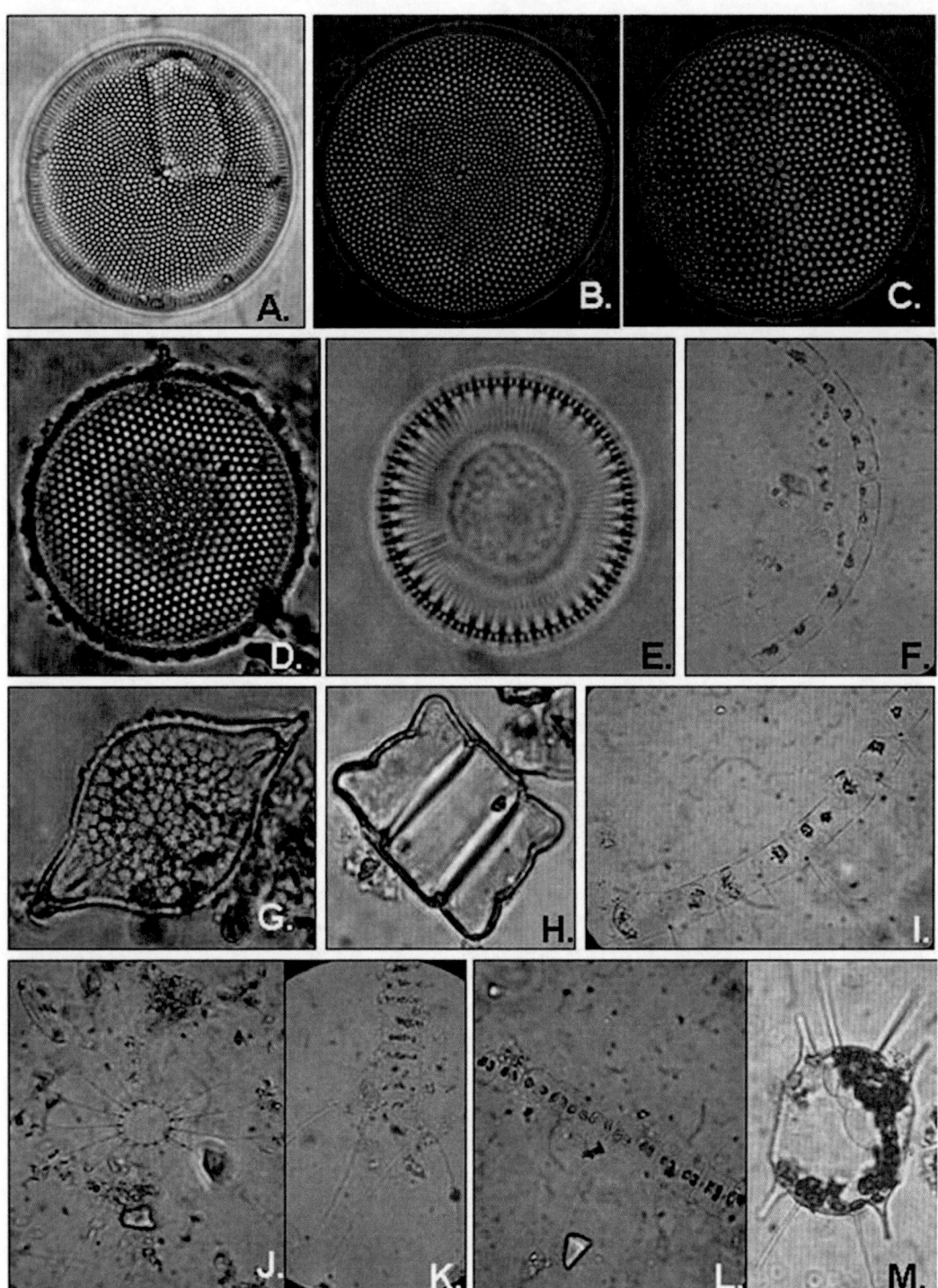

Plate 5.10 (**A**) *Coscinodiscus granii* Gough, (**B**) *Coscinodiscus excentricus* Ehrenberg, (**C**) *Actinocyclus normanii* f. *subsala* (Juhl.-Dannf.) Hustedt, (**D**) *Thalassiosira decipiens* (Grunow) Jørgensen, (**E**) *Cyclotella meneghiniana* Kutzing, (**F**) *Chaetoceros curvisetus* Cleve, (**G**) *Odontella rhombus* (Ehrenb.) Kützing, (**H**) *Biddulphia alternans* (Bailey) van Heurck, (**I**) *Chaetoceros massenensis* Castracane, (**J**) *Bacteriastrum varians* Lauder, (**K**) *Chaetoceros diversus* Cleve, (**L**) *Chaetoceros wighamii* Brightwell, (**M**) *Odontella mobiliensis* (J. W. Bailey) Grunow

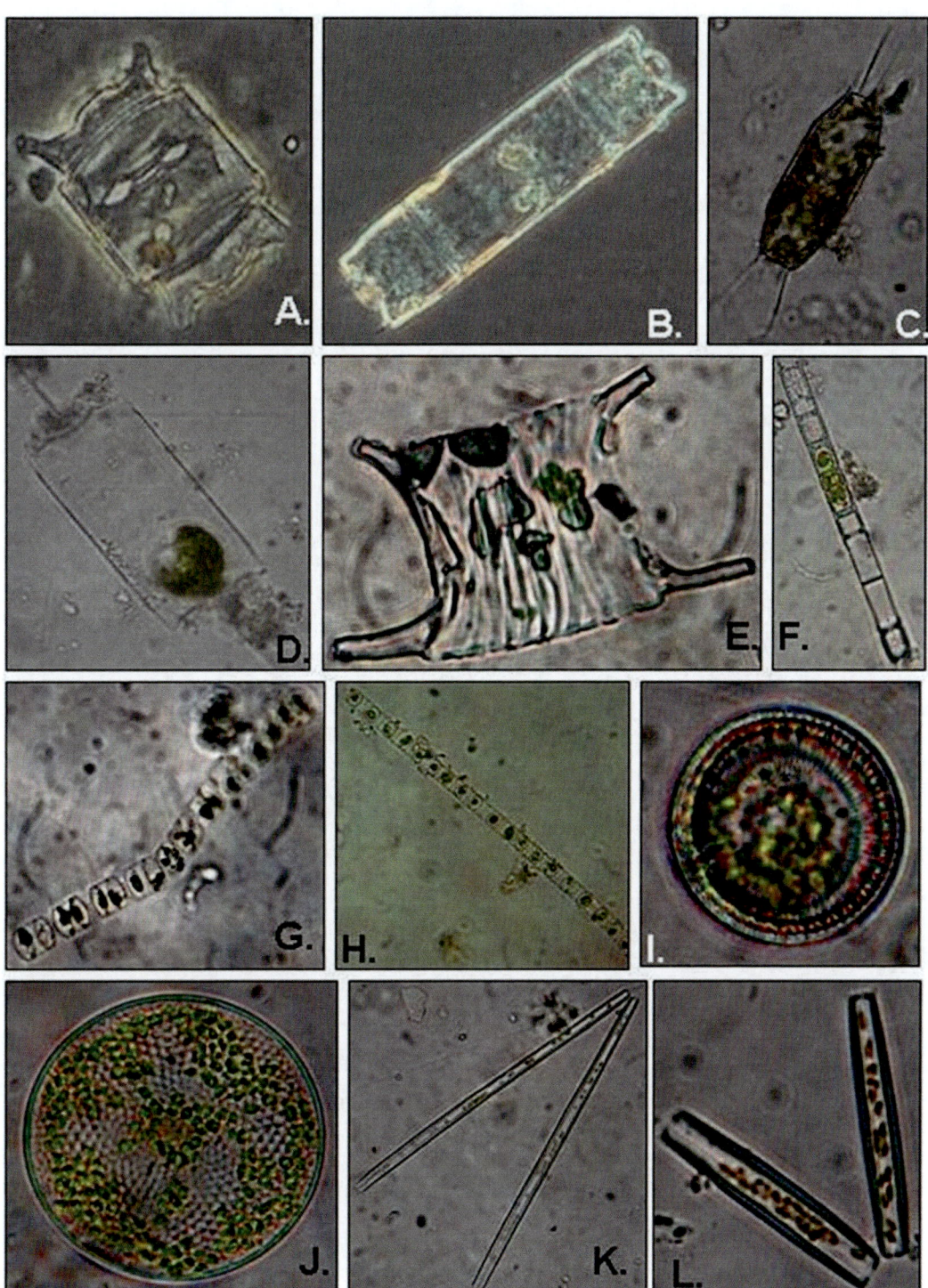

Plate 5.11 (**A**) *Odontella aurita* (Lyngbye) Brébisson and Godey, (**B**) *Biddulphia dubia* (Brightwell) Cleve, (**C**) *Biddulphia heteroceros* Grunow, (**D**) *Ditylum brightwelli* (West) Grunow, (**E**) *Eucapmia zoodiacus* Ehrenberg, (**F**) *Leptocylindrus danicus* Cleve, (**G**) *Skeletonema costatum* (Greville) Cleve, (**H**) *Aulacoseira granulata* (Ehr.) Ralfs var. *Angustissima*, (**I**) *Cyclotella striata* (Kutzing) Grunow, (**J**) *Coscinodiscus excentricus* Ehrenberg var. *fasciculata* Hustedt, (**K**) *Thalassiothrix frauenfeldii* Grunow, (**L**) *Thalassionema nitzschoides* Grunow

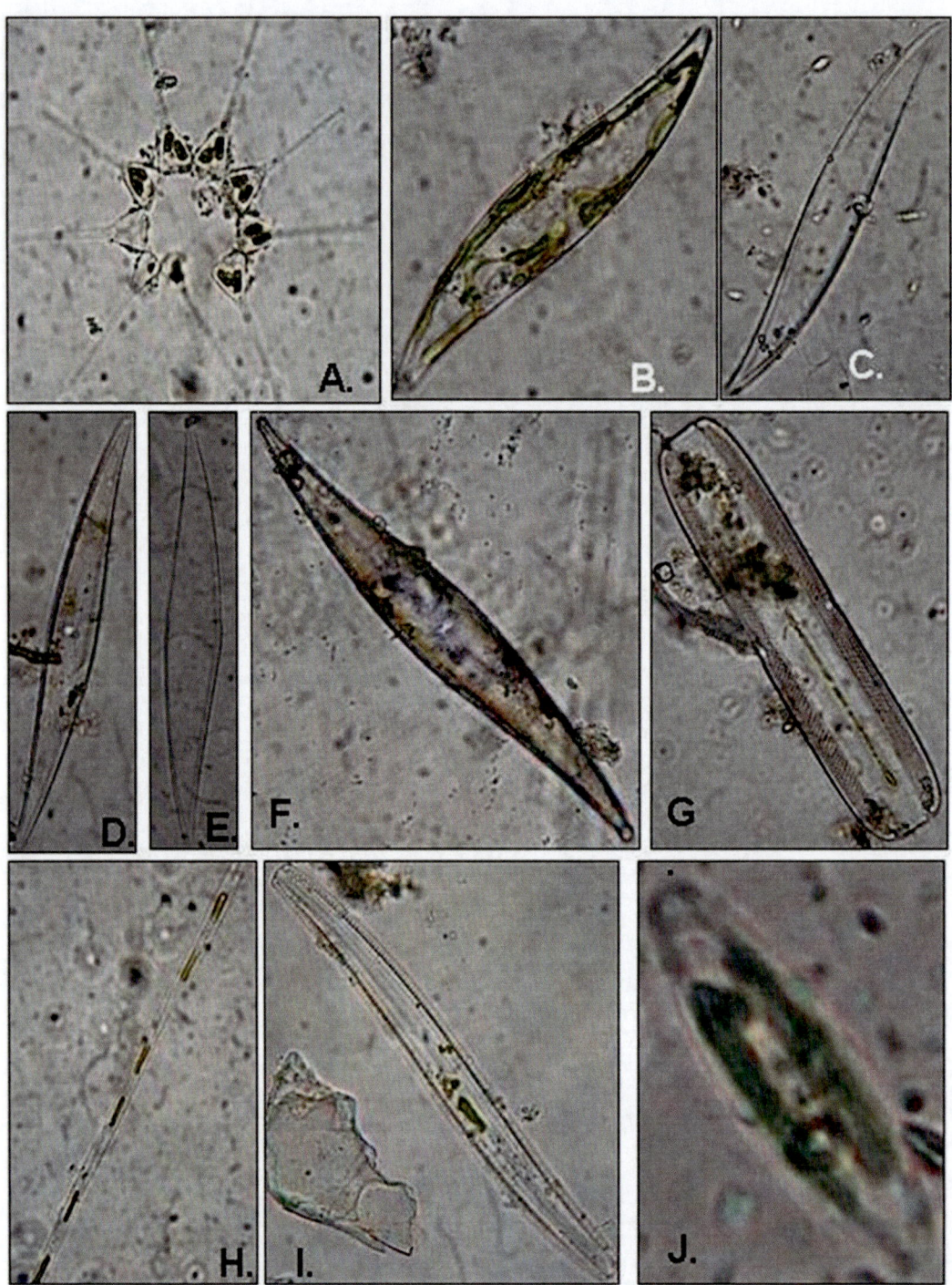

Plate 5.12 (**A**) *Asterionella japonica* Cleve, (**B**) *Gyrosigma obtusatum* (Sull. & Wormley) Boyer, (**C**) *Gyrosigma acuminatum* (Kütz.) Rabenh., (**D**) *Pleurosigma salinarum* Grunow, (**E**) *Pleurosigma normanii* Ralfs, (**F**) *Gyrosigma beautifortianum* Hust., (**G**) *Amphiprora gigantea* Grunow, (**H**) *Nitzschia delicatissima* Cleve, (**I**) *Nitzschia sigmoidea* Ehrenberg, (**J**) *Navicula peregrina* (Ehrenberg) Kutzing

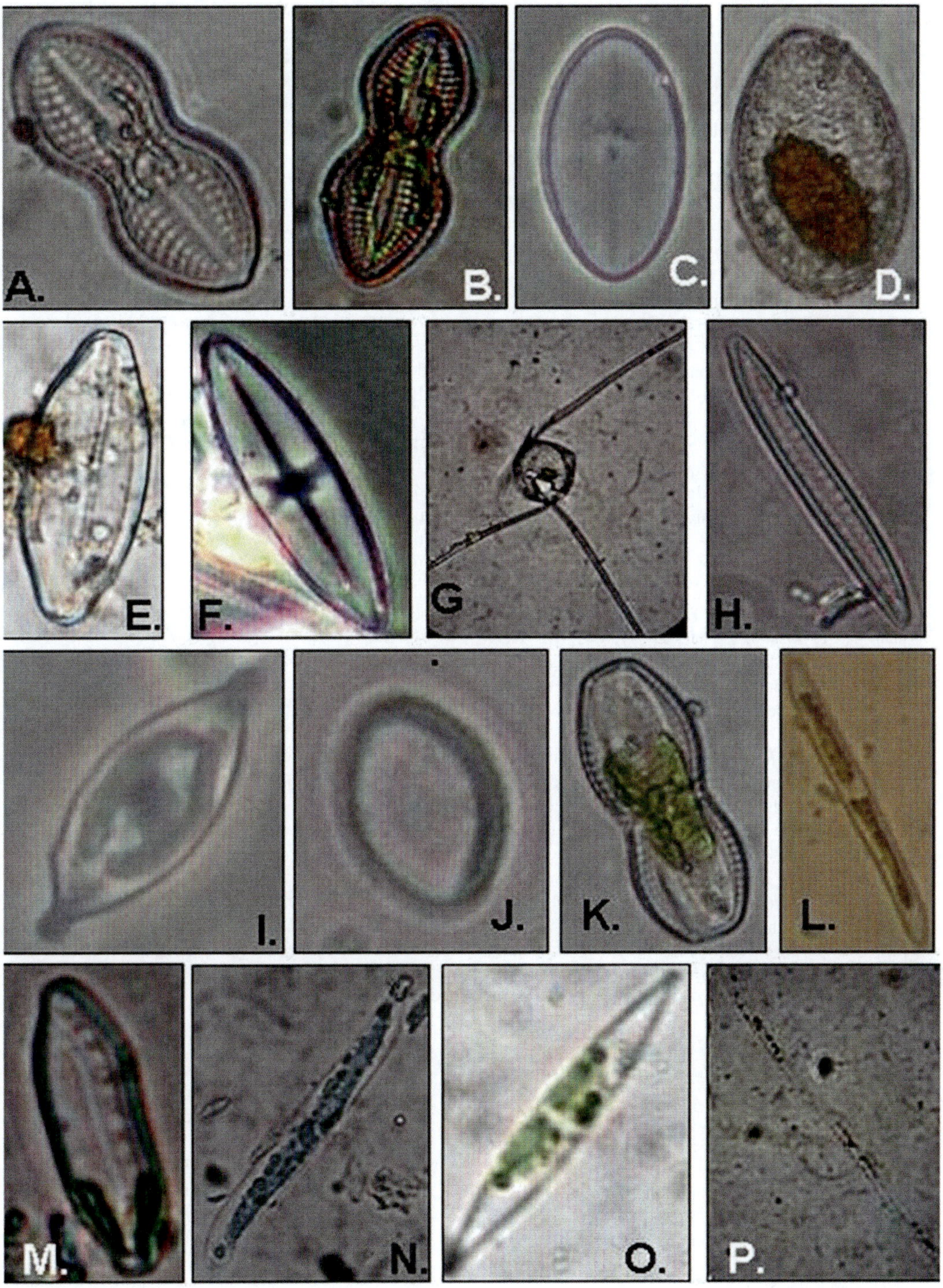

Plate 5.13 (**A**) *Diploneis weissflogii* (A. Schmidt) Cleve, (**B**) *Diploneis interrupta* (Kutzing) Cleve, (**C**) *Cocconeis dirupta* Greg., (**D**) *Surirella fastuosa* var. *recedens* (A. Schmidt) Cleve, (**E**) *Navicula quadripartita* Hustedt, (**F**) *Navicula minima* Grunow, (**G**) *Bacteriastrum delicatulum* Cleve (*top view*), (**H**) *Nitzschia ignorata* Kresske, (**I**) *Cymbella naviculiformis* Allerswald, (**J**) *Navicula mutica* Kutz. fo. *Cohni*, (**K**) *Nitzschia bilobata* var. *minor* Grunow, (**L**) *Nitzschia obtusa* Smith, (**M**) *Nitzschia punctata* W. Smith, (**N**) *Bacillaria paxillifer* (O. F. Müller) Hendy, (**O**) *Nitzschia amphibia* Grunow, (**P**) *Nitzschia pacifica* n. sp

References

Ayukai, T. (1992). Picoplankton dynamics in Davies Reef lagoon, the Great Barrier Reef, Australia. *Journal of Plankton Research, 14*, 1593–1606.

Azam, F., Fenchel, T., Field, J. G., Gray, J. S., Mayer-Reil, L. A., & Thingstad, T. (1983). The ecological role of water-column microbes in the sea. *Marine Ecology Progress Series, 10*, 257–263.

Azov, Y. (1986). Seasonal patterns of phytoplankton productivity and abundance in nearshore oligotrophic waters of the Levant Basin (Mediterranean). *Journal of Plankton Research, 8*(1), 41–53.

Badylak, S., & Phlips, E. J. (2004). Spatial and temporal patterns of phytoplankton composition in a subtropical lagoon, the Indian River Lagoon, Florida, USA. *Journal of Plankton Research, 26*, 1229–1247.

Berman, T., Azov, Y., & Townsand, D. W. (1984a). Understanding oligotrophic oceans: Can the Eastern Mediterranean be a useful model? In O. Holm-Hansen, L. Bolis, & R. Gilles (Eds.), *Marine phytoplankton and productivity, lecture notes on coastal and estuarine studies* (Vol. 8, pp. 101–111). Berlin: Springer.

Berman, T., Townsand, D. W., El-sayed, S. Z., Trees, C. C., & Azov, Y. (1984b). Optical transparency, chlorophyll and primary productivity in the Eastern Mediterranean near the Israeli Coast. *Oceanologica Acta, 7*(3), 367–372.

Bonin, D. J., Bonin, M. C., & Berman, T. (1989). Mise en evidence experimentale des facteurs nutritifs limitants de la productiondu microplancton et de l'ultraplancton dans une eau cotierede la mediterranee orientale. *Aquatic Sciences, 51*, 129–152.

Børsheim, K. Y., & Bratbak, G. (1987). Cell volume to cell carbo conversion factors for a bacterivorous *Monas* sp. enriched from seawater. *Marine Ecology Progress Series, 36*, 171–175.

Bradford-Grieve, J. M., Boyd, P. W., Chang, F. H., Chiswell, S., Hadfield, M., Hall, J. A., James, M. R., Nodder, S. D., & Shushkina, E. A. (1999). Pelagic ecosystem structure and functioning in the subtropical front region east of New Zealand in austral winter and spring 1993. *Journal of Plankton Research, 21*(3), 405–428.

Brussaard, C. P. D., Gast, G. J., van Duyl, F. C., & Riegman, R. (1996). Impact of phytoplankton bloom magnitude on a pelagic microbial food web. *Marine Ecology Progress Series, 144*, 211–221.

Caron, D. A., Peele, E. R., Lim, E. L., & Dennett, M. R. (1999). Picoplankton and nanoplankton and their trophic coupling in surface waters of the Sargasso Sea south of Bermuda. *Limnology and Oceanography, 44*, 259–272.

Caroppo, C. (2000). The contribution of picophytoplankton to community structure in a mediterranean brakish environment. *Journal of Plankton Research, 22*, 381–397.

Christaki, U., Van Wambeke, F., & Dolan, J. R. (1999). Nanoflagellates (mixotrophs, heterotrophs and autotrophs) in the oligotrophic eastern Mediterranean: Standing stocks, bacterivory and relationships with bacterial production. *Marine Ecology Progress Series, 181*, 297–307.

Christaki, U., Giannakourou, A., Van Wambeke, F., & Gregori, G. (2001). Nanoflagellate predation on auto- and heterotropic picoplankton in the oligotrophic Mediterranean Sea. *Journal of Plankton Research, 23*, 1297–1310.

Cupp, E. E. (1943). *Marine plankton diatoms of the west coast of North America* (Bulletin of the Scripps Institute of Oceanography, pp. 1–237). Berkeley/Los Angeles: University of California Press.

Curl, H. C., Jr. (1959). The phytoplankton of Apalachee Bay and the Northeastern Gulf of Mexico. *Publications of the Institute of Marine Science, 6*, 311–320.

Dangeard, P. A. (1902). Euglena flava. *Le Botaniste, 8*, 180, pl 5.

Dam, H. G., & Peterson, W. T. (1988). The effect of temperature on the gut clearance rate constant of planktonic copepods. *Journal of Experimental Marine Biology and Ecology, 123*, 1–14.

De, T. K., Ghosh, S. K., Choudhury, A., & Jana, T. K. (1994). Plankton community organization and species diversity in the Hugli estuary, north east coast of India. *Indian Journal of Marine Sciences, 23*(3), 152–156.

De Jonge, V. N., & van Beusekom, J. E. E. (1995). Wind- and tide-induced resuspension of sediment and microphytobenthos from tidal flats in the Ems estuary. *Limnology and Oceanography, 40*, 766–778.

Desikachary, T. V. (1959). *Cyanophyta*. New Delhi: IARI.

Dowidar, N. M. (1984). Phytoplankton biomass and primary productivity of the south-eastern Mediterranean. *Deep-Sea Research, 31*(6–8A), 983–1000.

Foged, N. (1979). *Diatoms in New Zealand. The North Island* (Bibliotheca Phycologica, Band 47). Vaduz: J. Cramer.

Gopalakrishnan, P. (1972). Studies on marine planktonic diatoms off Port Okha in the Gulf of Kutch. *Phykos, 12*, 37–49.

Gopalakrishnan, V. V., & Sastry, J. S. (1985). Surface circulation over the shelf off the coast of India during the southwest monsoon. *Indian Journal of Marine Sciences, 14*, 62–66.

Gregory, W. (1857). On new forms of marine Diatomaceae found in the Firth of Clyde and in Loch Fyne, illustrated by numerous figures drawn by R.K. Greville, LL.D., F.R.S.E.. *Transactions of the Royal Society of Edinburgh, 21*, 473–542, pl. 9–14.

Ha, K., Hyun-Woo, K., & Gea-Jae, J. (1998). The phytoplankton succession in the lower part of hypertrophic Nakdong River (Mulgum), South Korea. *Hydrobiologia, 369–370*, 217–227.

Hangovan, G. (1987). A comparative study on species diversity distribution and ecology of Dinophyceae from Vellar estuary and nearby Bay of Bengal. *Journal of the Marine Biological Association of India, 29*(1–2), 280–285.

Hernandez-Becerril, D. U. (1987). Vertical distribution of phytoplankton in the central and northern part of the Gulf of California (June 1982). PS.Z.N.I. *Marine Ecology, 8*(3), 237–251.

Hinga, K. R. (2002). Effects of pH on coastal marine phytoplankton. *Marine Ecology Progress Series, 238*, 281–300.

Hustedt, F. (1930a). Bacillariophyta (Diatomee). In *A. Pascher's Die Süsswasser – Flora Mitteleuropa.* Jena: Gustav Fisher.

Hustedt, F. (1930b, 1931, 1932). Die Kieselalgen in Dr. L. Rabenhorst's *Kryptogamen Flora* von Deutschlands, Osterreichs und der Schweiz, bd. 7, Teil 1, Teil 2, Lief. pp. 1–4.

Ignatiades, L. (1998). The productive and optical status of the oligotrophic waters of the Southern Aegean Sea (Cretan Sea), Eastern Mediterranean. *Journal of Plankton Research, 20*(5), 985–995.

Ignatiades, L. (2005). Scaling the trophic status of the Aegean Sea, eastern Mediterranean. *Journal of Sea Research, 54*(1), 51–57.

Ignatiades, L., Karydis, M., & Vounatsou, P. (1992). A possible method for evaluating oligotrophy and eutrophication based on nutrient concentration scales. *Marine Pollution Bulletin, 24*(5), 238–243.

Ignatiades, L., Psarra, S., Zervakis, V., Pagou, K., Souvermezoglou, E., Assimakopoulou, G., & Gotsis-Skretas, O. (2002). Phytoplankton size-based dynamics in the Aegean Sea (Eastern Mediterranean). *Journal of Marine Systems, 36*, 11–28.

Jacobsen, B. A., & Simonsen, P. (1993). Disturbance events affecting phytoplankton biomass, composition and species diversity in a shallow, eutrophic, temperate lake. *Hydrobiologia, 149*, 9–14.

Jerlov, N. G. (1997). Classification of sea water in terms of quanta irradiance. *ICES Journal of Marine Science, 37*, 281–287.

Jumars, P. A. (1976). Deep sea species diversity: Does it have a characteristic scale? *Journal of Marine Research, 34*(2), 217–246.

Karydis, M., & Tsirtsis, G. (1996). Ecological indices: A biometric approach for assessing eutrophication levels in the marine environment. *Science of the Total Environment, 186*, 209–219.

Kilham, P. (1971). A hypothesis concerning silica and the freshwater planktonic diatoms. *Limnology and Oceanography, 16*, 10–18.

Kirchman, D. L. (1993). Leucine incorporation as a measure of biomass production by heterotrophic bacteria. In P. F. Kemp, B. F. Sherr, E. B. Sherr, & J. J. Cole (Eds.), *Handbook of methods in aquatic microbial ecology* (pp. 509–512). Ann Arbor: Lewis Publishers.

Kommarek, J. (2005). Studies on the cyanophytes (Cyanobacteria, Cyanoprokaryota) of Cuba; (11) Fresh water *Anabaena* species. *Preslia, 77*, 211–234.

Krishna Kumari, L., et al. (2002). Primary productivity in Mandovi-Zuari estuaries in Goa. *Journal of the Marine Biological Association of India, 44*(1&2), 1–13.

Krishnamurthy, V. (2000). *Algae of India and neighbouring countries. I. Chlorophycota* (pp. 1–125). Oxford & IBH.

Krom, M. D., Brenner, S., Kress, N., Neori, A., & Gordon, L. I. (1993). Nutrient distributions during an annual cycle across a warm-core oceanic eddy from the E. Mediterranean Sea. *Deep-Sea Research, 40*, 805–825.

Krom, M. D., Groom, S., & Zohary, T. (2003). The eastern Mediterranean. In K. D. Black & G. B. Shimmield (Eds.), *The biogeochemistry of marine systems* (pp. 91–126). Oxford: Blackwell.

Lee, S., & Fuhrman, J. A. (1987). Relationships between biovolume and biomass of naturally derived marine bacterioplankton. *Applied and Environmental Microbiology, 53*, 1298–1303.

Lemmermann, E. (1904). Das Plankton schwedischer Gewasser. *Arkiv för Botanik, 2*(2),1–209. Pls. 1, 2. McComb, A. J., & Lukatelich.

Li, K. W., Zohary, T., Yacobi, Y. Z., & Wood, A. M. (1993). Ultraphytoplankton in the eastern Mediterranean Sea: Towards deriving phytoplankton biomass from flow cytometric measurements of abundance, fluorescence and light scatter. *Marine Ecology Progress Series, 102*, 79–87.

Livingston, R. J. (Ed.). (2001). *Eutrophication processes in coastal systems.* Boca Raton: CRC Press.

Madhupratap, M., Gauns, M., Ramaiah, N., Prasanna Kumar, S., Muraleedharan, P. M., de Sousa, S. N., Sardessai, S., & Muraleedharan, U. (2003). Biogeochemistry of the Bay of Bengal: Physical, chemical and primary productivity characteristics of the central and western Bay of Bengal during summer monsoon 2001. *Deep-Sea Research Part II, 50*, 881–896.

Mallin, M. A. (1994). Phytoplankton ecology of North Carolina estuaries. *Estuaries, 17*, 561–574.

Malone, T. C., Garside, C., & Neale, P. J. (1980). Effects of silicate depletion on photosynthesis by diatoms in the plume of the Hudson River. *Marine Biology, 58*, 197–204.

Malone T. C., Pike, S. E., & Conley, D. J. (1993). Transient variations in phytoplankton productivity at JGOFS Bermuda times series station. Deep sea research part 1. *Journal of Oceanography, 40*(5), 903–924.

Matondkar, S. G. P., Dwivedi, R. M., Parab, S., Pednekar, S., Desa, E. S., Mascarenhas, A., Raman, M., & Singh, S. K. (2006). Satellite and ship studies of phytoplankton in the Northeastern Arabian during 2000–2006 period. *Proceedings of SPIE, 6406*(64061I) 1–10.

Mazzocchi, M. G., Christou, E., Fragopoulu, N., & Siokou-Frangou, I. (1997). Mesozooplankton distribution from Sicily to Cyprus (Eastern Mediterranean): I. General aspects. *Oceanologica Acta, 20*(3), 521–535.

Mercado, J. M., Ramírez, T., Cortés, D., Sebastián, M., & Vargas-Yáñez, M. (2005). Seasonal and inter-annual variability of the phytoplankton communities in an upwelling area of the Alborán Sea (SW Mediterranean Sea). *Scientia Marina, 69*(4), 451–465. doi:10.3989/scimar.2005.69n4451.

Nielsen, T. G., & Hansen, B. (1995). Plankton community structure and carbon cycling on the western coast of Greenland during and after the sedimentation of a diatom bloom. *Marine Ecology Progress Series, 125*, 239–257.

Nielsen, T. G., Lokkegaard, B., Richardson, K., Pedersen, F. B., & Hansen, L. (1993). Structure of plankton communities in the Dogger Bank area (North Sea) during a stratified situation. *Marine Ecology Progress Series, 95*, 115–131.

Padisak, J. (1993). The influence of different disturbance frequencies on the species richness, diversity and equitability of phytoplankton in shallow lakes. *Hydrobiologia, 249*, 135–156.

Pegler, K., & Kempe, S. (1988). The carbonate system of the North Sea: Determination of alkalinity and TCO_2 and calculation of PCO_2 and Sical (Spring 1986). *Mitt Geol-Paläont Inst, 65*, 35–87.

Pochman, A. (1942). Synopsis der Gattung *Phacus*. *Archiv fur Protistenkunde, 95*(2), 81–252; Figs.1-170.

Prescott, G. W. (1982). *Algae of the great western lakes area* (pp. 1–997). Dubuque: W.C. Brown Co.

Psarra, S., Tselepides, A., & Ignatiades, L. (2000). Primary productivity in the oligotrophic Cretan Sea (NE Mediterranean): Seasonal and interannual variability. *Progress in Oceanography, 46*, 187–204.

Qasim, S. Z. (1977). Biological productivity of the Indian Ocean. *Indian Journal of Marine Sciences, 6*, 122–137.

Radhakrishna, K., Devassay, V. P., Bhargava, R. M. S., & Bhattathiri, P. M. A. (1978). Primary production in the northern Arabian Sea. *Indian Journal of Marine Sciences, 7*, 271–275.

Ramette, A. (2007). Multivariate analyses in microbial ecology. *FEMS Microbiology Ecology, 62*, 142–160.

Redekar, P. D., & Wagh, A. B. (2000). Planktonic diatoms of the Zuari estuary, Goa (west coast of India). *Seaweed Research and Utilization Association, 22*(1&2), 107–112.

Redfield, A. C. (1958). The biological control of chemical factors in the environment. *American Scientist, 46*(3), 205–221.

Reynolds, C. S. (1984). *The ecology of freshwater phytoplankton* (pp. 1–396). Cambridge: Cambridge University Press.

Richardson, K., Nielsen, T. G., Pedersen, F. B., Heilmann, J. P., Lokkegaard, B., & Kaas, H. (1998). Spatial heterogeneity in the structure of the planktonic food web in the North Sea. *Marine Ecology Progress Series, 168*, 197–211.

Robarts, D. R., Zohary, T., Waiser, M. J., & Yacobi, Y. Z. (1996). Bacterial abundance, biomass, and production in relation to phytoplankton biomass in the Levantine Basin of south-eastern Mediterranean Sea. *Marine Ecology Progress Series, 137*, 273–281.

Roman, M. R., Caron, D. A., Kremer, P., Lessard, E. J., Madin, L. P., Malone, T. C., Napp, J. M., Peele, E. R., & Youngbluth, M. J. (1995). Spatial and temporal changes in the partitioning of organic carbon in the planktonic community of the Sargasso Sea off Bermuda. *Deep-Sea Res Part I, 42*(6), 973–992.

Sasamal, S. K., Panigrahy, R. C., & Misra, S. (2005). *Asterionella* bloom in the north western Bay of Bengal during 2004. *International Journal of Remote Sensing, 26*(17), 3853–3858.

Schelske, C. L., & Stoermer, E. F. (1971). Eutrophlcation, silica depletion and predicted changes in algal quality in Lake Michigan. *Science, 173*, 423–424.

SenGupta, R., De Sousa, S. N., & Joseph, T. (1977). On nitrogen and phosphorous in the western Bay of Bengal. *Indian Journal of Marine Sciences, 6*, 107–110.

Sherr, E., & Sherr, B. (1988). Role of microbes in pelagic food webs: A revised concept. *Limnology and Oceanography, 33*(5), 1225–1227.

Siokou-Frangou, I., Bianchi, M., Christaki, U., Christou, E., Giannakourou, A., Gotsis-Skretas, O., Ignatiades, L., Pagou, K., Pitta, P., Psarra, S., Souvermezoglou, E., Van Wambeke, F., & Zervakis, V. (2002). Differential carbon transfer along a gradient of oligotrophy in the Aegean Sea (Mediterranean). *Journal of Marine Systems, 33–34*, 335–353.

Skirrow, G. (1975). The dissolved gases – Carbon dioxide. In J. P. Riley & G. Skirrow (Eds.), *Chemical oceanography*. New York: Academic.

Skvortzow, B. W. (1928). Die Euglenacaengathung *Phacus* Dujardin. Ibid., *46*, 105–125;Pl.2.

Sournia, A. (1973). La production primaire planctonique en Mediterranee, Essai de mise en jour. *Bulletin Etude en Commun de la Méditerranée 5* (no sp.), 128 pp.

Stefanou, P., Tsirtsis, G., & Karydis, M. (2000). Nutrient scaling for assessing eutrophication: The development of a stimulated normal distribution. *Ecological Applications, 10*, 303–309.

Subrahmanyan, R. (1946). A systematic account of the marine plankton diatoms of the Madras coast. *Proceedings of the Indian Academy of Sciences, 24B*, 85–197.

Swirenko, D. O. (1915). Materialy K floria vordoroslei Rossii. Niepotoryia dannyia K sistematikie I geografii Euglenaceae. *Obschchestvo Ispytatelei Prirody, Khar'kov. Turdy, 48*(1), 67–148.

Trigueros, J. M., & Orive, E. (2001). Seasonal variations of diatoms and dinoflagellates in a shallow, temperate estuary, with emphasis on neritic assemblages. *Hydrobiologia, 444*, 119–133.

Tsirtis, G. E. (1995). A simulation model for the description of a eutrophic system with emphasis on the microbial processes. *Water Science and Technology, 32*, 189–196.

Turley, C. M., Bianchi, M., Christaki, U., Conan, P., Harris, J. R. W., Psarra, S., Ruddy, G., Stutt, E. D., Tselepides, A., & Van Wambeke, F. (2000). Relationship between primary producers and bacteria in an oligotrophic sea – The Mediterranean and biogeochemical implications. *Marine Ecology Progress Series, 193*, 11–18.

Venkataraman, G. (1939). A systematic account of some south Indian diatoms. *Proceedings of the Indian Academy of Sciences, 10*, 293–368.

References

Venkataraman, K., & Wafar, M. (2005). Coastal and marine biodiversity of India. *Indian Journal of Marine Sciences, 34*(1), 57–75.

Verlencar, X. N., & Qasim, S. Z. (1985). Particulate organic matter in the coastal and estuarine waters of Goa and its relationship with phytoplankton production. *Estuarine, Coastal and Shelf Science, 21*, 235–242.

Vollenweider, R. A., & Kerekes, J. J. (1982). *Eutrophication of waters: Monitoring, assessment and control*. Paris: OECD.

Vollenweider, R. A., Giovanardi, F., Montanari, G., & Rinaldi, A. (1998). Characterization of the trophic conditions of marine coastal waters with special reference to the NW Adriatic Sea: Proposal for a trophic scale, turbidity and generalized water quality index. *Environmetrics, 9*, 329–357.

Wondie, A., et al. (2007). Seasonal variation in primary production of a large high altitude tropical lake (Lake Tana, Ethiopia): Effects of nutrient availability and water transparency. *Aquatic Ecology, 41*, 195–207.

Yacobi, Y. Z., Zohary, T., Kress, N., Hecht, A., Robarts, R. D., Waiser, M., & Wood, A. M. (1995). Chlorophyll distribution throughout the southeastern Mediterranean in relation to the physical structure of the water mass. *Journal of Marine Systems, 6*(3), 179–190.

Glossary

Algae Algae are polyphyletic group of simple, oxygenic photosynthetic organisms that have chlorophyll as their main photosynthetic pigment and lack sterile covering of cells around the reproductive cells.

Algal bloom Dense population of planktonic algae or cyanobacteria that distinctly colours the water and may form scum on the surface.

Algal trophic index Quantitative expression of algal species counts, providing a measure of the trophic (nutrient) status of the aquatic environment.

Allochthonous Materials (usually organic) produced within a water body.

Apochlorotic Colourless or without chlorophyll.

Autotroph Organism capable of synthesizing organic matter by means of photosynthesis.

Bathyal zone Ocean water over continental slope.

Biofilm Community of microorganisms occurring at a physical (e.g. water/solid) interface that is typically present within a layer of extracellular polysaccharide that is secreted by the community.

Bioluminescence Emission of light by a living organism.

Biovolume Volume of single algae and algal populations in a particular population. It may be measured either for a single genus or for a mixed population (volume of individual taxa should be measured).

Brackish Saline water with a salinity less than that of seawater (33 $0/_{00}$).

Calcification Deposition of calcium carbonate, usually in association with smaller amounts of other carbonate.

Carotenoid Yellow, orange or red hydrocarbon or fat-soluble pigment.

Chlorophyll Fat-soluble, green, porphyrin-type pigment.

Chloroplast Plastid with chlorophyll.

Chloroplast endoplasmic reticulum or chloroplast ER One or two membranes surrounding the chloroplast envelope; ribosomes are usually attached to the outside of the outer membrane.

Chromatic adaptation Change in the proportion of different photosynthetic pigments enabling optimum absorption of the available wavelengths of light.

Circadian rhythm Repeated sequence of metabolic activities that occur at about 24 h intervals.

Coccoid Spherical structure.

Coccolith Calcified scale in a coccolithophorid (Prymnesiophyceae).

Coenobium Spherical colony of algal cells with central hollow and number that is fixed at the time of origin and is not subsequently augmented.

Compensation depth Depth of water at which sufficient light is penetrated so that photosynthesis equals respiration over a 24 h period.

Compensation point Particular light intensity at which respiration equals photosynthesis over a 24 h period of a specific area.

Coralline Calcified algae.

Cryptomonads Group of unicellular motile eukaryote algae of the division Cryptophyta.

Cyanelle Endosymbiotic blue-green alga that gave rise to chloroplast.

Cyanome Host cell containing a cyanelle that gave rise to eukaryotic algae.

Cyanophage Virus that infects the cells of the Cyanophyceae.

Cyanophycin granule Polypeptide storage granules within the cells of Cyanophyceae.

Diatom indices Use of diatom species counts to assess the trophic status of a water body.

Dystrophic Brown or yellow coloured waters rich in organic matter where the rate of decay of that organic matter is slow having a low pH.

Ecosystem Self-regulating biological community living in a defined habitat.

Environmental stress factor External change that impairs biological function at the level of individual organisms and molecular systems of an ecosystem.

Epilithic Organisms living on rock surfaces.

Epipelic Organisms growing on mud.

Epiphyte One plant living on other plant.

Epontic Organisms living on the bottom of ice.

Estuary The junction of a river and ocean where tidal effects are evident and where freshwater and seawater mix.

Euphotic or photic zone Regions of water body above the compensation depth.

Eurysaline (euryhaline) Organisms tolerant of a wide salinity range.

Eutrophic A body of water that receives large amounts of nutrients, usually resulting in a large growth of algae.

Eutrophication An increase in the concentration of soluble inorganic nutrients such as phosphates and nitrates in aquatic ecosystem.

Gas vacuole Gas-filled vacuum found in some aquatic blue-green algae and bacteria that increases buoyancy. It is composed of gas vesicles which are made of protein.

Habitat The living place of an organism or community, characterized by its physicochemical and biotic properties.

Holoplanktonic Aquatic organisms which are present in the water column over most of the annual cycle.

Hydrology All aspects of water flow connected with an aquatic system, including inflow and outflow of water.

Hypertrophic Water body with extremely high levels of dissolved inorganic nutrients, also called hypereutrophic.

Hypolimnion Region of water body beneath the thermocline in thermally stratified water bodies with low light intensity.

Intertidal Occurring between the low and high tide marks.

Iridescence The play of colours caused by refraction and interference of light waves at the surface.

K-selected species (K-strategist) Organisms adapted to high levels of competition in a crowded environment where they survive, grow and reproduce.

Lentic Related to a pond or lake habitat.

Limnology Study of aquatic systems in relation to physicochemical and biotic factors.

Littoral zone Peripheral shoreline at the edge of lakes and rivers.

Lotic Related to rivers or stream habitat.

Macroplankton Planktons larger than 75 µm in diameter. Also called net plankton.

Meroplanktonic Algae with only a limited planktonic existence in the water column. Most of the annual cycle is spent on sediments as a resting stage.

Mesotrophic Water body with moderate levels of inorganic nutrients and moderate primary productivity – intermediate state between oligotrophic and eutrophic condition.

Microplankton Unicellular and multicellular planktonic organisms in the size range of 20–200 µm.

Mixotrophy Organisms having the ability to combine both autotrophic (using inorganic carbon sources) and heterotrophic (organic carbon sources including phagotrophy) nutrition.

Nannoplankton or nanoplankton Plankton smaller than 75 µm but larger than 2 µm.

Nephelometer Submerged instrument used to measure the particulate concentration (turbidity) of water by collecting light scattering from suspended matter.

Oligotrophic Water body with less dissolved inorganic nutrients (particularly nitrogen and phosphorous) resulting in low levels of biological productivity.

Organotroph (osmotrophy, saprotrophy) Organisms that either use reduced organic compounds as its sources of electrons or carry out organotrophy.

Pelagic All organisms normally present in the water column of water bodies like plankton and nekton.

Pelagic zone The central main part of a lake.

Periphyton Plantlike organisms present in a community mainly on underwater substrata – including algae, bacteria and fungi.

Photic zone Upper part of water column on aquatic ecosystem in which net photosynthesis can occur (also known as the euphotic zone).

Phototroph Organisms that use solar energy to fix inorganic carbon to organic compounds by photosynthesis.

Phragmoplast Wall formation by the coalescence of Golgi vesicles between spindle microtubules.

Phytoplankton Free-floating plants that float aimlessly or swim too feebly to maintain a constant position against water current.

Picoplankton Plankton with a diameter of 0.2–2 µm.

Plankton Organisms that float aimlessly or swim too feebly to maintain a constant position against water current.

Plankton sedimentation Gravitational force-induced sinking of nonmotile plankton in the water column.

Primary production Synthesis of biomass by photosynthetic organisms – higher plants, algae and photosynthetic bacteria.

Productivity The rate of increase in biomass (growth rate) in a population of organism. Can be expressed as $mgC/m^2/day$.

r-selected species (r-strategist) Organism adapted to an uncrowded environment, with low competition.

Saline lakes Lakes with highly concentrated salts often resulting in white salt 'crusts' round their margins where evaporation from the surface greatly exceeds the inputs.

Saprobic pollution High concentration of soluble organic nutrients.

Secchi depth A particular depth of a water column at which a suspended sectored plate (Secchi disc) can no longer just be seen indicating the measure of water turbidity and phytoplankton biomass.

Sedgwick rafter cell counter A grooved slide with counting chamber commonly used for phytoplankton samples.

Stratification Vertical structuring of static or very slow moving water bodies into three distinct layers – epilimnion, metalimnion and hypolimnion. Determined by temperature and circulatory differences with the water column.

Sublittoral zone In the freshwater region, the zone from the end of rooted vegetation (about 6 m) to the compensation depth and in the marine ecosystem the zone from the lowest low tide mark to 200 m depth.

Succession Temporal sequence of organism that occurs in a developing community such as biofilm or lake pelagic community.

Supralittoral zone In marine ecosystem the zone above the high tide mark in the ocean and in freshwater region above the standing water mark, which receives splash during windy periods.

Trophic The term 'trophic' is used to describe the inorganic nutrient status of different water bodies (oligotrophic to eutrophic) and the feeding relationships (trophic interactions) of freshwater biota.

Turbidity Opacity of water caused by suspended particulate matter used as measure of phytoplankton biomass.

Tychoplankton Organisms circumstantially carried into the plankton often from plant or rock surfaces. Also referred to as 'accidental plankton' or 'pseudoplankton'.

Water bloom See algal bloom.

Zooplankton Assemblage of invertebrate planktonic organism.

Bibliography

Alley, R. B. (2007). Wally was right; predictive, ability of the North Atlantic "Conveyer Belt" hypothesis for abrupt climate change. *Annual Reviews of Earth and Planetary Science, 35*, 241–272.

Falkowski, P. G., & Kolber, Z. (1995). Variations in chlorophyll fluorescence yields in phytoplankton in the world oceans. *Australian Journal of Plant Physiology, 22*(2), 341–355.

Hecky, R. E. (1998). Low N: P ratios and the nitrogen fix: Why watershed nitrogen removal will not improve the Baltic. In *Effects of nitrogen in the aquatic environment* (KVA Report 1998: 1, 85–115 pp). Stockholm: Kungl Vetenskapsakademien [Royal Swedish Academy of Science].

Jeffrey, S. W., & Vesk, M. (1997). Introduction to marine phytoplankton and their pigment signatures. In S. W. Jeffrey, R. F. C. Mantoura, & S. W. Wright (Eds.), *Phytoplankton pigments in oceanography: guidelines to modern methods* (pp. 37–84). Paris: UNESCO Publishing.

Lewitus, A. J., et al. (1995). Discovery of the "Phantom" dinoflagellate in Chesapeake Bay. *Estuaries, 18*(2), 373–378.

Moss, B. (1968). The chlorophyll a content of some benthic algal communities. *Archives of Hydrobiology, 65*, 51–62.

Steeman-Nielsen, E. (1952). The use of radioactive carbon (^{14}C) for measuring organic production in the sea. *Journal du Conseil/Conseil Permanent International pour l'Exploration de la Mer, 18*, 117–140.

Weckström, K., & Juggins, S. (2005). Coastal diatom-environment relationships from the Gulf of Finland, Baltic Sea. *Journal of Phycology, 42*, 21–35.

Whitten, B. (1970). Biology of Cladophora in freshwaters. *Water Research, 4*, 457–476.

Willen, E. (1991). Planktonic diatoms – An ecological review. *Algological Studies, 62*(Suppl), 69–106.

Preface

Tremendous effort was made in thermoelectric materials research in the late 1950s and 1960s after Ioffe first proposed the investigation of semiconductor materials for utilization in thermoelectric applications. Alloys based on either the Bi_2Te_3 or $Si_{1-x}Ge_x$ system soon became some of the most widely studied thermoelectric materials. These materials were extensively studied and optimized for their use in thermoelectric applications (solid state refrigeration and power generation; Goldsmid, 1986; Rowe, 1995) and remain the state-of-the-art materials for their specific temperature use. By the 1970s, research on thermoelectric materials had begun a steady decline and essentially vanished by the 1980s in the United States. However, since the early 1990s there has been a rebirth of interest in the field of thermoelectric materials research, and over the past few years many new classes of materials have been investigated for their potential for use in thermoelectric applications. Much of this was brought about by the need for new alternative energy materials, especially solid-state energetic materials. Many new concepts of materials, including bulk and thin-film materials, complex structures and geometry, materials synthesis, theory, and characterization have been advanced over the past decade of work. These three volumes of *Semiconductors and Semimetals* are dedicated to identifying the efforts of research in this past decade and preserving them in a concise and relatively complete overview of these efforts. It is hoped that this will provide future generations a significant added advantage over the current generation, who have worked hard to revive this field of research.

The first two volumes are focused primarily on bulk materials, with one chapter on transport through interfaces. The first volume contains an overview of the field, including an introduction by Julian Goldsmid, who is credited with discovering the Bi_2Te_3 materials. Volumes 69 and 70 contain reviews of theoretical, synthesis, and characterization methods and directions, as well as in-depth reviews of some of the most active areas of bulk materials research. The third volume in this series (Volume 71) is dedicated primarily to low-dimensional and thin-film thermoelectric materials, including both theory and experimental work.

Thermoelectric energy conversion utilizes the Peltier heat transferred when an electric current is passed through a thermoelectric material to provide a temperature gradient with heat being absorbed on the cold side and rejected at the sink, thus providing a refrigeration capability. Conversely, an imposed temperature gradient, ΔT, will result in a voltage or current, that is, small-scale power generation (Tritt, 1996, 1999). This aspect is widely utilized in deep space applications. A radioactive material acts as the heat source in these RTGs (radioactive thermoelectric generators) and thus provides a long-lived energy supply. The advantages of thermoelectric solid-state energy conversion are compactness, quietness (no moving parts), and localized heating or cooling, as well as the advantage of being "environmentally friendly." Applications of thermoelectric refrigeration include cooling of CCDs (charge coupled devices), laser diodes, infrared detectors, low-noise amplifiers, computer processor chips, and biological specimens.

The essence of defining a good thermoelectric material lies primarily in determining the material's dimensionless figure of merit, $ZT = \alpha^2 \sigma T/\lambda$, where α is the Seebeck coefficient, σ the electrical conductivity, λ the total thermal conductivity ($\lambda = \lambda_L + \lambda_E$; the lattice and electronic contributions, respectively), and T is the absolute temperature in kelvins. The Seebeck coefficient, or thermopower, is related to the Peltier effect by $\Pi = \alpha T = Q_P/I$, where Π is the Peltier coefficient, Q_P is the rate of heating or cooling, and I is the electrical current. The efficiency (η) and coefficient of performance (COP) of a thermoelectric device are directly related to the figure of merit of the thermoelectric material or materials. Both η and COP are proportional to $(1 + ZT)^{1/2}$.

Narrow-gap semiconductors have long been the choice of materials to investigate for potential thermoelectric applications because they satisfy the necessary criteria better than other materials. Material systems that exhibited complex crystal structures and heavy atoms, to facilitate low thermal conductivity, yet were easy to dope to tune the electronic properties, were of primary interest. Currently, the best thermoelectric materials have a value of $ZT \approx 1$. This value, $ZT \approx 1$, has been a practical upper limit for more than 30 years, yet there is no theoretical or thermodynamic reason why it cannot be larger. But recently many new materials and concepts of materials have been introduced, as you will see in the following chapters. The development of rapid synthesis and characterization techniques, coupled with much-advanced computational models, provides the ability to more rapidly investigate a class of materials for potential for thermoelectric applications. The need for higher performance energetic materials (providing alternative energy sources) for refrigeration applications such as cooling electronics and optoelectronics and power generation applications such as waste heat recovery are of great importance. One of the goals of the current research is to achieve $ZT \approx 2-3$ for many applications. Such values of ZT would make thermoelectric refrigeration competitive with vapor compression refrigeration systems and would make high-temperature materials feasible for utilization in many waste heat recovery applications, such as waste heat from automobile engines and exhaust.

Over the past decade, much of the recent research in bulk materials for thermoelectric applications has revolved around the concept of the "phonon glass electron crystal" model (PGEC) developed by Slack (1979, 1995). This paradigm suggests that a good thermoelectric material should have the electronic properties of a crystalline material and the thermal properties of a glass. The "kickoff talk" given by Glen Slack in Symposium Z at the 1998 Fall Materials Research Society (MRS) was entitled "Holey and Unholey Semiconductors as Thermoelectric Refrigeration Materials" (Tritt *et al.*, 1998). The chapters in Volume 69 such as that on skutterudites by Uher, clathrates by Nolas *et al.*, and Chapter 1 in Volume 70 on the use of ADP parameters by Sales *et al.*, discuss the concept of "holey" semiconductors or cage-structure materials that use "rattling" atoms to scatter phonons and reduce the lattice thermal conductivity of a material. In Volumes 69 and 70, other materials are discussed, such as the half-Huesler alloys (Poon), BiSb (Lenoir *et al.*), and quasicrystals (Tritt *et al.*) are more typical of the "unholey" materials, which have to depend on more typical scattering mechanisms, such as mass fluctuation scattering, to reduce lattice thermal conductivity in a material. The PGEC paradigm is also prevalent in much of the research focused on thin-film and superlattice materials and electrical and heat transport through interfaces.

It is my strong belief that a new, higher performance thermoelectric material will be found and it will truly have a large impact on the world around us. The advances that I have seen over the past 5 or 6 years give me great optimism. However, I am always reminded just how good the Bi_2Te_3 materials really are. The aspect of low-temperature refrigeration ($T < 200\,K$) of electronics and optoelectronics would yield a revolution in the electronics industry. The possibility of superconducting electronics cooled below their superconducting transition by a solid-state and compact thermoelectric device is very enticing. Where will the breakthrough be? Will it be in the bulk materials, either "holey" or "unholey"? Will it be in the new exotic structures, such as superlattice or thin-film materials, or will it be in using thermionic refrigeration? In these new exotic structures we are learning much about interface scattering of the phonons as well as the electrical transport in these "confined structures." Added to this is an even greater challenge than in the bulk materials—characterizing the figure of merit of such complex geometries. Measurements on these structures have proven to be quite challenging. Hopefully, one or possibly more of the next generation thermoelectric materials will have been identified and discussed in one of these three volumes of *Semiconductors and Semimetals*. There are many possibilities and much work is left to do.

I came into the field of thermoelectric materials research in 1994 while working as a research physicist at the Naval Research Laboratory, NRL, in Washington, D.C. We had decided to start a program in thermoelectric materials early that year at NRL. This program was headed by A. C. Ehrlich and included others at NRL such as David Singh, who had already been working in the field. I attended the 1994 International Conference on Thermoelectrics (ITC), which was held in Kansas City, MO. From the very first meeting I knew I had much to learn. Over the period from late 1970 until 1996, most of the research on thermoelectric materials was published and archived in the proceedings of these ITC conferences. The measurements necessary to evaluate thermoelectric materials were certainly nontrivial and the interplay of the electrical and thermal transport was indeed a challenge. At the 1994 meeting, I heard the term *thermoelectrician* for the first time, used by Cronin Vining, then president of the International Thermoelectrics Society. Much of the meeting was centered around Bi_2Te_3 alloys and incremental improvements to these state-of-the-art materials, as well as more efficient design of devices based on these materials. There were talks about a new class of materials called skutterudites, which were viewed as very promising materials. The year before, 1993, Hicks and Dresselhaus had published a paper in which they predicted that much higher ZT values were possible in quantum well structures. This enhanced ZT is due to an enhanced density of states and thus higher mobility and also a higher thermopower as the quantum well width decreased from a "bulklike" term. The excitement that something new and promising might be happening in the field of thermoelectrics was apparent. Around this same time a program was developed by John Pazik at the Office of Naval Research (ONR) to investigate the possibilities of finding and developing higher performance thermoelectric materials. Then in late 1996, another new program on high performance thermoelectric materials was started by DARPA (Defense Applied Research Projects Agency), which was headed by Stuart Wolf. There were also a few programs funded by the Army Research Office (ARO), most of which were managed by John Prater. The coordination and cooperation of the ONR, ARO, and DARPA thermoelectric programs was very impressive and continues to be. The goals were lofty and still remain a challenge: "Find a material with a $ZT \approx 3-4$!" These DOD programs were the "heartblood" of the rebirth of research in thermoelectric materials in the 1990s in the United States. Volume 70 contains a chapter on "Military Applications of Enhanced Thermoelectric Materials."

Much of the work that is highlighted in these three volumes has direct ties to that original ONR program, and most were supported by one or more of the DOD programs. Without the vision of these program managers as well as DARPA, ARO, and ONR, these volumes would certainly not have been possible. I take this opportunity to acknowledge them and thank them for their support.

As we underwent the "rebirth" of thermoelectrics research in the 1990s we had a distinct advantage. Many of the "great minds of thermoelectrics" such as J. Goldsmid, G. A. Slack, G. Mahan, M. Dresselhaus, and T. Harman were still very active; thus I am pleased to say that most of these researchers have contributions in these volumes. (Note: Ted Harman was invited to write a chapter but declined due to time constraints. However, some of his work on quantum dot superlattices is some of the most exciting work in the field.) Their contributions to this field of research are impressive, with some of them dating back to the "early days" of thermoelectrics in the late 1950s. The work and vision of Raymond Marlow and Dr. Hylan Lyon, Jr., of Marlow Industries and their contributions related to the rebirth of this field of research are also worth noting.

Over the course of development of *Recent Trends in Thermoelectric Materials Research* it became apparent that the work would have to be divided initially into two and finally into three volumes. I decided to divide the volumes between two primary themes: *Overview and Bulk Materials* (Volumes 69 and 70) and *Thin-Film/Low Dimensional Materials: Theory and Experiment* (Volume 71). In the end, I think that the division of the volumes works quite well and will make it easier for the reader to follow specific areas of interest. Some chapters may seem somewhat out of place; this is due primarily to the timing of receiving manuscripts and to space constraints, and was also left somewhat to the discretion of the editor.

First and foremost, I express great thanks to the authors who contributed to these volumes for their hard work and dedication in producing such an excellent collection of chapters. They were very responsive to the many deadlines and requirements and they were a great group of people to work with. I want to personally acknowledge my many conversations with Glen Slack, Julian Goldsmid, Jerry Mahan, Hylan Lyon, Jr., Ctirad Uher, Al Ehrlich, Cronin Vining, and others in the field, as I grasped for the knowledge necessary to personally advance in this field of research. Their contributions to me and to others in the field are immeasurable. Thanks also to my many other colleagues in the thermoelectrics community. I acknowledge the support of DARPA, the Army Research Office, and the Office of Naval Research in my own research. I also acknowledge the support of my own institution, Clemson University, during the editorial and manuscript preparation process. I am truly indebted to my graduate students for their contributions to these volumes, their hard work, and for the patience and understanding they exemplified during the editorial and writing process. I especially acknowledge A. L. Pope and R. T. Littleton IV for their help. A special thanks goes to my publisher, Greg Franklin, for his encouragement in all stages of the development of these manuscripts for publication. Thanks also to his assistant, Marsha Filion, for her help and contributions. I am especially indebted to my assistant at Clemson University, Lori McGowan, whose attention to detail and hard work (copying, reading, filing, corresponding with authors, etc.) really made these volumes possible. Without her dedication and hard work, I would not have been able to tackle the mountain of paperwork that went into these volumes. I also wish to acknowledge my wife, Penny, and my wonderful kids, Ben, Karen, Kristin, and Mary, for their patience and understanding during the many hours I spent on this work.

References

H. J. Goldsmid, *Electronic Refrigeration.* Pion Limited Publishing, London, 1986.
D. M. Rowe, ed., *CRC Handbook of Thermoelectrics.* CRC Press, Boca Raton, FL, 1995.
G. A. Slack, in *Solid State Physics*, Vol. 34 (F. Seitz, D. Turnbull, and H. Ehrenreich, eds.), p. 1. Academic Press, New York, 1979.
G. A. Slack, in *CRC Handbook of Thermoelectrics* (D. M. Rowe, ed.), p. 407. CRC Press, Boca Raton, FL, 1995.
Terry M. Tritt, *Science* **272**, 1276 (1996); **283**, 804 (1999).
Terry M. Tritt, M. Kanatzidis, G. Mahan, and H. B. Lyon, Jr., eds. *Thermoelectric Materials — The Next Generation Materials for Small Scale Refrigeration and Power Generation Applications*, MRS Proceedings Vols. 478 (1997) and 545 (1998).

TERRY M. TRITT

List of Contributors

Numbers in parentheses indicate the pages on which the authors' contribution begins.

VALERIE M. BROWNING (25), *U.S. Naval Research Laboratory, Washington, D.C.*

T. CAILLAT (101), *Jet Propulsion Laboratory, California Institute of Technology, Pasadena, California*

H. JULIAN GOLDSMID (1), *Department of Physics, University of Tasmania, Hobart, Tasmania, Australia*

MERCOURI G. KANATZIDIS (51), *Department of Chemistry, Michigan State University, East Lansing, Michigan*

B. LENOIR (101), *Ecole des Mines, Laboratoire de Physique des Matériaux, Nancy, France*

GEORGE S. NOLAS (255), *R&D Division, Marlow Industries, Inc., Dallas, Texas*

H. SCHERRER (101), *Ecole des Mines, Laboratoire de Physique des Matériaux, Nancy, France*

SANDRA B. SCHUJMAN (255), *Department of Physics, Rensselaer Polytechnic Institute, Troy, New York*

GLEN A. SLACK (255), *Department of Physics, Rensselaer Polytechnic Institute, Troy, New York*

TERRY M. TRITT (25), *Department of Physics and Astronomy, Clemson University, Clemson, South Carolina*

CTIRAD UHER (139), *Department of Physics, University of Michigan, Ann Arbor, Michigan*

CHAPTER 1

Introduction

H. Julian Goldsmid

SCHOOL OF MATHEMATICS AND PHYSICS
UNIVERSITY OF TASMANIA
HOBART, TASMANIA, AUSTRALIA

I. THE THERMOELECTRIC EFFECTS 1
II. SEMICONDUCTOR THERMOELEMENTS 9
III. THERMOELECTRIC MATERIALS 13
IV. LIMITS ON THE FIGURE OF MERIT 18
V. NEW DEVELOPMENTS 20
    REFERENCES 22

I. The Thermoelectric Effects

The thermoelectric effects provide a means by which thermal energy can be converted into electricity and by which electricity can be used for heat pumping or refrigeration. The effects were discovered early in the 19th century but it was only in the second half of the 20th century that thermoelectric generation became reasonably efficient and thermoelectric refrigeration became even practicable.

Seebeck discovered the effect that bears his name in 1821. In this effect a voltage appears when two different conductors are joined together and the junction is heated. The phenomenon is familiar to all scientists and engineers, as it has long been used in the measurement of temperature. The Peltier effect, discovered in 1834, is not so well known. When an electric current passes through the junction between two conductors, that junction becomes heated or cooled according to the direction of the current. This reversible effect is usually masked by the irreversible phenomenon of Joule heating, and it is also opposed by heat flow along the conductors. As we shall see, it is possible to choose the conductors so that the Peltier effect can be used to produce worthwhile cooling and so that the efficiency of generation through the Seebeck effect becomes large enough for certain purposes.

It is not surprising that the Seebeck and Peltier effects are closely related to one another. A thermodynamic relationship between them was established by Kelvin in 1855. Lord Kelvin (who was then known as William Thomson) predicted that there should be a third thermoelectric effect. This, the Thomson effect, consists of reversible heating or cooling when an electric current passes along a single conductor that is also subject to a temperature gradient. It was on the basis of Kelvin's theory that Altenkirch in 1911 was able to determine the conditions under which efficient thermoelectric energy conversion could be achieved. Altenkirch introduced the concept of a figure of merit, which has ever since assisted researchers in the development of thermocouple materials. In particular, the figure of merit, as it is defined today, was used by Abram Ioffe (1957) in his original treatise on the principles of thermoelectric energy conversion employing semiconductor thermoelements.

Qualitatively, it is obvious that the thermocouples should be made from materials that have large thermoelectric coefficients, high electrical conductivities, and low thermal conductivities. The electrical conductivity, σ, and the thermal conductivity, λ, will be familiar to all readers, but the thermoelectric coefficients may require definition. The two that are of most concern to us are the Seebeck coefficient, α, and the Peltier coefficient, π. Both coefficients are defined for junctions between two conductors. Thus, if the conductors a and b, shown in Fig. 1, meet at junctions 1 and 2 that differ in temperature by an amount ΔT, the voltage that appears when an opening is introduced in either conductor is

$$V = \alpha_{ab} \Delta T. \tag{1}$$

The Seebeck coefficient is assigned a positive value if the voltage tends to drive a current from 1 to 2 through the conductor a when 1 is at a higher temperature than 2.

The Peltier coefficient is defined, for the same pair of conductors, by the

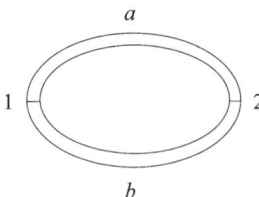

FIG. 1. Definition of the thermoelectric effects. The conductors a and b meet at junctions 1 and 2. If these junctions are at different temperatures, a Seebeck voltage appears across an opening in one of the conductors. If a current is passed, there is Peltier cooling at one junction and Peltier heating at the other.

relation

$$\pi_{ab} = \frac{q}{I}. \tag{2}$$

Here q is the rate of heating or cooling at one of the junctions when an electric current I passes round the circuit. The Peltier coefficient is regarded as positive if junction 1 becomes heated and junction 2 is cooled when the electric current through conductor a passes from 1 to 2.

Although it is only of secondary importance, the Thomson coefficient, τ, should also be defined since it is related to the Peltier and Seebeck coefficients. It is a property of a single conductor and is given by

$$\tau = \frac{dq/dx}{I\, dT/dx}, \tag{3}$$

where dq/dx is the rate of heating per unit length and dT/dx is the temperature gradient.

The Kelvin relationships are

$$\tau_a - \tau_b = T \frac{d\alpha_{ab}}{dT} \tag{4}$$

and

$$\pi_{ab} = \alpha_{ab} T. \tag{5}$$

The second of these equations is particularly important, since it relates the Peltier coefficient to the more easily measured Seebeck coefficient.

Although the Peltier and Seebeck effects only become apparent at junctions, the two coefficients are essentially properties of the bulk materials. Thus, it is correct to regard π_{ab} as being equal to the difference between the absolute coefficients π_a and π_b. Similarly, α_{ab} is equal to $\alpha_a - \alpha_b$.

It may be noted that the differential Peltier and Seebeck coefficients are zero for the junction between any two superconductors and it is, therefore, reasonable to assume that all such materials have zero absolute coefficients. It is then possible to determine the absolute Peltier or Seebeck coefficient for any normal conductor by joining it to a superconductor. The absolute coefficients at temperatures above that at which superconductivity disappears can be found with the aid of Eq. (4), after the Thomson coefficient has been measured.

We now outline the procedures for evaluating the efficiency of a thermoelectric generator and the coefficient of performance of a thermoelectric refrigerator. Although any practical device will use many thermocouples acting thermally in parallel, it is sufficient for our purposes to consider the behavior of a single couple. We assume that the only thermal flow between the heat source and the heat sink is through the two thermoelements. We also suppose that there is no temperature difference between the source and one of the junctions and between the sink and the other junction.

For the case of a generator the thermocouple is connected to an electrical load of resistance R_L, as shown in Fig. 2. In this context, it is noteworthy that an additional conductor can be introduced into the circuit without any effect on the thermoelectric behavior, provided that the points of insertion are at the same temperature. If the temperature difference between the source and the sink is $T_H - T_C$, the electromotive force is $\alpha_{ab}(T_H - T_C)$ and this drives a current I, which is equal to $\alpha_{ab}(T_H - T_C)/(R + R_L)$, around the circuit, R being the electrical resistance of the thermocouple. The useful power is given by

$$w = I^2 R_L. \tag{6}$$

The heat, q, that is drawn from the source is partly lost by conduction through the thermoelements. The remainder is used to compensate for the Peltier cooling of the hot junction when a current is flowing. Thence,

$$q = K(T_H - T_C) + \alpha_{ab} I T_H. \tag{7}$$

where K is the thermal conductance of the thermoelements in parallel. The efficiency is

$$\eta = \frac{w}{q}. \tag{8}$$

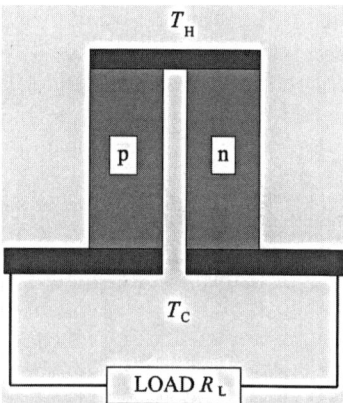

FIG. 2. A single thermocouple used as a thermoelectric generator. When the source and sink are at different temperatures, an electric current is driven through the load.

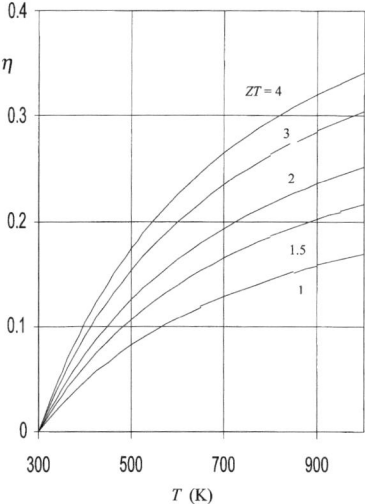

FIG. 3. Plots of efficiency against source temperature for thermoelectric generators having various values of the dimensionless figure of merit, ZT. The sink is supposed to be at 300 K.

The efficiency for a given thermocouple varies with the resistance of the load. Ioffe (1957) showed that the highest efficiency is given by

$$\eta = \frac{(T_H - T_C)}{T_H} \frac{(1 + ZT)^{1/2}}{T_C/T_H + (1 + ZT)^{1/2}}, \tag{9}$$

where T is taken to be equal to $(T_H + T_C)/2$, that is, the mean temperature over which the device operates. It is noted that $(T_H - T_C)/T_H$ is the efficiency of an ideal thermodynamic machine and this value is only approached by the thermoelectric generator if Z becomes very large. The quantity Z, which is known as the figure of merit of the thermocouple, is given by

$$Z = \frac{\alpha_{ab}^2}{KR}. \tag{10}$$

Figure 3 shows how the efficiency varies with T_H for various values of ZT.

Z itself is not a fixed quantity for a given pair of thermoelectric materials, but depends on the relative dimensions of the two branches of the thermocouple. The efficiency reaches its optimum value if the lengths l_a and l_b and the cross-section areas S_a and S_b of the branches satisfy the relation

$$\frac{l_a S_b}{l_b S_a} = \left(\frac{\sigma_a \lambda_a}{\sigma_b \lambda_b}\right)^{1/2}. \tag{11}$$

When this equation is satisfied, the figure of merit is given by

$$Z = \frac{(\alpha_a - \alpha_b)^2}{[(\sigma_a \lambda_a)^{1/2} + (\sigma_b \lambda_b)^{1/2}]^2}. \tag{12}$$

It is important to note that this figure of merit is a function of the properties of both the materials that form the thermocouple. Nevertheless, it is a common practice to define the figure of merit for a single material as

$$z = \frac{\alpha^2 \sigma}{\lambda}. \tag{13}$$

It often happens that the figure of merit, Z, for a couple is approximately equal to the average of the values of z for the two branches, but the reader should be warned that this is not always the case. However, it is convenient to use z rather than Z when one is searching for new thermoelectric materials.

We turn now to the problem of refrigeration using the Peltier effect. We suppose that a current I is passed around the thermoelectric circuit shown in Fig. 4 in such a direction that junction 2, at the temperature T_H, is heated and junction 1, at the temperature T_C, is cooled at the rate q. The cooling rate is given by

$$q = \alpha_{ab} T I - I^2 R/2 - K(T_H - T_C). \tag{14}$$

Here use has been made of the Kelvin relation (5) to eliminate the Peltier

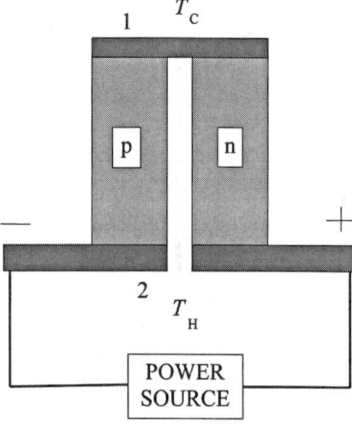

FIG. 4. Essential features of a thermoelectric refrigerator.

coefficient in terms of the Seebeck coefficient. It is assumed, as can easily be proved, that half of the Joule heating flows to each junction. The third term on the right-hand side of Eq. (14) represents the heat loss due to thermal conduction.

The electrical power, w, is consumed partly in overcoming the Seebeck voltage and partly in forcing current through the electrical resistance, R. Thence,

$$w = \alpha_{ab}(T_H - T_C)I + I^2 R. \tag{15}$$

The coefficient of performance is

$$\phi = \frac{q}{w}. \tag{16}$$

Not surprisingly, there is a current for which the coefficient of performance reaches its highest value. When this current is applied,

$$\phi = \frac{T_1[(1 + ZT)^{1/2} - T_H/T_C]}{(T_H - T_C)[(1 + ZT)^{1/2} + 1]}. \tag{17}$$

It will be seen once again that the quantity Z has an important part to play in the performance of the device. In other words, the same figure of merit applies for thermoelectric refrigeration as for generation.

For refrigeration purposes, an important quantity is the maximum temperature difference that can be achieved when the thermal load is zero. It is found that

$$\Delta T_{max} = \frac{ZT_C^2}{2}. \tag{18}$$

In Fig. 5, this maximum temperature difference is plotted against ZT, when the sink temperature, T_H, is set at 300 K.

The absolute Seebeck and Peltier coefficients can be of either positive or negative sign. It is obvious from Eq. (12) that the figure of merit of a couple will generally be largest if materials *a* and *b* have coefficients of opposite sign. Henceforth, we shall replace the subscript *a* by *p* and the subscript *b* by *n* to emphasize that a thermocouple used in energy conversion will normally have a positive and a negative branch. Occasionally, however, one may use a passive material for one of the branches (Goldsmid *et al.*, 1988). Such a material has absolute thermoelectric coefficients that are very small or zero, combined with a very large value for the ratio σ/λ. The figure of merit of the couple is then nearly equal to *z* for the active branch.

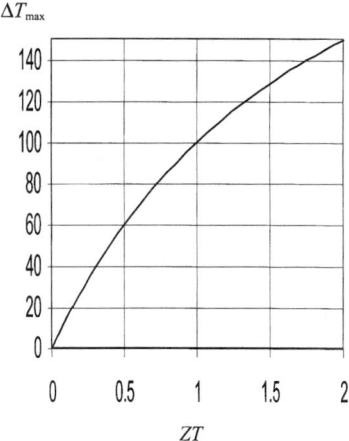

FIG. 5. Maximum temperature difference for a single stage thermoelectric refrigerator plotted against ZT. The heat sink is set at 300 K.

The figures of merit, Z and z, have the dimensions of inverse temperature, but in any of the expressions that describe the performance of a thermoelectric energy converter they invariably appear multiplied by the absolute temperature, T. It is usual, therefore, to find that researchers employ the dimensionless figures of merit, ZT and zT, instead of Z and z, when describing their work. Another common practice is to define the figure of merit z as the ratio of the so-called power factor to the thermal conductivity, λ. The power factor is given by $\alpha^2\sigma$ and thus depends solely on the electronic properties of the material (Bhandari and Rowe, 1994). The thermal conductivity, on the other hand, is often dominated by the contribution from the lattice vibrations, although it must not be forgotten that there is always a substantial electronic component of the thermal conductivity in any good thermoelectric material.

In some low-temperature applications, the figure of merit can be improved by applying a magnetic field (Wolfe, 1964), though such a field usually has only a very small effect at and above ordinary temperatures. However, when a magnetic field is applied to a conductor a number of new effects occur. In particular, when a longitudinal electric current is accompanied by a transverse magnetic field, the result is a flow of heat in the mutually perpendicular direction. This, the Ettingshausen effect, is an analog of the Peltier effect and can be used for refrigeration. Likewise, a longitudinal flow of heat in a transverse magnetic field produces a mutually perpendicular electromotive force. This is the Nernst effect and could, in principle, form the basis for an electrical generator. The Ettingshausen and Nernst effects are illustrated in Fig. 6 and are related to one another in more

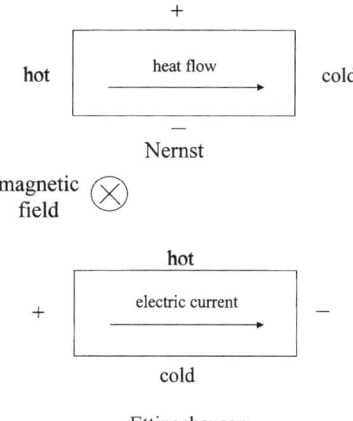

FIG. 6. Definition of the Ettingshausen and Nernst effects. In the Ettingshausen effect, a longitudinal electric current gives rise to a transverse temperature gradient, and in the Nernst effect, a longitudinal temperature gradient leads to a transverse electric field. In both cases, a mutually perpendicular magnetic field is applied.

or less the same way as are the Peltier and Seebeck effects. It is possible to define a thermomagnetic figure of merit, Z_E, which has the same significance as Z (Kooi *et al.*, 1963) insofar as the efficiency and coefficient of performance are concerned. There are certain practical advantages that derive from the use of the thermomagnetic effects, but, with the values of magnetic field strength that are readily available, these advantages can be utilized only at temperatures of the order of that of liquid nitrogen or less.

II. Semiconductor Thermoelements

When Altenkirch formulated his theory of thermoelectric energy conversion in 1911, the only thermocouple materials were made from metals or metallic alloys. These substances generally have Seebeck coefficients of less than some tens of microvolts per degree. They also invariably have ratios of the electrical to thermal conductivity that lie close to the value given by the Wiedemann–Franz law. When these factors are taken into account, one finds that zT is always going to be very much less than unity. It was not until work began on semiconductor thermoelements in the middle of the 20th century that worthwhile figures of merit appeared.

It was found that some semiconductors displayed Seebeck coefficients of the order of millivolts per degree but, unfortunately, the same materials also had very small values for the ratio of electrical to thermal conductivity. This was due to the dominance of heat conduction by the lattice in substances that have small electrical conductivities. Eventually, though, it was found that some semiconductors with more modest values of the Seebeck coefficient had ratios of the electrical to thermal conductivity that are not much smaller than the Wiedemann–Franz value, and it is these materials that form the present-day thermoelements.

Semiconductors can display conduction by negative electrons, positive holes, or both. The Seebeck coefficient is negative when the charge carriers are electrons and positive when the carriers are holes. In discussing the selection and optimization of semiconductors for thermoelectric applications, we shall suppose that there is only one type of charge carrier. This is a desirable condition since the thermoelectric effects from the two carriers in mixed and intrinsic semiconductors act in opposition to one another.

According to the modern theory of solids one must generally employ Fermi–Dirac statistics rather than classical statistics when determining the properties of the charge carriers. One of the fundamental differences between metals and semiconductors is that the former satisfy the degenerate approximation, whereas the latter may often be treated using the classical approximation. In fact, it turns out that the best thermoelectric materials are semiconductors in which the carrier concentration is high enough for the classical approximation to break down, but we shall nevertheless, for the sake of simplicity, assume that this approximation applies.

In a classical semiconductor the Seebeck coefficient is given by

$$\alpha = \pm \frac{k}{e}(\eta - (r + \tfrac{5}{2})). \tag{19}$$

Here k is Boltzmann's constant and e is the electronic charge. η is the Fermi energy divided by kT, and r is a scattering parameter that describes the way in which the relaxation time for the carriers varies with energy. It is usually supposed that r lies between $-\tfrac{1}{2}$ and $\tfrac{3}{2}$, the extremes corresponding to acoustic-mode lattice scattering and ionized-impurity scattering. In Eq. (19) the upper sign applies when the charge carriers are holes and the lower sign when they are electrons.

The classical expression for the electrical conductivity is

$$\sigma = ne\mu, \tag{20}$$

where μ is the carrier mobility and n is the carrier concentration, which is given by

$$n = 2\left(\frac{2\pi m^* kT}{h^2}\right)^{3/2} \exp \eta. \tag{21}$$

The quantity m^* is known as the effective mass of the carriers, and h is Planck's constant.

The thermal conductivity is the sum of a lattice component λ_L and an electronic component λ_e,

$$\lambda = \lambda_L + \lambda_e, \tag{22}$$

where

$$\lambda_e = L\sigma T \tag{23}$$

and

$$L = \left(\frac{k}{e}\right)^2 (r + \tfrac{5}{2}). \tag{24}$$

An advantage of the classical approximation is that it leads to a simple expression for the figure of merit z that can be used to determine the optimum value for the reduced Fermi energy η. We find (Goldsmid, 1986) that this optimum value satisfies the equation

$$\eta_{opt} + 2(r + \tfrac{5}{2})\beta \exp \eta_{opt} = r + \tfrac{1}{2}. \tag{25}$$

Here the quantity β, which was introduced by Chasmar and Stratton (1959), is defined by

$$\beta = \left(\frac{k}{e}\right)^2 \frac{\sigma_0 T}{\lambda_L} \tag{26}$$

where

$$\sigma_0 = 2e\mu \left(\frac{2\pi m^* kT}{h^2}\right)^{3/2}. \tag{27}$$

If we express μ in the units $m^2\,V^{-1}\,s^{-1}$ and λ_L in $W\,m^{-1}\,K^{-1}$, we find that

$$\beta = 5.745 \times 10^{-6} \frac{\mu}{\lambda_L} \left(\frac{m^*}{m}\right)^{3/2} T^{5/2}$$

where m is the mass of a free electron.

The dimensionless figure of merit zT is given by

$$zT = \frac{[\eta - (r + 5/2)]^2}{(\beta \exp \eta)^{-1} + (r + 5/2)}. \tag{28}$$

This equation shows that β should be as high as possible and, incidentally, this remains true even if we use the full Fermi–Dirac statistics. However, it turns out that $\beta < 0.5$ for all known materials. This means that η_{opt} is almost certainly going to exceed -2 and, in this case, classical statistics will be invalid. Nevertheless, our simple approach is correct in indicating that the optimum Fermi level must be quite close to the appropriate band edge. The material will certainly not satisfy the degenerate approximation. The optimum Seebeck coefficient can be calculated using Fermi–Dirac statistics. Its precise value will depend on β and r, but it can be stated with some confidence that it will lie close to $\pm 200\,\mu\text{V K}^{-1}$ unless materials with greatly improved values of β are found in the future.

At any particular temperature β is proportional to $\mu(m^*/m)^{3/2}/\lambda_L$. This shows that the preferred semiconductors are those with a high carrier mobility, a large effective mass, and a low lattice conductivity. It is also necessary that the energy gap, E_g, be large enough for the concentration of the minority carriers to be negligible when the Seebeck coefficient is optimized. It can be shown quite simply that the maximum value of the Seebeck coefficient at a given temperature, for a classical semiconductor, is given by $E_g/2eT$ (Goldsmid and Sharp, 1999).

One of the most useful rules for the selection of materials has been based on the observation that, in a given series of elements or compounds, the lattice thermal conductivity falls with increasing mean atomic weight (Ioffe and Ioffe, 1956). The semiconductors with the highest carrier mobilities are those that have predominantly covalent binding, but these also tend to have the highest values for the lattice conductivity. Thus, it is probably not too much of a disadvantage to have some measure of ionic binding. There is a tendency for semiconductors of high atomic weight to have large mobilities, but these are often associated with small values for the energy gap. Thus, some high atomic weight materials are satisfactory for use in thermoelectric refrigeration but cannot be used at the higher temperatures that are usually employed in thermoelectric generation.

The high atomic weight rule was used in the selection of bismuth telluride for Peltier cooling (Goldsmid and Douglas, 1954). Subsequent work on this compound showed that it has a multivalley band structure for both the electrons and holes, and it was realized that this is also a contributing factor to its good performance.

Ioffe and some of his colleagues suggested that any given element or compound might be improved by alloying it with an isomorphous substance (Ioffe et al. 1956). It was pointed out that the lattice thermal conductivity should be reduced on the formation of a solid solution, whereas the carrier mobility might not be changed. It was predicted that the phonons would be scattered by the disturbances in the short-range order, but the preservation of long-range order would prevent additional scattering of the electrons and holes. For example, the figure of merit of bismuth telluride can be signifi-

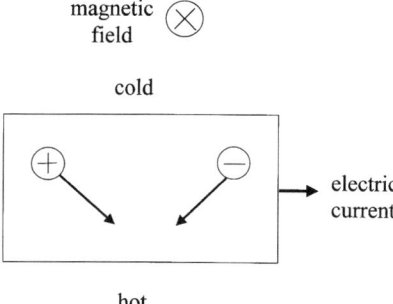

FIG. 7. Origin of the bipolar Ettingshausen effect. The partial flows of the electrons and holes are in opposite directions, but the magnetic field deflects both types of carrier in the same direction.

cantly improved by alloying it with antimony telluride or bismuth selenide. Likewise, lead–tin telluride is superior to lead telluride, while silicon–germanium alloys are far superior as generator materials to either silicon or germanium on their own.

Mention has already been made of the need to avoid significant minority carrier conduction in thermoelectric materials. However, it turns out that bipolar effects are beneficial when the transverse thermomagnetic effects are being employed. For example, when an intrinsic semiconductor is carrying a longitudinal current, the electrons and holes travel in opposite directions. A transverse magnetic field then drives both types of carrier in the same mutually perpendicular direction, as shown in Fig. 7. The electron–hole pairs carry their activation energy with them and this enhances the Ettingshausen effect. The Nernst effect, too, is improved when bipolar conduction occurs though, as yet, no practical generator based on this effect has been proposed.

III. Thermoelectric Materials

We now review the materials that have become established over the past half-century for thermoelectric generation and for refrigeration using both the Peltier and Ettingshausen effects.

The most useful materials have been those based on bismuth telluride, though nowadays this compound is rarely used other than as a solid solution with antimony telluride or bismuth selenide or both. These compounds all have a hexagonal structure with the formula X_2Y_3, where X is bismuth or antimony and Y is tellurium or selenium. The crystals of these compounds are formed of layers of atoms that follow the sequence Y-X-Y-X-Y. The atoms within this sequence are subject to mixed covalent–ionic binding, but the linkage to the next sequence is through weak van der Waals bonds. Consequently, the crystals are easily cleaved perpendicular to the c-axis. This leads to manufacturing problems if the thermoelements are made from single crystals, or polycrystals in which the cleavage planes all lie parallel to a common axis. However, it is just these materials that are preferred if the highest figure of merit is desired (Yim and Rosi, 1972). The lattice conductivity is certainly less by a factor of about 2 in a direction perpendicular to the cleavage planes, but the electrical conductivity displays an even stronger anisotropy. This is the case particularly for n-type alloys of bismuth telluride. Nevertheless, thermoelements are sometimes made from polycrystalline material containing randomly oriented grains, since these have much better mechanical properties than material with aligned grains.

Aligned material is usually produced from the melt by, for example, the Bridgman method, while polycrystalline samples are usually made by a powder metallurgy, or sintering, technique. Some degree of alignment can be introduced in sintered samples if the powders are hot-pressed rather than cold-pressed (Kajihara et al., 1998). Even better alignment results from the use of an extrusion method (Seo et al., 1998), and it seems possible that extruded material may eventually be the equal of that prepared by melt growth.

Bismuth telluride is a semiconductor with the relatively small energy gap of about 0.15 eV. This allows the Seebeck coefficient to reach a value of $\pm 260\,\mu\text{V}\,\text{K}^{-1}$ at room temperature when appropriate doping agents are added. However, the preferred Seebeck coefficient is closer to $\pm 200\,\mu\text{V}\,\text{K}^{-1}$ as suggested by theory. When grown from the melt, the compound is p-type and the hole concentration can be increased if an acceptor impurity such as lead is added. On the other hand, it can be made less strongly p-type or, eventually, n-type if a donor impurity, such as iodine, is added. At the optimum Seebeck coefficient, for either type of material, the electrical conductivity is about $10^5\,\Omega^{-1}\,\text{m}^{-1}$.

The thermal conductivity of the optimized material is of the order of $2\,\text{W}\,\text{m}^{-1}\,\text{K}^{-1}$, of which about $1.5\,\text{W}\,\text{m}^{-1}\,\text{K}^{-1}$ is the lattice contribution. However, the lattice conductivity becomes less on the formation of a solid solution. Thus, for the best p-type alloy, $Bi_{0.5}Sb_{1.5}Te_3$, the lattice conductivity falls to below $1\,\text{W}\,\text{m}^{-1}\,\text{K}^{-1}$ and, since there is also a marginal improvement in the power factor, the figure of merit rises from about $2\times 10^{-3}\,\text{K}^{-1}$ for Bi_2Te_3 to $3.3\times 10^{-3}\,\text{K}^{-1}$ for the alloy. There is a similar improvement for n-type material when the alloy $Bi_2Te_{2.7}Se_{0.3}$ is used instead of Bi_2Te_3. It will be seen, then, that a ZT value of about unity can be achieved with a thermocouple made from alloys of bismuth telluride.

The alloys of bismuth telluride can be used for thermoelectric generation if the heat source is not at too great a temperature, but above, say, 200°C these materials are no longer the most suitable. The fact that the energy gap is so small means that mixed conduction is increasingly difficult to avoid, and there is also the question of chemical instability to be considered.

Bismuth telluride alloys can be used below room temperature. However, for both conductivity types, z becomes less as the temperature is reduced, largely because of an increase in the lattice conductivity. Of course, zT becomes smaller even more rapidly. Nevertheless, the p-type alloys continue to be employed down to the lowest temperatures that can be reached with a multistage thermoelectric cooler, for want of any better material. On the other hand, there is an alternative n-type material at low temperatures.

The element bismuth is a semimetal. This means that the energy gap is negative but the overlap between the conduction and valence bands is small enough for the carrier concentration to be much less than in a true metal and capable of change by doping with donor or acceptor impurities. When the isomorphous semimetal antimony is alloyed with bismuth, there is a small range over which the energy gap becomes positive (Lenoir et al., 1995). The gap is too small to allow the Bi–Sb alloys to become better than bismuth telluride alloys at room temperature, but they are superior as n-type materials below about 200 K.

It might be thought, then, that multistage cooling units, with some of the stages maintained below 200 K, would invariably employ Bi–Sb alloys for their negative branches, as shown schematically in Fig. 8. There is, however, a practical problem associated with bismuth and Bi–Sb. The power factor is significantly greater, and the lattice thermal conductivity significantly smaller if the current flow lies in the c-direction rather than the plane of the a-axes. This means that the thermoelements should ideally be made from single crystals with the cleavage planes normal to the length direction. Such thermoelements are too fragile for use in most devices. Nevertheless, it is possible that extruded Bi–Sb will have good enough thermoelectric and mechanical properties to allow its use in low temperature cooling (Martin-Lopez et al., 1997).

Bismuth and its alloys have exceptionally high values for the electron mobility. This means that the thermogalvanomagnetic effects can become quite large at relatively small magnetic field strengths, particularly at low temperatures. In fact, the figure of merit of bismuth itself is improved when a transverse field is applied, and a similar improvement is found for the Bi–Sb alloys. For example, Wolfe and Smith (1962) observed an increase in z by a factor of more than 2 when a field of no more than about 0.5 T was applied to $Bi_{0.88}Sb_{0.12}$ at a temperature of 160 K. Yim and Amith (1972) found that zT for $Bi_{0.85}Sb_{0.15}$ was in excess of unity at 100 K in a field of only 0.15 T. Unfortunately, although p-type samples can be produced by the addition of tin, their figure of merit is much lower than that of the n-type material, though at temperatures on the order of 100 K it is possible to use high-T_c superconductors as passive branches in conjunction with n-type Bi–Sb thermoelements.

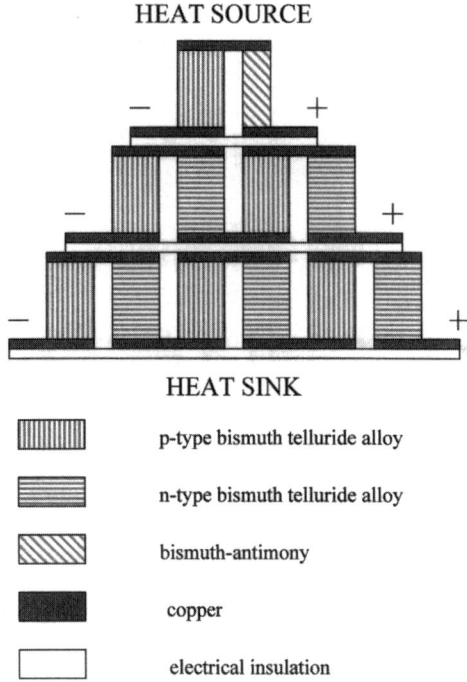

Fig. 8. Schematic diagram of a high-performance multistage cooling unit. The positive branches and some of the negative branches are made from alloys of bismuth telluride. The low-temperature negative thermoelements are made from bismuth–antimony. The connectors to each stage are not shown.

Bismuth and Bi–Sb alloys are the best materials for thermomagnetic energy conversion. This is essentially a low-temperature application, since even in these materials the mobility of the carriers is not particularly large above, say, 200 K. Although the theoretical thermomagnetic figure of merit in a high magnetic field can be quite high at ordinary temperatures, the required field is then too large for practical applications. The mobility is higher for bismuth than for the alloys, but the smaller lattice conductivity of the latter makes them preferable for Ettingshausen cooling as well as Peltier cooling. The bipolar effects on which a large thermomagnetic figure of merit depends really require both the electron and hole mobilities to be high. The hole mobility in bismuth and its alloys is not nearly so large as the electron mobility, but even so, quite substantial values of the Ettingshausen figure of merit, z_E, can be obtained. For example, Horst and Williams (1980) have observed a value of $z_E T$ equal to about unity at 150 K for $Bi_{0.97}Sb_{0.03}$, the required magnetic field being about 1 T.

Turning now to materials for thermoelectric generation, we realize that they must be chemically stable at the highest temperature of operation. They must also have an energy gap that is wide enough for a Seebeck coefficient of the order of $\pm 200\,\mu\text{V}\,\text{K}^{-1}$ to be achieved in the extrinsic region of conduction. However, although this will mean that z may be smaller than for bismuth telluride or bismuth-based materials, it is quite possible for the dimensionless figure of merit, zT, to be as large or even larger. Generally we find that a given material is only suitable for use over a particular range of temperature.

It is worth mentioning that the thermocouples that have been used for a great many years for the measurement of temperature are really no more than thermoelectric generators, albeit very inefficient ones. There are also other applications of thermoelectric generation in which cheap, robust materials are preferable to those that have a higher figure of merit but that are expensive or fragile. Thus, iron disilicide is still of some interest (Birkholz, 1989).

At temperatures that are just above those for which bismuth telluride can be used, the preferred materials are based on lead telluride, PbTe. This compound was, in fact, one of the first semiconductors to be studied with a view to thermoelectric applications (Ioffe, 1957). It is similar to bismuth telluride in that it has a high mean atomic weight and a multivalley band structure. However, it differs in having the cubic rock salt structure. It makes up for somewhat inferior properties compared with bismuth telluride at room temperature by having an energy gap of 0.32 eV at 300 K, which allows one to obtain Seebeck coefficients in excess of $300\,\mu\text{V}\,\text{K}^{-1}$. Both p-type and n-type samples can be produced either by making the compound nonstoichiometric or by adding alkali metals as acceptors or halogens as donors (Fano, 1994).

Although the electron mobility in lead telluride is higher than that in bismuth telluride, the effective mass is somewhat smaller. Also, the lattice conductivity at room temperature, equal to $2.0\,\text{W}\,\text{m}^{-1}\,\text{K}^{-1}$, is greater than that of bismuth telluride. In consequence, the figure of merit is no more than about $1 \times 10^{-3}\,\text{K}^{-1}$ at 300 K (Wright, 1970) but it eventually becomes larger than that of bismuth telluride, as the temperature is raised.

Lead telluride forms solid solutions with isomorphous compounds, such as lead selenide and tin telluride. In doing so, the lattice thermal conductivity is lowered, with a beneficial effect on the figure of merit. It was once thought possible that PbTe–SnTe was best for p-type thermoelements while, for n-type material, PbTe–PbSe was to be preferred. However, later work suggests that PbTe–SnTe is suitable for both the negative and positive branches (Rowe and Bhandari, 1983). Within this system, any fall in the carrier mobility is more than offset by the reduction in the lattice conductivity. Because of a band reversal effect, the energy gap in the PbTe–SnTe system actually becomes zero at the composition $Pb_{0.4}Sn_{0.6}Te$, even though the compounds PbTe and SnTe are both true semiconductors. For this reason, the concentration of tin telluride should not be too large and the composition $Pb_{0.75}Sn_{0.25}Te$ has been recommended (Rosi et al., 1961).

The power factor in the lead telluride alloys is greater for n-type material than for p-type material. For example, at 700 K, zT is equal to about 1.0 for the former and only 0.7 for the latter. However, at this temperature, zT can also be raised to about unity for the positive thermoelements, if they have the so-called TAGS formulation. TAGS materials are alloys between $AgSbTe_2$ and GeTe (Skrabek and Trimmer, 1994). They are closely related to lead telluride in that they possess the rock salt structure over part of the compositional range. However, there is a phase transition to a rhombohedral structure when the concentration of germanium telluride is less than about 80% and, in fact, about 70% GeTe is preferred. It is possible that the strain of the lattice associated with the phase transition may assist in the reduction of the lattice conductivity.

At still higher temperatures, the thermoelectric materials most frequently used are the alloys of silicon with germanium. This is remarkable since neither of the elements is at all promising for thermoelectric applications, both having rather large values of the lattice thermal conductivity. It so happens, however, that the lattice conductivity falls by about an order of magnitude within the Si-Ge alloy system. Thus, at 300 K silicon and germanium have lattice conductivities of 113 and 63 $W m^{-1} K^{-1}$ respectively, whereas a typical solid solution, $Si_{0.7}Ge_{0.3}$, has a value of only 10 $W m^{-1} K^{-1}$.

There is a complete range of solid solution between silicon and germanium, with solidus temperatures lying between 1693 and 1231 K. $Si_{0.7}Ge_{0.3}$ has a solidus temperature of about 1500 K and can be operated for long periods at a heat source temperature of 1300 K (Vining, 1994). The energy gap of silicon is 1.15 eV, while that of germanium is 0.65 eV, so there is an advantage in using a composition at the silicon-rich end of the system. Then, even at high temperatures, there is not too great a problem with minority carriers. In spite of the strong alloy scattering on the phonons, there is little effect on the carrier mobility, and for both n-type and p-type material it is possible to obtain values for the dimensionless figure of merit, zT, of about 0.5 or greater, at temperatures above 600 K (Rowe and Bhandari, 1983).

IV. Limits on the Figure of Merit

From our review of thermoelectric materials it is evident that values of zT of the order of unity can be obtained over a wide range of temperature. It is therefore tempting to propose that there is some physical limitation that prevents the dimensionless figure of merit from reaching higher values. In this section we consider the factors that may set a limit on zT, and we give reasons for expecting that this quantity will eventually become significantly greater than unity.

It is evident that the figure of merit of an optimally doped thermoelectric material depends on both electronic and lattice properties. The power factor is determined primarily by the product $\mu(m^*/m)^{3/2}$, though the width of the energy gap must also be considered. The other quantity of importance is the lattice thermal conductivity, λ_L. Thus, the ideal material is one in which λ_L is as small as possible and $\mu(m^*/m)^{3/2}$ is as large as possible, provided that this is consistent with sufficient separation of the conduction and valence bands at the temperature of operation. It is convenient to discuss these requirements separately.

The lattice conductivity is largest for crystals and smallest for glasslike or amorphous materials. Certain types of crystal have relatively large values for λ_L while others have values that can approach those that are typical of glasses. Slack (1979) has proposed that the smallest possible lattice conductivity can be predicted by setting the mean free path of the phonons equal to that which it would have in the amorphous state.

Slack suggested that the minimum thermal conductivity is obtained when each phonon has a free path length equal to its wavelength. An alternative view is that the smallest free path length is equal to half this value (Cahill et al., 1992). A compromise, based on the simple assumption that all the phonons have a free path length equal to the lattice spacing, results in a thermal conductivity that is intermediate between the values given by these more sophisticated theories. In any case, one expects the minimum lattice conductivity to be about $0.2\,\mathrm{W\,m^{-1}\,K^{-1}}$.

Next, one must consider the electronic properties. It is noteworthy that the highest value for $\mu(m^*/m)^{3/2}$ in any known material is found for electrons in bismuth. Thus, one approach that has been adopted is to suppose that a material can be found in which $\mu(m^*/m)^{3/2}$ has the value, $0.075\,\mathrm{m^2\,V^{-1}\,K^{-1}}$, that is characteristic of electrons in bismuth at 300 K, and in which λ_L is equal to $0.2\,\mathrm{W\,m^{-1}\,K^{-1}}$. It is also supposed that the energy gap is large enough to prevent bipolar conduction after the optimization of the carrier concentration. The preferred Seebeck coefficient would be higher, at some $\pm 350\,\mu\mathrm{V\,K^{-1}}$, than it is for existing materials. This is because the small lattice conductivity makes the electronic contribution to the transfer of heat more important.

The preceding assumptions yield a dimensionless figure of merit equal to about 4 (Slack, 1994). However, it is now thought that this estimate of an upper limit for zT may be too optimistic. It may be argued that the exceptionally favorable electronic properties of bismuth are the result of the fact that this material is a semimetal. It certainly seems to be true that as the energy gap becomes larger, so the value of $\mu(m^*/m)^{3/2}$ becomes less. It therefore seems to be unreasonable to suppose that the combination of properties just given will ever be realized. Rather, it seems more likely that the limiting value of zT will turn out to be nearer 2 than 4. Of course, even this lower value would yield a substantial improvement in performance over that of existing devices.

When we consider the possibilities at higher temperatures, we come to the same conclusions. At these temperatures, smaller values of z would suffice, but larger energy gaps would be needed, and a detailed study suggests that the limit on zT would remain the same for generation materials as well as those used for refrigeration.

V. New Developments

There has been a renewed interest in thermoelectricity during the past few years. Several new classes of material have been studied, and some of these show great promise. There have been proposals for novel configurations, such as two-dimensional and one-dimensional thermoelements, and they also look promising. Thus, in the last section of this introductory chapter, these developments are reviewed.

Central to the research on new materials has been Slack's concept of a minimum lattice conductivity and his ideas for its realization. The first semiconductors to be studied in detail, such as the group IV elements and the III–V compounds, had simple crystal structures. Most of the present-day thermoelectric materials are alloys based on elements or binary compounds. It is, perhaps, significant, however, that the most useful of the current materials, namely those based on bismuth telluride, have a rather more complex structure than the semiconductors generally used in electronic applications. Slack, in fact, pointed out that crystalline materials that have lattice conductivities approaching his predicted minimum value are usually quite complex in their structure. These materials include YB_{68}, Tl_3AsSe_3, and H_2O. In such crystals, there are atoms or molecular groups of atoms that do not have precisely defined positions or orientations. There is then no long-range correlation and the atoms or molecules are said to "rattle." Slack stated that the lattice conductivity in this type of system is generally smaller than in mixed crystals even though λ_L for the latter is less than in pure elements or compounds.

The low lattice conductivity must be combined with good electronic properties if the material is to be useful. Thus, one needs what may be termed a phonon glass and electron crystal (PGEC). Several of the substances that are now being studied fall into this category. Later chapters in this book describe these systems in detail, but we mention one or two of them here so that the reader can appreciate the trend that current research is taking.

Work has been in progress for several years on the skutterudites. Binary skutterudites have the formula AB_3 where A = Co, Ir, Rh and B = P, As, Sb. A major feature of the crystal structure is two large empty spaces within the unit cell. Some skutterudites have reasonably large Seebeck coefficients of the order of $\pm 200\,\mu\text{V K}^{-1}$, but the thermal conductivity is generally unacceptably large. However, λ_L can be considerably reduced by introducing heavy atoms into the spaces in the lattice. This has been done, for example, by Nolas et al. (1998a), who added La to $CoSb_3$. In this case, it was found that the lattice conductivity was lowered by an order of magnitude at room temperature. It was stated that part of the reduction was due to mass-defect scattering of the phonons and part due to rattling of the loosely bound atoms in their "cages." A dimensionless figure of merit in excess of unity has been observed for such materials at 700 K.

Perhaps more promising at lower temperatures are the clathrates. The germanium clathrates, for example, are based on compounds with the formula A_8Ge_{46}, where A represents atoms within voids in the germanium network. Some of the properties of these materials have been described by Nolas et al. (1998b). One particular material with the composition $Sr_8Ga_{16}Ge_{30}$ has a lattice conductivity at room temperature that is less than twice that of amorphous germanium. Similarly low thermal conductivities have been reported for germanium clathrates containing europium rather than strontium (Cohn et al., 1999). There are also silicon clathrates, but, as expected, these have higher lattice conductivities than their germanium counterparts. On the other hand, we would expect lower thermal conductivities for the tin clathrates, though, somewhat surprisingly, $Cs_8Zn_4Sn_{42}$ has a value of λ_L, which, though small, is greater than that of Cs_8Sn_{44}, and this material does not appear to behave as a phonon glass (Nolas et al., 1999). Some of the clathrates have sufficiently high Seebeck coefficients for thermoelectric applications and zT approaches unity at about 700 K, but so far the figure of merit at ordinary temperatures is less than half that of the best bismuth telluride alloys.

There are other systems that also show promise, and the number of possible compounds is so large that it would be most disappointing if none of them become superior to present-day materials.

One of the most exciting recent developments originated in a proposal by Hicks and Dresselhaus (1993) that the figure of merit might be improved if two-dimensional structures were used. It is necessary to modify conventional three-dimensional band theory when one or two of the crystal dimensions become comparable with the lattice spacing. The confinement of the carriers within a low-dimensional quantum well leads to the modification of the density-of-states function. The most likely benefit, then, is an increase in the number of carriers for a given Fermi energy, that is, for a particular Seebeck coefficient. It has become possible to fabricate two-dimensional crystals using such techniques as molecular beam epitaxy

(MBE). Harman et al. (1999) have shown that the power factor in two-dimensional lead telluride layers exceeds that of bulk material. It does not appear that the limitation on the figure of merit for bulk material will apply for low-dimensional systems.

It is possible that a one-dimensional system would be better still, but it is not yet clear how low-dimensional systems could be incorporated in practical devices. Nevertheless, there is an abundance of good ideas. One problem is the loss of heat through the passive layers, used in the creation of the barriers that contain the thin active layers. However, it is possible that this will be more than compensated by a reduction of the thermal conductivity in the active material. This has been demonstrated convincingly by Venkatasubramanian et al., (1999), who, indeed, may well have succeeded in producing a system with a zT value substantially in excess of unity.

The object in a low-dimensional thermoelectric system is to confine the charge carriers by means of barriers. However, it is also possible to design a thermoelectric material in which conduction takes place through the barriers. When the thickness of the semiconductor between the barriers is not large compared with the mean free path of the carriers, thermionic transport occurs. There is a reversible heating or cooling effect associated with this flow, just as there is in a vacuum diode. It has been suggested that the thermionic effect in a solid state system may actually be more efficient than the thermoelectric effect (Mahan et al., 1998). However, Vining and Mahan (1999) have now shown that this will not normally be the case unless the lattice thermal conductivity happens to be reduced.

Finally, it may be mentioned that vacuum diodes could eventually prove to be superior to thermocouples for refrigeration. Mahan (1994) has shown that they would become effective at around 500 K, if a thermionic emitter with a work function of 0.7 eV were used. There seems to be a distinct possibility that materials with even smaller values for the work function may be available and, if this quantity could be reduced to about 0.3 eV, thermionic refrigeration becomes most attractive (Nolas and Goldsmid 1999). This is because of the absence of conduction of heat by the lattice. Thermal radiation must be taken into account when thermionic vacuum diodes are used for energy conversion at high temperatures, but its effect is quite small near 300 K.

It remains to be seen which of these ideas will become dominant, but there are surely good prospects for an improved performance before many years have passed.

References

C. M. Bhandari and D. M. Rowe, Optimization of carrier concentration, in *CRC Handbook of Thermoelectrics*, Ed. D. M. Rowe, Boca Raton: CRC Press, 43 (1994).

U. Birkholz, Iron disilicide as a thermoelectric generator material, Proceedings 8th International Conference on Thermoelectrics, Nancy, France, INPL, 98 (1989).

D. G. Cahill, S. K. Watson, and R. O. Pohl, Lower limit to the thermal conductivity of disordered crystals, *Phys. Rev. B*, **46**, 6131 (1992).

R. P. Chasmar and R. Stratton, The thermoelectric figure-of-merit and its relation to thermoelectric generators, *J. Electron. Control*, **7**, 52 (1959).

J. L. Cohn, G. S. Nolas, V. Fessatidis, T. H. Metcalf, and G. A. Slack, Glasslike heat conduction in high-mobility crystalline semiconductors, *Phys. Rev. Lett.*, **82**, 779 (1999).

V. Fano, Lead telluride and its alloys, in *CRC Handbook of Thermoelectrics*, Ed. D. M. Rowe, Boca Raton: CRC Press, 257 (1994).

H. J. Goldsmid, *Electronic Refrigeration*, London: Pion (1986).

H. J. Goldsmid and R. W. Douglas, The use of semiconductors in thermoelectric refrigeration, *Brit. J. Appl. Phys.*, **5**, 386 (1954).

H. J. Goldsmid and J. W. Sharp, Estimation of the thermal band gap of a semiconductor from Seebeck measurements, *J. Electron. Mats.*, **28**, 869 (1999).

H. J. Goldsmid, K. K. Gopinathan, D. N. Matthews, K. N. R. Taylor, and C. A. Baird, High-T_c superconductors as passive thermo-elements, *J. Phys. D: Appl. Phys.*, **21**, 344 (1988).

T. C. Harman, D. L. Spears and M. P. Walsh, PbTe/Te superlattice structures with enhanced thermoelectric figures of merit, *J. Electron. Mats.*, **28**, L1 (1999).

L. D. Hicks and M. S. Dresselhaus, Effect of quantum-well structures on the thermoelectric figure of merit, *Phys. Rev. B*, **47**, 12727 (1993).

R. B. Horst and L. R. Williams, Application of solid state cooling to spaceborne infrared focal planes, Proceedings 3rd International Conference on Thermoelectrics, Arlington, Texas, IEEE, 183 (1980).

A. F. Ioffe, *Semiconductor Thermoelements and Thermoelectric Cooling*, London: Infosearch (1957).

A. F. Ioffe, S. V. Airapetyants, A. V. Ioffe, N. V. Kolomoets and L. S. Stil'bans, On increasing the efficiency of semiconducting thermocouples, *Dokl. Akad. Nauk.*, SSSR, **106**, 981 (1956).

A. V. Ioffe and A. F. Ioffe, Thermal conductivity of semiconductors, *Izv. Akad. Nauk.*, SSSR, Ser. Fiz., **20**, 65 (1956).

T. Kajihara, K. Fukuda, Y. Sato, and M. Kikuchi, Improvement of crystal orientation and thermoelectric properties for n- and p-type Bi_2Te_3 compounds, Proceedings 17th International Conference on Thermoelectrics, Nagoya, Japan, IEEE, 129 (1998).

C. F. Kooi, R. B. Horst, K. F. Cuff, and S. R. Hawkins, Theory of the longitudinally isothermal Ettingshausen cooler, *J. Appl. Phys.*, **34**, 1735 (1963).

B. Lenoir, A. Dauscher, H. Scherrer, S. Scherrer, M. Cassart and Yu. I. Ravich, Highest figure of merit in undoped Bi–Sb alloys, Proceedings 14th International Conference on Thermoelectrics, St. Petersburg, Russia (1995).

G. D. Mahan, Thermionic refrigeration, *J. Appl. Phys.*, **76**, 4362 (1994).

G. D. Mahan, J. O. Sofo, and M. Bartkowiak, Multilayer thermionic refrigerator and generator, *J. Appl. Phys.*, **83**, 4683, (1998).

R. Martin-Lopez, A. Dauscher, X. Devaux, B. Lenoir, H. Scherrer, and M. Zandona, Influence of the consolidation technique on the thermoelectric properties of mechanically alloyed Bi–Sb, Proceedings 16th International Conference on Thermoelectrics, Dresden, Germany, IEEE, 184 (1997).

G. S. Nolas and H. J. Goldsmid, A comparison of projected thermoelectric and thermionic refrigerators, *J. Appl. Phys.*, **85**, 4066 (1999).

G. S. Nolas, J. L. Cohn, and G. A. Slack, Effect of partial void filling on the lattice thermal conductivity of skutterudites, *Phys. Rev. B*, **58**, 164 (1998a).

G. S. Nolas, G. A. Slack, J. L. Cohn, and S. B. Schujman, The next generation of thermoelectric materials, Proceedings 17th International Conference on Thermoelectrics, Nagoya, Japan, IEEE, 294 (1998b).

G. S. Nolas, T. J. R. Weakley, and J. L. Cohn, Structural, chemical, and transport properties of a new clathrate compound $Cs_8Zn_4Sn_{42}$, *Chemistry of Materials*, **11**, 2470 (1999).

F. D. Rosi, E. F. Hocking, and N. E. Lindenblad, Semiconducting materials for thermoelectric power generation, *RCA Review*, **22**, 82 (1961).

D. M. Rowe and C. M. Bhandari, *Modern Thermoelectrics*, London: Holt, Rinehart and Winston (1983).

J. Seo, K. Park, and C. Lee, Fabrication and thermoelectric properties of n-type SbI_3-doped $Bi_2Te_{2.85}Se_{0.15}$ compounds by hot extrusion, *Mats. Res. Bull.*, **33**, 553 (1998).

E. A. Skrabek and D. S. Trimmer, Properties of the general TAGS system, *CRC Handbook of Thermoelectrics*, Ed. D. M. Rowe, Boca Raton: CRC Press, 267 (1994).

G. A. Slack, The thermal conductivity of nonmetallic crystals, in *Solid State Physics*, Ed. H. Ehrenreich, F. Seitz and D. Turnbull, New York: Academic Press, **34**, 1 (1979).

G. A. Slack, New materials and performance limits for thermoelectric cooling, in *CRC Handbook of Thermoelectrics*, Ed. D. M. Rowe, Boca Raton: CRC Press, 407 (1994).

R. Venkatasubramanian, E. Siivola, T. Colpitts, and B. C. O'Quinn, Phonon-blocking electron-transmitting structures, Proceedings 18th International Conference on Thermoelectrics, Baltimore, Maryland, IEEE, awaiting publication (1999).

C. B. Vining, Silicon germanium, in *CRC Handbook of Thermoelectrics*, Ed. D. M. Rowe, Boca Raton: CRC Press, 329 (1994).

C. B. Vining and G. D. Mahan, The B factor in multilayer thermionic refrigeration, *J. Appl. Phys.* **86**, 6852 (1999).

R. Wolfe, Magnetothermoelectricity, *Scientific American*, **210**, 70 (1964).

R. Wolfe and G. E. Smith, Effects of a magnetic field on the thermoelectric properties of bismuth–antimony alloys, *Appl. Phys. Lett.*, **1**, 5 (1962).

D. A. Wright, Materials for direct conversion thermoelectric generators, *Metallurgical Reviews*, **15**, 147 (1970).

W. M. Yim and A. Amith, Bi–Sb alloys for magneto-thermoelectric and thermomagnetic cooling, *Solid-State Electron.*, **15**, 1141 (1972).

W. M. Yim and F. D. Rosi, Compound tellurides and their alloys for Peltier cooling, *Solid-State Electron.*, **15**, 1121 (1972).

CHAPTER 2

Overview of Measurement and Characterization Techniques for Thermoelectric Materials

Terry M. Tritt

DEPARTMENT OF PHYSICS AND ASTRONOMY
CLEMSON UNIVERSITY
CLEMSON, SOUTH CAROLINA

Valerie M. Browning

U.S. NAVAL RESEARCH LABORATORY
WASHINGTON, D.C.

I. INTRODUCTION 25
II. THERMOELECTRIC MATERIALS 27
III. ESTIMATION OF THE EFFECT OF ERRORS IN CALCULATING ZT 28
IV. THERMOELECTRIC MEASUREMENTS 29
   1. *Issues* 29
   2. *Electrical Resistivity (and Electrical Conductivity)* 31
   3. *Seebeck Coefficient* 36
   4. *Thermal Conductivity* 39
   5. *Hall Coefficient, Carrier Concentration, and Mobility* 43
   6. *Z-Meters (or "Harman Technique")* 45
V. SUMMARY 47
   REFERENCES 48

I. Introduction

During the 1950s and 1960s, there was a considerable amount of research activity in the field of thermoelectric materials. Alloys based on the Bi_2Te_3 systems and $Si_{1-x}Ge_x$ systems were some of the most widely studied thermoelectric materials. These materials were extensively studied and optimized for their use in thermoelectric applications (solid state refrigeration and power generation) (Goldsmid, 1986; Rowe, 1995) and remain state-of-the-art materials for their specific temperature use. By the 1970s, research on thermoelectric materials had begun a steady decline. Universities, in particular, tended to discontinue their research efforts in thermoelectrics as it was considered to be a mature field, with little to gain by further study. Unfortunately, this resulted in somewhat of a void in the training of students and future researchers. The decline in interest in thermoelectrics meant that students were no longer learning many of the specific experimental techniques (and their limitations) necessary to characterize thermoelectric materials.

Recently there has been renewed interest in the field of thermoelectrics (Tritt *et al.*, 1997, 1998). This has been driven, in part, by new applications requiring materials that exhibit higher performance than existing materials (Hicks and Dresselhaus, 1993; Fleurial *et al.*, 1995; Sales *et al.*, 1996; Morelli *et al.*, 1995; Slack and Toukala, 1994; Nolas *et al.*, 1996; Tritt *et al.*, 1996). In addition, advances in thin film growth techniques and novel processes for forming bulk materials have allowed the exploration of a variety of new systems. Many new researchers have come into this field, often with new ideas and new concepts for materials. Multidisciplinary training in solid-state physics, solid-state chemistry, and materials science and engineering, as well as in extensive characterization of both the electrical and thermal transport of materials, is necessary for a successful research program in thermoelectrics. One distinction that is evident in this past decade of thermoelectric materials research is extensive collaboration between the solid-state chemists and experimental and theoretical physicists. This synergy makes for rapid advances in the development of new materials, from theoretical prediction to solid-state synthesis and subsequent characterization of a new material. A key factor to these advances is accurate and rapid measurements of the important properties that are related to a material's thermoelectric performance.

As stated previously, one of the difficulties in investigating thermoelectric materials lies in obtaining reliable and accurate measurements of their electrical and thermal properties. Tremendous efforts were expended in the late 1950s and 1960s in relation to the measurement and characterization of thermoelectric materials. These efforts were made by a generation of scientists who for the most part are no longer active, and this expertise will be lost to us unless we are aware of the great strides they made during their time. There are two recent papers that give excellent reviews of the issues related to accurate measurements of the electrical and thermal transport properties of thermoelectric materials, and the reader is encouraged to examine these papers (Tritt, 1997; Uher, 1996).

The purpose of this paper is to remind present-day researchers of these previous efforts and to acquaint new researchers with the difficulties of the measurements that they will encounter. The references provided (extensive, yet in no way complete) will serve as a useful resource for students as well as established scientists who are working in the area of thermoelectric materials research. This paper discusses the techniques employed to characterize thermoelectric materials and provides an overview of some of the systematic errors in these techniques that can be potential pitfalls to the researcher attempting to achieve reliable and accurate measurements. Statistical errors are not discussed in this paper, since understanding statistical errors is necessary for analyzing typical experimental data, and reference will be made to a couple of excellent general texts for analyzing statistical errors (Young, 1962; Taylor, 1982).

II. Thermoelectric Materials

Thermoelectric energy conversion utilizes the Peltier heat generated when an electric current is passed through a thermoelectric material to provide a temperature gradient with heat being absorbed on the cold side and rejected at the heat sink, thus providing a refrigeration capability. Conversely, an imposed ΔT will result in a voltage or current, i.e., small-scale power generation (Tritt, 1996). This aspect is widely utilized in deep-space applications. A radioactive material acts as the heat source in these RTGs (radioactive thermoelectric generators) and thus provides a long-lived energy supply. The advantages of thermoelectric solid-state energy conversion are compactness, quietness (no moving parts), and localized heating or cooling. Applications include cooling of CCDs, laser diodes, infrared detectors, low-noise amplifiers, computer chips, and biological specimens.

The essence of defining a good thermoelectric material lies primarily in determining the material's dimensionless figure of merit, $ZT = \alpha^2 \sigma T/\lambda$, where α is the Seebeck coefficient, σ the electrical conductivity, λ the total thermal conductivity ($\lambda = \lambda_L + \lambda_E$, the lattice and electronic contributions, respectively), and T is the absolute temperature in kelvins. The Seebeck coefficient, or thermopower, is related to the Peltier effect by $\Pi = \alpha T = Q_P/I$, where Π is the Peltier coefficient, Q_P is the rate of heating or cooling, and I is the electrical current.[1] The efficiency (η) and coefficient of performance (COP) of a thermoelectric device are directly related to the figure of merit of the thermoelectric material or materials. Both η and COP are proportional to $(1 + ZT)^{1/2}$. There are a number of excellent references that discuss the materials and measurement aspects and thoroughly discuss the field of thermoelectric materials (Goldsmid, 1986; Rowe, 1995; Wood, 1988; Rowe and Bhandari, 1983; Egli, 1960). These also contain many of the early references that can prove to be invaluable to the new researcher in this field.

Semiconductors have long been the materials of choice for thermoelectric applications. The most promising materials typically have carrier concentrations of approximately 10^{19} carriers/cm^3. The power factor, $\alpha^2 \sigma$, is typically optimized through doping to give the largest Z. High-mobility carriers are most desirable, thus yielding the highest electrical conductivity for a specific carrier concentration. In addition, to improve the figure of merit of these

---

[1] All the "Q-terms" discussed in this paper relate to rate of heat transfer or power related to that phenomenon.

TABLE I

PROPERTIES OF INTEREST FOR THERMOELECTRIC MATERIALS

Sample property	Relationship	Note
Seebeck coefficient	$\alpha = -\Delta V/\Delta T$	$Q_P = \alpha I T$
Electrical resistivity	$\rho = R_s \times A/l_0$	$Q_J = I^2 R$
Electrical conductivity	$\sigma = ne\mu = 1/\rho$	$Q_J = I^2 R$
Thermal conductivity	$\lambda = Q_K l_0 / A\Delta T$	$Q_K = \lambda A \Delta T / l_0$
Hall voltage	$V_H = R_H B I / t$	t: sample thickness
Hall coefficient (single carrier system)	$R_H = 1/ne$	Yields n and carrier sign
Carrier mobility	$\mu_H = \sigma R_H$	Want high μ for fixed n
Carrier concentration	$n = 1/R_H e$	$n \approx 10^{17} - 10^{19}\,\text{cm}^{-3}$

materials, attempts are made to lower the lattice thermal conductivity without decreasing the power factor proportionally and thus further increasing the figure of merit.

The measurements that are typically necessary to characterize a thermoelectric material are listed in Table I. Each of these is discussed later in this paper. These terms are defined as follows: μ is the carrier mobility, n is the carrier concentration, l_0 is the sample distance between the measuring leads, A is the cross-sectional area, R_H is the Hall coefficient, and e is the charge of the carrier, $-e$ (electrons) or $+e$ (holes). Of course, there are many other factors that go into the full understanding of the thermal and electrical transport in a material, but these aforementioned properties are given as a fundamental starting point.

III. Estimation of the Effect of Errors in Calculating ZT

Let us consider the importance of accurate measurements of the specific properties such as α, σ, and λ_T, by using an elementary example and performing a calculation of ZT using what will be defined as the real intrinsic numbers for the parameters and numbers, which include reasonable estimates of measurement errors. For the real ZT, we take $\alpha = 275\,\mu\text{V/K}$, $\sigma = 10^{+3}\,(\Omega\text{-cm})^{-1}$, $\lambda_T = 2$ watts/m-K, and $T = 300\,\text{K}$. These numbers would give $ZT = 1.13$ at 300K. If we perform the measurements with some systematic errors in our system, we can see how this would affect the calculation. The thermopower or Seebeck coefficient will be taken to be $\approx 10\%$ low ($\alpha' = 250\,\mu\text{V/K}$). The error could possibly come from a ΔT that is measured to be 10% higher than the real ΔT across the sample. This could be due to poor thermal anchoring of the thermocouples to the sample or inaccurate calibrations of the thermocouples. The conductivity [$\sigma' =$

$1.05 \times 10^{+3}$ $(\Omega\text{-cm})^{-1}$] will be taken to be 5% too high. This could result from inaccurate determination of the sample dimensions (3–4%) (recall $\rho = 1/\sigma = RA/l_0$) or determination of the sample current through the sample ($\approx$1–2%). Let the thermal conductivity ($\lambda'_T = 2.2$ watts/m-K) be 10% too high. This could result from inaccurate determination of the sample dimensions, errors in ΔT, or errors in determining the power through the sample, possibly due to radiation losses in the measurement. It should be stated that these are all very reasonable estimates of typical errors that can be made. Assume that there will be no error in determining the sample temperature ($T' = 300$ K). Given these systematic errors in our measurement, which are not unreasonable, unless one is very careful, a $(ZT)' = 0.89$ would be obtained which has an error of 21% from the real ZT value. It should be obvious, then, that accurate determination of these properties is very important.

IV. Thermoelectric Measurements

1. Issues

a. Standards and Samples

One of the first steps in improving the accuracy of a particular experimental setup is to identify and quantify any systematic errors in the measurement. One way to do this is to develop *standards* (materials with known and established thermoelectric properties) to check the apparatus and measurement techniques for accuracy and reproducibility. One can obtain standards for thermal conductivity or electrical resistivity from NIST, the National Institute of Standards and Technology (formerly the National Bureau of Standards). However, the lowest thermal conductivity standard that NIST currently has available is stainless steel, which has values around $\lambda_T = 15$ W/m-K at room temperature, an order of magnitude higher than that of a good thermoelectric material. A good standard for Seebeck measurements is even more difficult to obtain. Thus, very few good thermoelectric standards exist today. For this reason, the current thermoelectrics research community is working to develop laboratory standard materials to be shared between researchers. In addition, "round robin" or "blind measurement" studies are being conducted to assure the accuracy of results that are being reported. In order to appreciate the importance of accuracy in thermoelectric materials characterization, one need only consider the unfortunate circumstance of one research group that erroneously reported a very large ZT ($ZT \approx 2$). This report created quite a stir within the thermoelectrics community. However, it was later discovered that a subtle error in the resistivity measurement (sample decomposition at the surface) was yielding a false ZT value, and the result had to be withdrawn.

Another mistake that is often made results from calculating ZT using values of the various thermoelectric properties (α, σ, λ, n, and R_H) as measured from different samples. Thermoelectric materials have often proven to be notoriously inhomogeneous even among samples taken from the same batch. Therefore, it is best to take all measurements on the same sample. In addition, these measurements should be taken as closely together in time as possible in order to eliminate sample deterioration effects. Of course, in calculating ZT, it is also important to ensure that all of the measurements are taken at the same temperature in order not to be misled by any strong temperature dependence of the properties of the materials that are under investigation.

In general, electrical and thermal transport properties can have a very strong dependence on crystallographic direction, sometimes by orders of magnitude. Thus, when working with single crystals or oriented polycrystalline materials it is important to verify the crystallographic direction of the measurements being reported. Even "pressed pellet" polycrystalline samples can exhibit anisotropy in their properties, and sample orientation must be consistent when measuring and comparing the different thermoelectric properties. When possible, it is better to measure more than one sample of a given material. This will help to average the sample-to-sample differences, and possibly minimize discrepancies between various groups.

b. *Contacts and Contact Effects*

Establishing excellent electrical contacts to these thermoelectric materials is also an essential factor. Large contact resistances that result in Joule heating at the contacts $\{I^2(R_{C1} + R_{C2})\}$ can make these thermoelectric measurements extremely difficult. In some cases, Joule heating due to contact resistance can completely cancel the desirable "Peltier heat flow." In addition, if the contact resistances of the current leads differ significantly, an unwanted temperature gradient can develop because of the differences in Joule heating at the sample ends, $\Delta T \approx P \approx I^2(R_{C1} + R_{C2})$. For the Bi_2Te_3 class of materials, achieving good electrical contacts has always been an important issue. The fact that Cu or Au can readily diffuse into Bi_2Te_3 requires that these materials be plated with a diffusion barrier (typically Ni) prior to attaching contact leads. Once the sample has been plated with Ni, for example, a variety of contact techniques can be used to attach leads: solder (many different types and temperatures), arc welding–capacitor discharge (localized point contact—this requires a robust sample to avoid damage), metal-sputtering or evaporation (Au, Ag, or Cu, etc.), ion implantation or diffused contacts, Ag paints (e.g., Dupont 4929 Ag paint or SPI Ag coating for SEM), metal-plating or metal epoxies, and also needle pressure probes (e.g. "Pogo" contacts).[2]

Unfortunately, some of the materials of interest in thermoelectrics are susceptible to the formation of oxide layers or sample decomposition at the surface. Therefore, careful surface preparation is often one of the most crucial steps in achieving good adhesion and low resistance contacts. Also, most of the potential thermoelectric materials are semiconductors, thus requiring metal–semiconductor electrical contacts. Thus, this incorporates all the problems and issues that exist in making electrical contacts between metals and semiconductors. The readers are referred to two excellent texts that deal with this issue (Runyan and Shaffer, 1997; Streetman, 1995). In practice, each material will present a set of contact issues that will need to be resolved. Since poor contacts present the most likely source of error in measurement, the researcher who gives careful consideration to these issues minimizes the likelihood that he or she will be faced with the uncomfortable task of retracting an erroneous result.

2. Electrical Resistivity (and Electrical Conductivity)

a. Resistivity of Bulk Samples

For many materials, the determination of the electrical resistivity is a simple and straightforward measurement. A four-probe method is typically used in which current is injected through one set of current leads and voltage is measured using another set of voltage leads. This of course, eliminates the contributions of the leads or the contacts in the sample voltage measurement. However, measurements of resistivity in thermoelectric materials, which are typically semiconductors, can pose several significant challenges. As stated, since oxide layers can often form on surfaces, and these are typically metal–semiconducting interfaces, making good electrical contacts can be difficult. As discussed, because of the semiconducting nature of some of these materials, the contacts often form p–n junctions and can result in nonohmic voltages that lead to erroneous resistivity measurements.

Today's materials come in many different forms, from bulk materials (a few millimeters for each dimension) to either thick (≈ 1–$5\ \mu$m) or thin films (≈ 100–1000 Å). For all of these, accurate determination of sample dimensions and measurement current are very important since they factor into the calculation for resistivity ($\rho = VA/Il_0$). Accurate determination of the measurement current is typically accomplished by measuring the voltage

---

[2]DuPont 4929 is a product of Du Pont Corporation; SPI Ag coating for SEM is a product of Scientific Products Incorporated; Pogo contacts are a product of Augat-Pylon Corporation.

across a known precision resistor placed in series with the sample. With the exception of some thin film samples, the resistances measured in many thermoelectric samples is typically on the order of several milliohms; the value of the precision resistor should be chosen with this in mind.

Although a large Peltier effect is desirable for a good thermoelectric material, the presence of this effect can lead to significant errors in resistivity measurements. Thermoelectric materials exhibit relatively large Seebeck coefficients; therefore, the total voltage measured across the sample will be the sum of the Seebeck voltage, $V_{TE} = \alpha \Delta T$, plus the resistive or IR voltage, V_{IR}, [i.e., $V_{Total} = \alpha \Delta T + V_{IR}$]. Often, the Seebeck contribution to the total voltage is comparable to the resistive component, $V_{TE} \approx V_{IR}$. In order to minimize the effects of the Seebeck induced voltage, the measurement should be made relatively fast ($\approx 2-3$ sec). In addition, by switching the current direction, one can subtract out the Seebeck voltage ($V_{IR} = [(V(I+) + \alpha \Delta T) - (V(I-) + \alpha \Delta T)]/2$), where $I+$ is positive and $I-$ is negative. Therefore, resistivity measurements of thermoelectric materials should be performed using either AC or fast switching DC currents.

As mentioned previously, a precision resistor ($\approx 0.01-0.1\%$) in series with the sample is the best way to ensure accurate measurement of the current through the sample. The four-probe configuration eliminates the effects of the contact resistances; however, it is important to position the voltage leads away from the current leads in order to ensure uniform current flow through the sample at the points where the voltage is being measured. A good rule of thumb is to position the current and voltage leads such that $l - l_0 \geqslant 2w$, where l is the total length of the sample, l_0 is the distance between the voltage probes, and w is the thickness of the sample. Errors in measuring l_0 can be minimized (if the sample is sufficiently long) by using several sets of voltage leads and averaging resistivity measurements.

A setup for measuring the resistivity and Seebeck coefficient of bulk samples used by one the authors (TMT) is shown in Fig. 1. The typical sample size is 2–3 mm for the width and/or thickness and the total sample length is $\approx 10-12$ mm. A heater is attached to a small copper plate that is soldered to one end of the sample. Leads for the current ($+$), the upper Seebeck voltage lead and one end of the thermocouple are attached to this copper. The other end of the sample is attached to a copper base that is heat sunk to the thermometer and cooling stage. Similar leads ($-$) are attached to the copper base. Voltage leads for the resistivity measurements are attached onto the sample as shown. The thermocouples are embedded in the copper plates at the ends of the sample and the Seebeck voltage is measured at these points. This configuration yields the most reliable and consistent resistivity and Seebeck measurements. The sample current is reversed for the resistivity measurements to eliminate contributions of thermal voltages from the Seebeck effect.

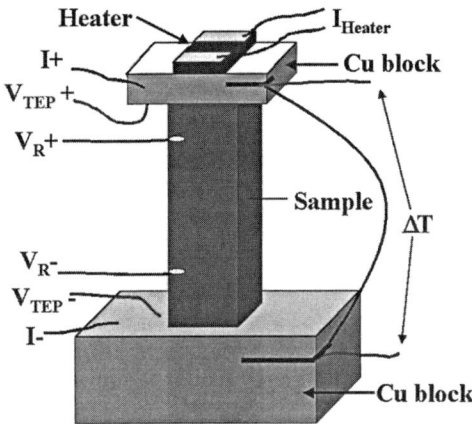

FIG. 1. A diagram of the setup for the measurement of the resistivity and Seebeck coefficient of bulk samples. A heater is attached to a small copper plate that is soldered to the top of the sample. Leads for the current (+), the upper Seebeck voltage lead, and one end of the thermocouple are attached to this copper. The other end of the sample is attached to a copper base that is heat sunk to the thermometer. Similar leads (−) are attached to the copper base. Voltage leads for the resistivity measurements, V_R, are attached to the sample as shown. Typical sample size is 2–3 mm for the width and/or thickness, and the total sample length is ≈ 10–12 mm.

b. Resistivity of Thin Disks or Thin Films

The issues previously discussed for bulk samples apply for thin films or disks, but additional factors are also important. We still have to consider issues related to metal–semiconductor junctions, and diffusion can be even more of a problem than for bulk samples. However, making good electrical contacts to semiconducting materials for devices has received a great deal of attention because of the microelectronics industry, and the reader is referred to Runyan and Shaffer (1997) and Streetman (1995) for further details.

Specific techniques are known for the measurement of the resistivity of thin disks or films or samples of arbitrary shape. The four-point probe technique for the measurement of a thin rectangular sample is given in Fig. 2 (Smits, 1958). The needle point probes are colinear and evenly spaced along the sample. The important dimensions are given by the sample dimension parallel to the contact line (*a*); the sample dimension perpendicular to the contact line (*b*); spacing between the contact probes (*s*), and the thickness of the sample (*w*). The relationship between the resistivity, ρ, and the sheet resistivity, ρ_S, is given by $\rho = \rho_S w$. The resistivity for this

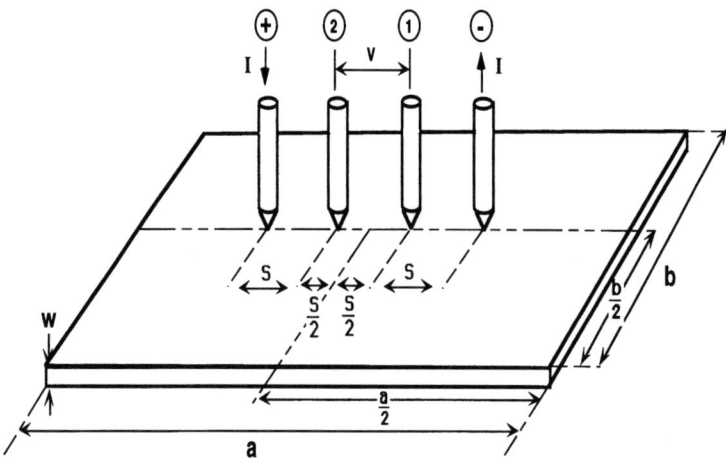

FIG. 2. Diagram of a four-point probe configuration for the measurement of resistivity of a thin disk. The sample dimensions are length, a, width, b, thickness, w, and spacing between the contact probes, s. (After Smits, 1958.) Copyright © 1958 AT&T. All rights reserved. Reprinted with permission.

configuration is given by

$$\rho = \{(Vw)/I\}\{C_1(a/d)C_2(w/s)\}. \tag{1}$$

The terms in Eq. (1) are $C_1(a/d)$, a correction factor for the planar dimensions, e.g., for an infinite sheet ($C_1(a/d) = \pi/\ln 2$) and $C_2(w/s)$, a correction factor for the ratio of thickness to the contact spacing (e.g., $C_2(w/s) = 0.9995$ for $w/s = 0.4$ and $C_2(w/s) = 0.9214$ for $w/s = 1$). Values for these correction factors are taken from Smits (1958) and are given in Tables II and III. If $C_1(a/d) = \pi/\ln 2$, then the thickness dependence is given by

$$\rho = \{(Vw)/I\}\{(\pi/\ln 2)C_2(w/s)\}. \tag{2}$$

A technique for measuring samples of arbitrary shape was developed by van der Pauw in the late 1950s (van der Pauw, 1958, 1961). An illustration of the sample and lead configuration is shown in Fig. 3. Many others have elaborated on this technique with additional corrections in latter years (Wasscher, 1961; Koon et al., 1989; Koon, 1989). The resistivity is given by

$$\rho = F\{(\pi w)/\ln 2\}\{R_{AB-CD} + R_{BC-DA}\}/2, \tag{3}$$

where $F = Fnc\,(R_{AB-CD}/R_{BC-DA})$, and R_{AB-CD} and R_{BC-DA} are the resistances measured in the different contact configurations (I^+, I^-) and (V^+, V^-) rotated by 90° with respect to each other. The designations of A, B, C, and D are shown in Fig. 3. The function F has a value of $F \approx 1$ for (R_{AB-CD}/R_{BC-DA}) ≤ 1.5. The van der Pauw technique assumes the contacts are

TABLE II
CORRECTION FACTOR FOR PLANAR DIMENSIONS, $C_1(a/d)$

d/s	Circle diam. d/s	$a/d = 1$	$a/d = 2$	$a/d = 3$	$a/d \geqslant 4$
1.0				0.9988	0.9994
1.3				1.2467	1.2248
1.5			1.4788	1.4893	1.4893
1.8			1.7196	1.7238	1.7238
2.0			1.9454	1.9475	1.9475
2.5			2.3532	2.3541	2.3541
3.0	2.2662	2.4575	2.7000	2.7005	2.7005
4.0	2.9289	3.1137	3.2246	3.2248	3.2248
5.0	3.3625	3.5098	3.5749	3.5750	3.5750
7.5	3.9273	4.0095	4.0361	4.0362	4.0362
10.0	4.1716	4.2209	4.2357	4.2357	4.2357
15.0	4.3646	4.3882	4.3947	4.3947	4.3947
20.0	4.4364	4.4516	4.4553	4.4553	4.4553
40.0	4.5076	4.5120	4.5129	4.5129	4.5129
∞	4.5324	4.5324	4.5324	4.5325	4.5324

placed on the extreme edges of the sample. A correction also exists for the placement of the contacts onto the sample (van der Pauw, 1961). If d is the distance from the edge of the sample to the contact point and D is the diameter of the sample, then the error in the resistivity is given by

$$\Delta\rho/\rho = -\ln\{1 + (d/D)^2/(1 - d/D)^2\}/2(\ln 2). \tag{4}$$

TABLE III
CORRECTION FACTOR FOR RATIO OF THICKNESS/CONTACT SPACING, $C_2(w/s)$

w/s	$C_2(w/s)$
0.4000	0.9995
0.5000	0.9974
0.5555	0.9948
0.6250	0.9898
0.7143	0.9798
0.8333	0.9600
1.0000	0.9214
1.1111	0.8907
1.2500	0.8490
1.4286	0.7938
1.6666	0.7225
2.0000	0.6336

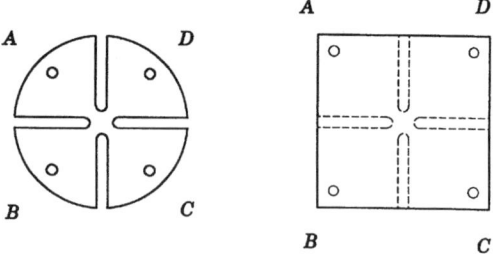

FIG. 3. An illustration of the sample and lead configuration for the van der Pauw technique is shown. (After van der Pauw, 1958, 1961; also I. A. Nishida, p. 159 in Rowe, 1995.) Reprinted with permission.

If d is less than $0.1D$ ($d < 0.1D$), then the error in $\Delta\rho/\rho$ is less than 1%. Care must be taken for oxide layers on the samples or surface decomposition of the sample. Needle probes are usually able to break through ordinary oxide layers. Also, another technique known as the Montgomery technique is used to determine the resistivity tensor of single crystal samples (Montgomery, 1971). It utilizes a series of probe configurations on each face of the sample and then various current and voltage combinations.

3. SEEBECK COEFFICIENT

The Seebeck coefficient or thermopower, α, is an intrinsic property of a material related to the material's electronic structure, much like the resistivity. Errors related to determining the sample dimensions are not present as evident from Eq. (1); however, the more common error exists in the correct determination of the temperature gradient of the sample. The thermopower yields information about the sign of the charge carrier and is essentially the entropy per carrier divided by the charge of the carrier (Chaikin, 1990; Blatt et al., 1976; MacDonald, 1962). The temperature dependence of the thermopower can be quite complicated and difficult to interpret and is typically used in conjunction with other measurements in the understanding of various physical phenomena. Many other contributions, such as phonon drag, can be involved in combination with the simple diffusion thermopower (linear in T) that is typical of metals. The thermopower is not geometry specific (as is the sample resistance) and is given by a measurement of the ratio of the sample voltage (electric field) to the temperature difference (temperature gradient) along the sample,

$$\alpha_{AB} = \Delta V/\Delta T = (V_H - V_L)/(T_H - T_L), \tag{5}$$

where $\alpha_{AB} = \alpha_B - \alpha_A$ is the measured value of the thermopower, which includes both the sample contribution, α_A, as well as the lead contribution α_B. The lead contribution, typically Au, Cu, or Chromel, must be known and subtracted from each data point at each temperature. In principal, thermopower appears to be a relatively easy measurement. However, in practice there are a number of sources for error.

It is very easy to get the sign of the thermopower wrong, and there are many published papers where this is the case, so a word of caution is appropriate. To avoid making this mistake, it is essential to use consistency in defining the direction of your temperature gradient with respect to your voltage measurement. Keep in mind that the sample or absolute thermopower will typically be opposite in sign from the measured or relative thermopower. This is due to the fact that one has to subtract out the lead contribution ($\alpha_A = \alpha_B - \alpha_{AB}$), which is usually small compared to the sample thermopower.

Another potential source of error is the placement of the thermocouples in relation to the voltage leads. Accuracy in thermopower measurements relies on determining the temperatures T_H and T_L precisely at the location of the voltage probes V_H and V_L, respectively. If one is using a differential thermocouple, great care must be taken to thermally anchor the ends of the thermocouple as closely as possible to the voltage contacts. This is the basis for the placement of the thermocouples and Seebeck voltage leads as shown and described in Fig. 1. By anchoring each of these to the high thermal conductivity copper plates (where the thermal conductivity of the copper is much greater than that of the sample), one is assured of each being at the same temperature. A better approach, although less practical, is to use separate thermocouples for measuring the temperature gradient. In this configuration, one leg of each thermocouple can be used as a sample voltage lead. Of course, this configuration requires that the thermocouples be in good thermal *and* electrical contact with the sample. This technique is used in the labs at Clemson (TMT) for the measurement of the Seebeck coefficients at high temperatures, where solders melt and thermal contact is more difficult (Littleton *et al.*, 1998). In bulk samples, it is often desirable to make a small hole in which the thermocouple bead can be placed. This maximizes the surface area over which the sample and thermocouple are in contact. Unfortunately, although this method is generally reliable, attaching thermocouples directly to the sample often requires the use of a new thermocouple for each measurement, making this technique somewhat tedious and impractical for general use.

For faster sample throughput, it is often desirable to use a permanently mounted thermocouple and thermally anchor the sample and the thermocouple through a common medium using varnish or some other contacting adhesive. This has advantages but is more likely to cause a thermal anchoring problem, resulting in a wrong ΔT reading. Kopp and Slack discussed the subject of thermal anchoring and the associated errors quite thoroughly in an article on the subject (Kopp and Slack, 1971). In any case, one should definitely test any technique by measuring known standards and comparing values. There are also a number of systematic "checks" that can be performed to rule out errors due to poor thermal anchoring. A significant time lag between changes in sample voltage and changes in the temperature gradient suggests a problem with thermal anchoring. Since the sample is its own best thermometer, one can compare changes in the thermocouple voltage to that of the sample voltage as the ΔT is varied by using an oscilloscope to measure the time differences in their response. A poor thermal contact between the thermocouple and the sample will also typically be revealed by a change of pressure in the sample space. As the gas or air is pumped out, the thermal link between the sample and the thermocouple typically becomes worse and the measured thermopower will change. A large difference between measurements taken under vacuum and in a gas atmosphere usually indicates poor thermal contact between sample and thermocouple. At low temperatures, calibration and subtraction of the lead contributions can also be a source of error. When calibration of lead wires at low temperatures is uncertain, then they should be recalibrated against a high-T_C superconductor (with transition temperatures T_C as high as $T_C \approx 130\,\mathrm{K}$). Since the thermopower of a superconductor below T_C is zero, the lead contribution is the only contribution to the measured value.

There are essentially two techniques that are used for measuring thermopower. The simplest of these involves fixing the sample temperature, T, and varying a small temperature gradient across the sample. The sample voltage is recorded as a function of sample gradient and the slope yields α_{AB}. This technique is somewhat slow and cumbersome, particularly if one is interested in looking at the temperature dependence of the thermopower over a wide range of temperatures. An alternative approach involves applying a small fixed temperature gradient (where $\Delta T/T \approx$ a few percent) across the sample and slowly varying T. At each temperature of interest, several voltage measurements are made and α_{AB} is calculated from the relationship $\alpha_{AB} = \Delta V/\Delta T$. To ensure that ΔV is linear in ΔT and goes through $\Delta V = 0$ at $\Delta T = 0$, one should periodically perform ΔV vs ΔT sweeps, and compare the slope of these measurements to the fixed measurement. Although simpler and somewhat faster than other thermopower measurements, this technique is particularly susceptible to errors due to high resistance contacts that can exhibit AC pickup resulting in a rectified DC offset voltage.

Measurement of thermopower in thin film samples requires special consideration. The thermal conductivity of the film substrate is typically quite high and can be a source of unwanted temperature gradients. Thermally anchoring the thermocouple bead to the film surface is also difficult. Only a very small fraction of the bead surface can actually be placed in contact with the film surface. To ensure that the remainder of the bead is at the same temperature, one can use solder or a thermally conducting epoxy to thermally anchor the entire bead to the film. If the film temperature is being measured using thermocouples that are permanently mounted into the sample holder, then it is extremely important to ensure that the film is in excellent thermal contact with the sample holder. Checking for measurement consistency in various atmospheres is a good way to verify good thermal contacts.

As stated earlier, the Seebeck coefficient or thermopower is conceptually the easiest of the three parameters to measure, but we emphasize, only conceptually. Thermal contact errors must be understood and minimized. The Seebeck voltage leads must be in excellent thermal contact to the point at which the temperature is being measured. This is difficult in bulk materials, and one of us (TMT) uses the configuration shown in Fig. 1 for bulk sample measurements. In a recent panel discussion on thermoelectric measurements at the 1999 International Conference on Thermoelectrics, J. Goldsmid agreed that this is the best method for measuring the Seebeck coefficient in bulk materials. Measurement of the Seebeck coefficient in thin films is even more of a challenge. Again, great care is encouraged.

4. Thermal Conductivity

Thermal conductivity measurements are by far the most difficult to make with relatively high accuracy, $\approx 5\%$. There are many excellent texts and techniques available that discuss in detail many of the corrections and potential errors one must consider, and we refer the reader to a few of these (Tritt, 1997; Uher 1996; Slack, 1979; Tye, 1969; Berman, 1976; Marton, 1959). Many other excellent references also exist in the literature. The thermal conductivity, λ_T, of good thermoelectric materials is very low, typically $\lambda_T \leqslant 2$ W/m-K. This makes the measurement even more difficult, since the heat will flow through other paths of higher thermal conductivity, such as down lead wires and conduction by any gases or air flow around the sample. These result in an error in determination of the power input into the sample. Thus, calculating the heat loss corrections and proper thermal shielding techniques to minimize these corrections and radiation effects is critical for these thermoelectric materials. The thermal conductivity for a typical "steady-state" method is given by

$$\lambda_T = Q_T l_0 / A \, \Delta T, \qquad (6)$$

where Q_T is the heating power through the sample, l_0 is the length between the thermocouple leads. A typical sample setup is shown in Fig. 4. A small heater (strain gauge) is placed on top of the sample and the heating power is given by $I^2 R$ through the heater. Number 38 (0.004″) phosphor bronze

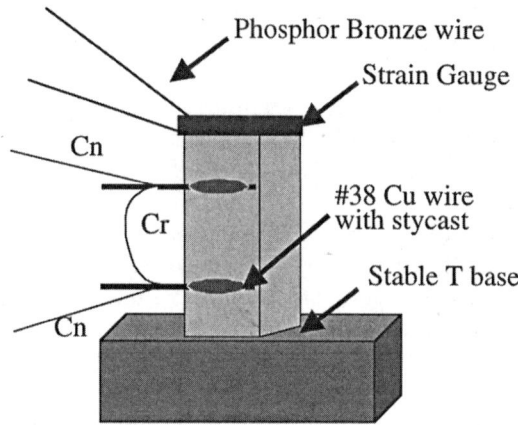

FIG. 4. Diagram of the steady-state thermal conductivity method used in one of our laboratories (TMT). Small copper wires (flags) are attached to the samples on which the Cn–Cr thermocouples (0.001" diameter) are attached. A thin resistor (strain gauge) is attached to the top of the sample for the power source, I^2R. Phosphor bronze current leads [#38 (0.004")] are attached to the strain gauge resistor heater.

leads are attached to the heater. These will result in small resistive contributions and small thermal conduction losses. Small copper flags (#38 gauge) are attached to the sample with thermal epoxy. Small Cn–Cr thermocouples (0.001" diameter) are attached to these flags to determine the temperature gradient. One can also attach small Cernox (Lake Shore Cryotronics), carbon glass, or other semiconducting thermometers to the copper flags to determine ΔT. These are important for low-temperature measurements, $T < 10$ K, where thermocouples are rapidly losing their sensitivity. Using a temperature controller that controls the temperature stabilizes the base temperature. Power sweeps at fixed temperature yield power vs ΔT curves from which the slope is calculated, yielding the thermal conductance of the sample. This is coupled with the sample dimension measurements to yield the thermal conductivity of the sample at a given temperature. This system is described in detail elsewhere (Pope et al., 2000). Errors due to radiation loss or gain between the surroundings and the sample or to convection and conduction through any lead wires can be substantial. The radiation loss is given by

$$Q = \varepsilon \sigma_{S-B} A (T_0^4 - T_S^4), \qquad (7)$$

where T_0 (T_S) is the temperature of the sample and the surroundings, respectively, σ_{S-B} is the Stephan–Boltzmann constant ($\sigma_{S-B} = 5.7 \times 10^{-8}$ W/m^2-K^4), and $\varepsilon (0 < \varepsilon < 1)$ is the emissivity. Proper thermal shielding and thermal anchoring are essential for reliable and accurate measurements. Heat losses can also be due to convection or circulating gas flow around the sample. The best way to minimize these convection losses is to operate the measurement with the sample in a moderate vacuum (10^{-4}–10^{-5} torr). This will also reduce the heat loss due to conduction through the gaseous medium. The other substantial heat loss mechanism is due to conduction. This can be due to loss from the thermocouple or other leads attached to sample for temperature measurement. Long lead lengths of small diameter (small A) with sufficient thermal anchoring, so essentially no ΔT arises between the sample and shield, are important for minimizing this effect. One must accurately determine the power through the sample by considering the various loss mechanisms. Thermal resistance of leads, heaters, etc., as well as interface anchoring between the sample, the heater, and the heat sink, is also important.

At times the radiation losses may be estimated by determining the temperature dependence of the lattice thermal conductivity between 50 and 150 K. If there is a distinct temperature dependence of the lattice thermal conductivity, this can be extrapolated to higher temperature, say, 300K. The difference between the calculated and measured lattice thermal conductivity can be calculated. If this difference has a T^3 temperature dependence, it can usually be attributed to radiation losses. This is illustrated in Fig. 5 for a research sample (quasicrystal) measured in the labs at Clemson. Even if one minimizes or effectively measures many of these losses, the sample length and cross-section must be accurately determined, a challenge that can still yield 5–10% uncertainty. Again, measuring known standards and thoroughly calibrating the apparatus is essential. It is suggested that one use a number of different standards with different thermal conductivity. Pyrex and Pyroceram are suggested as low thermal conductivity standards.

Many techniques other than the standard steady-state method are valid. In the *comparative technique*, a known standard is put in series between the heater and the sample. This technique is best when the thermal conductivity of the standard is comparable to that of the sample. Also, the same type of errors and corrections must be considered as for the steady-state technique. The power through the standard (1) is equal to the power through the sample (2), and given the thermal conductivity of the standard, λ_1, the thermal conductivity of the sample, λ_2, is given by

$$\lambda_2 = \lambda_1 \{A_1 \Delta T_1 L_2 / A_2 \Delta T_2 L_1\}. \tag{8}$$

Another technique that is becoming popular for thermoelectric materials, as well as for many nonconducting low thermal conductivity systems, is the 3-ω technique (Cahill, 1990; Cahill et al., 1989). This technique was

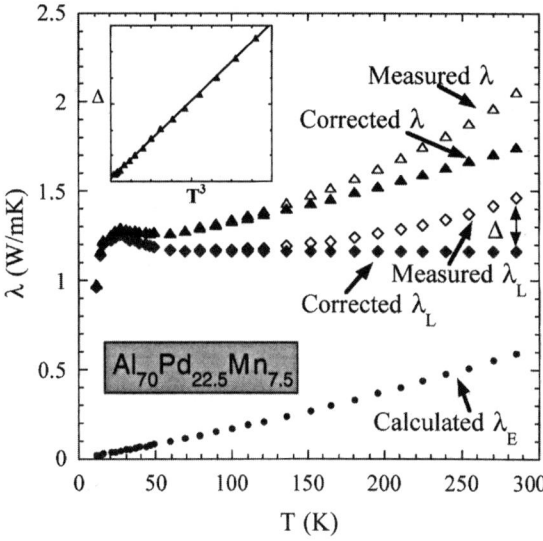

FIG. 5. A plot of the thermal conductivity as a function of temperature is shown. The total, lattice, and electronic contributions are shown. The difference between the extrapolated and measured lattice thermal conductivity, Δ, is plotted as a function of T^3 in the inset, illustrating losses due to radiation effects.

originally developed for measuring the thermal conductivity of glasses and other amorphous solids. More recently, it has been used to measure thermal conductivity in thin film samples. In this technique a thin metal strip (typically Au or Pt) is evaporated onto the sample. If the sample is an electrical conductor, then an insulating layer must be deposited prior to evaporating the metal strip. Because of Joule heating, an AC current applied to the film provides an oscillating heat source ($P = I^2R \approx \Delta T$) of frequency 2ω. The known temperature dependence of the resistivity of the thin metal film is used to calibrate the temperature rise associated with the Joule heating. Thus, the strip serves as both a heat source and a thermometer.

In order to measure thermal conductivity, the AC voltage is monitored as a function of the frequency of the AC applied current. The measured voltage, $V = IR$, will have both an ω component and a 3ω component. This is due to the Joule heating of the film, which manifests itself in the film's resistance as a small perturbation in temperature with frequency 2ω, that is, $V = IR = I_0 e^{i\omega t} (R_0 + dR/DT \times \Delta T) = I_0 e^{i\omega t}(R_0 + \text{const.} \times e^{i2\omega t})$. The thermal conductivity, λ, is determined by the linear slope of the ΔT vs $\log(\omega)$ curve.

The 3ω technique has many advantages. The temperature dependence of the thermal conductivity can be acquired much more readily than by the steady-state technique. In addition, radiation effects are minimized with this method because of the AC nature of the measurement. Although the 3ω technique requires some level of expertise in thin film patterning and microlithography, it is generally user-friendly and inexpensive. For these reasons, the 3ω technique is probably best "pseudocontact" method available.

Another way to measure the thermal properties of both thin film and bulk samples is the *laser-flash thermal diffusivity method* (Taylor, 1995). In this technique one face of a sample is irradiated by a short ($\leqslant 1$ msec) laser pulse. Using an IR detector, the temperature rise of the opposite side of the sample is monitored. The thermal diffusivity is calculated from the temperature rise vs time profile. A number of algorithms exist for correcting for the various losses that are typically present in this measurement. The thermal conductivity is related to the thermal diffusivity. Therefore, in principle, this technique can be used to measure thermal conductivity. However, the utility of this method requires fairly stringent sample preparation requirements. In order to prevent "flash-throughs" to the IR detector, there is very little flexibility in the required sample geometry (typically thin disks or plates). In addition, the sample surfaces must be highly emissive to maximize the amount of thermal energy transmitted from the front surface and to maximize the signal observed by the IR detector. Usually this requires the application of a thin coating of graphite to the sample surfaces. If good adhesion is not achieved, this coating procedure can potentially be a source of significant error.

Commercial units are available that allow measurement of thermal diffusivity at temperatures that range from 77 K up to ~ 2300 K. These units are typically automated and reasonably easy to use. Since the thermal diffusivity is related to thermal conductivity through the specific heat and sample density, the laser flash method is sometimes used to determine thermal conductivity indirectly when the specific heat and density have been measured in separate experiments. However, these systems require a relatively large sample size, a 2-inch disk for some systems. This can be difficult to obtain for a research sample.

5. Hall Coefficient, Carrier Concentration, and Mobility

The power factor of a thermoelectric material is typically optimized around some carrier concentration, n, where $n \approx 10^{17}$–10^{19} cm^{-3}. Thus, it is necessary to measure the carrier concentration by measuring the Hall effect.

$$V_H = BIb/neA = R_H BI/t, \tag{9}$$

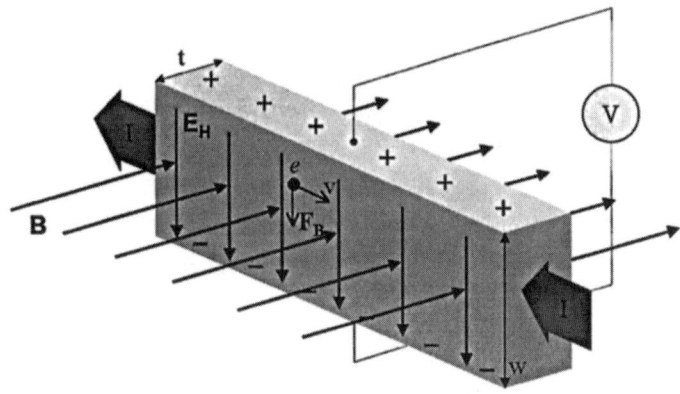

FIG. 6. Configuration for measurement of the Hall voltage and Hall effect to determine the carrier concentration.

where the Hall coefficient is given by $R_H = 1/ne$ for a single-carrier system and is much more complicated for a two-carrier system. The Hall voltage, V_H, is due to the Lorentz force ($F = qE = qV/b$) acting on a charged particle moving with a velocity, v, in a magnetic field B, where b is the width of the sample and t the thickness. The configuration for the measurement of the Hall voltage and Hall effect is shown in Fig. 6. Notice from Eq. (9) that the magnitude of the Hall voltage is proportional to B and I and inversely proportional to the sample thickness and the carrier concentration. The parameters I, B, and V_H are all perpendicular to each other, as shown in Fig. 6. For example, given a rectangular sample, B is the magnetic field (B_Z), I is the current (I_X), and V_H, the Hall voltage, is measured in the y direction. The Hall effect and Hall measurements in metals and semiconductors are discussed in many texts (see Runyan and Streetman) and other references (McKelvey, 1966; Runyan, 1975; Putley, 1960; Chien and Westgate, 1979; Hurd, 1972; Ehrlich, 1993). The electrical conductivity is related to the carrier mobility, $\mu = R_H \sigma$ ($\sigma = ne\mu = 1/\rho$) and for a fixed n, μ must be large for a good thermoelectric ($\mu \approx 1000$ cm^2/V-s for Bi$_2$Te$_3$ and skutterudites). For a semiconductor of one carrier type, $R_H = 3\pi/8ne$. In general, $R_H = r/ne$, where $1 \leqslant r \leqslant 2$ and r is known as the Hall factor, which depends on magnetic field, temperature, and scattering mechanism. Measurements of V_H, $R_H = 1/ne$, and μ are important for fully characterizing and understanding a thermoelectric material. Again there are errors that must be considered. Misalignment of the leads (IR voltage) or magnetic field can lead to errors unless corrected and terms must be canceled that are not proportional to BI. Sometimes five-wire Hall measurements are performed. This uses three Hall voltage leads that are balanced with a bridge at zero magnetic field to eliminate any resistive contributions. There are also other corrections such as the thermoelectric voltages, since temperature gradients will lead to voltage differences. Care should be taken to eliminate any unnecessary temperature gradients. Other configurations such as the van der Pauw configurations (for thin films) are also used for measuring the Hall effect. In addition to the Hall voltage, there are other magneto-thermoelectric effects that must be considered and corrected for, such as the Nernst effect, the Ettingshausen effect, and the Ridgi-Leduc effect. These are not discussed here because of space constraints, but are discussed in detail elsewhere (McKelvey, 1966; Runyan, 1975; Putley, 1960; Chien and Westgate, 1979; Hurd, 1972; Ehrlich, 1993). The Hall voltage is given by

$$V_H \approx \{V(I^+, B^+) + V(I^-, B^-) - V(I^-, B^+) - V(I^+, B^-)\}/4, \quad (10)$$

canceling out all the terms discussed earlier except the Ettingshausen effect, which should only be a small correction to the Hall voltage (except, of course, where $R_H \approx 0$). An AC Hall measurement may prove the most applicable with few corrections for these thermoelectric materials.

6. Z-Meters (or "Harman Technique")

Another technique that is widely used for characterizing thermoelectric materials is the so-called Harman technique or Z-meter (Tritt, 1996; Egli, 1960; Harman, 1959; Harman et al., 1959; Penn, 1964; Bowley et al., 1961; Buist, 1992). This is a direct method for obtaining the figure of merit or ZT of a material or a device. Consider the voltage as a function of time for a thermoelectric material under various conditions. If there is no ΔT and $I = 0$, then $V_S = 0$, where V_S is the sample voltage. A current, I, is applied and the voltage increases by IR_S, where R_S is the sample resistance, V_{IR}. Recall that when a current is applied to a thermoelectric material a ΔT arises from the Peltier effect ($Q_P = \alpha I T$) and a voltage, V_{TE}, will add to the IR_S voltage. The sample voltage as a function of time is shown in Fig. 7 for a thermoelectric material. Under steady-state or adiabatic conditions, the heat pumped by the Peltier effect will be equal to heat carried by the thermal conduction:

$$(\alpha I T) = (\lambda A \Delta T/L). \quad (11)$$

One can derive a relationship between ZT and the adiabatic voltage ($V_A = V_{IR} + V_{TE}$) and the IR sample voltage, V_{IR} (Egli, 1960; Harman, 1959; Harman et al., 1959):

$$ZT = (V_A/V_{IR}) - 1 = (\alpha^2 T/\rho\lambda). \quad (12)$$

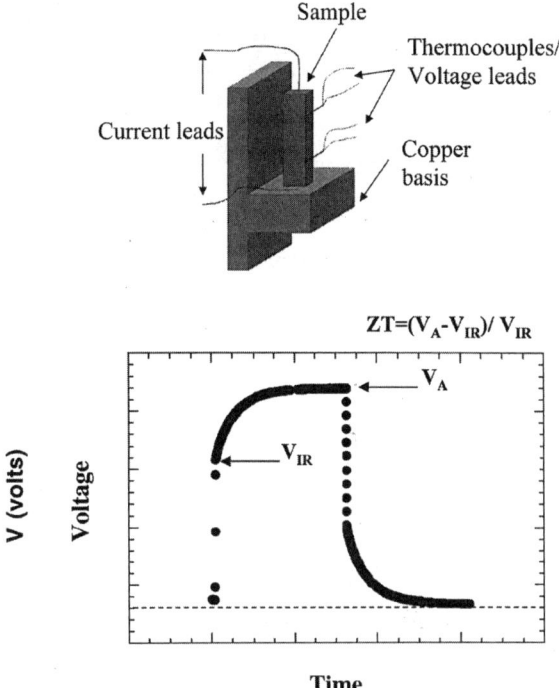

FIG. 7. (Top) Sample setup for the Harman technique. (Bottom) Plot of the sample voltage as a function of time, illustrating the utilization of the Harman technique for the measurement of the thermoelectric figure of merit, ZT.

This relationship is a reasonable approximation to ZT, but assumes ideal conditions unless a number of corrections are accounted for such as contacts, radiation effects, and losses. The first criterion is that the sample typically needs to possess a $ZT \geqslant 0.1$. Also, contact effects, sample heating from the contacts, and the sample resistance should be negligible, and ΔT effects from contact resistance differences can also be negligible. The thermal conductivity can be estimated from the Harman technique in two ways: first measure R and α, and then measure ZT from Eq. (12), and the thermal conductivity can then be determined. Another way is to use Eq. (11) as given in the form

$$I = \lambda(A/\alpha TL)\Delta T. \tag{13}$$

Then at a constant temperature, $I = \lambda(A/\alpha TL)\Delta T = \lambda C_0 \Delta T$, where C_0 is a constant at a given T. Thus, the linear part of the slope of an I–ΔT plot will yield the thermal conductivity. The ZT determined from the Harman method is essentially an "effective ZT," one that yields the operating figure of merit of the device. This technique requires essentially no contact effects, (recall $I^2\{R_{C1} - R_{C2}\} = \Delta P_C \approx \Delta T$ across the sample from I^2R heating). Information from this technique should be compared to the measurements of the individual parameters that go into ZT. It should not be a substitute for knowing the individual parameters.

V. Summary

It is certainly no easy task to come up with highly reliable and accurate measurements, given all the parameters that must be measured to effectively determine ZT of a thermoelectric material. Many special errors and corrections for these thermoelectric materials must be taken into consideration. Thus, reports of materials that are only a few percent different from previous materials are not so convincing and of course are really not that much of a technological advancement in material performance. Many new researchers are coming into this field, and we remind them that much information concerning thermoelectric materials and the many excellent techniques for measuring their properties was developed in the late 1950s and 1960s. Careful investigation of the literature can provide valuable information and insight into the measurement of thermoelectric materials. The errors that can be made are sometimes very subtle and if we are to achieve much higher ZT materials, $ZT \approx 2-3$, then we will have to be able to validate these numbers. Measuring standards, careful and accurate calibrations, and potential sharing of samples through "round robin" measurements are very important to eliminate erroneous information from being reported. There are, of course, other measurement techniques and apparatus that were not discussed in this overview, such as the slow AC technique for thermopower measurements (Chaikin and Kwak, 1975) and other techniques that use slowly varying temperature methods (Maldonado, 1992) just to name a couple of the more common methods. Researchers will have to judge for themselves which of the various techniques is the most appropriate for a given set of materials and for the equipment available to perform these measurements. We hope that this paper is informative both in terms of the issues involved in measuring thermoelectric materials and in terms of an extensive (yet incomplete) reference set of literature that is available. We hope it is especially useful to people new to the field of thermoelectric materials research, while it may serve as an elementary but important reminder to the established experts in the field. The accurate measurement and characterization of any new material will remain a challenge, and only correct care and attention to these measurements will rapidly advance the field of thermoelectric materials research.

Acknowledgments

One of us (TMT) acknowledges some of the many people who have imparted some of their vast wisdom to me concerning the measurements of thermoelectric materials. These include Dr. Glen Slack, Dr. Ctirad Uher, Dr. Julian Goldsmid, Dr. Ted Harman, Dr. Al Ehrlich, and Dr. Cronin Vining, to mention just a few. We also acknowledge A. L. Pope, R. T. Littleton IV, M. Kaeser, N. Lowhorn, and Dr. B. M. Zawilski in the preparation of the figures in this manuscript, and A. L. Pope for critically reading the manuscript. Each of us (TMT and VMB) would like to acknowledge the support of our institutions during the development of this manuscript.

References

Berman, R. (1976). *Thermal Conduction in Solids*. Clarendon Press, Oxford.
Blatt, F., Schroeder, P., Foiles, C., and Grieg, D. (1976). *Thermoelectric Power of Metals*. Plenum Press, New York.
Bowley, A. E., Cowles, L. E. J., Williams, G. J., and Goldsmid, H. J. (1961). *J. Sci. Instrum.* **38**, 433.
Buist, R. (1992). Proceedings of the XI International Conference on Thermoelectrics, AIP.
Cahill, D. (1990). *Rev. Sci. Instrum.* **61**, 802.
Cahill, D., Fischer, H. E., Klitsner, T., Swartz, E. T., and Pohl, R. O. (1989). *J. Vac. Sci. Technol. A* **7**, 1260.
Chaikin, P. M. (1990). In *Organic Superconductivity* (Kreisin, V. Z., and Little, W. A., Eds.). Plenum Press, New York.
Chaikin, P. M., and Kwak, J. F. (1975). *Rev. Sci. Instrum.* **46**, 218.
Chien, C. L., and Westgate, C. R. (1979). *The Hall Effect and Its Application*. Plenum Press, New York.
Egli, P. H., Ed. (1960). *Thermoelectricity*. John Wiley and Sons, New York.
Ehrlich, A. C. (1993). In *The Electrical Engineering Handbook* (Dorf, R. C., Ed.), p. 1106. CRC Press, Boca Raton, FL.
Fleurial, J. P., Caillat, T., and Borschchevsky, A. (1995). Proceedings of the XIII International Conference on Thermoelectrics, AIP, pp. 40–44.
Goldsmid, H. J. (1986). *Electronic Refrigeration*. Pion Limited Publishing, London.
Harman, T. C. (1959). *J. Appl. Phys.* **30**, 1373.
Harman, T. C., Cahn, J. H., and Logan, M. J. (1959). *J. Appl. Phys.* **30**, 1351.
Hicks, L. D., and Dresselhaus, M. S. (1993). *Phys. Rev. B* **47**, 12727.
Hurd, C. (1972). *The Hall Effect in Metals and Alloys*. Plenum Press, New York.
International Conference on Thermoelectrics (1999). Panel Session on Thermoelectric Measurements, "$ZT = 2-3$! Fact or Fiction?," chaired by Terry M. Tritt.
Koon, D. W. (1989). *Rev. Sci. Instrum.* **60**, 271.
Koon, D. W., Bahl, A. A., and Duncan, E. O. (1989). *Rev. Sci. Instrum.* **60**, 275.
Kopp, J., and Slack, G. A. (1971). *Cryogenics*, Feb., p. 22.
Littleton, R. T., IV, Jeffries, J., Kaeser, M. A., and Tritt, T. M. (1998). In *New Materials for Small Scale Thermoelectric Refrigeration and Power Generation Applications*, Vol. 545 (Tritt, T. M., Kanatzidis, M., Mahan, G., and Lyon, H. B., Jr., Eds.), Proceedings of the 1998 Materials Research Society.
MacDonald, D. K. C. (1962). *Thermoelectricity: An Introduction to Principles*. John Wiley and Sons, New York.

Maldonado, O. (1992). *Cryogenics* **32**, 908.
Marton, L., Ed. (1959). *Methods of Experimental Physics: Solid State Physics*, Vol. 6. Academic Press, New York.
McKelvey, J. P. (1966). *Solid State and Semiconductor Physics*. Harper and Row, New York.
Montgomery, H. C. (1971). *J. Appl. Phys.* **42**, 2971.
Morelli, D. T., Caillat, T., Fleurial, J.-P., Borshchevsky, A., Vandersande, J., Chen, B., and Uher, C. (1995). *Phys. Rev. B* **51**, 9622.
Nolas, G., Slack, G., Morelli, D. T., Tritt, T. M., and Ehrlich, A. C. (1996). *J. Appl. Phys.* **79**, 4002.
Penn, A. W. (1964). *J. Sci. Instrum.* **41**, 626.
Pope, A. L., Zawilski, B. M., and Tritt, T. M. Submitted to *Rev. Sci. Instrum.* (2000).
Putley, E. H. (1960). *The Hall Effect and Related Phenomena*. Butterworth, London.
Rowe, D. M., Ed. (1995). *CRC Handbook of Thermoelectrics*. CRC Press, Boca Raton, FL.
Rowe, D. M., and Bhandari, C. M. (1983). *Modern Thermoelectrics*. Reston Publishing Co., Reston, VA.
Runyan, W. R. (1975). *Semiconductor Measurements and Instrumentation*, 1st. ed. McGraw-Hill, New York.
Runyan, W. R., and Shaffer, T. J. (1997). *Semiconductor Measurements and Instrumentation*, 2nd ed. McGraw-Hill, New York.
Sales, B. C., Mandrus, D., and Williams, R. K. (1996). *Science* **272**, 1325.
Slack, G. A. (1979). *Solid State Physics*. Academic Press, New York.
Slack, G. A., and Toukala, V. G. (1994). *J. Appl. Phys.* **76**, 1635.
Smits, F. M. (1958). *Bell Syst. Tech. J.*, Vol. 37, May, p. 711.
Streetman, B. G. (1995). *Solid State Electronic Devices*. Prentice-Hall, Englewood Cliffs, NJ.
Taylor, J. R. (1982). *An Introduction to Error Analysis: The Study of Uncertainties in Physical Measurement*. University Science Books, Mill Valley, CA.
Taylor, R. (1995). In *CRC Handbook of Thermoelectrics* (Rowe, D. M., Ed.), p. 165. CRC Press, Boca Raton, FL. This paper gives a good overview of many techniques used for measuring thermal properties of thermoelectric materials.
Tritt, T. M. (1996). *Science* **272**, 1276.
Tritt, T. M. (1997). In *Thermoelectric Materials—New Directions and Approaches*, Vol. 478 (Tritt, T. M., Kanatzidis, M., Mahan, G., and Lyon, H. B., Jr., Eds.), Proceedings of the 1997 Materials Research Society, p. 25.
Tritt, T. M., Nolas, G. S., Slack, G. A., Ehrlich, A. C., Gillespie, D. J., and Cohn, J. L. (1996). *J. Appl. Phys.* **79**, 8412.
Tritt, T. M., Kanatzidis, M., Mahan, G., and Lyon, H. B., Jr., Eds. (1997). *Thermoelectric Materials—New Directions and Approaches*, Vol. 478, Proceedings of the 1997 Materials Research Society.
Tritt, T. M., Kanatzidis, M., Mahan, G., and Lyon, H. B., Jr., Eds. (1998). *New Materials for Small Scale Thermoelectric Refrigeration and Power Generation Applications*, Vol. 545, Proceedings of the 1998 Materials Research Society.
Tye, R. P., Ed. (1969). *Thermal Conductivity*, Vols. 1 and 2. Academic Press, New York.
Uher, C. (1996). In *Thermoelectric Materials*, Vol. 48, Naval Research Reviews, p. 48.
van der Pauw, L. J. (1958). *Philips Res. Rep.* **13**, 1.
van der Pauw, L. J. (1961). *Philips Res. Rep.* **16**, 187.
Wasscher, J. D. (1961). *Philips Res. Rep.* **16**, 301.
Wood, C. W. (1988). *Rep. Prog. Phys.* **51**, 459.
Young, H. D. (1962). *Statistical Treatment of Experimental Data*. McGraw-Hill, New York.

CHAPTER 3

The Role of Solid-State Chemistry in the Discovery of New Thermoelectric Materials

Mercouri G. Kanatzidis

DEPARTMENT OF CHEMISTRY
MICHIGAN STATE UNIVERSITY
EAST LANSING, MICHIGAN

I. INTRODUCTION	51
II. NECESSARY CRITERIA FOR THERMOELECTRIC MATERIALS	54
III. SOLID-STATE CHEMISTRY AS A TOOL FOR TARGETED MATERIALS DISCOVERY	55
IV. THERMOELECTRIC MATERIALS DISCOVERY	57
V. NEW MATERIALS	60
1. The Sulfides $KBi_{6.33}S_{10}$ and $K_2Bi_8S_{13}$	61
2. The Selenides β-$K_2Bi_8Se_{13}$ and $K_{2.5}Bi_{8.5}Se_{14}$	64
3. Materials Related to β-$K_2Bi_8Se_{13}$	74
4. The Family $A_{1+x}Pb_{4-2x}Bi_{7+x}Se_{15}$ ($A = K, Rb$)	75
5. The Telluride Systems A/Bi/Te and A/Pb/Bi/Te	82
VI. OTHER CLASSES OF COMPOUNDS	92
VII. CONCLUSIONS AND OUTLOOK	96
REFERENCES	98

I. Introduction

The challenge in any effort to discover new thermoelectric (TE) materials lies in achieving simultaneously high electronic conductivity, high thermoelectric power, and low thermal conductivity in the same solid (Tritt *et al.*, 1997, 1998; Kanatzidis *et al.*, 1996; Mahan, 1998; DiSalvo, 1999; Tritt, 1996; Rowe, 1999). These properties define the thermoelectric figure of merit $ZT = (S^2\sigma/\kappa)T$, where S is the thermopower, σ the electronic conductivity, κ the thermal conductivity, and T the temperature. The first three quantities are determined by the details of the electronic structure and scattering of charge carriers (electrons or holes) and thus are not independently controllable parameters. The thermal conductivity κ has a contribution from lattice vibrations, κ_l, which is called the lattice thermal conductivity. Thus, $\kappa = \kappa_e + \kappa_l$, where κ_e is the carrier thermal conductivity.

There seem to be several possible ways one may go about increasing ZT. One is to minimize κ_l while retaining good electronic and thermopower properties. Although there are several approaches to minimizing κ_l, the most intriguing is that associated with the concept of a "phonon glass electron crystal" (PGEC) that was suggested by Slack as the limiting characteristic for a superior thermoelectric (Slack, 1995, 1997). A material that is a PGEC features cages (or tunnels) in its crystal structure inside which reside atoms small enough to "rattle." This situation produces a phonon damping effect that results in dramatic reduction of the solid's lattice thermal conductivity. In this picture a loosely bound atom with a large thermal parameter scatters phonons much more strongly than electrons, thus permitting a glasslike thermal conductivity to coexist with the high electron mobilities found in crystals. The thermal conductivity can also be decreased by introducing large nonperiodic mass fluctuations in the crystal lattice, through solid solutions (i.e., alloy scattering), or by increasing the lattice period (i.e., large unit cell parameters), thus providing short mean path lengths for the heat-carrying phonons.

Another way to maximize ZT would be to increase the thermopower and electronic conductivity without drastically increasing the total thermal conductivity. This is a difficult problem because if the electronic conductivity is greatly increased according to the Wiedemann–Franz (WF) law κ also increases. Based on WF law the electronic conductivity scales linearly with the carrier thermal conductivity κ_e, and thus very high electronic conductivities (i.e., >2000 S/cm) are ineffectual, even undesirable, in a good thermoelectric material. What is the maximum conductivity that can be tolerated, for a room temperature TE material, without causing the κ_e to be overly high? Let us first consider the total thermal conductivity at room temperature of the best currently known thermoelectrics, that is, $Bi_2Te_{3-x}Se_x$ and $Bi_{2-x}Sb_xTe_3$ alloys, which is 1.5 W/m-K. About 0.7 W/m-K of this value is due to the κ_e that arises from an electronic conductivity of ~ 1000 S/cm. Therefore, for every 1000 S/cm increase in conductivity, the total thermal conductivity is burdened with an additional 0.7 W/m-K (at room temperature). This places severe constraints on the magnitude of the lattice thermal conductivity if the total amount is to remain in the neigborhood of 1.0–2.0 W/m-K. Based on these considerations, we estimate that an optimum electronic conductivity between ~ 1000 and ~ 1400 S/cm is required. For a room temperature $ZT \sim 2$ and a κ value of ~ 1.5 W/m-K we obtain a target range for the thermopower values $260 \leqslant |S| \leqslant 440\,\mu$V/K. Currently, bulk materials with these physical characteristics remain elusive. The sought-after materials almost certainly have to be novel with new compositions or structures and therefore they have to be discovered through exploratory chemical synthesis. By every assessment the solution to this problem seems to be essentially a chemical one.

The least understood problem is how to increase the thermopower of a material without depressing the electronic conductivity and how to predict precisely which materials will have very large thermopower. Boltzmann transport theory describes both electronic and thermal transport in the vast majority of solids. This theory provides a general understanding of the thermopower that is expressed in the Mott equation (Mott and Jones, 1958):

$$S = \frac{\pi^2}{3} \cdot \frac{k^2 T}{e} \cdot \frac{d \ln \sigma(E)}{dE}\bigg|_{E=E_f}. \qquad (1)$$

$\sigma(E)$ is the electronic conductivity determined as a function of band filling or Fermi energy, E_F. If the electronic scattering is independent of energy, then $\sigma(E)$ is just proportional to the density of states (DOS) at E. In the general case, S is a measure of the variation in $\sigma(E)$ above and below the Fermi surface, specifically through the logarithmic derivative of σ with E. Since the thermopower of a material is a measure of the asymmetry in electronic structure and scattering rates near the Fermi level, we should aim to produce complexities in either or both in a small energy interval (a few kT) near E_F. What insights can be gleaned from considering the implications of the Mott equation? How can the logarithmic derivative of $\sigma(E)$ (essentially DOS) be maximized in a real chemical compound? Although there is no straightforward answer, crude insights include the search of compounds with complex structures and compositions so that they have a good chance of possessing complex electronic structure. Another class of materials with a good probability of possessing a high $d\ln\sigma/dE$ is mixed-valent compounds, particularly those of f elements (Chung et al., 1998; Kanatzidis et al., 1996; Meng et al., 2000; Proctor et al., 1999b; Jones et al., 1999).

Additional guidance to the experimenter comes from electronic band structure considerations. For example, the following parameters are critical: actual bandgap, the shape and width of the bands near the Fermi level, and the carrier effective masses and mobilities, as well as the degeneracy of band extrema (i.e., the number of valleys in the conduction band or peaks in the valence band) (Bandari, 1995). It is often emphasized that the presence of a large number of valleys in the bands tends to significantly increase the ZT (mainly through the thermopower). This is true if the carrier scattering between valleys is minimized or absent. Compounds with a large number of valleys are typically those with high symmetry crystal structures. Systems with many degenerate extrema, characterized by a degeneracy parameter γ, have higher thermoelectric power than those with a single extremum. This is because, for the same total carrier concentration, the concentration in each pocket is smaller for larger γ. This increases the value of S associated with each pocket compared to the value obtained for the single band case because S increases with decreasing pocket carrier concentration. The amount of increase, however, depends on γ, the temperature, bandgap, and other band parameters. If the carrier mobilities associated with each pocket are the same, then the total conductivity is independent of γ, but the conductivity weighted thermopower for the multiband case is larger than for the single-band case (Hicks and Dresselhaus, 1993; Larson et al., 2000).

For an anisotropic three-dimensional single-band case and band degeneracy of γ, when the thermal and electrical currents travel in the same direction (x), Hicks and Dresselhaus have shown that the figure of merit ZT increases with a parameter B defined as

$$B = \gamma \frac{1}{3\pi^2}\left(\frac{2k_B T}{h^2}\right)^{3/2} \sqrt{m_x m_y m_z} \frac{k_B^2}{e\kappa_1} \mu_x, \qquad (2)$$

where m_i is the effective mass of the carriers (electrons or holes) in the ith direction, μ_x is the carrier mobility along the transport direction, and κ_1 is the lattice contribution to the thermal conductivity. Thus, in order to increase the value of Z, large effective masses, high carrier mobility, and low lattice thermal conductivity are necessary. It has been shown that semiconductors with a bandgap of approximately $10\, k_B T$ best satisfy this criterion (Mahan et al., 1997), a property that can be easily determined from appropriate band-structure calculations. A detailed analysis of the electronic structure of a semiconducting compound (obtained through appropriate quantum mechanical calculations) can give information about the gap, the degeneracies of the conduction and valence band extrema (i.e., γ), and the effective mass parameters. Equation (2) also suggests that while high mobility along the current flow direction is needed, high effective masses need not occur along the same crystallographic directions. Therefore, highly anisotropic structures may be highly suitable for high ZT by combining high μ along the "good" direction with very masses associated with other directions.

The challenge in TE research is to incorporate all the desirable features associated with the charge and thermal transport in a single solid-state material. Next we review the criteria and the types of compounds that might be suitable.

II. Necessary Criteria for Thermoelectric Materials

Based on the preceding discussion, materials suitable for TE applications fall mainly into two categories: semiconductors and mixed-valent compounds, although some semimetals may also be viable. Many research groups are also investigating whether multiple quantum wells in semiconductors will improve thermoelectric behavior (Dresselhaus et al., 1999; Chen, 1997; Venkatasubramanian et al., 1999; Hicks and Dresselhaus, 1993).

The improvements for the electronic properties so far seem small, but quantum wells may have potentially a much smaller thermal conductivity than bulk counterparts, which may eventually make them useful for devices.

Metals generally have small Seebeck coefficients, but S has a large contribution from spin fluctuations near the Kondo temperature. Several intermetallic mixed-valent compounds such as $CePd_3$ and $YbAl_3$ have the largest value of the "power factor" $\sigma \cdot S^2$ among all known materials (Mahan and Sofo, 1996; Proctor et al., 1999a). Their large thermal conductivity prevents them from having a large ZT. Yet mixed-valent materials have the ideal shape of the density of states for producing large values of S, according to the Mott equation given earlier.

For semiconductors, the best materials have the following properties:

1. Electronic bands near the Fermi level with many valleys preferably away from the Brillouin zone boundaries. This requires high symmetry.
2. Elements with large atomic number with large spin-orbit coupling (Larson et al., 2000).
3. Compositions with more than two elements (i.e., ternary, quaternary compounds).
4. Low average electronegativity differences between elements (Tritt, 1996; Rowe, 1999).
5. Large unit cell sizes.
6. Energy gaps equal to 10 $k_B T$, where T is the operating temperature of the thermoelectric. For room temperature operation this should be $0 < E_g < 0.30$ eV.

These basic criteria, if satisfied, should give rise to high carrier mobility (criteria 1 and 4), low thermal conductivity (criteria 2, 3, and 5) and large thermopower (criteria 1 and 6). The last criterion (6) suggests that low-temperature thermoelectrics ($T < 300$ K) have very small bandgaps. For operation at higher temperatures (e.g., power generation applications) higher band gaps must be used. A more extensive discussion of criteria for TE compounds has been elaborated in the past by Slack (1995, 1997).

III. Solid-State Chemistry as a Tool for Targeted Materials Discovery

The critical role of solid-state chemistry and physics in modern technology is in little doubt. Solid-state compounds have been the foundation of the chemical industry, and metallurgy as well as the entire electronics industry for many years, and many emerging technologies such as thermoelectrics will hinge on developments in the discovery of new materials with new or enhanced properties. In this context, the importance of exploratory solid-state synthesis cannot be overemphasized.

Traditionally, the use of high temperatures to directly combine elements or simpler compounds into more complex ones has been quite successful in providing new materials, however, they often give rise to important synthetic limitations. For example, the reactions almost always proceed to the and most thermodynamically stable products and the high energies involved often leave little room for kinetic control. These thermodynamically stable products are typically the simplest of binary or ternary compounds, and because of their high lattice stability, they become synthetic obstacles. Second, the high reaction temperatures also dictate that only the simplest chemical building blocks can be used — that is, elements on the atomic level. Attempts to synthesize using molecules of known structure are doomed because the high temperatures used sunder all bonds and reduce the system to atoms rushing to a thermodynamic minimum. Hence, multinary compounds can be more difficult to form, the preference lying with the more stable binary and ternary compounds. Being almost totally at the mercy of thermodynamics, the solid-state chemist has traditionally relied on experience and intuition, rather than a set of predictable rules.

Over the past decade, work has been underway on many fronts to move away from the high temperatures of classical solid state synthesis, toward techniques that take advantage of lower reaction temperatures. The expectation here is that moderate temperature will favor the stabilization of more complex compounds that otherwise may be unstable with respect to disproportionation to simpler compounds. One technique is chemical vapor deposition (CVD) or molecular beam epitaxy. Here the synthesis of solid-state compounds in various technologically useful forms or shapes proceeds by the intimate gas-phase mixing of volatile precursors leading to deposition of solid-state materials on various substrates (Venkatasubramanian et al., 1997). This technique is intended for the synthesis of known solid-state compounds rather than the discovery of new ones, although there is no real reason why new ones (including metastable compounds) could not be discovered in this way (Johnson, 1998). Also noteworthy is the low-temperature solid state metathesis technique, which achieves rapid synthesis of pure binary and ternary chalcogenides (Treece et al., 1995; Parkin, 1996). The hydro(solvo)thermal technique has proven advantageous in that polyatomic building blocks have been used in the synthesis of solid-state frameworks. This method uses solvents heated in closed containers above their boiling point (but below their critical point) as reaction media (Sheldrick and Wachhold, 1998). Solubility of reactants is increased by virtue of the unusually high temperature and pressure within the container. Diffusion and crystal growth are further enhanced by the inclusion of mineralizers, species that can aid in the solvation and reprecipitation of solid state reactants and products, analogous to the function of the transport agent in chemical vapor transport crystal growth. Solvothermal synthesis has been the method of choice for the synthesis of a large variety of solid-state materials and eventually may find uses in TE materials discovery.

A highly successful approach, and most relevant in this article, has been performing reactions using molten salts as solvents. Such media have been employed for well over 100 years for high-temperature single crystal growth (Elwell and Scheel, 1975; Scheel, 1974; Sanjines et al., 1988; Garner and White, 1970). Although many salts are high-melting species, eutectic combinations of binary salts and salts of polyatomic species often have melting points well below the temperatures of classical solid-state synthesis, making possible their use in the exploration of new chemistry at intermediate temperatures. In many cases, such salts act not only as solvents, but also as reactants, providing species and building blocks that can be incorporated into the final product. In the search for new TE materials we have found this molten salt method to be most suitable for exploratory synthesis involving heavy elements such as Ba, Sr, Bi, Pb, Sn, Se, and Te and alkali atoms. Since the known materials that are used in TE devices are mainly chalcogenide compounds, we are searching for more complex semiconducting compounds of this type.

We are not merely interested in new compounds that are substitutions and variations of known structures, but in entirely *new structure types*. If significantly enhanced TE properties are to be found, new materials must become available. Therefore, novel types of syntheses must be explored that allow for higher ZTs. Since the electrical properties of solids are directly dependent on their crystal structure, we are motivated to look for new materials with new lattice structures.

IV. Thermoelectric Materials Discovery

Structural and compositional complexity can result in corresponding complexities in the electronic structure that may produce the required large asymmetry in DOS (see Eq. (1)) to obtain large thermopower. The phonon contribution to the thermal conductivity can also be lowered by such structural complexity, by choosing heavy elements as components of the material and by choosing combinations of elements that normally make moderate to weak chemical bonds.

The concept that certain materials can conduct electricity like a crystalline solid but heat like a glass is very useful. In these materials a weakly bound atom or molecule called a "rattler" is used to lower the thermal conductivity of the solid without severely affecting electronic conduction. This can lead to improved thermoelectric efficiency (Sales, 1997; Nolas et al., 1996, 1998, 1999).

The class of chalcogenide materials described here tends to satisfy this description because, as will become apparent later, they are made of three-dimensional or two-dimensional bismuth–chalcogenide frameworks, stabilized by weakly bonded alkali atoms that reside in cavities, tunnels, or galleries of the framework. These electropositive atoms almost always possess the highest thermal displacement parameters in the structure, which is evidence that a certain degree of "rattling" is present (Sales et al., 1999). This feature is very important in substantially suppressing the thermal conductivities of these materials. These materials also incorporate other beneficial features such as large unit cells and complex compositions.

For chalcogenide materials discovery, the use of molten alkali metal polychalcogenides, of the type A_2Q_x (A = alkali metal, Q = S, Se, Te) as solvents is very appropriate as we have demonstrated already (Kanatzidis and Sutorik, 1995). As solvents for intermediate temperature reactions, A_2Q_x salts are especially well suited because the melting points range between 200 and 600°C. Most alkali polytelluride salts melt between 300 and 500°C. Low-melting A_2Q_x fluxes remain nonvolatile over a wide temperature range, and so once above the melting point, reaction temperatures can be varied considerably without concern for solvent loss. Polychalcogenide fluxes are highly reactive toward metals because they are strong oxidants. Reactions between metals and molten A_2Q_x are performed in situ. The powdered reagents (polychalcogenide and metal or metal chalcogenide) are mixed under inert atmosphere and loaded into reaction vessels of either Pyrex or silica. Once evacuated, the tubes are sealed under vacuum and subjected to the desired heating program in a computer-controlled furnace. In a given A-M-Q system, large composition ranges are explored simultaneously (many synthesis runs in a given time) so that all possible phases can be found. At the end of the experiment and before product isolation, the tubes can be inspected with an optical microscope to assess the existence of promising reactions. In this sense this is a "poor man's" combinatorial method. Because this is a flux technique it means that only the part of the phase space that is rich in flux can be explored. Metal-rich compositions will have to be investigated in other ways.

To synthesize new compounds, one or more metals are added directly to the molten A_2Q/Q reaction mixture and heated in a sealed pyrex or silica container. Crystalline products either precipitate from the melt or form on slow cooling of the melt, depending on the specific stoichiometric and processing conditions. Presumably, the nucleated species are in equilibrium with the soluble intermediates, especially if the flux is present in excess, and hence a solvation/reprecipitation effect (often referred to as the mineralizer effect) occurs. This aids in the growth of single crystals because the flux can redissolve small or poorly formed crystallites and then reprecipitate the species onto larger, well-formed crystals. The advantage of the flux method is that one allows the system to end up "where it wants" in the kinetic or thermodynamic sense without attempting to force upon it a certain stoichiometry or structure. Provided the temperature and time are appropriate, the reaction systems have all the ingredients and freedom to form a new phase. The benefit of this becomes apparent from the unusual compositions often found in the new materials that most certainly could not have been predicted *a priori*.

The molten flux approach is suitable for quick and wide explorations of phase space but is most practical when carried out in small (<2 g) quantities. In this sense it is a *discovery* method, not a production method. Of course, once a new material has been discovered by such techniques, the next step is to attempt to prepare it on a large scale. Therefore, once the composition of a new material has been established, other synthesis methods amenable to scale-up must then be employed.

What are the empirical guidelines with which we have to design TE materials, and how do we go about choosing a particular system for exploration? We know that heavy atoms are desirable because they tend to vibrate slowly, giving rise to low-frequency lattice phonons that help slow down heat transfer through a material, leading to low thermal conductivity. The fact that Bi_2Te_3 is the best material known to date suggests that it combines many of the features necessary for high a figure of merit. We note here that even Bi/Sb alloys show some promise for TE applications. Therefore, we reason that if there is anything particular (or special) about bismuth that gives rise to simultaneously high electronic conductivity and thermoelectric power, it could be manifested in other compounds of bismuth as well. Because of this, we think a sensible approach is a research direction exploring other multinary chalcogenides of bismuth (and by extension antimony) in the hope that these elements will impart some (or all) of the key properties needed for superior TE performance. Furthermore, structurally and compositionally more complex bismuth chalcogenides would, most likely, have a low lattice thermal conductivity for the reasons outlined earlier. Based on these considerations, we decided to perform exploratory chemical synthesis involving Bi as one of the elements. As our results show, promising new materials can be found. The other elements employed in the synthesis are chalcogens such as S, Se, and Te, as well as alkali metals.

The presence of alkali metals in the structures of ternary and quaternary bismuth chalcogenides induce the stabilization of covalently bonded Bi–chalcogen frameworks with cages or tunnels that accommodate the charge-balancing alkali atoms. The interactions of alkali metals and the Bi–chalcogen frameworks are considered to be mainly electrostatic in nature. Alkaline earth metals such as Sr and Ba (and by extension Eu) with a $+2$ charge are also desirable as substitutes for alkali metals. They tend to have similar properties to alkali metals, but their greater charge tends to stabilize different structures.

V. New Materials

The considerations and criteria outlined earlier led us to investigate the pseudoternary phase space between $A_2Q/Bi_2Q_3/PbQ$ systems. The reactivity of Bi_2Q_3 and PbQ separately or together was investigated over a wide variety of ratios and conditions in molten A_2Q_x and over a wide variety of x. In chemical terms, this is a system of enormous size given all the possible combinations that can be investigated. Figure 1 shows a phase-space diagram with the most promising thermoelectric phases identified. Many lie on the A_2Q–Bi_2Q_3 and Bi_2Q_3–PbQ axes, while others (i.e., quaternary) are found in the body of the triangle. Selected phases are discussed later.

A common feature that has emerged during our investigations of these compounds is that they are constructed from simple building blocks. These blocks are regarded as excised fragments from the basic archetypal lattice of NaCl (or PbS-galena). These fragments formally derive from "cutting" the NaCl structure along various directions generating blocks of various sizes and shapes. The theme in all structures seems to be that these building blocks vary in size and width while they are usually infinite in at least one dimension. In addition to size and width, the particular cut (i.e., crystallo-

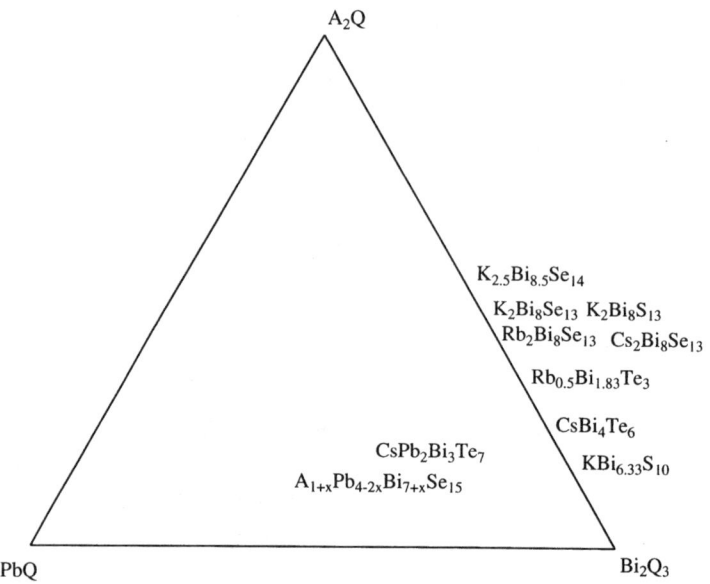

FIG. 1. New phases discovered in the pseudoternary phase $A_2Q/Bi_2Q_3/PbQ$ systems. Note: This is not a thermodynamic phase diagram, but a graph intended to give a better perspective on some of the associated compounds in this system.

graphic orientation) is also important and seems to vary from compound to compound.

The systems $(AE)Q/Bi_2Q_3/PbQ$ (where $AE = Sr$ and Ba; $Q = Se$ and Te) are related to those containing alkali metals and deserve an equal level of scrutiny, but we have only very recently begun to investigate these.

1. THE SULFIDES $KBi_{6.33}S_{10}$ AND $K_2Bi_8S_{13}$

One of the first materials discovered in our investigations was $KBi_{6.33}S_{10}$, and the closely related $K_2Bi_8S_{13}$ followed (Kanatzidis et al., 1996). The compounds belong to the series $(A_2Q)_n(Bi_2Q_3)_m$ (A = alkali metal; Q = S, Se) with $n = 1$ and $m = 6.33, 4$, respectively, and they lie in the A_2Q–Bi_2Q_3 axis of the triangle of Fig. 1. Although these compounds continue to be under investigation, the physicochemical characteristics that make them worthy of consideration for TE investigations are presented. Before we go into the discussion of these characteristics, we wish to explain why the metal sulfides are attractive for consideration. Metal sulfides are lighter and less dense than selenides and tellurides. If a metal sulfide compound were found with a $ZT \sim 1$ at room temperature, this would be a significant development from the point of view of device weight, despite the fact that the material may be only as good as the Bi_2Te_3 alloys. Other advantages include lower materials costs and possible environmental reasons, that is, sulfur is less toxic. Lightweight TE devices could provide a significant benefit when they are placed on moving vehicles such as trucks, cars, or ships.

The orthorhombic structure of $KBi_{6.33}S_{10}$ is closely related to that of the mineral cosalite $Pb_2Bi_2S_5$ (Srikrishnan and Nowacki, 1974); see Fig. 2. The structure of $KBi_{6.33}S_{10}$ is a defect cosalite type in which the high coordination positions in the lattice are occupied by a mixture of K and Bi atoms. This can be understood if the formula is doubled and written as $Pb_2Pb_2Bi_4S_{10}$ and then compared to the rearranged formula of $(*K_1Bi_{0.33})Bi_2Bi_4S_{10}$ ($* = $ vacancy). The mixed occupancy between K/Bi on certain sites in the structure and the presence of vacancies are special defects that are responsible for the observed very low thermal conductivity (see later discussion). The structure of $K_2Bi_8S_{13}$ is similar but the Bi_2Te_3-type (NaCl-type) blocks and CdI_2-type fragments are arranged differently; see Fig. 3.

In both compounds and in those described later, the alkali atoms interact with the metal chalcogenide framework via ionic bonds, whereas the Bi and S atoms form stronger covalent bonds to form a three-dimensional $[Bi_xS_y]^{z-}$ semiconducting framework. The sizes of the tunnels in which the alkali metals reside are often larger than their ionic size, and this generates either vibrational motion ("rattling") or slight positional disorder. These two potassium bismuth sulfides have promising electrical properties with

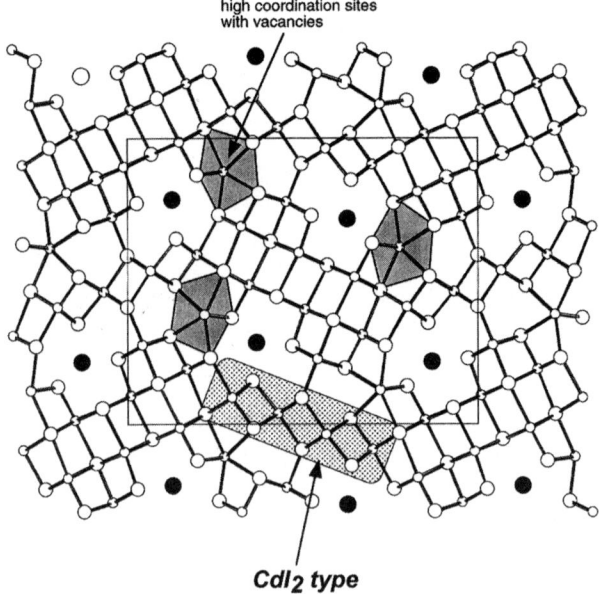

FIG. 2. The structure of $KBi_{6.33}S_{10}$. This is essentially the cosalite structure with K atom vacancies in the high coordination sites.

maximum conductivity and thermopower of 200 S/cm and $\sim 90\,\mu V/K$, respectively. These are unoptimized values from samples that have not been deliberately doped. Controlled doping can optimize the power factor ($\sigma \cdot S^2$) in $KBi_{6.33}S_{10}$, rendering it a good thermoelectric material. Additional information regarding the transport properties, including carrier concentrations and mobility, is needed. Preliminary results with polycrystalline samples indicate that $KBi_{6.33}S_{10}$ have carrier concentrations of 3×10^{19} cm^{-3}. Doping experiments with SbI_3 indicate that the conductivity type of $K_2Bi_8S_{13}$ can be controlled to be n-type with room temperature thermopower reaching $-100\,\mu V/K$; see Fig. 4. The data show a rise in thermopower at >300 K suggesting that maximum performance is reached at higher temperatures. The electronic conductivity of this sample at room temperature was ~ 200 S/cm. We believe they can be greatly improved by further crystal processing and systematic doping studies.

$KBi_{6.33}S_{10}$ has a very low thermal conductivity of ~ 1.3 W/m·K, lower than that of optimized $Bi_2Te_{3-x}Se_x$ (~ 1.5 W/m·K). This pleasant surprise comes from the rather low thermal conductivity they possess, because $KBi_{6.33}S_{10}$, being a sulfide, is expected to possess higher thermal conductivity compared to the heavier tellurides. The temperature dependence is shown in Fig. 5. The absence of a low-temperature (Umklapp process)

3 THE DISCOVERY OF NEW THERMOELECTRIC MATERIALS 63

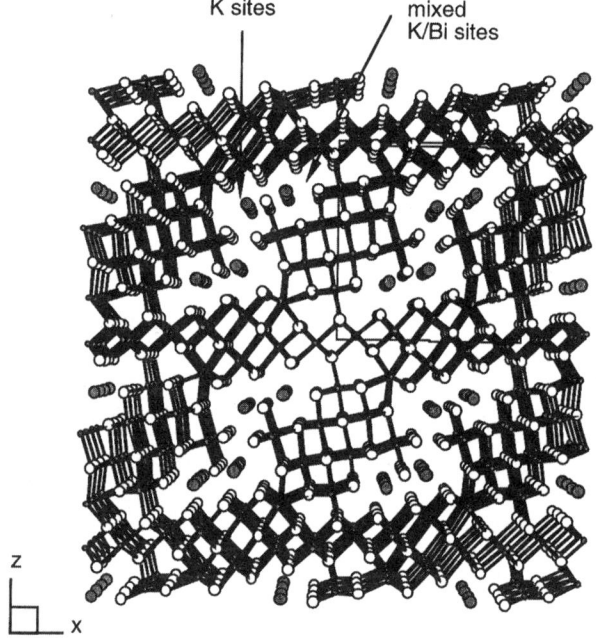

FIG. 3. The structure of $K_2Bi_8S_{13}$.

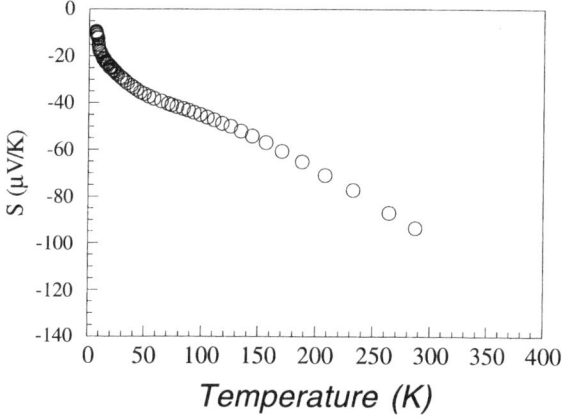

FIG. 4. Thermopower of a single crystal of $K_2Bi_8S_{13}$ as a function of temperature.

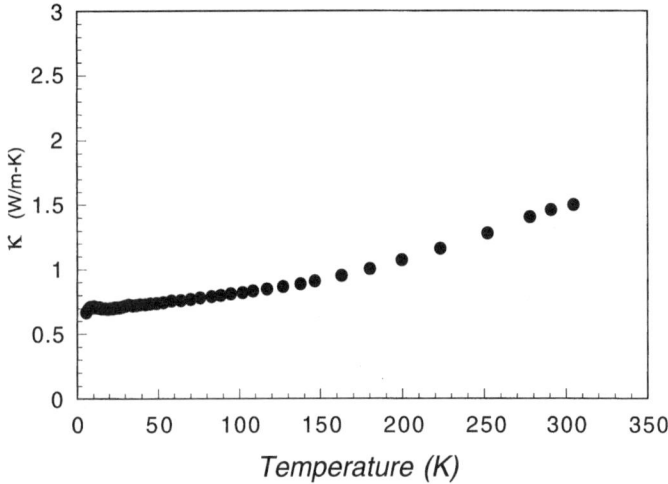

FIG. 5. Thermal conductivity as a function of temperature of a polycrystalline sample of $KBi_{6.33}S_{10}$.

maximum in the thermal conductivity is consistent with a phonon-glass-like behavior (Slack, 1995, 1997), which could be attributed to the presence of vacancies in the lattice. Using the measured values of the electronic conductivity in conjunction with the WF law, we estimate the maximum possible value of the electronic thermal conductivity contribution to be below 10% of the total thermal conductivity. Thus, essentially all heat in these compounds is carried by lattice phonons. This result suggests that certain sulfide materials possess low enough thermal conductivity to be viable candidates for TE research. In contrast, $K_2Bi_8S_{13}$, which does not possess lattice vacancies, shows a higher thermal conductivity of ~ 2.8 W/m·K. The temperature dependence is very different from that of $KBi_{6.33}S_{10}$ and more typical of that of a crystalline solid judging from the presence of the Umklapp maximum; see Fig. 6.

Both $KBi_{6.33}S_{10}$ and $K_2Bi_8S_{13}$ melt with no decomposition at 710°C and 713°C, suggesting they will be amenable to thermoelectric element fabrication and processing similar to that currently used in Bi_2Te_3 technology.

2. THE SELENIDES β-$K_2Bi_8Se_{13}$ AND $K_{2.5}Bi_{8.5}Se_{14}$

In moving to the corresponding selenides we found that the Se analog of $KBi_{6.33}S_{10}$ could not be prepared; however, β-$K_2Bi_8Se_{13}$ (isostructural to $K_2Bi_8S_{13}$) and the related $K_{2.5}Bi_{8.5}Se_{14}$ are very stable. As was found in the

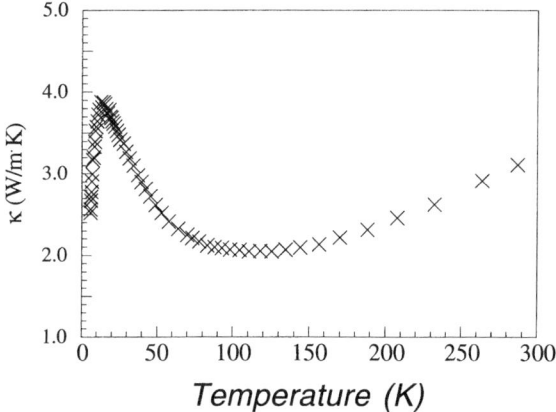

FIG. 6. Thermal conductivity as a function of temperature of a polycrystalline sample of $K_2Bi_8S_{13}$.

isostructural sulfide, in β-$K_2Bi_8Se_{13}$, the high coordination sites in the lattice (i.e., those with coordination number 7 or higher) are occupied by both K^+ and Bi^{3+} ions. The octahedral Bi sites have a smaller cavity size and do not accept K^+ ions. A previous version for this composition was found in α-$K_2Bi_8Se_{13}$, but the two structure types are completely different (Chung et al., 1997). Although both are three-dimensional, α-$K_2Bi_8Se_{13}$ has a more open structure than β-$K_2Bi_8Se_{13}$. Overall, the structure of β-$K_2Bi_8Se_{13}$ is slightly more dense than that of the α-form, because in the latter 25% of the Bi atoms are found in a trigonal pyramidal geometry, whereas in the former all Bi atoms are in an octahedral or higher coordination geometry; see Fig. 7. The origin of the structural and property differences between these two forms lies partly in the ability of the $6s^2$ lone pair of Bi^{3+} to stereochemically express itself.

$K_{2.5}Bi_{8.5}Se_{14}$ possesses a complex three-dimensional anionic framework related to that of β-$K_2Bi_8Se_{13}$; see Fig. 8 for comparison. Compositionally, $K_{2.5}Bi_{8.5}Se_{14}$ derives from β-$K_2Bi_8Se_{13}$ by addition of 0.5 equiv of $KBiSe_2$. The main difference between the two structures is that in $K_{2.5}Bi_{8.5}Se_{14}$ only NaCl- and Bi_2Te_3-type blocks exist. $K_{2.5}Bi_{8.5}Se_{14}$ forms by addition of half "$BiSe_2$" atoms to CdI_2-type fragments in β-$K_2Bi_8Se_{13}$. This small structural modification preserves the same connectivity of the NaCl-type fragments as well as the size and shape of this K site as in β-$K_2Bi_8Se_{13}$. Although the width of the NaCl block in the structure of $K_{2.5}Bi_{8.5}Se_{14}$ is also that of three Bi polyhedra, the width of its Bi_2Te_3 block is five Bi polyhedra, which is an important difference with β-$K_2Bi_8Se_{13}$; see Figs. 7 and 8.

The high coordination sites in the lattice (i.e., ~ 8) exhibit mixed occupancy between K and Bi atoms, and it could be explained on the basis of

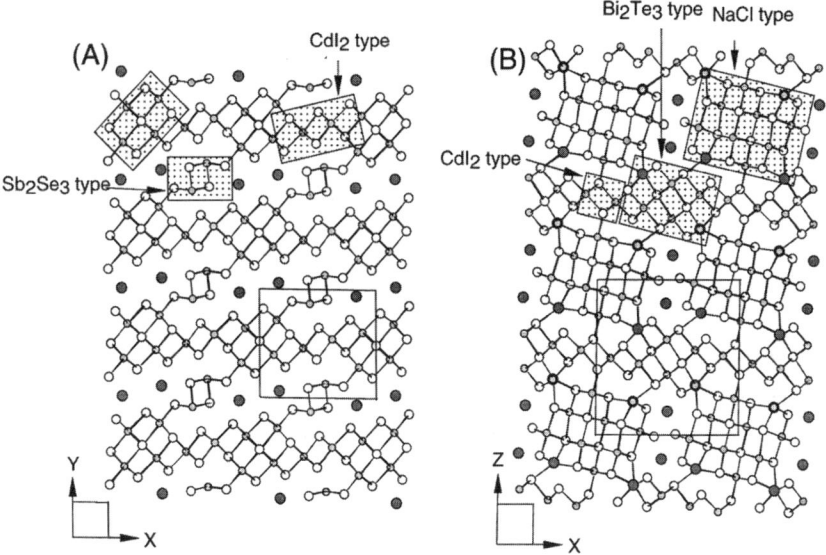

FIG. 7. Comparison of the structures of (A) α-$K_2Bi_8Se_{13}$ and (B) β-$K_2Bi_8Se_{13}$.

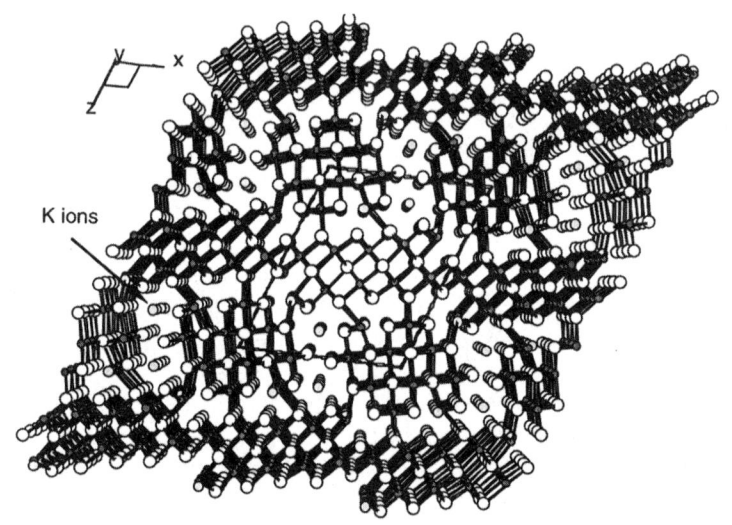

FIG. 8. The crystal structure of $K_{2.5}Bi_{8.5}Se_{14}$, viewed down the b axis.

similar ionic sizes of K^+ and Bi^{3+} in high coordination. This phenomenon is actually desirable in thermoelectrics because it contributes substantially to the very low thermal conductivity values found in these materials. In octahedral lattice sites, K^+/Bi^{3+} disorder is much less common.

The electrical properties of β-$K_2Bi_8Se_{13}$ and $K_{2.5}Bi_{8.5}Se_{14}$ were measured from undoped single crystal samples and polycrystalline ingots. A typical room temperature conductivity value obtained for single crystals of β-$K_2Bi_8Se_{13}$ is ~ 250–300 S/cm with a weak negative temperature dependence consistent with a semimetal or a narrow bandgap semiconducting material; see Fig. 9. In general $K_{2.5}Bi_{8.5}Se_{14}$ seems to be obtained from the synthesis in a more highly doped state than β-$K_2Bi_8Se_{13}$ since its thermopower tends to be smaller and its electronic conductivity greater. Its room temperature values vary between 300 and 1000 S/cm.

β-$K_2Bi_8Se_{13}$ and $K_{2.5}Bi_{8.5}Se_{14}$ are narrow bandgap semiconductors with room temperature bandgaps of 0.59 and 0.56 eV, respectively. In the case of the Sb analogs, we found greater bandgaps at 0.78 eV for $K_2Sb_8Se_{13}$ and 0.82 eV for $K_{2.5}Sb_{8.5}Se_{14}$. Tunability in the bandgaps and consequently electrical properties could be achieved by preparing solid solutions of the type $K_2Bi_{8-x}Sb_xSe_{13}$ and $K_{2.5}Bi_{8.5-x}Sb_xSe_{14}$. Figure 10 shows the evolution of the energy gap in the $K_2Bi_{8-x}Sb_xSe_{13}$ system as a function of x. Because of the narrower bandgap of $K_{2.5}Bi_{8.5}Se_{14}$, we expect better properties than β-$K_2Bi_8Se_{13}$, provided a good handle on its doping behavior can be obtained.

a. Undoped Samples

Thermopower measurements on typical "as prepared" β-$K_2Bi_8Se_{13}$ and $K_{2.5}Bi_{8.5}Se_{14}$ samples indicate that they always form n-type, suggesting electrons as the carriers with the Seebeck coefficient typically varying from -50 to $-250\,\mu V/K$. In general, the thermopower of $K_{2.5}Bi_{8.5}Se_{14}$ samples is lower than that of β-$K_2Bi_8Se_{13}$, depending on doping. The origin of n-type conductivity is not understood, but it may be due to a slight excess of Se atoms occupying Bi sites in the structure. However, it could also arise from a slight Se deficiency. Annealing β-$K_2Bi_8Se_{13}$ under vacuum at a temperature of $\sim 50°C$ below its melting point causes the electronic conductivity to rise substantially from ~ 200 S/cm to ~ 900 S/cm at room temperature. The metal-like slope to the data as a function of temperature is enhanced. At the same time the thermopower decreases from $-200\,\mu V/K$ to $-100\,\mu V/K$ or even lower. These results suggest that the number of n-type carriers in the material has increased. This could happen through the creation of Se vacancies in the lattice, which results in electron injection into the material's conduction band (each Se atom generates two electrons); however, other mechanisms could be responsible as well.

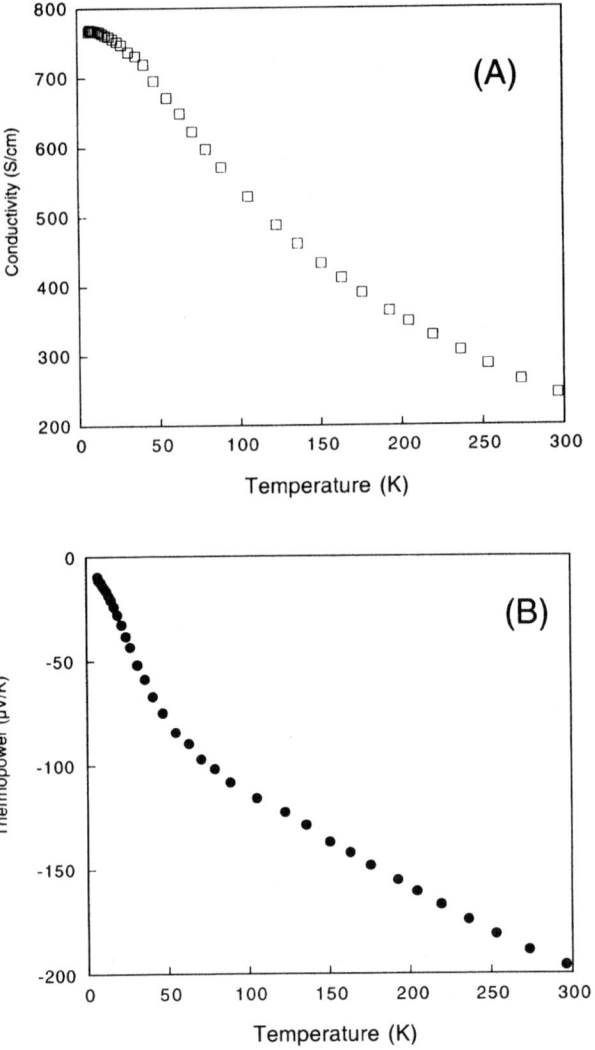

FIG. 9. Temperature dependence of (A) the electrical conductivity and (B) thermopower for a single crystal of β-$K_2Bi_8Se_{13}$.

At room temperature the total thermal conductivities of both β-$K_2Bi_8Se_{13}$ and $K_{2.5}Bi_{8.5}Se_{14}$ are very low and comparable at 1.28 and 1.24 W/m·K, respectively. These are slightly lower than that of optimized solid solution $Bi_2Te_{3-x}Se_x$ ($\kappa_{tot} \sim 1.5$ W/m·K). A plot of the thermal conductivity of β-$K_2Bi_8Se_{13}$ as a function of temperature is shown in Fig. 11. The

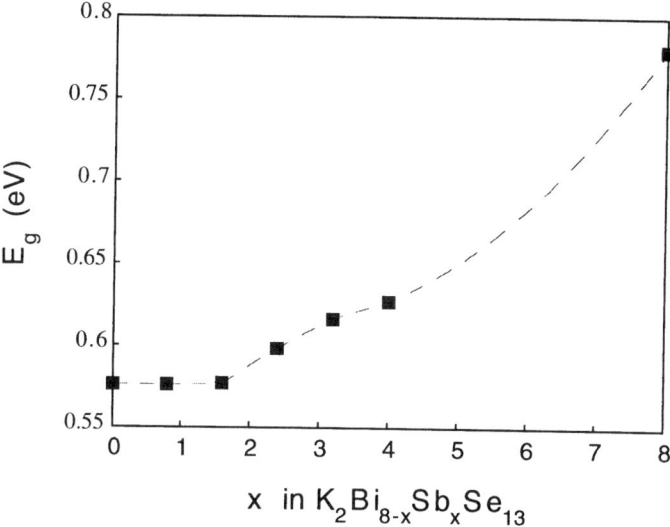

FIG. 10. Dependence of the energy gap as a function of x in the solid solution system $K_2Bi_{8-x}Sb_xSe_{13}$.

very low thermal conductivity of these compounds is attributed to a number of factors such as the low crystal symmetry (monoclinic), large unit cells, and the presence of "rattling" alkali atoms in tunnels that are only electrostatically interacting with Se atoms on the tunnel walls. The thermal conductivity of β-$K_2Bi_8Se_{13}$ in the temperature range of 4–300 K is significantly lower than that of the isostructural compound $K_2Bi_8S_{13}$, consistent with the fact that the heavier Se atoms reduce the frequency of the lattice phonons, thereby suppressing heat transport in the material. Using the measured values of the electronic resistivity and based on WF law, the maximum possible values of the κ_e contribution in both cases were estimated to be less

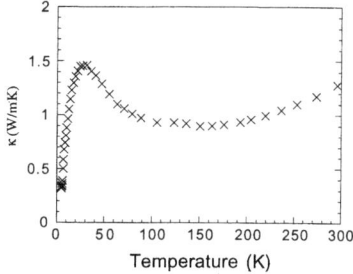

FIG. 11. Thermal conductivity as a function of temperature of a polycrystalline sample of β-$K_2Bi_8Se_{13}$.

b. Doped Samples

β-$K_2Bi_8Se_{13}$ doped with varying concentrations of $SbBr_3$ and Sn were synthesized. The electronic conductivity for the Sn-doped samples generally decreased as the doping level increased, shown in Fig. 12. The room temperature conductivity was increased when the material was doped with 0.5% Sn, but the conductivities steadily decreased from 542 to 90.7 S/cm as the doping concentration increased from 0.5 to 3.0% Sn. The temperature dependence also showed a transition from a weakly metallic behavior to semiconducting behavior. With the exception of the 2.0% Sn sample, which showed metallic behavior, the electronic conductivity data suggest the activation energy increases with doping level.

The electronic conductivity of the entire range of $SbBr_3$-doped samples showed a weak metallic dependence, that is, decreasing as temperature increased from 4.2 to 340 K. The electronic conductivity was maximized when the sample contained 2.2% $SbBr_3$, with a steady reduction in conductivity values as the doping level moved away from 2.2%. Room temperature values are listed in Table I.

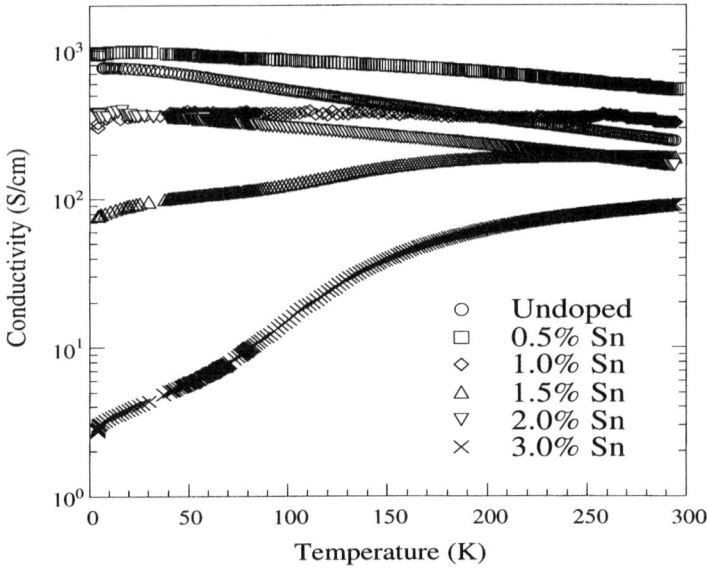

FIG. 12. Temperature dependence of the electrical conductivity for the series of tin-doped β-$K_2Bi_8Se_{13}$ samples.

TABLE I

TE Properties and Power Factors for β-K$_2$Bi$_8$Se$_{13}$ at 295 K for Sn and SbBr$_3$ Dopants

Dopant	Concentration (%)	Conductivity (S/cm)	Seebeck (μV/K)	$S^2\sigma$ (μW/cm·K^2)
Undoped	—	244	−222	12.1
Sn	0.5*	544	−266	**38.5**
	1.0	327	−284	26.3
	1.5	190	−280	14.9
	2.0	168	−295	14.6
	3.0	90.9	−333	10.1
SbBr$_3$	0.5	129	−211	5.74
	1.0	54.7	−272	4.03
	1.5	157	−198	6.17
	1.8	136	−262	9.34
	2.0	191	−276	14.6
	2.2*	258	−276	**19.7**
	2.5	—	−350	—
	3.0	133	−268	9.53

Thermopower data, shown in Fig. 13 for the Sn dopant, exhibited similar behavior to the undoped β-K$_2$Bi$_8$Se$_{13}$ material. The n-type thermoelectric response tended toward zero as the temperature approached 0 K. The thermopower values become very large at higher temperatures and seem to maximize outside of the range accessible by our measurements. Based on the approximate relationship $E_g \sim 2S_{max} \times T_{max}$ (Goldsmid and Sharp, 1999) and the optically measured $E_g \sim 0.54$ eV, the thermopower is expected to maximize in the region 500–700 K between 400 and 550 μV/K. This temperature range is also the range of maximum ZT for this material. Thermopower values showed a steady increase with doping concentration, with the room temperature value steadily increasing from $-220\ \mu$V/K for the undoped β-K$_2$Bi$_8$Se$_{13}$ to $-333\ \mu$V/K for the sample doped with 3.0% Sn. The thermopower for the SbBr$_3$-doped sample, Fig. 14, showed a minimum value for the 1.5% doped sample, and a maximum value with the 2.2% doped sample.

The conductivity and thermopower data were used to calculate the power factors, $S^2\sigma$, for each dopant versus temperature and doping concentration. The power factor data are a good indication of the optimal doping level for each dopant, as well as the temperature where the power factor is a maximum. The most promising sample was 0.5% Sn-doped β-K$_2$Bi$_8$Se$_{13}$, which shows a maximum power factor of 38.5 μW/cm·K^2 at 295 K; see Fig. 15. This value is approximately a threefold increase from the room temperature power factor of undoped β-K$_2$Bi$_8$Se$_{13}$. Further increase in doping levels resulted in a steady reduction of the power factor.

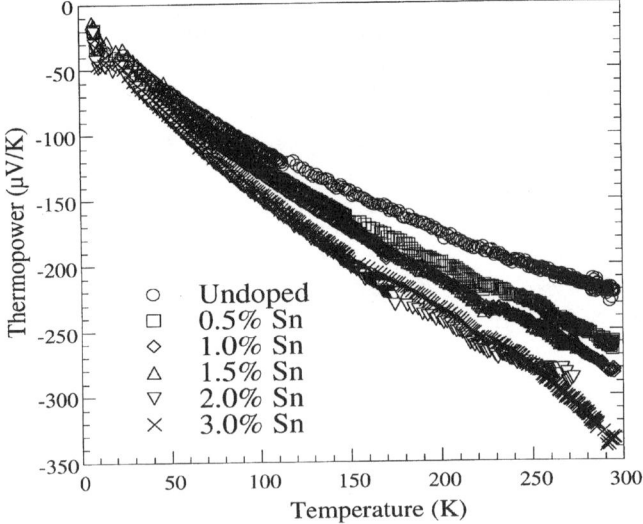

FIG. 13. Temperature dependence of the thermopower for the series of Sn-doped β-$K_2Bi_8Se_{13}$ samples.

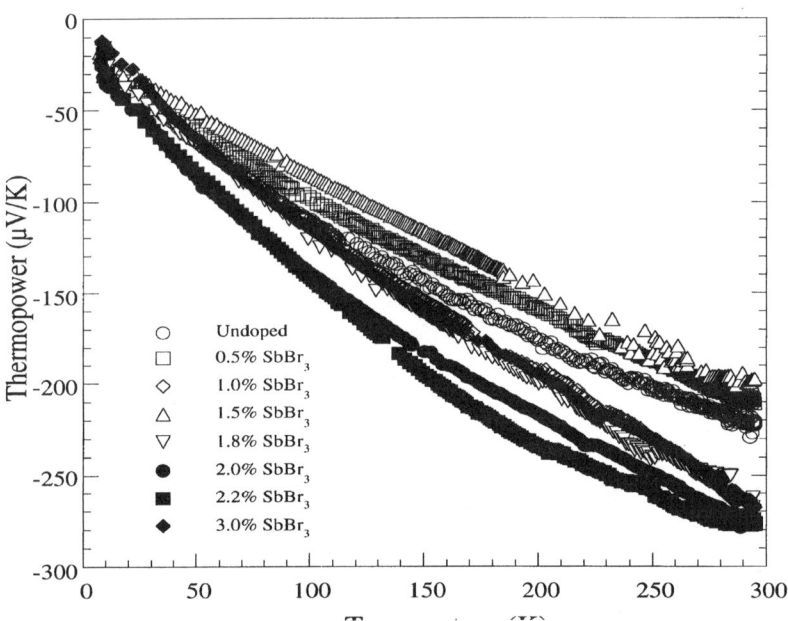

FIG. 14. Temperature dependence of the thermopower for the series of $SbBr_3$-doped β-$K_2Bi_8Se_{13}$ samples.

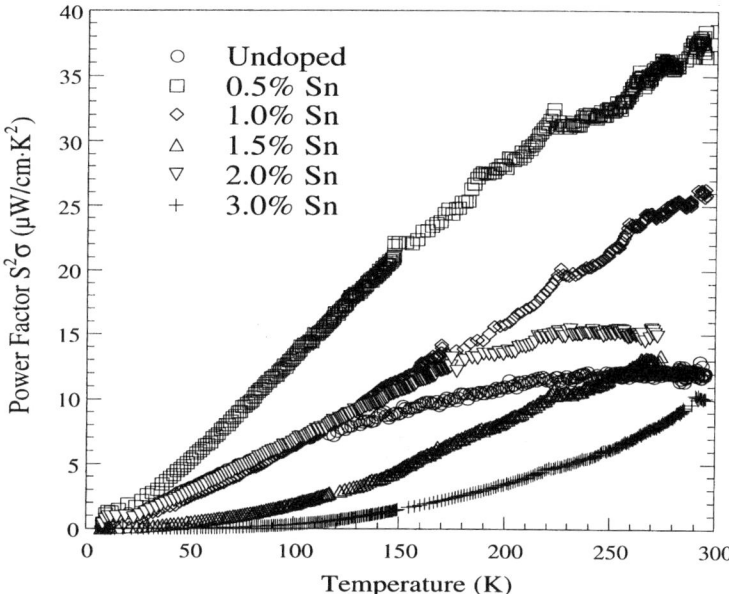

FIG. 15. Power factor versus temperature for the series of tin-doped β-$K_2Bi_8Se_{13}$ samples.

The power factor values for the $SbBr_3$-doped samples, Fig. 16, showed a sharp drop as doping was raised to 1.0%, then increased to the highest power factor found for the series: 19.7 $\mu W/cm \cdot K^2$ at 295 K for the 2.2% doped sample. The power factor again begins to decrease as doping is increased to 3.0%. Although the enhancement of power factor was not as dramatic as for the 0.5% Sn-doped sample, the 2.2% $SbBr_3$-doped sample still exhibited a 63% increase compared to the room temperature power factor of undoped β-$K_2Bi_8Se_{13}$. Both the maximum and room temperature power factors for all doped samples are shown in Table I.

It is apparent from these studies that power factors that exceed those of undoped β-$K_2Bi_8Se_{13}$ by approximately a factor of 3 can be obtained with a relatively small or no increase in thermal conductivity. Greater enhancements are expected with further investigations of dopants.

A reasonable approach to improve the thermoelectric figure of merit of these ternary compounds could be sulfur alloying or solid solutions of $K_xBi_y(Se,S)_z$ and $K_x(Bi,Sb)_ySe_z$. This type of alloying is expected to lower the κ_l further and increase the thermopower by increasing the bandgap. An outstanding challenge is to achieve p-type doping in these materials. Although such efforts are only beginning, so far only n-type transport has been observed.

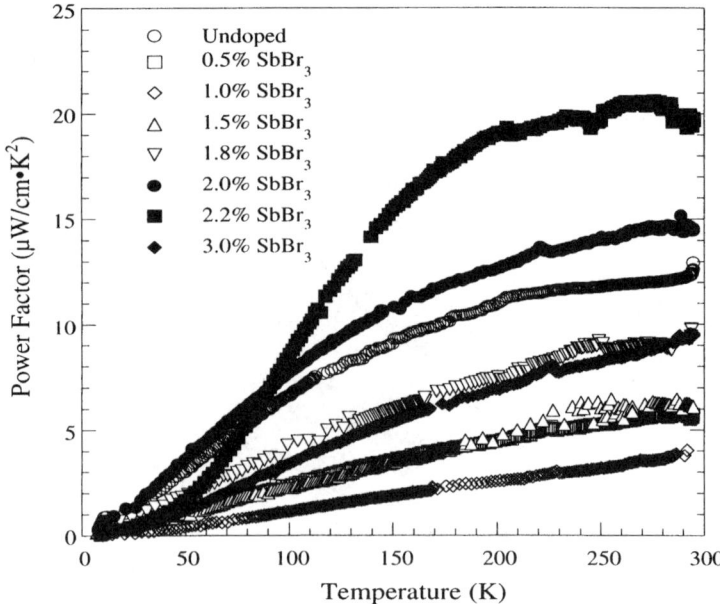

FIG. 16. Power factor versus temperature for a series of $SbBr_3$-doped β-$K_2Bi_8Se_{13}$ samples.

3. Materials Related to β-$K_2Bi_8Se_{13}$

The structure of β-$K_2Bi_8Se_{13}$ is the same as that of $Sr_4Bi_6Se_{13}$ (Cordier et al., 1985) and it derives by replacing two Sr^{2+} ions with two K^+ ions and the remaining two with Bi^{3+} ions. These substitutions are isoelectronic on average and do not require compositional changes in the "Bi_6Se_{13}" part of the compound. Therefore, an alternative way of representing the β-phase is $(K, Bi)_4Bi_6Se_{13}$. Once again the K and high coordination sites of Bi in the structure of β-$K_2Bi_8Se_{13}$ are distinct from the rest of the metal sites, which are essentially octahedral. The high coordination sites are susceptible to a substantial degree of chemical substitution provided cations of similar size are used and electroneutrality is preserved. An interesting substitution here is the replacement of the four Sr^{2+} atoms in $Sr_4Bi_6Se_{13}$ with two Ba^{2+} ions and two Pb^{2+} ions, or the replacement of two Sr^{2+} ions with Pb^{2+} or Eu^{2+} ions. This results in the isostructural compounds $Ba_2Pb_2Bi_6Se_{13}$, $Eu_2Pb_2Bi_6Se_{13}$, and $Sr_2Pb_2Bi_6Se_{13}$, as well as the solid solution compounds $Ba_{2-x}Pb_{2+x}Bi_6S_{13}$ and $Sr_{2-x}Pb_{2+x}Bi_6Se_{13}$. Many other interesting derivatives may be envisioned by such substitutions. Interestingly, the selenide analog $Ba_{2-x}Pb_{2+x}Bi_6Se_{13}$ has not been observed yet; instead, a different structure type is adopted when these elements are combined, namely that of $A_{1+x}Pb_{4-2x}Bi_{7+x}Se_{15}$ (see later discussion).

The compounds $Ba_4Bi_6Se_{13}$ and $Eu_2Pb_2Bi_6Se_{13}$ are two such derivatives. In $Ba_4Bi_6Se_{13}$ the K^+ and Bi^{3+} cations are replaced with Ba^{2+} cations, whereas in $Eu_2Pb_2Bi_6Se_{13}$ they are replaced with Eu^{2+} and Pb^{2+} cations. Such modifications are useful because they provide for considerable manipulation of the TE properties. Eventually they should help us learn more about structure–property relationships.

Preliminary charge transport data shows that $Ba_4Bi_6Se_{13}$ tends to form in a highly doped form as judged from its room temperature conductivity of ~ 2500 S/cm; see Fig. 17A. The samples are n-type with corresponding Seebeck coefficient between -32 and $-40\,\mu V/K$; see Fig. 17B. The analogous $Eu_2Pb_2Bi_6Se_{13}$ is also not optimally doped with room temperature conductivity and Seebeck coefficient of ~ 400 S/cm and $-40\,\mu V/K$. A related system worth mentioning here is $Eu_2Pb_2Bi_4Se_{10}$ (isostructural to cosalite), which shows greater promise with corresponding values of ~ 500 S/cm and $-140\,\mu V/K$; see Fig. 18. All these compounds show n-type charge transport, which is consistent with a number of possibilities, including Se deficiency, Se occupation of Bi sites, or presence of accidental electron donor impurities. The class of materials related to the structure of β-$K_2Bi_8Se_{13}$ present a fertile ground in which to explore new combinations for novel TE compounds. Much work needs to be done to identify these, but the fact that so many members already look promising is a very positive hint that additional member could be even more interesting.

4. The Family $A_{1+x}Pb_{4-2x}Bi_{7+x}Se_{15}$ (A = K, Rb)

Lead is an attractive element in thermoelectrics research because, like Bi, it is heavy, influenced by relativistic effects, and forms covalent bonds with Se or Te atoms. The covalency arises because the electronegativity difference with Se and Te is quite small. Lead also has a similar size to Bi and it is likely to occur in similar crystal environments with it. A distinguished electronic feature of Bi and Pb is the $6s^2$ lone pair of electrons. The stereochemical expression of the $6s^2$ lone pair of electrons influences the structure type and electronic structure, and consequently the electrical properties of Bi and Pb compounds. Because they exhibit very similar coordination geometry and crystallographic behavior, this chemistry produces diverse combinations of NaCl-type fragments. The incorporation of lead into the A–Bi–Se system increases its complexity at all levels from structure to composition.

One phase of interest in this system is $K_{1+x}Pb_{4-2x}Bi_{7+x}Se_{15}$ (Choi et al., 2000). In fact there is a series of compounds with this formula in which the ratio of K/Pb/Bi varies. The general formula $A_{1+x}M_{4-2x}M'_{7+x}Se_{15}$ (A = K, Rb; M = Pb, Sn; M' = Bi, Sb) derives from a large degree of variability in composition that is expressed in terms of mixed occupancy among

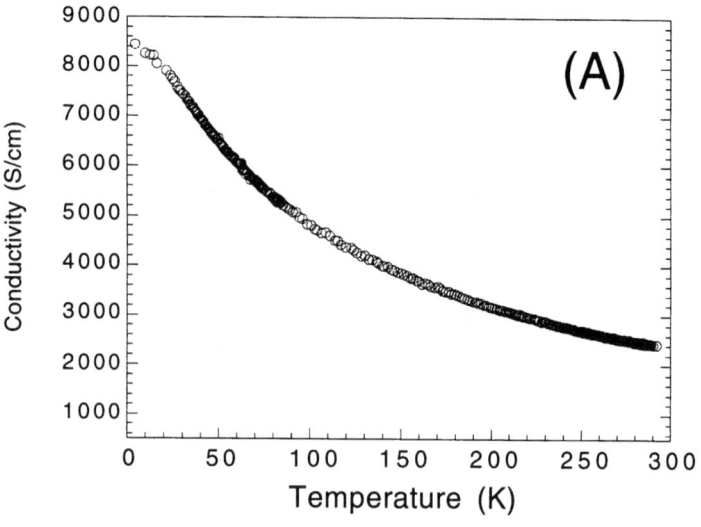

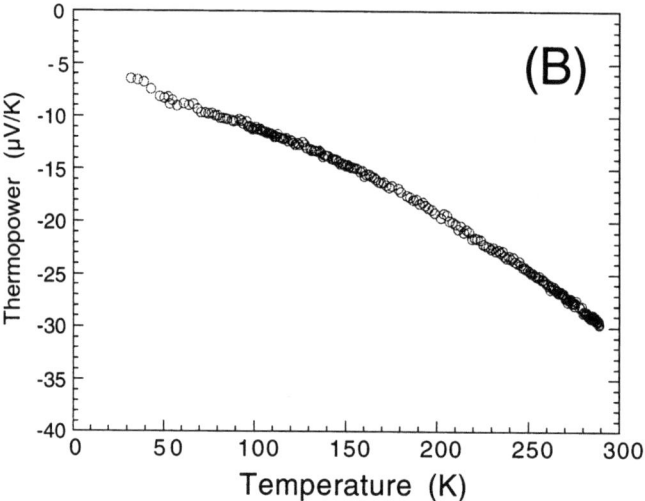

FIG. 17. Temperature dependence of (A) the electrical conductivity and (B) thermopower for a single crystal of $Ba_4Bi_6Se_{13}$.

A^+, M^{2+}, and M'^{3+} atoms. This variability has two important consequences: First, it generates a continuum of isostructural compositions with regularly varying energy gap, and second, by virtue of its internal mixed occupancy disorder, it sets the stage for materials with very low thermal conductivity. The former consequence is advantageous because it provides

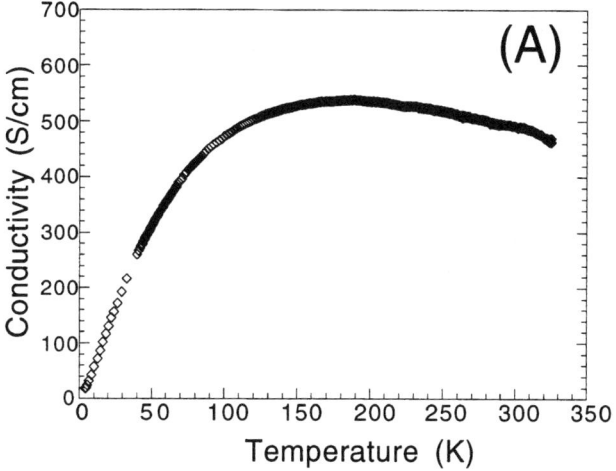

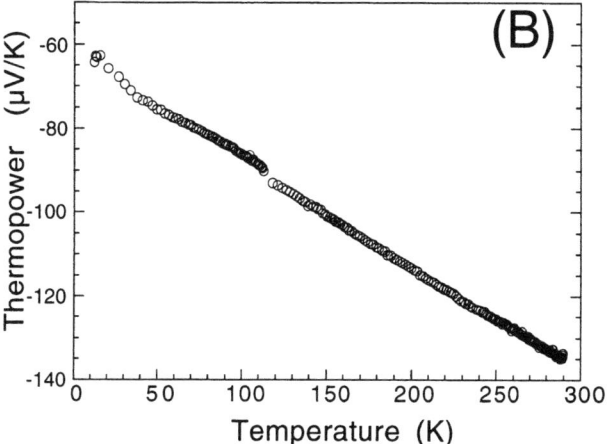

FIG. 18. Temperature dependence of (A) the electrical conductivity and (B) thermopower for a single crystal of $Eu_2Ba_2Bi_6Se_{13}$.

a mechanism for controlling the electrical properties of these materials, whereas the latter is a necessary condition for any viable thermoelectric material. Therefore, certain members of this, apparently large, class of isostructural compounds may hold high potential for thermoelectric applications.

The structure type has a three-dimensional framework assembled from NaCl- and Bi_2Te_3-type building units; see Fig. 19. The framework features narrow tunnels filled with K^+ or Rb^+ ions. The NaCl- and Bi_2Te_3-type

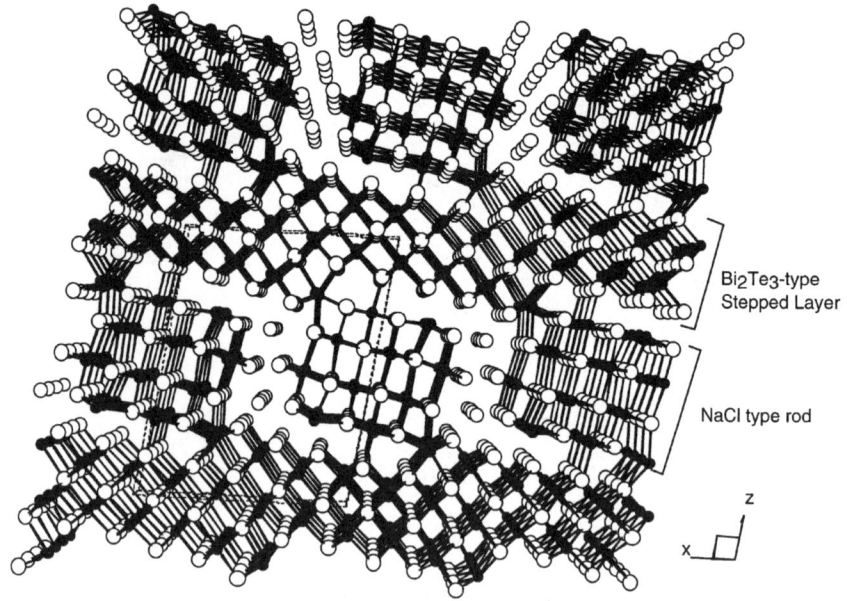

FIG. 19. The crystal structure of $K_{1+x}Pb_{4-2x}Bi_{7+x}Se_{15}$, viewed down the b axis.

units are composed of edge shared distorted Bi^{3+}/Sb^{3+} octahedra. The Pb/Sn atoms are stabilized in 8-coordinated bicapped trigonal prismatic sites at the connecting points of NaCl-type blocks and Bi_2Te_3-type blocks. There exists considerable occupancy disorder among Pb/Sn, Bi/Sb, and K/Rb in the structure.

This structure is closely related to those of β-$K_2Bi_8Se_{13}$ and $K_{2.5}Bi_{8.5}Se_{14}$ (Chung et al., 1997). The local environment of alkali metal ions and the size of the NaCl-type block in $K_{1.25}Pb_{3.50}Bi_{7.25}Se_{15}$ are also shared by β-$K_2Bi_8Se_{13}$ and $K_{2.5}Bi_{8.5}Se_{14}$, see Figs. 7 and 8. Only the size of the Bi_2Te_3-type unit in each compound is different. The relationship between these three compounds can easily be seen if the formulas are broken down into two parts, the anionic framework and the alkali metal cations in the tunnels. For example, in β-$K_2Bi_8Se_{13}$, one equivalent of K^+ is stabilized in the tunnel and the other equivalent is disordered with Bi atoms in the anionic framework. Therefore, the formula can be described as $K^+[KBi_8Se_{13}]^-$ or $K^+[M_9Se_{13}]^-$ (M = K + Bi in the anionic framework). In the same way, $K_{2.5}Bi_{8.5}Se_{14}$ can be described as $K^+[M_{10}Se_{14}]^-$ and $K_{1+x}Pb_{4-2x}Bi_{7+x}Se_{15}$ as $K^+[M_{11}Se_{15}]^-$ (M = K + Bi + Pb in the anionic framework). Therefore, $K_{2.5}Bi_{8.5}Se_{14}$ and $K_{1+x}Pb_{4-2x}Bi_{7+x}Se_{15}$

derive by successively adding neutral "MSe" units to $K_2Bi_8Se_{13}$ as follows.

$$K^+[M_9Se_{13}]^- \xrightarrow{[MSe]} K^+[M_{10}Se_{14}]^- \xrightarrow{[MSe]} K^+[M_{11}Se_{15}]^- \rightarrow ?$$

Therefore, $\beta\text{-}K_2Bi_8Se_{13}$, $K_{2.5}Bi_{8.5}Se_{14}$, and $K_{1+x}Pb_{4-2x}Bi_{7+x}Se_{15}$ are all members of a *homologous* series in both a compositional and structural sense. This structural homology is flexible enough to preserve the basic framework through successive addition of "MSe" equivalents by adjusting the width of the Bi_2Te_3-type blocks. It would be interesting to investigate if higher order homologs in the series exist. The modular construction of this type of compound is a major underlying characteristic and facilitates the understanding of such compounds' structure and interrelationships. It is also expected to be useful in the prediction of new compounds based on different sizes and shapes of the NaCl modules.

The electrical properties of these compounds were measured from single crystal samples (see Fig. 20). The room temperature conductivity of $K_{1.25}Pb_{3.50}Bi_{7.25}Se_{15}$ was obtained around 260 S/cm. The conductivity shows a metallic trend between 300 K and 50 K and it turns semiconducting below 50 K. The thermopower at room temperature is quite large and negative at $-150\ \mu V/K$ and it increases at higher temperatures. The

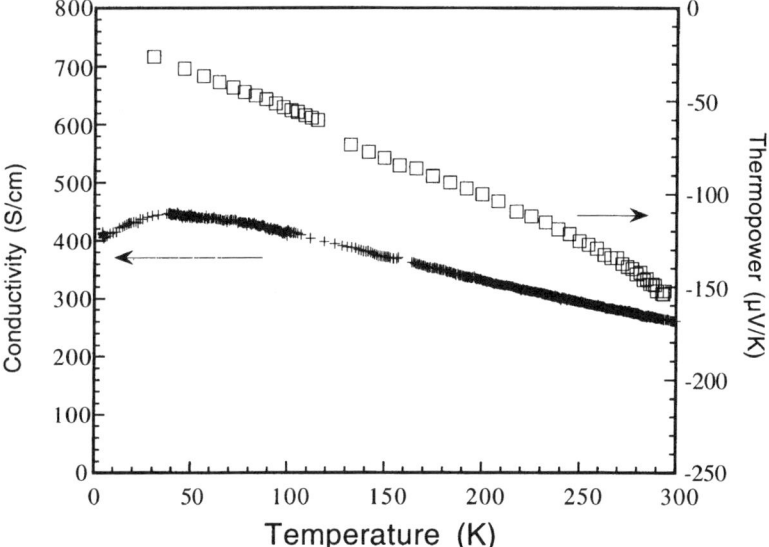

FIG. 20. Temperature dependence of the electrical conductivity and thermopower for a single crystal of $K_{1+x}Pb_{4-2x}Bi_{7+x}Se_{15}$.

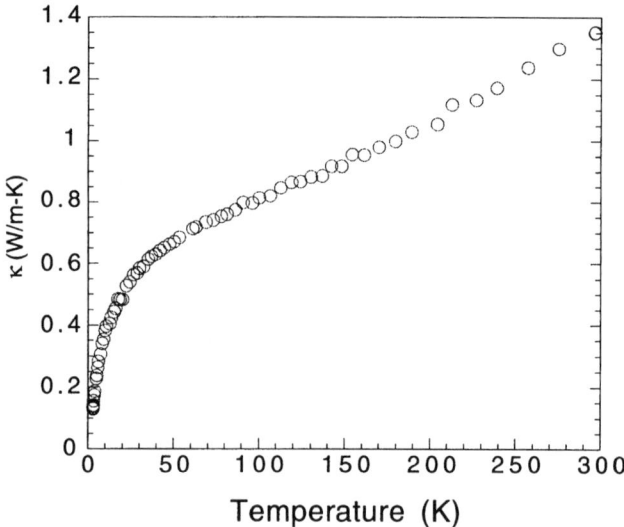

FIG. 21. Thermal conductivity as a function of temperature of a polycrystalline ingot sample of $K_{1+x}Pb_{4-2x}Bi_{7+x}Se_{15}$.

negative sign indicates that charge carriers are electrons (n-type). The electronic conductivities of Sb compounds are 10^5 times smaller than those of the Bi analogs. Therefore, the Sb compounds are not viable for TE, though Sb can served as a suitable modifier of the electrical properties of the Bi compounds.

Unfortunately, it is not feasible to measure the thermal conductivity of small single crystals; for this measurement, large ingots of the material are required. The thermal transport properties of $K_{1.25}Pb_{3.50}Bi_{7.25}Se_{15}$ were measured on elongated ingots (see Fig. 21). The thermal conductivity κ was estimated at 1.26 W/m·K after correcting for radiative losses. This value is comparable to those obtained for the ternary β-$K_2Bi_8Se_{13}$ and $K_{2.5}Bi_{8.5}Se_{14}$ and lower than that of optimized $Bi_2Te_{3-x}Se_x$ (*Thermoelectric Semiconductors*, 1986). As we have done earlier, here again we conclude that the structural and compositional complexity, the rattling effect of alkali metal and Pb^{2+} ions in cavities, and the occupancy disorder among the K, Pb, and Bi in $K_{1.25}Pb_{3.50}Bi_{7.25}Se_{15}$ all combine to result in a very low lattice thermal conductivity indeed. At this point a figure of merit (ZT) for single crystals of undoped $K_{1.25}Pb_{3.50}Bi_{7.25}Se_{15}$ is estimated at ~ 0.2 at room temperature, suggesting that this material merits serious experimental efforts at optimization.

The isostructural tin compound $K_{1+x}Sn_{4-2x}Bi_{7+x}Se_{15}$ has a room temperature electronic conductivity (measured on an ingot) of ~ 100 S/cm. The thermopower measurement reveals n-type behavior with a room temperature Seebeck coefficient of $-85\ \mu V/K$ that rises to $\sim -120\ \mu V/K$ at 400 K.

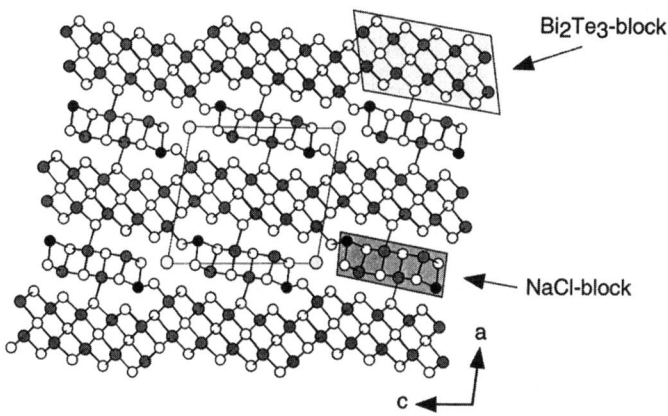

FIG. 22. The structure of $K_{1-x}Sn_{5-x}Bi_{11+x}Se_{22}$ ($x \sim 0.33$).

This material also shows a very low thermal conductivity at room temperature of ~ 1.5 W/m·K.

$K_{1-x}Sn_{5-x}Bi_{11+x}Se_{22}$ ($x \sim 0.33$) (Mrotzek et al., 2000) is another related compound, but not a member of the same homologous series, which crystallizes in a low symmetry monoclinic space group. Its structure is shown in Fig. 22, and it, too, contains Bi_2Te_3-type and NaCl-type building units connected to form a new kind of three-dimensional anionic framework with K^+ filled tunnels. The building units are assembled from distorted, edge-sharing (Bi, Sn)Se_6 octahedra. Bi and Sn atoms are disordered over all metal sites of the chalcogenide network while the K site is only one-third occupied. The potassium ions fill the tunnels created by the linkage of the two different structure units. The tricapped trigonal prismatic sites are only 33% occupied. The high atomic displacement parameters (ADPs) of this site suggests that the potassium ions "rattle" in their positions, although, probably, static disorder along the tunnel direction is also present.

The charge transport measurements of $K_{1-x}Sn_{5-x}Bi_{11+x}Se_{22}$ ($x \sim 0.33$) at room temperature reveal an electronic conductivity of 450 S/cm and a thermopower of -43μV/K. This is consistent with a narrow bandgap semiconductor. Because of the same type of extensive disorder of the metal atoms and the presence of "rattling" potassium atoms, as described earlier, $K_{1-x}Sn_{5-x}Bi_{11+x}Se_{22}$ has similarly low thermal conductivity of ~ 1.5 W/m·K. The material has naturally n-type doping and this may be amenable to manipulation with changes in the value of x. Synthetic investigations of the system K/Sn/Bi/Se should lead to additional quaternary bismuth selenide compounds.

The systems just discussed possess some of the lowest thermal conductivities reported for crystalline compounds. The results presented here demonstrate that it is possible to achieve lower thermal conductivity in quaternary compounds with complex composition and crystal structures compared to simpler high-symmetry binary compounds. The fact that there exist Sb and Sn analogs of $K_{1.25}Pb_{3.5}Bi_{7.25}Se_{15}$ makes systematic Bi–Sb or Pb–Sn solid solution studies possible without disrupting the crystal structure.

5. The Telluride Systems A/Bi/Te and A/Pb/Bi/Te

In sharp contrast to the selenide compounds, the isostructural Te analogs are not stable. Rather, the A/Bi/Te and A/Pb/Bi/Te systems give rise to different structure types. Several interesting new compounds have been identified. These include $CsBi_4Te_6$, $RbBi_{3.66}Te_6$ and $APb_2Bi_3Te_7$ (A = Cs, Rb).

a. $CsBi_4Te_6$

In an attempt to make the Te analog of β-$K_2Bi_8Se_{13}$ with Cs, we obtained $CsBi_4Te_6$ instead of the expected $Cs_2Bi_8Te_{13}$. From a chemical point of view, this amounts to a reduction of a Bi_2Te_3 unit by a half equivalent of electrons. This reduction results in a complete restructuring of the Bi_2Te_3 framework so that a new structure forms. This compound is a possible new candidate for low-temperature TE applications. When doped appropriately, it achieves a maximum ZT of ~ 0.8 at 225 K, making it the best bulk thermoelectric material below room temperature (Chung et al., 2000). The material is air and water stable and melts without decomposition at 545°C. The crystals grow with long needlelike morphology (Fig. 23). The direction

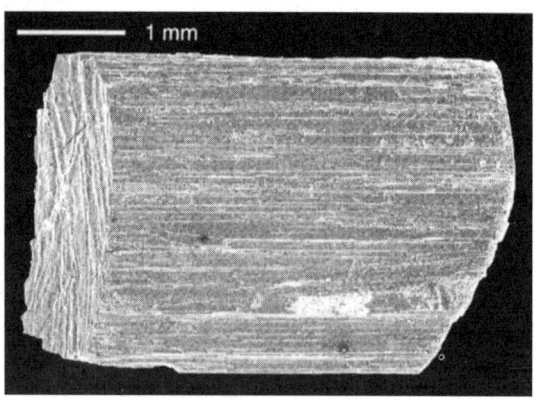

Fig. 23. A crystal of $CsBi_4Te_6$.

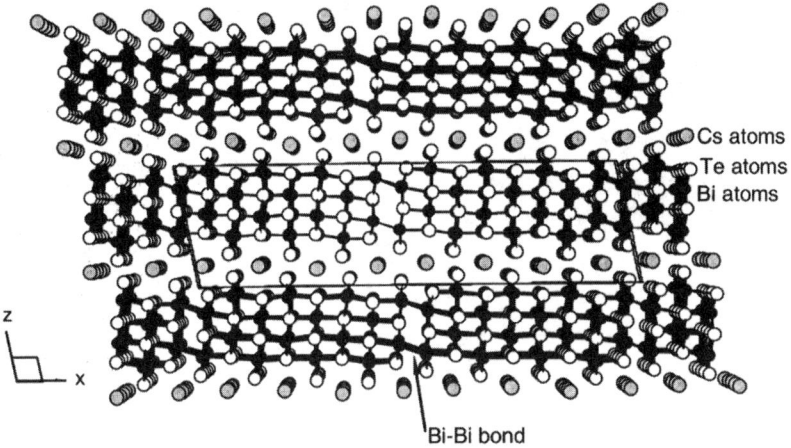

FIG. 24. The structure of $CsBi_4Te_6$. Open white circles are Te atoms.

of rapid growth along the needle axis is also the direction of maximum TE performance.

$CsBi_4Te_6$ has a layered anisotropic structure. It is composed of anionic $[Bi_4Te_6]$ slabs alternating with layers of Cs^+ ions (Fig. 24). The added electrons localize on the Bi atoms to form Bi–Bi bonds that are 3.238(1) Å long. The presence of these bonds is unusual in bismuth chalcogenide chemistry and it is not clear whether they play a role in the enhanced TE properties of the material (Larson, Mahanti, Chung, and Kanatzidis, work in progress). The $[Bi_4Te_6]$ layers are strongly anisotropic, as they consist of one-dimensional (1D) $[Bi_4Te_6]$ lathlike ribbons running parallel to the b-axis. The width and height of these laths is 23 Å by 12 Å, (Fig. 25). The laths arrange side by side and are connected via the Bi–Bi bonds mentioned earlier. This structural feature is responsible for the strongly 1D needlelike appearance of the $CsBi_4Te_6$ crystals. The Bi atoms are octahedrally surrounded either by six Te atoms or by five Te atoms and one other Bi atom. The degree of distortion around the Bi atoms is relatively small. The longest and shortest Bi–Te bonds are 3.403(1) Å and 2.974(1) Å, respectively, with an average distance of 3.18 Å. The Cs^+ ions lie between the $[Bi_4Te_6]$ layers and their ADPs are 1.6 times greater than those of the Bi and Te atoms, which suggests they undergo considerable "rattling" motion. Such a dynamic motion in the lattice can be responsible for strong scattering of heat-carrying phonons and leads to low κ values (see later discussion). The immediate environment of Cs is a square prismatic arrangement of Te atoms.

As obtained directly from the synthesis (with no deliberate attempt at doping), crystals of $CsBi_4Te_6$ have high room-temperature σ ranging from

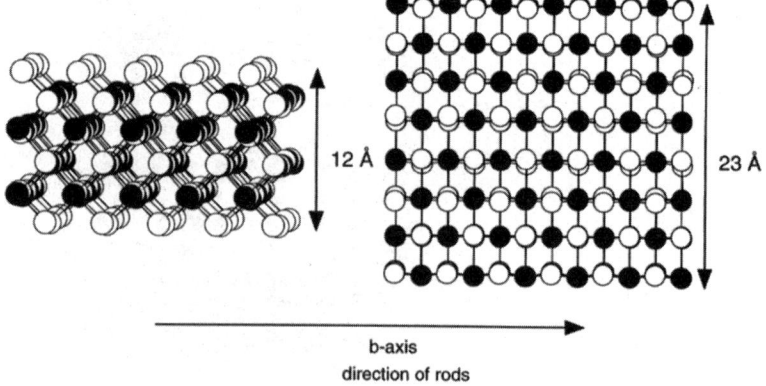

FIG. 25. Two different views of segments of the [Bi$_4$Te$_6$] lathlike ribbons. The laths are joined side by side via Bi–Bi bonds at 3.238(1) Å. Open white circles are Te atoms. Dark solid circles are Bi atoms.

900 to 2500 S/cm and S values from 90 to 120 μV/K (Fig. 26). At lower temperature, S typically exhibits a maximum of 120 μV/K at $\sim$240 K and then slopes toward zero at 0 K. The κ measurements on a large number of pressed pellets (>97% theoretical density) or oriented ingots show values between 1.25 and 1.85 W/m·K, for lightly and heavily doped samples, respectively. These values give rise to a relatively high room temperature ZT of 0.2 to 0.5, suggesting that the material deserves further optimization.

Already we have pursued doping studies of this material with various chemical doping agents, such as SbI$_3$, BiI$_3$, and In$_2$Te$_3$, in amounts varying from 0.02 to 4 mol%. Doping with SbI$_3$ and BiI$_3$ produces p-type samples, whereas In$_2$Te$_3$ produces n-type samples. Depending on the type and degree of doping, room temperature S values between $+175$ μV/K and -100 μV/K were observed. CsBi$_4$Te$_6$ is amenable to considerable doping manipulation, much like Bi$_2$Te$_3$, and thus higher ZT values may be expected.

The best power factors, $S^2 \cdot \sigma$, in this study were obtained with SbI$_3$. Figure 27A shows the evolution of power factor as a function of SbI$_3$ added. From these data, the optimal concentration seems to lie at 0.05% SbI$_3$. This sample achieved a maximum power factor of $\sim$51.5 μW/cm·K^2 at 184 K. The temperature dependence of σ and S of the best sample are shown in Fig. 27B. The S maximum was found at $\sim$250 K. The total κ of the doped samples of the material, $\sim$1.48 W/m·K, Fig. 27C, is considerably smaller than that of Bi$_2$Te$_3$ (at $\sim$1.85 W/m·K) and more comparable to that of the optimized Bi$_{2-x}$Sb$_x$Te$_{3-y}$Se$_y$ alloy (at $\sim$1.56 W/m·K). Perpendicular to the growth axis (b-axis), κ was dramatically lower ($\sim$0.6 W/m·K), reflecting the highly anisotropic nature of CsBi$_4$Te$_6$.

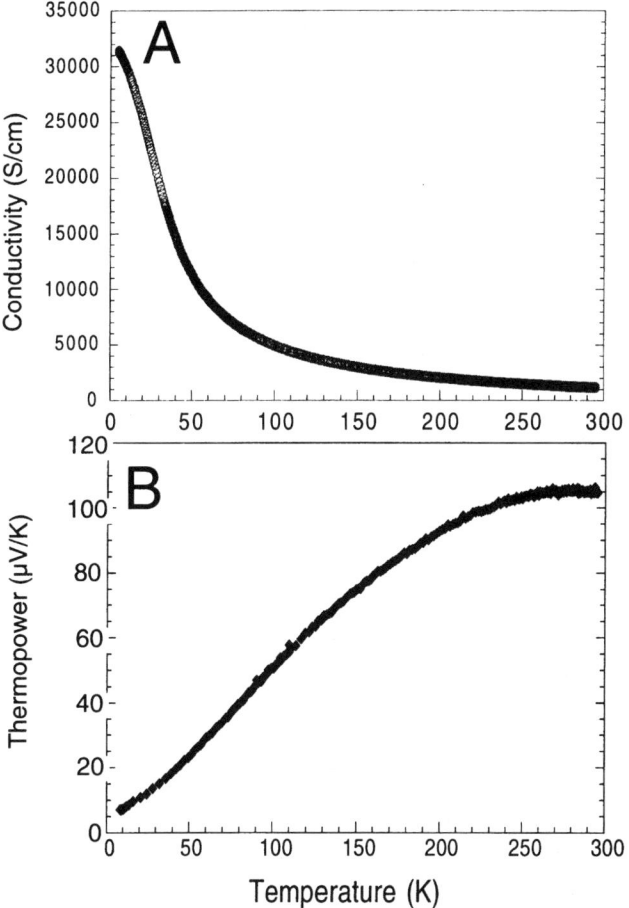

FIG. 26. Temperature dependence of (A) the electrical conductivity and (B) thermopower for a single crystal of $CsBi_4Te_6$.

The ZT for this sample (0.05% SbI_3-doped $CsBi_4Te_6$) and that of the optimized commercial $Bi_{2-x}Sb_xTe_3$ are compared as a function of temperature in Fig. 27D. Along the growth axis, the calculated ZT values for $CsBi_4Te_6$ reach a maximum of 0.82 at 225 K and 0.65 at room temperature. In contrast, optimized $Bi_{2-x}Sb_xTe_3$ p-type alloy has a peak of $ZT \sim 0.95$ at room temperature, whereas at 225 K its ZT drops to 0.58. Both ZT curves are similar in shape, but that of $CsBi_4Te_6$ is shifted significantly (by ~ 70 K) to lower temperatures. Because the $CsBi_4Te_6$ samples reach optimum performance well below that of $Bi_{2-x}Sb_xTe_3$, we expect that this new material could be exploited for low-temperature applications, particularly at

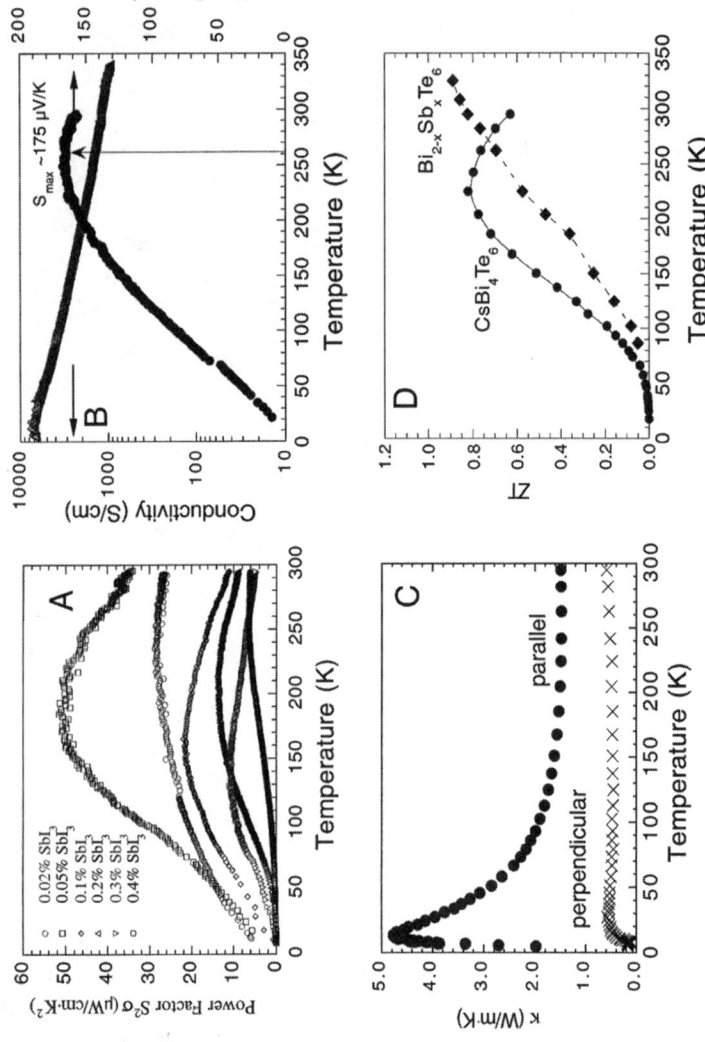

FIG. 27. (A) Power factors ($S^2\sigma$) as a function of SbI_3 doping. (B) Electrical conductivity and thermopower data for a single crystal of 0.05% SbI_3-doped $CsBi_4Te_6$. (●) Heat flow parallel to the growth axis (*b* axis) direction. (X) Heat flow perpendicular to the growth axis (*b* axis) direction. (D) Comparison of ZT versus temperature for 0.05 mol% SbI_3 doped single crystal $CsBi_4Te_6$ and optimally doped $Bi_{2-x}Sb_xTe_3$ *p*-type alloy (commercial sample from Marlow Industries). The peak maximum for $CsBi_4Te_6$ occurs at ~225 K.

temperatures where $Bi_{2-x}Sb_xTe_3$ alloy is ineffective. Recent optimization work on p-type $Bi_{2-x}Sb_xTe_{3-y}Se_y$ alloys claimed low-temperature (~ 210 K) ZT_{max} values of 0.64 (Vedernikov et al., 1997).

Hall effect measurements for SbI_3-doped $CsBi_4Te_6$ samples show that carrier concentrations are on the order of 3×10^{18} to 10^{19} cm^{-3} for samples doped at 0.1% and 0.2% SbI_3. Hole mobilities calculated from the electronic conductivity and Hall data show exponentially decreasing mobility as the temperature increases. The hole mobilities in doped $CsBi_4Te_6$ samples range between 700 and 1000 cm^2/V·s at room temperature. These are significantly greater than those typically found in the optimized p-type bismuth telluride alloy (~ 380 cm^2/V·s) (Süssman and Heiliger, 1982). At low temperatures, the mobility rises to > 5000 cm^2/V·s. The very high hole mobilities could be due to the 1D character of $CsBi_4Te_6$ and the lack of atomic disorder in its crystal lattice.

A TE cooling device needs both a p-type and an n-type version of a material to operate. Thus, an important issue to be addressed in future studies with $CsBi_4Te_6$ is whether n-type doping is possible. So far we have been able to demonstrate that In_2Te_3 doping leads to n-type charge transport. Though optimum levels have not been reached yet, Fig. 28 shows not only that n-type behavior is achievable, but the maximum TE power of -100 μV/K occurs at ~ 160 K, a temperature that has important implications for the development of even lower temperature TEs. Probably, further improvements in TE performance in this material are expected with continued exploration of doping agents and the study of solid solutions such as $CsBi_{4-x}Sb_xTe_6$, $CsBi_4Te_{6-x}Se_x$ and $Cs_{1-x}Rb_xBi_4Te_6$. The latter could result in substantially lower thermal conductivities (as much as 40–50% lower) giving projected values of $ZT > 1.5$. Band structure calculations for $CsBi_4Te_6$ now in progress, should be insightful in helping to understand at a deeper level the electronic properties of this material (Larson, Mahanti, Chung, and Kanatzidis, in preparation).

b. $RbBi_{3.66}Te_6$ and $APb_2Bi_3Te_7$

Attempts to obtain the Rb analog of $CsBi_4Te_6$ have been unsuccessful so far. Instead, the new material $RbBi_{3.66}Te_6$ was obtained. The compound is charge-balanced assuming Rb^{+1}, Bi^{3+}, and Te^{2-} formal oxidation states and therefore, unlike the Cs compound, $RbBi_{3.66}Te_6$ does not contain reduced Bi atoms. The structure of $RbBi_{3.66}Te_6$ is composed of $[Rb_{0.34}Bi_{3.66}Te_6]$ layers separated by Rb atoms; see Fig. 29. The Rb and Bi atoms in this layer are statistically disordered (i.e., mixed occupancy). The $[Rb_{0.34}Bi_{3.66}Te_6]$ layer has a similar array of $[MTe_6]$ octahedra to that of Bi_2Te_3 and can be considered a fragment excised from the NaCl structure. The excision is made along the (110) plane of the NaCl-type lattice, whereas

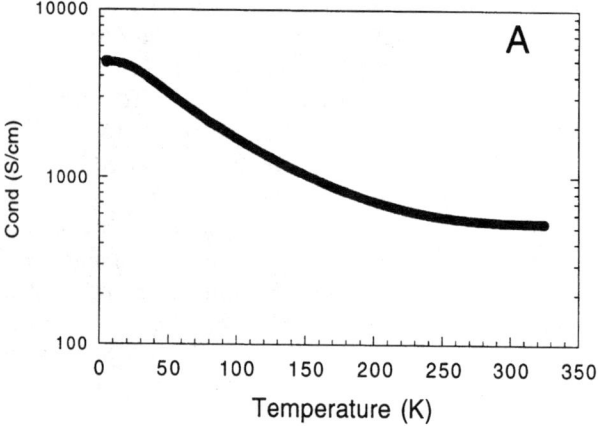

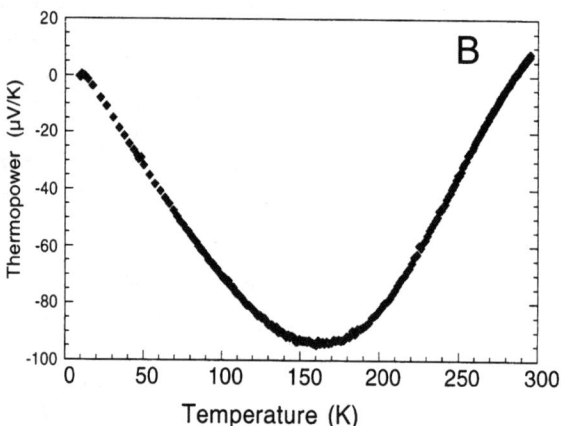

FIG. 28. (A) Electrical conductivity data for In_2Te_3-doped $CsBi_4Te_6$. (B) Thermopower data from a single crystal of In_2Te_3-doped $CsBi_4Te_6$.

the Bi_2Te_3 layer is formed by excising along the (111) plane. For comparison the structure of Bi_2Te_3 is shown in Fig. 29B. These two types of layers have the same number of $[BiTe_6]$ octahedra and thus both have the stoichiometry of $[M_2Te_3]$. The Rb atoms are sandwiched between the $[Rb_{0.34}Bi_{3.66}Te_6]$ layers and their sites are filled only at the 66.7% level. Therefore, this compound contains vacancies and can be formulated as $Rb_{0.66}[Rb_{0.34}Bi_{3.66}Te_6]$.

The statistical disorder of Rb and Bi atoms inside the $[Rb_{0.34}Bi_{3.66}Te_6]$ layer and Rb site vacancies in the interlayer space are novel structural features of this compound. The Rb atoms between the layers display some

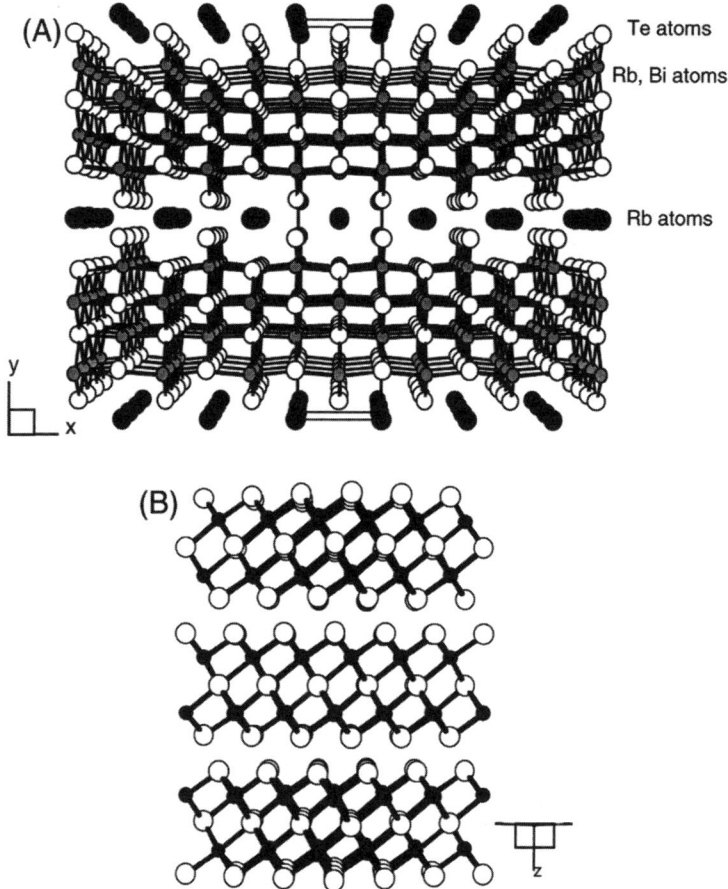

FIG. 29. The structure of (A) $RbBi_{3.66}Te_6$ viewed down the c axis in comparison with that of (B) Bi_2Te_3 viewed down the a axis. Open circles are Te atoms, light gray circles Bi atoms.

mobility due to the available vacant space, which is responsible for the high ADP of 4.0 compared to ~1.6 for the Bi and Te atoms. This situation indicates that Rb atoms indeed "rattle" between the layers and may favorably affect the thermal conductivity of $RbBi_{3.66}Te_6$. On the other hand, the Rb atoms that are part of the layers are expected to create discontinuities in the covalent Bi–Te bonding of the $[Rb_{0.34}Bi_{3.66}Te_6]$ framework.

Single crystals of $RbBi_{3.66}Te_6$ show n-type charge transport behavior with conductivity of ~925 S/cm and thermopower of $-41\,\mu V/K$ at room temperature; see Fig. 30A. Because the compound is valence precise, it is likely to be a narrow-gap material or a semimetal. A semimetal could result from near contact or even partial overlap of conduction and valence bands

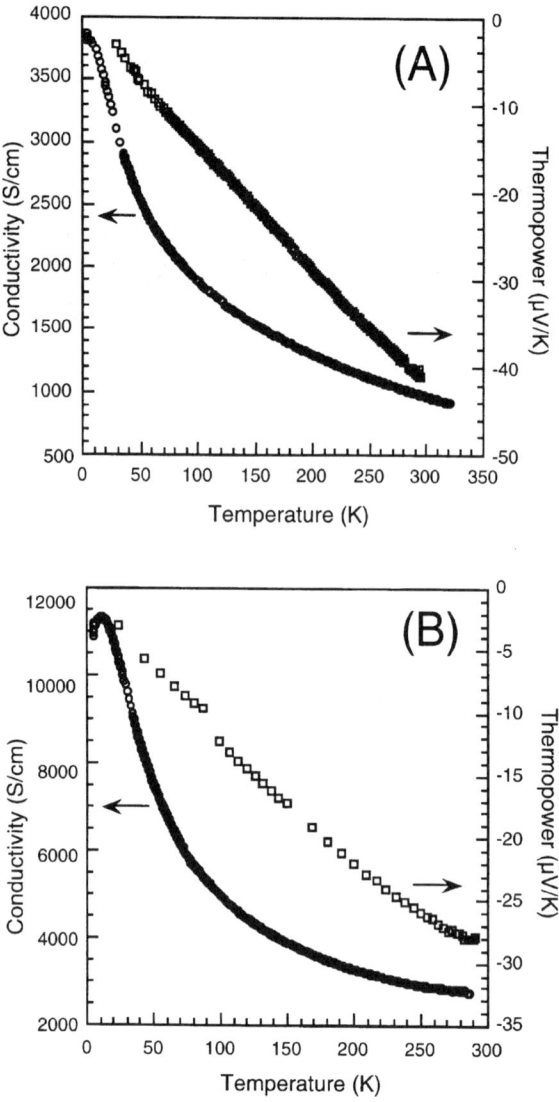

FIG. 30. Variable temperature electronic conductivity and thermopower data for a single crystal of (A) $RbBi_{3.66}Te_6$ and (B) $RbBi_{3.30}Sb_{0.36}Te_6$.

at the Fermi level. The exact details of the electronic structure near the Fermi level are not known at this time.

To examine whether the bandgap responds to partial substitution of Sb atoms in the Bi sites or Se atoms in the Te sites, we prepared solid solutions of Sb and Se with 5 and 10 mol%, respectively, which correspond to compositions $RbBi_{3.48}Sb_{0.18}Te_6$, $RbBi_{3.30}Sb_{0.36}Te_6$, $RbBi_{3.66}Te_{5.70}Se_{0.30}$, and $RbBi_{3.66}Te_{5.40}Se_{0.60}$. The room temperature conductivities of these solid solutions are 738, 2950, 1872, and 1106 S/cm, and thermopowers -41, -28, -31, $-31\,\mu V/K$, respectively; see Fig. 30B. Although the thermopower values of the solid solutions are still low and remain almost unchanged, greater variations are observed in the electronic conductivity. The substitution of Se for Te is expected to increase the bandgap of the compound and result in lower electronic conductivity. Since this was not observed, it may be that the material has no significant bandgap (i.e., a semimetal) or the amount of Se is not nearly enough to bring about significant widening of the energy gap. An increase in the Se/Te ratio is needed to check this. The extent of this type of Se substitution is known, but it is noteworthy that the isostructural $Rb_{0.66}[Rb_{0.34}Bi_{3.66}Se_6]$ compound is not known.

Including Pb and Sn in this system produced a new set of materials $AM_2Bi_3Te_7$ (A = Cs, Rb; M = Sn, Pb). The compounds $APb_2Bi_3Te_7$ possess a very similar layer to that observed in $RbBi_{3.66}Te_6$; see Figs. 29 and 31. In $APb_2Bi_3Te_7$, the alkali metal atoms reside only between the layers. The $[Pb_2Bi_3Te_7]$ (M_5Te_7; M = Bi, Pb) layer consists of a $[2M_2Te_3]$

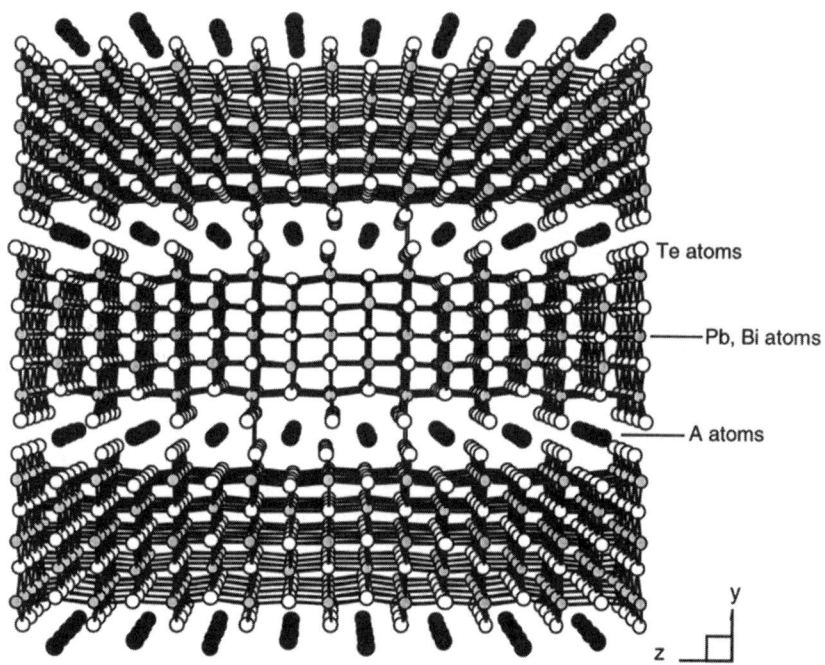

FIG. 31. The orthorhombic crystal structure of $APb_2Bi_3Te_7$.

(M_4Te_6) layer, which is the exactly same structural framework as [$Rb_{0.34}Bi_{3.66}Te_6$] (M_2Te_3) in $RbBi_{3.66}Te_6$, with an additional [MTe] monolayer attached. Therefore, the [$Pb_2Bi_3Te_7$] layer is one [BiTe] unit thicker than [$Rb_{0.34}Bi_{3.66}Te_6$], as depicted in Fig. 31. The distinction between Pb and Bi atoms in Pb/Bi mixed structures was not possible because of their almost identical X-ray scattering power ($Z_{Bi} = 83$, $Z_{Pb} = 82$). In $APb_2Bi_3Te_7$, we tentatively assume the Pb atoms to be statistically disordered over all Bi atom sites; however, this may not actually be the case.

The temperature dependence of conductivity and thermopower of $RbPb_2Bi_3Te_7$ and $CsPb_2Bi_3Te_7$ are similar and reveal n-type transport. The room temperature values are ~ 970 S/cm and $-20\,\mu V/K$ for $CsPb_2Bi_3Te_7$, and 1270 S/cm and $-20\,\mu V/K$ for $RbPb_2Bi_3Te_7$. First attempts to prepare solid solutions were also carried out for $APb_2Bi_3Te_7$ (A = Cs, Rb) with Se of 10 mol%. Both the electronic conductivity and thermopower remained essentially unaffected in this system (at room temperature, 520 S/cm and $-18\,\mu V/K$ for $CsPb_2Bi_3Te_{6.3}Se_{0.7}$, and 1100 S/cm and $-25\,\mu V/K$ for $RbPb_2Bi_3Te_{6.3}Se_{0.7}$). Preliminary band structure calculations on $APb_2Bi_3Te_7$ indicate that they may be metals, that is, $E_g < 0$ (Larson, Mahanti, Chung, and Kanatzidis, work in progress).

VI. Other Classes of Compounds

Our experience with the metal chalcogenides, discussed earlier, naturally drew our interest in other main group framework held by heavy elements. For example, clathrate-I structure A_8X_{46} members (A = alkali or alkaline earth metal; X = group 12, 13, 14 elements), are currently being pursued as possible TE materials (Blake et al., 1999). Could new materials with group 13 and 14 elements, instead of group 15 and 16, be expected to have similar structural and compositional characteristics worth exploring for TE applications? To explore this possibility we turned to Zintl phases with complex polyanionic frameworks. Zintl phases are built from group 13, 14, and 15 elements and alkali or alkaline earth metals, and they often possess large cages or channels (Huang and Corbett, 1999; Henning and Corbett, 1999; Fassler and Kronseder, 1998; Todorov and Sevov, 1998; Xu and Guloy, 1998; Kauzlarich et al., 1989; Vidyasagar et al., 1996; Young and Kauzlarich, 1995; Cordier et al., 1984). The great majority of Zintl compounds, however, are reported to be air and moisture sensitive. For TE applications environmentally stable compounds are necessary, and this may be achieved with the use of heavier elements and divalent electropositive cations. If heavy element analogs of the Zintl phases were synthesized, they should be narrow-gap semiconductors and may have interesting TE properties.

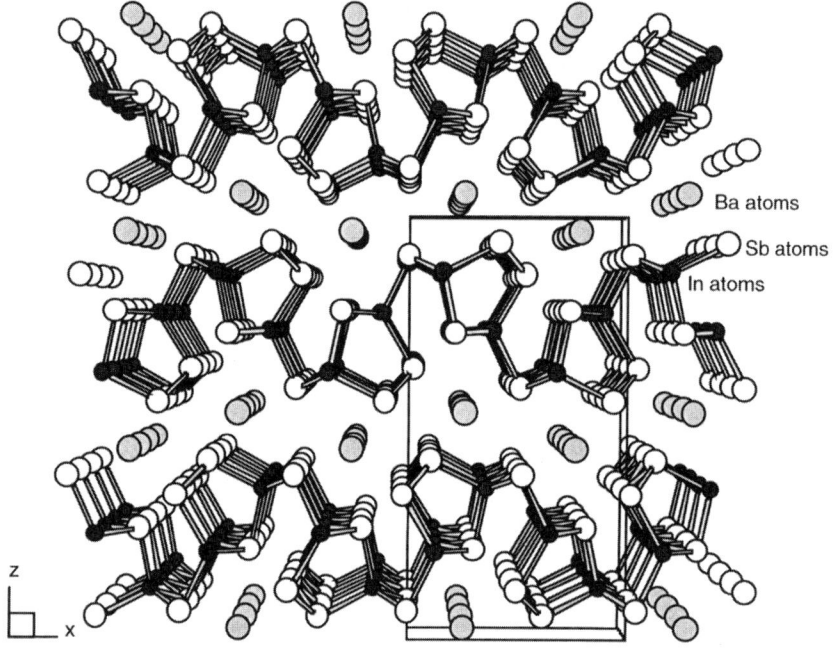

FIG. 32. The crystal structure of $Ba_4In_8Sb_{16}$.

One Zintl phase with narrow bandgap is $Ba_4In_8Sb_{16}$ (Kim et al., 1999), which possesses an unprecedented two-dimensional network structure and infinite zigzag Sb chains. This compound was not discovered with the use of the flux technique discussed earlier, but with more conventional techniques typically used in solid-state chemistry.

$Ba_4In_8Sb_{16}$ has a structure with two-dimensional character in which undulating puckered $(In_8Sb_{16})^{8-}$ anionic layers are separated with Ba^{2+} ions. The structural complexity of $Ba_4In_8Sb_{16}$ is reminiscent of those found in the ternary A/Bi/Q phases discussed previously.

The projection along the b axis illustrates a simplified view and shows small pentagonal tunnels composed of tetrahedral In and pyramidal Sb atoms; see Fig. 32. The building blocks of each pentagonal tunnel are $InSb_4$ tetrahedra and infinite Sb–Sb zigzag chains. All atoms in the structure lie on crystallographic mirror planes, that is, $z = \frac{1}{4}$ and $\frac{3}{4}$, and the $InSb_4$ tetrahedra are arranged in rows that run parallel to the b axis direction. The tetrahedra share three Sb atom corners to make infinite chains, while the fourth corner of Sb atoms make a close approach to chemically equivalent Sb atoms of a neighboring infinite chain to make Sb–Sb bonds and thus produce infinite zigzag chains of Sb. The Ba^{2+} ions are found between the corrugated $(In_8Sb_{16})^{8-}$ layers.

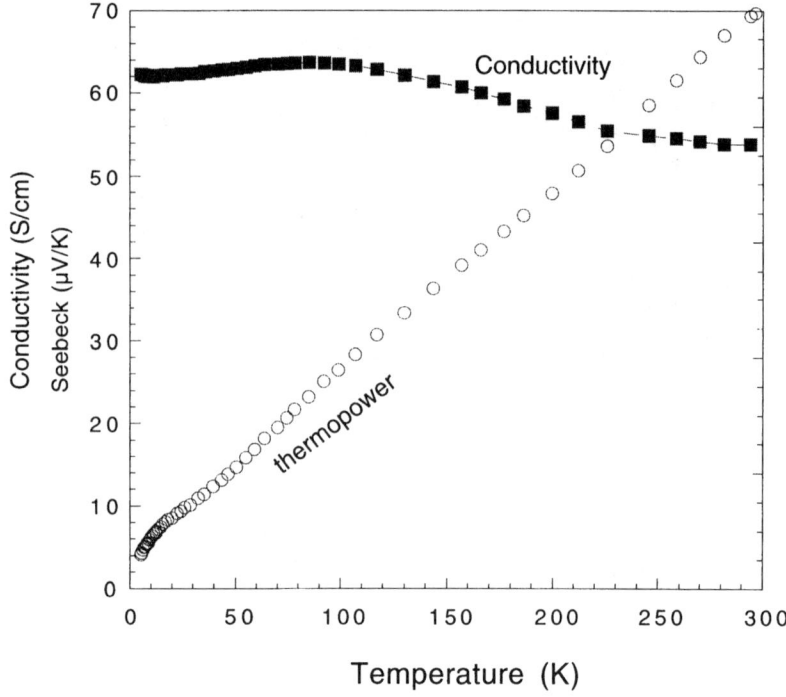

FIG. 33. Temperature dependence of the electrical conductivity and thermopower for a single crystal of $Ba_4In_8Sb_{16}$. Both properties share the same y axis.

On the basis of oxidation states, the three-coordinated Sb atoms can be assigned as Sb^{3-} ions and the Sb in Sb–Sb zigzag chains can be assigned as Sb^{1-}. The four coordinated In atoms can be assigned as In^{3+}. Therefore, the formula is best represented as $[8(In^{3+})8(Sb^{3-})8(Sb^{1-})]^{8-}$. The material possesses a narrow bandgap of ~ 0.1 eV.

The electronic conductivity at 300 K is ~ 135 S/cm and does not vary greatly with temperature; see Fig. 33. The electronic conductivity is low compared to the values of optimized semimetals such as Bi_2Te_3 (10^3 S/cm). The thermoelectric power of $Ba_4In_8Sb_{16}$ was about $+70\,\mu$V/K at room temperature and showed a linear decrease with decreasing temperature. The positive value of thermoelectric power indicates that $Ba_4In_8Sb_{16}$ is a p-type (hole) semiconductor. The values could likely be improved with better sample morphology and the introduction of appropriate dopants and dopant concentrations.

The most interesting aspect of $Ba_4In_8Sb_{16}$ is its total thermal conductivity (κ) of $Ba_4In_8Sb_{16}$, shown in Fig. 34 as a function of temperature. The total

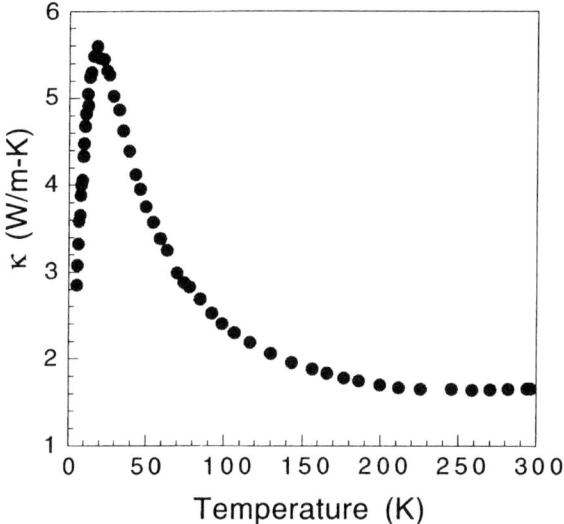

FIG. 34. Thermal conductivity of an ingot sample of $Ba_4In_8Sb_{16}$ as a function of temperature.

thermal conductivity is about 1.7 W/m·K at room temperature, which is comparable to the value of optimized Bi_2Te_3 alloy (1.4–1.6 W/m·K). The electronic thermal conductivity at 300 K is no more than about 5% of the total thermal conductivity and is even a smaller fraction at lower temperatures. In spite of an obviously short phonon mean free path, the overall character of the thermal conductivity is that of a crystalline dielectric solid. The ADPs of Ba^{2+} ions are comparable to those of the other atoms in the structure. From this, we can conclude that not much "rattling" motion of Ba atoms is occurring in $Ba_4In_8Sb_{16}$, indicative of a snag fit of Ba atoms in their cavity. Therefore, the total thermal conductivity is primarily due to the low symmetry and dimensionality of the structure as well as presence of heavy atoms. Because of the lack of thermoelectric data for other Zintl compounds, a thorough comparison was not possible. However, the thermoelectric properties of $Ba_4In_8Sb_{16}$ are encouraging and suggest a new category of possible thermoelectric materials in other similar Zintl phases.

It is likely that the total thermal conductivity can be depressed even farther lower by preparing solid solutions of the type $Ba_{4-x}Eu_xIn_8Sb_{16}$ or $Ba_{4-x}Sr_xIn_8Sb_{16}$. A "rattling" motion of the smaller Sr or Eu atoms, in cavities designed for Ba atoms, could cause additional heat-carrying phonon scattering, thereby further frustrating heat transport in this material.

Until recently, research on Zintl-type compounds concentrated predominantly on exploring the validity limits of the concept itself and as such focused mainly on systems with relatively light atoms in the framework.

Most currently known Zintl phases are air and moisture sensitive. However, the analogs with heavier elements, if environmentally stable, could be those with useful potential applications. Many more Zintl type phases with heavy main-group atoms and narrow bandgaps are anticipated. This class of solid state materials presents itself as a potentially rich source for new TE compounds.

VII. Conclusions and Outlook

In conclusion, the future of TE relies heavily on new materials and, consequently, solid state chemistry has an important role to play in the field of TE. The flux method is an excellent discovery technique that has produced a large number of interesting and promising new materials. This is why one of the most important long-term projects in the TE community should be the exploratory synthesis and study of a large range of heavy element multinary compounds. We note that maximum ZT will probably not be discovered in an as-made, unintentionally doped material (barring a stroke of luck). It will come from finding a new material with promising enough TE properties to begin with, so that optimization from those levels would result in great values of ZT. These new materials have not been optimized, whereas the best samples of $Bi_2Te_{3-x}Se_x$ and $Bi_{2-x}Sb_xTe_3$ have undergone decades of development. We believe some of the materials discussed here may indeed emerge as competitive thermoelectrics in the future, if their development is seriously pursued. The approach to performance improvement needs to be twofold: alloying and doping. To optimize a new material, such as $CsBi_4Te_6$, it must be kept in mind that one is dealing with a *multidimensional parameter space*. This is because not only must solid solutions be made, but also different dopants have to be tried, and different concentrations for each dopant. If, for example, there are 5 different solid solution members to be made and 5 different dopants to be investigated in 5 concentrations each, then in the $CsBi_4Te_6$ system alone we need to test at least 125 samples. At this stage measuring these samples seems to be the slowest step in the process of evaluation. Therefore, new high-throughput TE screening methods need to be devised that are capable of handling the number of samples available.

From the chalcogenide materials described here, we know most about $CsBi_4Te_6$ and $\beta\text{-}K_2Bi_8Se_{13}$. The rest need further investigation, and we hope that some of them will emerge as promising TE candidates. Since it is almost certain that the "as prepared" materials are not in their optimum doping state, we can expect substantial enhancements in the power factor with doping.

Where do we go from here? What other types of compounds can be targeted for high ZT? Based on the arguments presented earlier, systems with high γ or high B parameters are needed. While high effective masses could come from carriers residing in the flat (low dispersion) regions of bands as well as in heavy relativistic atoms, large degeneracy in the band extrema (i.e., large γ parameters) can come from compounds with highly symmetric crystal structures, ideally cubic or rhombohedral. The maximum degeneracy arises when the extrema (in valence or conduction bands) are found in the middle of the Brillouin zone rather than at the edges. One needs not only large degeneracy in the band extrema, but also a large number of extrema of similar energy in the same band near the Fermi level. Figure 35 shows a hypothetical band structure for such an ideal complex cubic material with the needed characteristics. None of the materials presented earlier have a high crystal symmetry. Therefore the γ parameter is low. Their promising TE properties presumably come from high carrier effective masses and inherently low thermal conductivity. One direction to pursue in the future is the development of cubic compounds in the A/Pb/Bi/Q (A = alkali metal; Q = S, Se, Te) system. Inroads have already been made, and we will be reporting on these results in the future (Sportouch *et al.*, 1999; Sportouch,

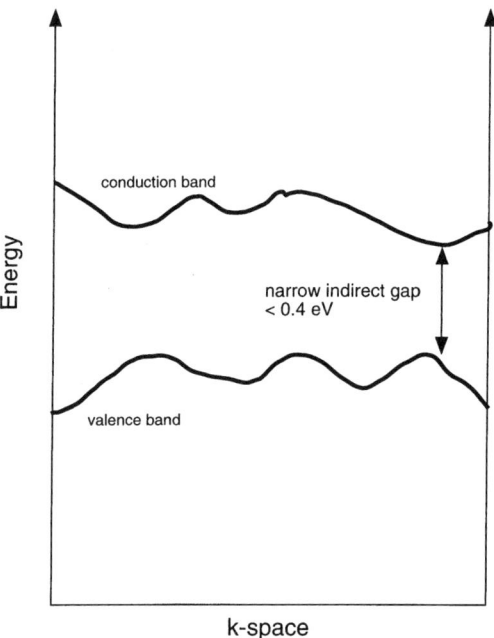

FIG. 35. One possibility of a generic yet simplistic band structure of a "nearly ideal" TE material. The high number of valleys in the conduction band or peaks in the valence band, coupled with a high band degeneracy parameter, γ (deriving from high crystal symmetry), could produce enormous power factors.

DeNardi, Brazis, Ireland, Kannewurf, Hogan, Uher, and Kanatzidis, work in progress).

As the number of atoms increases, the structural complexity increases and makes it difficult to achieve highly symmetric (e.g., cubic) structures. Nevertheless, large cubic cells of complex compounds have been observed, and their existence suggests that many more may be stable. One salutary example is $RE_{117}Fe_{52}Ge_{112}$ (RE = Sm, Gd, Tb, Dy, Ho, Er, Tm), which has a primitive cubic unit cell with $a = 29$ Å and space group $Fm3m$ (Morozkin et al., 1998). Other examples include phases that are single crystal approximants of quasicrystals (Saito et al., 2000; Sui et al., 1999; Liao et al., 1998). If the elemental composition is such that a narrow-gap semiconductor is obtained, the very complex compositions and heavy, electron-rich elements involved may well give rise to multiple band extrema and, we may hope, high ZT. Future investigations in the discovery and selection of candidate thermoelectric materials should take such considerations into account.

Interest in TE materials is not new, but the need for new materials is increasing and the next decade will be critical in the development of this field. Taking into account the difficulty of identifying the "right" compound and optimizing to $ZT > 1$, long-term sustainable planning is necessary, and close collaboration among chemists, physicists, and engineers is key to success. Such collaborative approaches have been the paradigm in which the present effort is based.

Acknowledgments

This work was supported at NU and MSU by the Office of Naval Research (N00014-98-1-0443) and DARPA through the Army Research Office (DAAG55-97-1-0184). The author expresses his deep appreciation for his long-standing collaborators Professors Carl R. Kannewurf, Tim Hogan, S. D. (Banhu) Mahanti, Ctirad Uher, and Dr. Duck-Young Chung. I also thank several very talented students and research associates, namely, Timothy J. McCarthy, Stephan DeNardi, Lykourgos Iordanidis, Kyoung-Shin Choi, Paul Brazis, Melissa Roci, John Ireland, and Drs. Li-Heng Chen, Sandrine Sportouch, Sung-Jin Kim, and Antje Mrotzek. Without their invaluable contributions and insight, none of this strongly interdisciplinary work would be possible.

References

Bandari, C. M. (1995). In *CRC Handbook of Thermoelectrics* (Rowe, D. M., Ed.), pp. 27–42. CRC Press, Boca Raton, FL.
Bensch, W., and Durichen, P. (1996). *Chem. Ber.* 129, 1207–1210.

Blake, N. P., Mollnitz, L., Kresse, G., and Metiu, H. (1969). *J. Chem. Phys.* 111, 3133–3144.
Chen, G. (1997). *J. Heat Transfer — Trans. ASME* 119, 220–229.
Choi, K.-S., Chung, D. Y., Mrotzek, A., Brazis, P., Kannewurf, C. R., Uher, C., Chen, C., Hogan, T., and Kanatzidis, M. G., *Chem. Mater.*, in press.
Chung, D.-Y., Choi, K.-Y., Iordanidis, L., Schindler, J. L., Brazis, P. W., Kannewurf, C. R., Chen, B., Hu, S., Uher, C., and Kanatzidis, M. G. (1997). *Chem. Mat.* 9, 3060–3071.
Chung, D.-Y., Kordanidis, L., Choi, K.-S., and Kanatzidis, M. G. (1998). *Bull Kor. Chem. Soc.* 19, 1283–1293.
Chung, D.-Y., Hogan, T., Brazis, P., Rocci-Lane, M., Kannewurf, C., Bastea, M., Uher, C., and Kanatzidis, M. G. (2000). *Science* 287, 1024–1027.
Cordier, G., Schafer, H., and Stelter, M. (1984). *Z. Anorg. Allg. Chem.* 519, 183–188.
Cordier, G., Schafer, H., Schwidetzky, C. (1985). *Rev. Chim. Miner* 22(5), 631–638.
DiSalvo, F. J. (1999). *Science* 285, 703–706.
Dresselhaus, M. S., Dresselhaus, G., Sun, X., Zhang, Z., Cronin, S. B., Koga, T., Ying, J. Y., and Chen, G. (1999). *Thermophys. Eng.* 3, 89–100.
Elwell, D., and Scheel, H. J. (1975). *Crystal Growth from High-Temperature Solutions*. Academic Press, New York.
Fassler, T. F., and Kronseder, C. (1998). *Angew. Chem., Int. Ed. Eng.* 37, 1571–1575.
Garner, R. W., and White, W. B. (1970). *J. Cryst. Growth* 7, 343.
Goldsmid, H. J., and Sharp, J. W. (1999). *Electron. Mat.* 28, 869.
Henning, R. W., and Corbett, J. D. (1999). *Inorg. Chem.* 38, 3883–3888.
Hicks, L. D., and Dresselhaus, M. S. (1993). *Phys. Rev. B* 47, 12727–12731.
Huang, B., and Corbett, J. D. (1999). *Solid State Sci.* 1, 555–565.
Johnson, D. C. (1998). *Curr. Opin. Solid State Mat.* 3, 159–167.
Jones, C. D. W., Regan, K. A., and DiSalvo, F. J. (1999). *Phys. Rev. B* 60, 5282–5286.
Kanatzidis, M. G., and DiSalvo, F. J. (1996). *ONR Quart. Rev.* XLVII, 14–22.
Kanatzidis, M. G., and Sutorik, A. (1995). *Prog. in Inorg. Chem.* 43, 151–265.
Kanatzidis, M. G., McCarthy, T. J., Tanzer, T. A., Heng, L.-C., Iordanidis, L., Hogan, T., Kannewurf, C. R., Uher, C., and Chen, B. (1996). *Chem. Mat.* 8, 1465–1474.
Kauzlarich, S. M., Kuromoto, T. Y., and Olmstead, M. M. (1989). *J. Am. Chem. Soc.* 111, 8041–8042.
Kim, S.-J., Hu, S., Uher, C., and Kanatzidis, M. G. (1999). *Chem. Mater.* 11, 3154–3159.
Larson, P., Mahanti, S. D., and Kanatzidis, M. G. (2000). *Phys. Rev. B* 61, 8162–8171.
Liao, X. Z., Sui, H. X., and Kuo, K. H. (1998). *Phil. Mag. A* 78, 143–156.
Mahan, G. D. (1998). *Solid State Phys.* 51, 81–157.
Mahan, G. D., and Sofo, J. O. (1996). *Proc. Natl. Acad. Sci. USA* 93, 7436–7439.
Mahan, G., Sales, B., and Sharp, J. (1997). *Phys. Today* 50, 42.
Meng, J. F., Polvani, D. A., Jones, C. D. W., DiSalvo, F. J., Fei, Y., and Badding, J. V. (2000). *Chem. Mat.* 12(1), 197–201.
Morozkin, A. V., Seropegin, Y. D., Portnoy, V. K., Sviridov, I. A., and Leonov, A. V. (1998). *Mat. Res. Bull.* 33, 903–908.
Mott, N. F., and Jones, H. (1958). *The Theory of the Properties of Metals and Alloys*. Dover Publications, New York.
Mrotzek, A., Chung, D. Y., Hogan, T., and Kanatzidis, M. G. (2000) *J. Mater. Chem.* 10:(7) 1667–1672.
Nolas, G. S., Slack, G. A., Morelli, D. T., Tritt, T. M., and Ehrlich, A. C. (1996). *J. Appl. Phys.* 79, 4002–4008.
Nolas, G. S., Cohn, J. L., Slack, G. A., and Schujman, S. B. (1998). *Appl. Phys. Lett.* 73, 178–180.
Nolas, G. S., Weakley, T. J. R., and Cohn, J. L. (1999). *Chem. Mat.* 11, 2470–2473.
Parkin, I. P. (1996). *Chem. Soc. Rev.* 25, 199.
Pell, M. A., and Ibers, J. A. (1997). *Chem. Ber.* 130, 1–8.

Proctor, K. J., Jones, C. D. W., and DiSalvo, F. J. (1999a). *J. Phys. Chem. Solids* 60, 663–671.
Proctor, K. J., Regan, K. A., Littman, A., and DiSalvo, F. J. (1999b). *J. Alloys Compounds* 292, 124–128.
Rowe, D. M. (1999). *Renew. Energ.* 16, 1–4.
Saito, K., Sugiyama, K., and Hiraga, K. (2000). *Phil. Mag. Lett.* 80, 73–78.
Sales, B. C. (1997). *Curr. Opin. Solid State Mat.* 2, 284–289.
Sales, B. C., Chakoumakos, B. C., Mandrus, D., and Sharp, J. W. (1999). *J. Solid State Chem.* 146, 528–532.
Sanjines, R., Berger, H., and Levy, F. (1988). *Mat. Res. Bull.* 23, 549.
Scheel, H. J. (1974). *J. Cryst. Growth* 24/25, 669.
Sheldrick, W. S., and Wachhold, M. (1998). *Coord. Chem. Rev.* 176, 211–322.
Slack, G. A. (1995). In *CRC Handbook of Thermoelectrics* (Rowe, D. M., Ed.), pp. 407–440. CRC Press, Boca Raton, FL.
Slack, G. A. (1997). In *Solid State Physics*, Vol. 34 (Ehrenreich, H., Seitz, F., and Turnbull, D., Eds.), p. 1. Academic Press, New York.
Sportouch, S., Bastea, M., Brazis, P., Ireland, J., Kannewurf, C. R., Uher, C., and Kanatzidis, M. G. (1998). In *New Materials for Small Scale Thermoelectric Refrigeration and Power Generation Applications*, Vol. 545 (Tritt, T. M., Kanatzidis, M. G., Mahan, G., and Lyon, H. B., Jr., Eds.), pp. 123–130. Proceedings of the 1998 Materials Research Society.
Srikrishnan, T., and Nowacki, W. Z. (1974). *Z. Kristallogr.* 140, 114–136.
Sui, H. X., Li, X. Z., and Kuo, K. H. (1999). *Phil. Mag. Lett.* 79, 181–185.
Süssman, H., and Heiliger, W. (1982). In *Proc. Conf. Transport in Compound Semiconductors*, p. 100, KTB series, MLT, Halle, Germany.
Thermoelectric Semiconductors: Encyclopedia of Materials Science and Engineering (1986). MIT Press, Cambridge, MA; Pergamon Press, Oxford.
Todorov, E., and Sevov, S. C. (1998). *Inorg. Chem.* 37, 3889.
Treece, R. E., Gillan, E. G., and Kaner, R. B. (1995). *Comment Inorg. Chem.* 16, 313–337.
Tritt, T. M. (1996). *Science* 272, 1276–1277.
Tritt, T. M., Kanatzidis, M. G., Mahan, G., and Lyon, H. B., Jr., Eds. (1997). *Thermoelectric Materials — New Directions and Approaches*, Vol. 478, Proceedings of the 1997 Materials Research Society.
Tritt, T. M., Kanatzidis, M. G., Mahan, G., and Lyon, H. B., Jr., Eds. (1998). *New Materials for Small Scale Thermoelectric Refrigeration and Power Generation Applications*, Vol. 545, Proceedings of the 1998 Materials Research Society, pp. 233–246.
Vedernikov, M. V., Kutasov, V. A., Luk'yanova, L. N., and Konstantinov, P. P. (1997). In *Proceedings of the 16th International Conference on Thermoelectrics*, p. 56.
Venkatasubramanian, R., Colpitts, T., Watko, E., Lamvik, M., and El-Masry, N. (1997). *J. Cryst. Growth* 170, 817–821.
Venkatasubramanian, R., Colpitts, T., O'Quinn, B., Liu, S., El-Masry, N., and Lamvik, M. (1999). *Appl. Phys. Lett.* 75, 1104–1106.
Vidyasagar, K., Honle, W., and von Schnering, H. G. (1996). *Z. Anorg. Allg. Chem.* 622, 518–524.
Xu, Z. H., and Guloy, A. M. (1998). *J. Am. Chem. Soc.* 120, 7349–7350.
Young, D. M., and Kauzlarich, S. M. (1995). *Chem. Mat.* 7, 206–209.

CHAPTER 4

An Overview of Recent Developments for BiSb Alloys

B. Lenoir and H. Scherrer

ECOLE DES MINES
LABORATOIRE DE PHYSIQUE DES MATÉRIAUX
NANCY, FRANCE

T. Caillat

JET PROPULSION LABORATORY
CALIFORNIA INSTITUTE OF TECHNOLOGY
PASADENA, CALIFORNIA

I. INTRODUCTION	101
II. BASIC FEATURES OF THE CRYSTAL STRUCTURE AND THE ELECTRONIC STRUCTURE OF $Bi_{1-x}Sb_x$ ALLOYS	103
1. *Crystal Structure*	103
2. *Electronic Structure of $Bi_{1-x}Sb_x$ Alloys*	106
3. *Carrier Mobility and Density*	111
III. SYNTHESIS OF Bi–Sb ALLOYS	113
1. *Single Crystals*	115
2. *Polycrystalline Bi–Sb*	117
IV. TRANSPORT PROPERTIES	120
1. *Single Crystals*	120
2. *Polycrystalline Samples*	129
V. CONCLUDING REMARKS	132
REFERENCES	132

I. Introduction

Among semimetals and narrow bandgap semiconductors, considerable attention has been devoted to bismuth–antimony solid solutions over the past century, not only because they possess unique physical properties from a fundamental point of view, but also because they present great interest as engineering materials. The impact of antimony substitution (Jain, 1959; Tanuma, 1959; Golin, 1968; Brandt *et al.*, 1982), temperature (Mendez, 1979), external pressure (Brandt and Chudinov, 1971; Mendez *et al.*, 1981), and magnetic field (Brandt and Svistova, 1970; Hiruma et al., 1980) on the electronic band structure of bismuth–antimony alloys stimulated extensive experimental and theoretical work. The possibility to observe a transition to a gapless state (a state intermediate between the metal and insulator state) in alloys has also received particular attention (Brandt et al., 1972a, 1972b, 1976).

The study of $Bi_{1-x}Sb_x$ alloys as possible materials for electronic refrigeration was undertaken by Smith and Wolfe in 1962 and extended some 10 years later by Yim and Amith (1972). Smith and Wolfe (1962) reported that undoped, Bi-rich, n-type Bi–Sb alloys have thermoelectric figures of merit significantly higher than those of Bi_2Te_3 alloys in the temperature range between 20 and 220 K. However, the practical use of these alloys in thermoelectric cooling at low temperatures has been constrained by the lack of p-type material having compatible thermoelectric properties.

The figure of merit Z_{np}, which is the thermoelectric measure of the efficiency of a thermocouple constituted of n and p legs (Ioffe, 1957), is defined as

$$Z_{np} = \frac{(\alpha_p - \alpha_n)^2}{[(\rho_p \lambda_p)^{1/2} + (\rho_n \lambda_n)^{1/2}]^2} \quad (1)$$

where $\alpha_{n,p}$ is the thermoelectric power, $\rho_{n,p}$ the electrical resistivity, and $\lambda_{n,p}$ the thermal conductivity of the n and p legs, respectively.

From relation (1), it is clear that if the p leg has far lower thermoelectric performance than the n leg, the figure of merit of the n–p couple will be decreased and thus the potential performance of solid state cooling devices will also be considerably reduced. Fortunately, the discovery of high-T_c superconductors has fundamentally modified the situation. Actually, a thermocouple constituted of an n-type Bi–Sb leg and of a p-type high-T_c superconductor leg, thermoelectrically passive, will possess a figure of merit close to that of the n leg (Goldsmid et al., 1988). Several workers have successfully demonstrated the possibility of using a high-T_c superconducting leg in solid-state cooling devices (Dashevskii et al., 1991; Vedernikov et al., 1991; Fee, 1993; Kuznetsov et al., 1994), which has resulted in a renewed interest in Bi–Sb alloys.

The advantage of using Bi–Sb single crystals with regard to conventional n-type materials is still more important in the presence of a transverse magnetic field, as first pointed out by Wolfe and Smith (1962) for a $Bi_{0.88}Sb_{0.12}$ single crystal. They reported a dimensionless figure of merit ZT greater than unity between 125 and 275 K by applying an optimum magnetic field. These values are at least twice the zero-field values. The reason for this large improvement is the presence of transverse thermomagnetic effects (Wolfe and Smith, 1962). These effects are particularly great for dilute Bi–Sb alloys, making them very attractive materials for

Ettingshausen cooling devices (Cuff et al., 1963; Horst and Williams, 1980; Jandl and Birkholz, 1994).

After briefly introducing the main physical properties of Bi–Sb alloys and their synthesis, the salient results pertaining to thermoelectric properties are discussed for both single crystalline and polycrystalline materials. Our attention is focused on the Bi–Sb alloys in the Bi-rich region. Transport properties are only described qualitatively because the band structure is strongly correlated to antimony content and temperature. Although the thermomagnetic effects present considerable interest in Bi–Sb solid solutions, we will essentially deal with the zero-field coefficients. The reader interested in the use of Bi–Sb alloys for thermomagnetic cooling may refer to this book or to Goldsmid (1995).

II. Basic Features of the Crystal Structure and the Electronic Structure of $Bi_{1-x}Sb_x$ Alloys

1. Crystal Structure

All of the group V semimetals (bismuth, antimony, and arsenic) belong to the rhombohedral A_7 structure of space group $R\bar{3}m$ (having one trigonal axis of threefold symmetry, three twofold or binary axes each normal to the trigonal direction and to a mirror plane, and a center of inversion). The unit cell contains two atoms situated on the trigonal axis at a distance u from each vertex, as Fig. 1 shows. This structure is defined by three parameters:

a_R, the side of the rhombohedral
α_R, the rhombohedral angle
u, the atomic parameter

The orthogonal set of coordinates (1, 2, 3) usually used is defined parallel to the binary, bisectrix and trigonal axes respectively (Fig. 1). The rhombohedral lattice parameters, listed in Table I for bismuth and antimony at different temperatures, are very close to those of the rhombohedral cell of a simple cubic (SC) lattice with one atom where $u = 0.25$ and $\alpha_R = 60°$, so that the A_7 structure deviates a few percent from a SC lattice (Fig. 2). The distortion can be best understood by considering the SC lattice as composed of two interpenetrating face-centered cubic (FCC) lattices of identical atoms (Abricosov and Falkovski, 1963). The first distortion in forming the A_7 crystal structure from the SC structure is generated by making a small translation of one FFC lattice relative to the other along the body diagonal of the cube, yielding a value of u slightly different from 0.25. The second

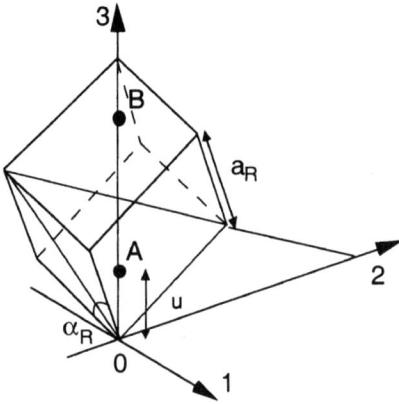

FIG. 1. Rhombohedral primitive unit cell of group V semimetals, described by three parameters: a_R, α_R, and u. The cell contains two atoms A and B located at (u,u,u) and $(1-u, 1-u, 1-u)$, respectively.

distortion is a rhomboedral shear along the same body diagonal that is involved in the displacement, reducing the angle α_R.

As in bismuth telluride compounds, the crystal structure is layered. The hexagonal representation, illustrated in Fig. 3, shows this characteristic. The structure consists of a staking of double layers γB, αC, and βA, separated by $(\frac{2}{3} - 2u)c_H$ within which atoms are bound by covalent bonds. Between the double layers the bonds are weaker. Crystals can then be cleaved along the basal planes (perpendicularly to the trigonal axis), but not so easily as in bismuth telluride.

TABLE I

SUMMARY OF THE CRYSTAL PARAMETERS OF BISMUTH AND ANTIMONY AT DIFFERENT TEMPERATURES (Schiferl and Barret, 1969)

	T (K)	Hexagonal			Rhombohedral		
		a_H (Å)	c_H (Å)	c_H/a_H	a_R (Å)	α_R (°)	u
Bi	4.2	4.5330	11.797	2.6025	4.7236	57.350	0.23407
	78	4.5350	11.814	2.6051	4.7273	57.280	0.23400
	298 ± 3	4.5460	11.862	2.6093	4.7458	57.230	0.23389
Sb	4.2	4.3007	11.222	2.6093	4.4898	57.233	0.23362
	78	4.3012	11.232	2.6114	4.4927	57.199	0.23364
	298	4.3084	11.274	2.6167	4.5067	57.110	0.23349

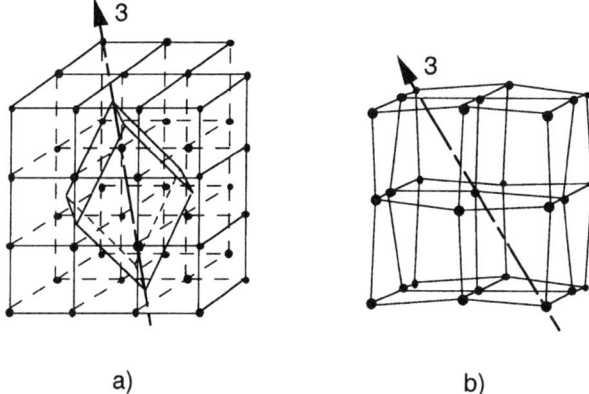

FIG. 2. (a) Representation of the rhombohedral cell in the simple cubic structure ($\alpha_R = 60°$ and $u = 0.25$). (b) Representation of the pseudocubic structure of group V semimetals.

Since bismuth and antimony have similar lattice parameters, it is not surprising that Bi–Sb alloys form a solid solution over the whole composition range. As expected, the lattice parameters for the alloys are between those of Bi and Sb. Cucka and Barrett (1962) reported that the lattice constants of the rhombohedral lattice of $Bi_{1-x}Sb_x$ alloys satisfy the Vegard's

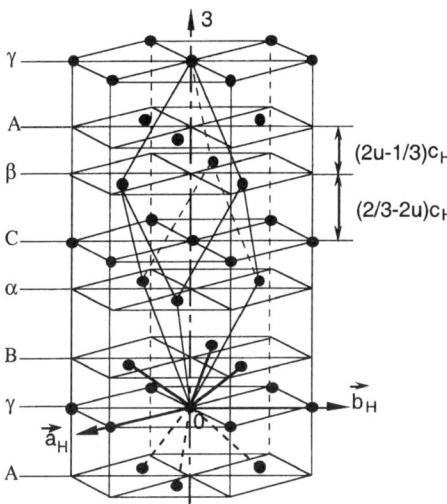

FIG. 3. Hexagonal representation of the A_7 structure. The atoms are arranged in double layers ...$\gamma B\alpha C\beta A$..., each atom (atom 0 for instance) having three near neighbors at a distance $(2u - \frac{1}{3})c_H$ in the double layer in which it is located, and three others at a greater distance $(\frac{2}{3} - 2u)c_H$ in an adjacent layer.

TABLE II

Rhombohedral Lattice Parameters of $Bi_{1-x}Sb_x$ ($0 \leqslant x \leqslant 0.3$) at Different Temperatures (Cucka and Barrett, 1962)

	298 ± 3 K	78 K
a_R (Å)	$4.746 - 21.94 \times 10^{-2}x$	$4.730 - 21.98 \times 10^{-2}x$
α_R (Å)	$57.24 - 34.5 \times 10^{-2}x$	$57.30 - 36.5 \times 10^{-2}x$
u	0.2340	0.2341

rule when $0 < x < 0.3$, as summarized in Table II. The substitution of Bi atoms by Sb atoms in the Bi lattice produces a decrease of a_R and α_R and leaves unchanged the atomic parameter u (at least up to $x = 0.3$) similarly to a decrease in temperature.

Any property of $Bi_{1-x}Sb_x$ alloys is described by a tensor in which the number of independent components is limited by crystallographic symmetry and the principle of microscopic reversibility. For example, the tensors of the three zero-field transport properties, that is, the electrical resistivity ρ, the thermoelectric power α, and the thermal conductivity λ, have only two independent components, one along the trigonal axis (index 33) and one in the basal plane (index 11 or 22).

2. Electronic Structure of $Bi_{1-x}Sb_x$ Alloys

Reviews concerning the electronic band structure of bismuth and antimony and the properties of electrons in bismuth have been reported by Dresselhaus (1971) and Edelman (1976), respectively. In the following section, we summarize the basic features of the band structure of the pure elements and alloys pertinent to the understanding of the transport properties.

The small distortion from the simple cubic lattice induces that bismuth and antimony are semimetals characterized by a small overlap of the fifth and sixth bands, leading to the presence of a small equal number of electrons (n) and holes (p) at all temperatures. Figure 4a represents the Brillouin zone for bismuth with the carrier pockets. The holes are located in two half-ellipsoidal pockets centered on T-points; these are revolution ellipsoids along the trigonal axis. The T valence band can be satisfactorily described by a simple parabolic model. The electrons are located in six half quasi-ellipsoidal pockets centered at the L-points of the Brillouin zone. These three quasi-ellipsoids are strongly elongated along a direction tilted by an angle φ_e out of the binary–bisectrix plane ($\varphi_e = 6 \pm 0.2°$ at 4.2 K, after Kao, 1963). This highly anisotropic shape leads to unusually very small effective masses along two directions, resulting in very high electron mobility.

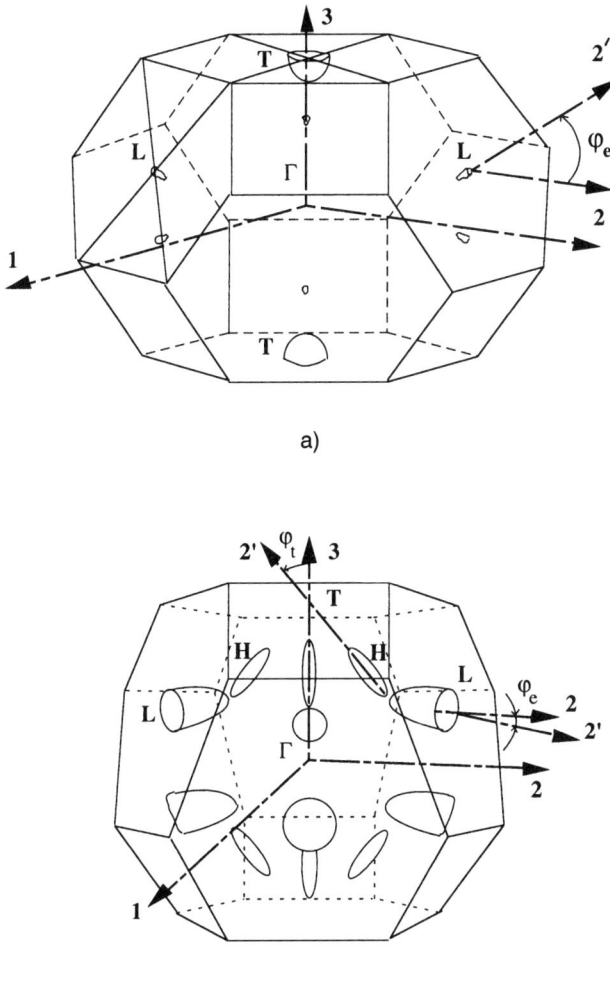

FIG. 4. Fermi surfaces of bismuth (a) and antimony (b) in the Brillouin zone. In bismuth, electrons are at L-points and holes at T-points. In antimony, electrons are at L-points and holes at H-points. Axes 1, 2, and 3 are parallel to the binary, bisectrix, and trigonal axes, respectively. φ_e and φ_t are the tilt angles of the electron and hole ellipsoids.

The same structure occurs for the "light" L hole band, which is separated from the conduction band by a narrow energy gap E_g ($E_g = 13.6$ meV at 0 K; Vecchi and Dresselhaus, 1974) and is coupled to it by the $\mathbf{k} \cdot \mathbf{p}$ interaction, causing a highly nonparabolic dispersion relationship of the two bands and a quasimomentum dependence of the Bloch amplitude. The simplest dispersion relation for the L-bands is the two-band model of Lax and Mavroides (1960), also called the ellipsoidal nonparabolic (ENP) model. Expressed in coordinates fixed to the ellipsoid axes, referred to as $(1', 2', 3')$, the energy dispersion law has the form

$$E\left(1 + \frac{E}{E_g}\right) = \frac{\hbar^2}{2m_0}\vec{k} \cdot \begin{pmatrix} m_{1'} & 0 & 0 \\ 0 & m_{2'} & 0 \\ 0 & 0 & m_{3'} \end{pmatrix}^{-1} \cdot \vec{k}, \qquad (2)$$

where E is the electron energy measured from the bottom of the conduction band, m_0 is the free electron mass, $m_{i'}$ the components of the effective mass tensor of electrons and L holes near the band edges, and $\vec{k}$ the wave vector. The nonparabolicity alters the effective masses. In the previous model, an effective mass tensor element m^* depends on energy E as

$$m^*(E) = m\left(1 + 2\frac{E}{E_g}\right). \qquad (3)$$

Other models that do not retain the ellipsoidal isoenergetic surfaces described by Eq. (2) have been proposed for the dispersion relation of the conduction band of Bi. A review is given by McClure and Choi (1977).

In antimony, the pockets of electrons are also located at the L points of the Brillouin zone, whereas holes are located at the six equivalent H points (Fig. 4b). The departure from cubic symmetry is more pronounced in antimony than in bismuth, resulting in a larger carrier density in antimony than in bismuth. Values for the principal band parameters of carriers in bismuth and antimony are reported in Table III after Dresselhaus (1971). It should be noted that L, T, and H hole bands differ greatly with regard to

TABLE III

BAND PARAMETERS FOR CARRIERS IN BISMUTH AND ANTIMONY AT 4.2 K[a] (Dresselhaus, 1971)

Carriers	$m_{1'}$ (m_0 units)	$m_{2'}$ (m_0 units)	$m_{3'}$ (m_0 units)	E_F (meV)	$n = p$	N_V
Bismuth						
L-electrons	0.00119	0.266	0.00228	27.2	2.7×10^{17}	3
T-holes	0.064	0.064	0.69	10.8	2.7×10^{17}	1
Antimony						
L-electrons	0.0093	1.14	0.088	93.1	3.74×10^{19}	3
H-holes	0.068	0.92	0.050	84.4	3.74×10^{19}	6

[a] E_F is the Fermi energy of electrons or holes, specified with respect to the energy extrema; N_V is the number of valleys in the Fermi surfaces; and the other symbols are explained in the text.

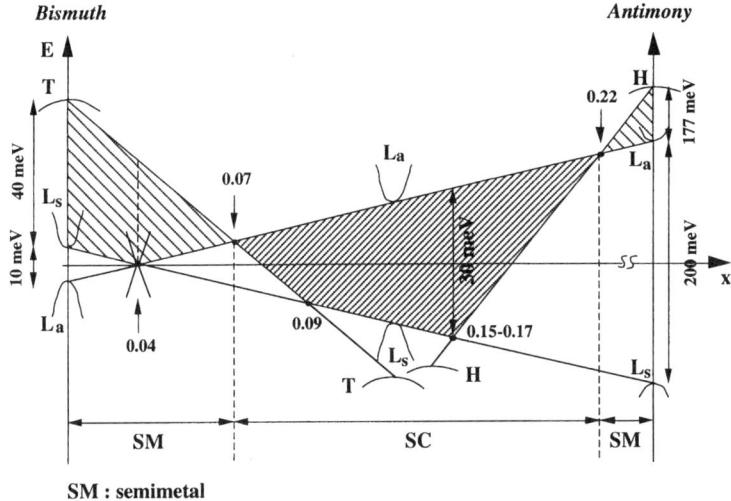

FIG. 5. Schematic diagram of the band edge configuration of $Bi_{1-x}Sb_x$ alloys as a function of x, at $T \cong 0$ K. For simplicity, the L-, T-, and H-point bands are drawn one on top of another. In the semiconducting range ($0.04 < x < 0.22$), the bandgap is maximum around $x = 0.15$–0.17.

the density of state effective mass m_d, which is equal to $m_{d,L} = 0.02\ m_0$, $m_{d,T} = 0.14\ m_0$, and $m_{d,H} = 0.5\ m_0$ for the L, T, and H hole bands, respectively.

The substitution of Bi atoms by Sb atoms in the Bi lattice drastically modifies the band structure of bismuth. Figure 5 represents the band structure variations for $Bi_{1-x}Sb_x$ alloys as a function of x, at low temperature. These results were obtained from powerful quantum oscillatory effects such as magnetoreflection or cyclotron resonance.

As can be seen in Fig. 5, alloying affects mainly three band parameters:

The overlap between L and T bands
The value of the direct energy gap E_g
The energy of the top of the H bands

In the range $0 < x < 0.04$, the band ordering is the same as in pure bismuth, but the overlap as well as the bandgap decrease. At $x \cong 0.04$, a gapless state appears (Tichovolsky and Mavroides, 1969; Oelgart and Herrmann, 1976). Beyond this composition, the bonding L_s and antibonding L_a bands are inverted and the bandgap increases with enhancing antimony concentration. At $x \cong 0.07$, there is no more overlap between the conduction band at the L point and the valence band at the T point. The material then loses its

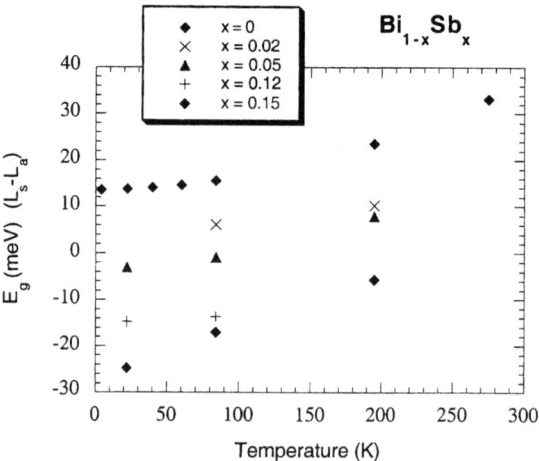

FIG. 6. Temperature dependence of the direct energy gap at the L-point for $Bi_{1-x}Sb_x$ alloys (from Vecchi and Dresselhaus, 1974, and Mendez, 1979).

semimetallic character and becomes a semiconductor (Ellet et al., 1966; Chao et al., 1974). This behavior prevails with increasing Sb concentrations up to $x \cong 0.22$ (Oelgart et al., 1976; Kraak et al., 1978), at which point the rising of the top of the H band restores the semimetallic behavior. Beyond this composition, the band structure of the alloys evolves up to the band structure of pure antimony. So, in the range $0.07 < x < 0.22$, $Bi_{1-x}Sb_x$ alloys are narrow-bandgap semiconductors with a maximal bandgap situated around $x = 0.15 - 0.17$.

Since the energy gap associated with the electron pockets at the L point is very small, it is not surprising that all the band parameters associated with the electron carriers are strongly temperature dependent. These observations were first reported by Vecchi and Dresselhaus (1974) for pure bismuth and were further extended to some Bi-rich alloys by Mendez (1979). Their results show drastic relative variations of the energy bandgap with regard to the value at 0 K compared to typical semiconductors, and this can be neglected in the analysis of transport properties of $Bi_{1-x}Sb_x$ alloys. For $T > 80$ K, extrema of L_s and L_a bands increase and decrease in energy, respectively, when the temperature is increased (Fig. 6). The effective masses at the L band extrema were also reported as being very sensitive to temperature. However, since the bisectrix mass $m_{2'}$ is much larger than the two other masses ($m_{1'}$ and $m_{3'}$), it is less accessible experimentally and it is necessary to make assumptions about the temperature dependence of $m_{2'}$. Nothing is a priori known about the temperature dependence of band parameters for T and H bands in Bi–Sb alloys.

All these unusual features considerably complicate the analysis of the transport properties of such alloys, the difficulties arising in part from the strong nonparabolicity and the large temperature dependence of the band parameters at the L points of the Brillouin zone. The situation is still more complex when several valence bands may contribute to transport properties. This becomes important when alloys are semiconductors, and it is precisely when the thermoelectric performance is the most attractive.

3. Carrier Mobility and Density

In isotropic monovalent metals or in extrinsic semiconductors, where only one type of carrier is present, the carrier density n and mobility μ may be easily experimentally determined by using the Hall coefficient, R_H,

$$R_H = \pm \frac{1}{ne}, \tag{4}$$

to first calculate n, and then using the measured electrical resistivity, ρ,

$$\rho = \frac{1}{ne\mu}, \tag{5}$$

to calculate the mobility. In $Bi_{1-x}Sb_x$ alloys, the simultaneous presence of at least two types of carriers leads to complicated relations between the measured coefficients and the electronic parameters, and more experimental investigations are necessary. One approach to determine the temperature variation of mobility and density turned out to be the low-field galvanomagnetic effects. The theory of such effects for the A_7 structure has been described in detail by many authors (see, for example, Juretschke, 1955). Briefly, if E_i is the electric field and J_i the current density, the resistivity $\rho_{ij}(\vec{B})$ is given by

$$E_i = \rho_{ij}(\vec{B})J_j. \tag{6}$$

In the limit of weak field ($\mu B \ll 1$, where B is the magnetic induction), only the terms in B and B^2 need to be retained in the expansion of this tensor as a power series. Experimentally, 12 coefficients should be measured: 2 zero-field resistivities, 2 Hall coefficients, and 8 magnetoresistance coefficients. Figure 7 schematically describes how the band parameters are deduced from these coefficients.

The analysis of the 12 independent coefficients performed on bismuth single crystals has been widely discussed in the literature (Hartman, 1969;

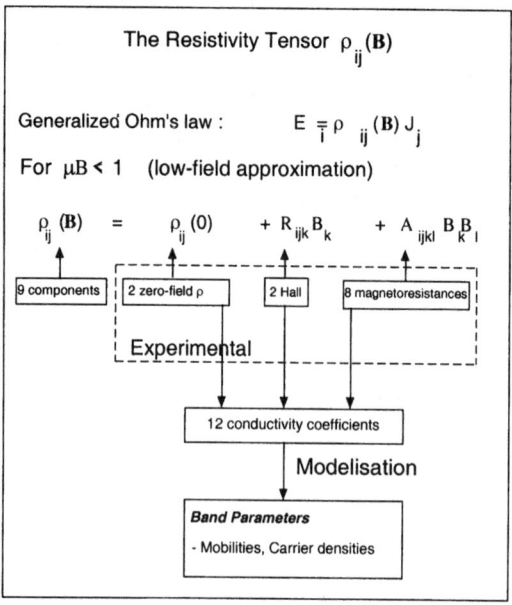

FIG. 7. Schematic representation of band parameter calculations from low-field galvanomagnetic measurements in the A_7 structure.

Michenaud and Issi, 1972; Nemchinskii and Ravich, 1991; Ravich and Rapoport, 1992; Ravich et al., 1995). Different attempts to obtain information about band parameters were also performed on semimetallic and semiconducting alloys at rather low temperatures: Rönnlund et al. (1965) investigated $Bi_{0.95}Sb_{0.05}$ and $Bi_{0.88}Sb_{0.12}$ at 4.2 K, Yazaki and Abe (1968a, 1968b) explored $Bi_{0.97}Sb_{0.03}$ and $Bi_{0.90}Sb_{0.10}$ at 77 and 300 K, Thomas and Goldsmid (1970) determined many of the weak-field galvanomagnetic coefficients of $Bi_{0.95}Sb_{0.05}$ with various additions of Te at 80 K, Brandt et al. (1972c) analyzed the components of the galvanomagnetic tensor of various $Bi_{1-x}Sb_x$ ($0 < x < 0.16$) alloys at 4.2 K, and Jacobson (1973) examined a dilute alloy ($Bi_{0.98}Sb_{0.02}$) at two fixed temperatures (4.2 and 89 K). When the experimental data were analyzed with a realistic model, the conclusions of the previous works point out that the carrier mobility is still high in alloys, as was first suggest by Jain (1959).

The knowledge of carrier density and mobility in the temperature range 80–200 K is of the greatest interest from the point of view of the use of Bi–Sb alloys in thermoelectric devices. However, the analysis of the galvanomagnetic data in this temperature range is complicated by the fact that it is not possible to assume both Hall and magnetoresistance factors to be close to unity. These factors are governed by the energy dependence of the relaxation time, and when the nonparabolicity of the conduction band is taken into account, they may considerably increased for electrons when there is no strong statistical degeneracy (Nemchinskii and Ravich, 1991). Ravich and Rapoport (1992) have clearly shown that these factors greatly differ from unity in bismuth for temperatures exceeding 100 K. The analysis of five weak-field galvanomagnetic coefficients in $Bi_{0.96}Sb_{0.04}$ alloy, performed by Demouge et al. (1995), showed that the difference from unity was greater in this alloy than in pure bismuth.

Lenoir et al. (1998a) have measured the complete set of the 12 independent galvanomagnetic coefficients in undoped $Bi_{0.96}Sb_{0.04}$ alloy within the temperature range 77–300 K under strict isothermal conditions, rigorously satisfying the low-field conditions. The analysis of these coefficients together with the two components of thermopower have been analyzed from a three-band model (L electrons, T and L holes) taking into account the nonparabolicity (model ENP), the temperature dependence of the band structure, and a complex mixed scattering model: the scattering of electrons and holes by acoustical phonons and interband (recombination) scattering due to electron transitions between the conduction band and the T valence band. It was found that the Hall and magnetoresistance factors significantly differ from unity for $T > 100$ K. Moreover, the contribution of light L holes to transport coefficients turned out to be very small in comparison with that of L electrons and T holes. The temperature variation of the carrier mobility (L-electrons and T-holes) and carrier density is shown in Fig. 8 and compared to pure bismuth. The important features of these curves are the retention of high carrier mobility on alloying, as was already observed at low temperatures, and a significant increase in carrier density (more than one order of magnitude) in the temperature range investigated.

III. Synthesis of Bi–Sb Alloys

Although bismuth and antimony are completely miscible in both the liquid and solid states, there is a large temperature difference between the liquidus and the solidus (Fig. 9). A similar situation is encountered for silicon–germanium alloys, which are the best thermoelectric materials for high-temperature applications. According to the phase diagram of the Bi–Sb system, a liquid of C_L composition will start to crystallize into a solid with the C_S composition, rich in Sb with regard to the melt, as seen in Fig. 9. Further lowering the temperature modifies both the compositions of the liquid along the liquidus line and of the grown crystal along the solidus line. Because of very low diffusion rates in the solid, the resulting ingot exhibits severe segregation. The preparation of homogeneous solids therefore requires either maintaining a constant composition of the melt during growth

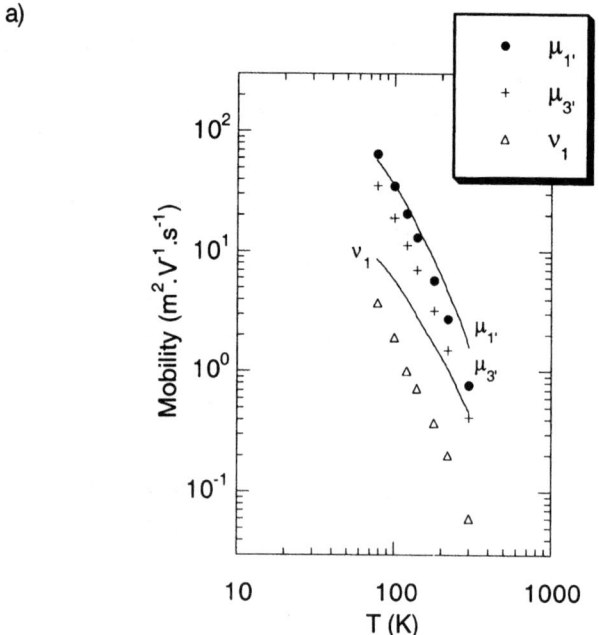

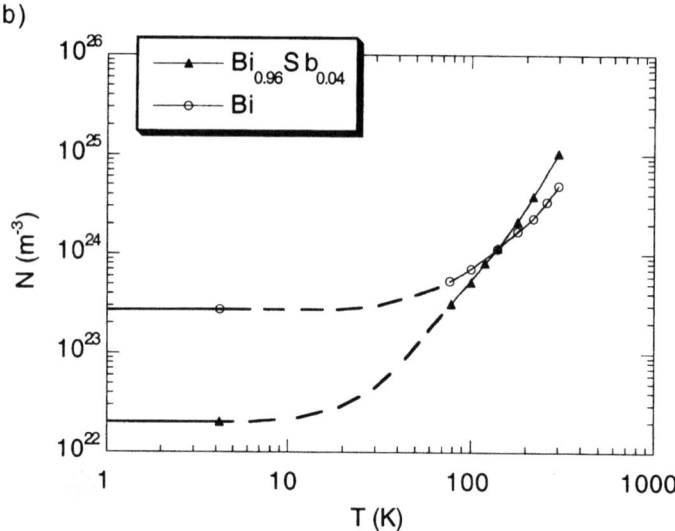

FIG. 8. Transport coefficients in the $Bi_{0.96}Sb_{0.04}$ alloy (after Lenoir et al., 1998a). (a) Temperature dependence of L-electron ($\mu_{1'}$ and $\mu_{3'}$) and T-hole (ν_1) mobility in the ellipsoid coordinate system. Comparison with bismuth (solid lines). (b) Temperature dependence of the carrier density in the alloy and in bismuth.

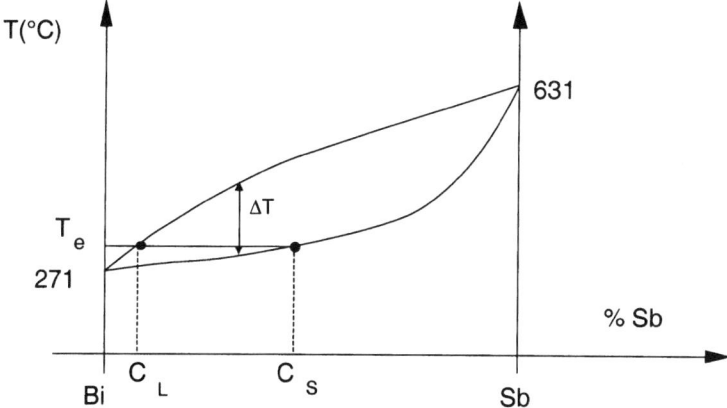

FIG. 9. Schematic representation of the phase diagram of the Bi-Sb system. The segregation coefficient of Sb (C_s/C_L) is greater than unity.

or using some process to homogenize the solid phase. Bi-Sb alloys have been extensively prepared in the single crystalline form and to a lesser extent in the polycrystalline form via powder metallurgy. Although Bi-Sb samples were mostly prepared without any intentional doping, effects on the thermoelectric properties of donor [Te (Ivanov et al., 1973; Belaya et al., 1994), Se (Ivanov et al., 1973)] and acceptor [Sn (Yim and Amith, 1972; Horst and Williams, 1980; Belaya et al., 1994), Pb (Smith and Wolfe, 1962; Tagiev et al., 1992), Ga (Okumura et al., 1991)] centers have been considered.

1. SINGLE CRYSTALS

Among the crystal growth processes, the repeated pass zone melting has been widely used with growth rates ranging from 0.1 to 51 mm h^{-1} (Jain, 1959; Smith and Wolfe, 1962; Brown and Heumann, 1964; Ivanov and Popov, 1964; Dugué, 1965; Short and Schott, 1965; Yazaki and Abe, 1968a; Ivanov et al.; 1973; Alekseeva et al., 1976). This elaboration method can overcome excessive macrosegregations in the alloys, whereas the microhomogeneity in composition can only be achieved from slow growth rates. Microinhomogeneities arise from the detrimental effect of constitutional supercooling. In order to eliminate constitutional supercooling or at least to reduce it substantially, Tiller et al. (1953) have shown that the growth rate R must satisfy the relation

$$R < DG/\Delta T \qquad (7)$$

where D is the binary alloy liquid diffusion constant (about $10^{-5}\,\text{cm}^2\,\text{s}^{-1}$ at 300°C as reported by Brown and Heumann, 1964), G the temperature gradient at the solidification interface, and ΔT the temperature difference between the liquidus and the solidus. It is, however, difficult to produce large temperature gradients because of the low melting temperatures of the alloys and their relatively high thermal conductivity. Consequently, small growth rates R are needed in conjunction with small G gradients and large ΔT. Further, it is not possible to produce homogeneous samples by annealing after growth because of the low melting temperature. Many early results taken from measurements on crystals grown without satisfying the relation (7) are questionable because of sample nonhomogeneity.

Contrarily to multipass zone melting, a single-pass zone leveling technique (Yim and Dismukes, 1966) or a traveling heater method (THM) (Lenoir et al., 1995) were successfully used to grow very homogeneous single crystals. In these processes, described in details elsewhere (see, for example, Borshchevsky, 1995), a precise knowledge of the phase diagram is critical for controlling the growth of the material. Despite the phase diagram of Bi–Sb alloys is relatively simple, the data available in the literature showed serious discrepancies concerning the solidus line on the Bi-rich side (Fig. 10) (Cook, 1992; Feutelais et al., 1992; Hansen and Anderko, 1958; Hüttner and

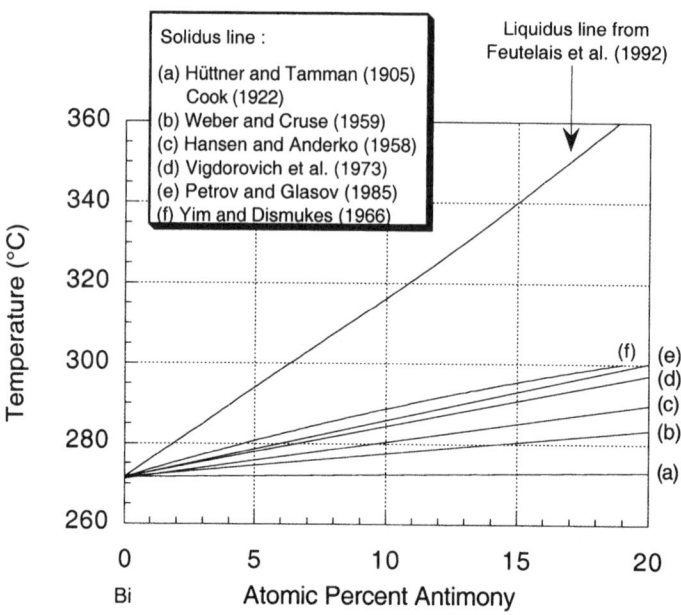

FIG. 10. Details of the phase diagram on the Bi-rich side. A serious discrepancy is observed for the solidus line.

Tamman, 1905; Petrov and Glasov, 1985; Vigdorovich et al. 1973; Weber and Cruse, 1959). It is only recently that confirmations of the experimental results of Yim and Dismukes (1966) were given by Lenoir et al. (1995) and Martin-Lopez et al. (1998a).

When low growth rates are used in the previous processes, Bi–Sb alloy crystals tend to grow in a direction perpendicular to the trigonal axis (Yim and Dismukes, 1966; Lenoir et al., 1995). Moreover, it is not necessary to use seed crystals, because even without seeding, a large portion of the grown ingot was single crystalline, especially for the alloys containing less than 15 at% Sb (Yim and Dismukes, 1966).

An example of longitudinal and radial homogeneities is reported in Fig. 11 for a $Bi_{0.96}Sb_{0.04}$ alloy (150 mm in length, 15 mm in diameter) elaborated by the THM method. As shown, the longitudinal variation in the antimony concentration is weak in the central part of the ingot. Radially, the composition is highly uniform (variation in composition are less than 0.3% Sb).

The Czochralski pulling technique was also successfully used to grow homogeneous doped (Te, Sn) and undoped Bi–Sb alloys up to 18 at% Sb at the Baikov Institute of Metallurgy in Moscow (Zemskov et al., 1985; Belaya et al., 1994). Large single crystals (20–25 mm in diameter and up to 100 mm long) of a high degree of crystalline perfection have been obtained. The melt-injection technique (Koh et al., 1995) was also developed to prepare these mixed crystals.

2. POLYCRYSTALLINE Bi–Sb

The use of Bi–Sb single crystals in low-temperature stages of solid-state coolers is limited because single crystals cannot withstand any bending along their trigonal axis. Unfortunately, it is precisely in that direction that the figure of merit is the best. Special attention was devoted to enhancing the mechanical properties of Bi–Sb single crystals. The influence of the growth parameters on the bending strength of Czochralski-grown ingots as well as the influence of the size and the surface quality of the samples has been investigated in detail to improve mechanical properties (Belaya et al., 1995). Significant work on extrusion of single crystals under high hydrostatic pressure has been performed by Sidorenko and coauthors (Kolomoetz et al., 1990; Sidorenko and Mosolov, 1992; Sidorenko, 1994) in an attempt to improve the mechanical properties without any significant decrease in the thermoelectric performance.

Another way to improve the durability of the material is to produce Bi–Sb polycrystalline alloys via powder metallurgy that are easier to synthesize, dimension, and handle than single crystals. The microstructure produced by powder metallurgy also eliminates the risk of catastrophic

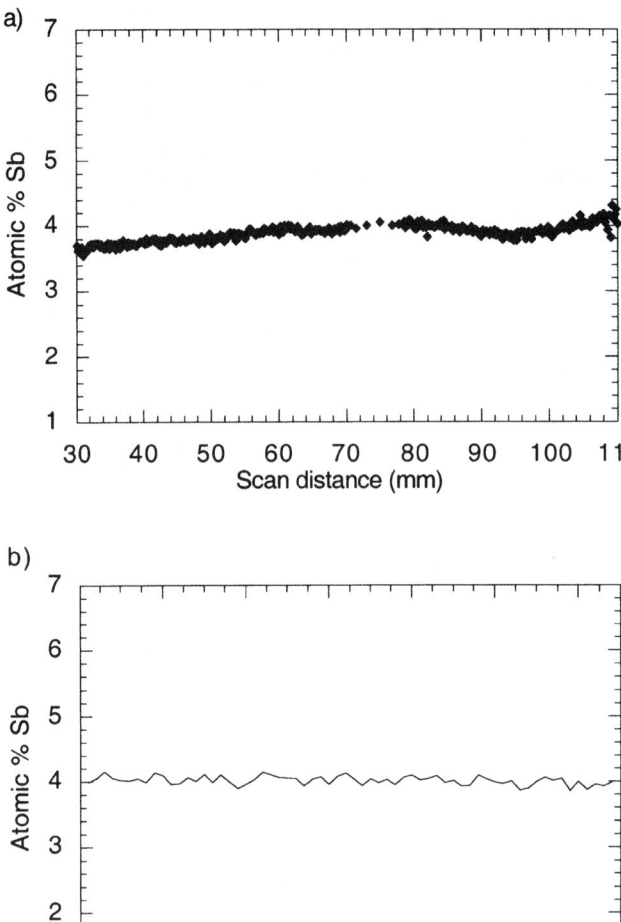

FIG. 11. Longitudinal (a) and radial (b) variation in Sb content as determined by electron probe X-ray microanalysis (EPMA) on a 150 mm in length and 15 mm in diameter $Bi_{0.96}Sb_{0.04}$ alloy grown by THM.

failure due to cleavage. Obtaining fine grains ($<10\,\mu$m) is also very important, in particular from the point of view of the processing of advanced thermoelectric materials. Grain refinement reduces the lattice thermal conductivity due to phonon scattering by grain boundaries (Goldsmid and Penn, 1968; Goldsmid et al., 1995), which is favorable from a thermoelectric point of view. The processing (hot pressing, cold pressing and annealing, extrusion, etc.) and its conditions may also strongly affect the microstructure and thus the transport properties. For instance, forming the material by extrusion may introduce a preferential orientation of the grains in the consolidated material. The presence of a strong texture in a layered material such as Bi–Sb alloys would be favorable from a thermoelectric point of view. Actually, if all the grains could be orientated so that their trigonal axes are nearly parallel, then their thermoelectric properties should approach those of the single crystals.

Among powder metallurgy processings, mechanical alloying (Martin-Lopez et al., 1995, 1997, 1998a), arc-plasma spraying (Suse et al., 1993; Devaux et al., 1997a), and the grinding of zone-melted rods (Cochrane and Youdelis, 1972; Banaga et al., 1986; Grabov et al., 1998) were applied to fabricate Bi-Sb alloys. Mechanical alloying of the elemental materials has proved to be a powerful technique to synthesize homogeneous $Bi_{1-x}Sb_x$ alloy powders (Fig. 12) with $x = 0.07, 0.1, 0.12, 0.15, 0.22$ with grain sizes about 10 μm, in relatively low milling times (4–15 hours, depending on the milling conditions) (Martin-Lopez et al., 1995, 1997, 1998a). The mechanical properties of a $Bi_{0.85}Sb_{0.15}$ alloy prepared by mechanical alloying and consolidated by hot extrusion were investigated at two fixed temperatures (77 and 300 K) by Martin-Lopez et al. (1998b). The three-point bending test (Fig. 13) is usually used in the course of this analysis because the stress state of sample in bending is very similar to the distribution of stresses in the thermocouple of a Peltier cooler device (Sidorenko and Mosolov, 1992; Belaya et al., 1995). The modulus of rupture of the polycrystalline material was roughly the same at the two studied temperatures with a value of about 100 MPa (Table IV). The improvement is significative comparatively to the

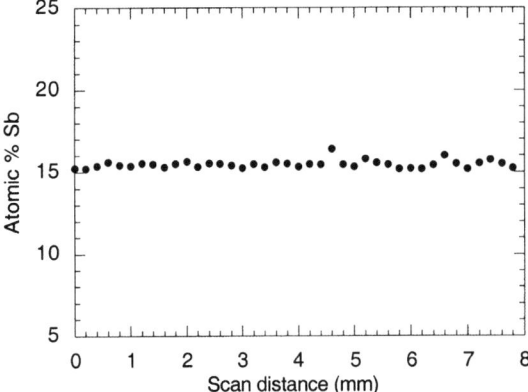

FIG. 12. Electron microprobe scanning along compacted $Bi_{0.85}Sb_{0.15}$ powders elaborated by mechanical alloying (ball to powder weight ratio: 10/1, particle sizes <100 mm, 15 h of milling time), after Martin-Lopez et al. (1995).

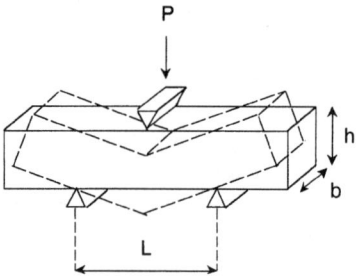

Fig. 13. Three-point bending setup. For a strip sample the modulus of rupture σ_b is equal to $\sigma_b = 1.5\, PL/bh^2$, where P is the load at failure, L is the length of the moment arm, b is the width of the strip, and h is the strip thickness (from Martin-Lopez et al., 1998b).

TABLE IV

Modulus of Rupture (Bending Strength) of a $Bi_{0.85}Sb_{0.15}$ Alloy, Either Polycrystalline or Single Crystalline, at 77 and 293 K (Martin-Lopez et al., 1998b)

	σ_b	
	77 K	293 K
Single crystal	10 MPa	20 MPa
Polycrystal	90 MPa	105 MPa

value obtained for a $Bi_{0.85}Sb_{0.15}$ single crystal prepared by the Czochralski method at 77 K.

The arc-plasma technique is particularly well adapted to provide fine powders of controlled stoichiometry and granulometry (0.025 to 0.8 μm) (Devaux et al., 1997a). By cold-pressing these powders, bulk materials with mean grain size ranging from 0.1 to 2.5 μm and with relative densities near 90% have been prepared.

IV. Transport Properties

1. Single Crystals

a. Electrical Resistivity

Many results have been published on the temperature dependence of the two components (ρ_{11} and ρ_{33}) of the electrical resistivity (Brandt et al., 1971; Jain, 1959; Smith and Wolfe, 1962; Brown and Silverman, 1964;

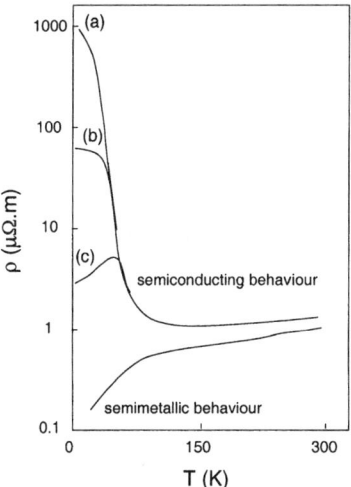

FIG. 14. Schematic temperature dependence of the electrical resistivity in Bi_{1-x}–Sb_x alloys ($0 < x < 0.25$).

Ivanov and Popov, 1964; Dugué, 1965; Yazaki, 1968; Brandt et al., 1969; Yim and Amith, 1972; Ivanov et al., 1973; Alekseeva et al., 1976; Kuhl et al., 1976; Rodionov et al., 1979; Probert and Thomas, 1979; Grabov et al., 1995; Lenoir et al., 1996). The general trends are similar in most of the studies except at low temperatures, where the crystal perfection and the concentration of residual impurities play an important role in semiconducting alloys. The different temperature variations of the electrical resistivity encountered in $Bi_x Sb_{1-x}$ alloys ($x < 0.25$) are illustrated in Fig. 14.

Above 150 K, the electrical resistivity increases almost linearly independently of the measurement direction and the alloy composition. At 300 K, the value of the resistivity is about $1.6\ \mu\Omega\,m$. This value, nearly identical to that for pure bismuth ($\cong 1.2\ \mu\Omega\,m$), is only two orders of magnitude higher than the value of typical metals. For bismuth, the low number of carriers (10^{17}–$10^{18}\ cm^{-3}$) (Ravich et al., 1995) is compensated by their very high mobility in some directions, as compared to metals, due to very small effective masses. Since the density of carriers in Bi-rich Bi–Sb alloys is certainly of the same order of magnitude as in pure bismuth, we can expect that the mobility is still very high in the solid solution at room temperature.

Below 150 K, a clear difference in the electrical resistivity behaviors appears. For $x < 0.07$, the electrical resistivity of the semimetallic alloys decreases with diminishing temperature, quite similarly to what happens for pure bismuth. For $0.07 < x < 0.20$, the resistivity first decreases down to a minimum and then three temperature dependence regimes have been reported as the temperature decreases (Fig. 14):

(a) A continuous increase of the resistivity
(b) An increase with a subsequent transition to saturation
(c) An increase up to a maximum and then a decrease down to a saturation limit

This unusual behavior at low temperature in the semiconducting range for undoped alloys can be understood qualitatively by referring to the expressions of both the first ionization energy E_i and Bohr radius a_B of a bound electron near an impurity ion,

$$E_i(\text{eV}) = 13.6 \left(\frac{m^*}{m_0}\right) \frac{1}{\varepsilon_r^2}$$

$$a_B(\text{cm}) = 0.53 \left(\frac{m_0}{m^*}\right) \varepsilon_r \times 10^{-8}, \tag{8}$$

where m^* is the effective mass, m_0 the free-electron mass, and ε_r the relative dielectric constant of the medium. Actually, these materials are characterized by large dielectric constant ($\cong 100$) and very small effective masses ($\cong 0.01 m_0$) leading to an unusually small ionization energy ($E_i \cong 10^{-5}$ eV) and a large Bohr radius ($a_B \cong 5 \times 10^{-5}$ cm) in comparison to typical semiconductors (Si: $E_i \cong 9 \times 10^{-2}$ eV and $a_B \cong 1.7 \times 10^{-7}$ cm). A large Bohr radius results in the formation of an impurity band for low concentrations of impurities n_i. In alloys, the overlap of this impurity band with the fundamental bands occurs when $n_i \cong 10^{12}$ cm^{-3} (Brandt et al., 1972b) and the heavy doping condition, namely, $a_B^3 n_i \gg 1$, is fulfilled when $n_i \cong 10^{14}$ cm^{-3} (Brandt et al., 1972b). So, even in very pure materials, a large residual conductivity can be expected at 4.2 K.

When the intrinsic carrier concentration is greater than the impurity density n_i, the resistivity decreases. In order to analyze the temperature-dependent resistivity for a given Sb concentration, Jain (1959) used the following exponential law to describe the semiconducting alloys:

$$\rho = \rho_0 \exp(\Delta E / 2kT). \tag{9}$$

Here, ρ_0 is a constant and ΔE is an effective bandgap. This relation is based on the assumptions that the contributing bands are parabolic with the same density of states, and that the carriers are scattered primarily by acoustic phonons. The thermal bandgap has been calculated by several authors using relation (9). The results are reported in Fig. 15. It is seen that, despite the approximations of such temperature dependence, the values obtained may be used to describe, at least qualitatively, the evolution of the band structure as a function of Sb concentration at low temperatures when the investigated alloys are uniform in composition. It results from the small value of the

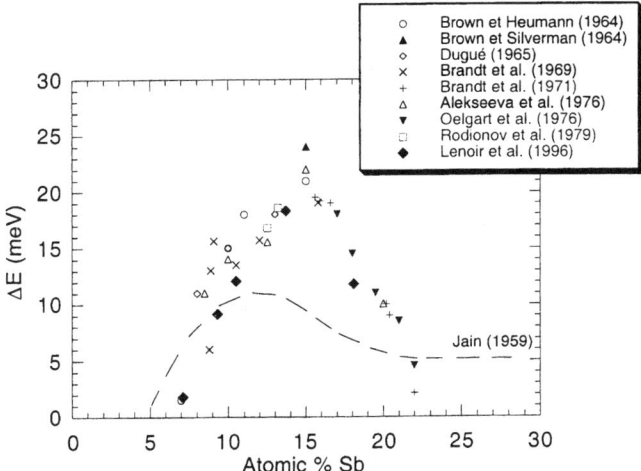

FIG. 15. Estimation of the thermal gap ΔE of Bi–Sb alloys from relation (9), according to several authors.

thermal gap that the electrical resistivity increases when the thermal energy kT is of the same order of magnitude as ΔE. The linear variation observed for $T > 150$ K, a sign of metallic behavior, is, however, purely fortuitous because both carrier densities and effective masses vary with temperature.

The anisotropy of the electrical resistivity of Bi-rich Bi–Sb alloys never exceed 30% when $50 < T < 300$ K, but depends on the antimony content (Yim and Amith, 1972; Lenoir et al., 1996): For $x < 0.07$, $\rho_{33} > \rho_{11}$, and for $x > 0.07$, it is the opposite.

b. Thermoelectric Power

The substitution of Bi atoms by Sb atoms results in larger absolute values of the thermoelectric power for both thermoelectric powers α_{11} and α_{33}, with regard to those of pure bismuth (Fig. 16). It is due to the decrease of the overlap between L and T bands, and to the appearance of a semiconducting state. It was shown experimentally by many authors that the thermoelectric power of semiconducting alloys is always negative in the region of intrinsic conductivity (Smith and Wolfe, 1962; Brown and Silverman, 1964; Yazaki, 1968; Yim and Amith, 1972; Rodionov et al., 1979, 1982; Probert and Thomas, 1979; Lenoir et al., 1996), whereas it can be either positive or negative in the region of extrinsic conductivity according to the type of doping (intentional or not).

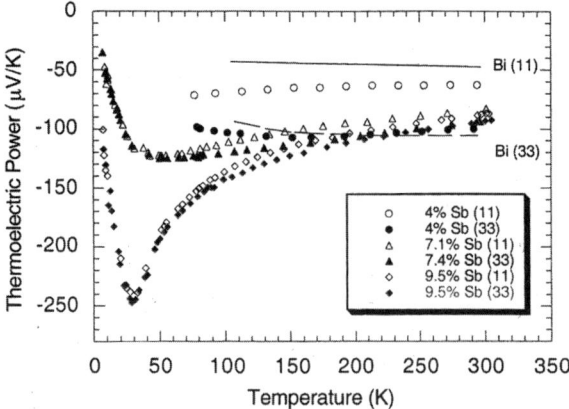

FIG. 16. Thermoelectric power of various Bi–Sb alloys as a function of temperature along directions parallel (α_{33}) or perpendicular (α_{11}) to the trigonal axis, after Lenoir et al. (1996). (—) represents the measurements for Bi (Gallo et al., 1963).

Many studies have been performed by Red'ko and coauthors (Rodionov et al., 1979, 1981; Gryaznov et al., 1982; Red'ko et al., 1986) on the measurement of the thermoelectric power on p or n $Bi_{1-x}Sb_x$ ($0.085 < x < 0.17$) alloys at low temperatures ($T < 100\,K$) with the attempt to correlate their results with the complex valence band structure. By varying the impurity carrier densities, they have investigated the carrier scattering mechanisms in a single-band state (only L-light holes participate in the transport phenomena) or in a many-band state (both the light holes and the heavy holes are involved in the transport effects). Very large values of the thermoelectric power were reported at very low temperatures in a series of samples (Fig. 17). Because of the narrow bandgap of the semiconducting alloys, the transition from extrinsic to intrinsic conductivity takes place at low temperatures. An increase of the carrier concentration at low temperatures only leads to a shift of the transition temperature toward higher temperatures, as can be seen in Fig. 17.

The negative sign of the thermoelectric power in the intrinsic range reflects only the higher mobility of electrons compared to that of holes. Actually, the total diffusion thermoelectric power α_{ii} is expressed in terms of the corresponding partial contributions as

$$\alpha_{ii} = \frac{\alpha^e \sigma_{ii}^e + \alpha^h \sigma_{ii}^h}{\sigma_{ii}^e + \sigma_{ii}^h}, \quad (10)$$

where σ_{ii}^j is the partial contribution of carriers to the total electrical conductivity in the considered direction (index j refers to electrons or holes)

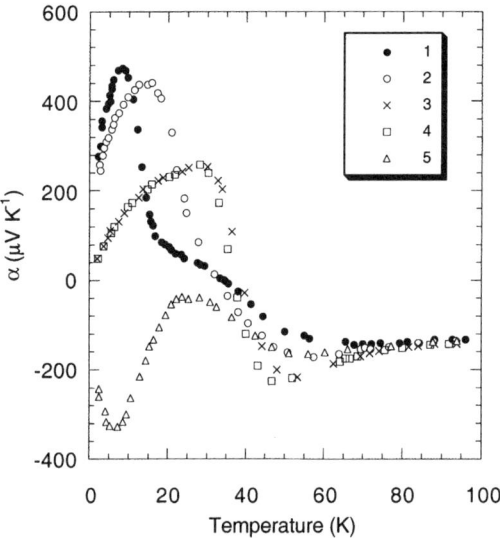

FIG. 17. Temperature dependence of the thermoelectric power as a function of hole and electron density p and n, respectively: (1) α_{22}, $p = 3.4 \times 10^{13}$ cm^{-3}; (2) α_{22}, $p = 2.1 \times 10^{14}$ cm^{-3}; (3) α_{22} and (4) α_{33}, $p = 4.6 \times 10^{14}$ cm^{-3}; (5) α_{33}, $n = 4.1 \times 10^{13}$ cm^{-3}. After Rodionov et al. (1979).

and α^j are the partial diffusion thermopowers. For holes, L, T, and H bands may contribute to the partial contributions.

As $\alpha^e < 0$ and $\alpha^h > 0$, the expression for α_{ii} is

$$\alpha_{ii} = \frac{\alpha^e \left(1 - \left|\frac{\alpha^h}{\alpha^e}\right| \frac{\sigma_{ii}^h}{\sigma_{ii}^e}\right)}{1 + \frac{\sigma_{ii}^h}{\sigma_{ii}^e}}. \qquad (11)$$

Since the contribution of light and heavy holes to the intrinsic conductivity is small in comparison with that of electrons, because the mobility of electrons is much higher than that of holes, the ratio of the partial conductivity is essentially less than unity. The difference between electron and hole partial thermopowers is less than that between partial conductivity. Therefore, in first approximation, the thermopower $\alpha_{ii} \cong \alpha^e$, and it is negative.

c. *Thermal Conductivity*

In bismuth–antimony alloys, the thermal conductivity λ is principally the sum of two terms: an electronic contribution λ_E (including both unipolar and bipolar terms) and a contribution associated with the lattice λ_L:

$$\lambda = \lambda_E + \lambda_L. \tag{12}$$

In pure bismuth, in the range 2–20 K, the thermal conductivity λ_{11} is only due to the lattice contribution. The relative contribution of λ_E increases with increasing temperature. Uher and Goldsmid (1974) have reported that λ_E contributes to 14% of the total conductivity at 35 K and to 42% at 140 K. The knowledge of the relative contributions of carriers and phonons has also been studied in the alloys (Yazaki, 1968; Chaudhuri and Dey, 1975; Probert and Thomas, 1979; Red'ko, 1990; Kagan and Red'ko, 1991, 1995). Significant results have been obtained by Kagan and Red'ko (1991, 1995), who succeeded in separating the relative contributions by applying a strong magnetic field ($\mu B > 1$) (Fig. 18). As might be expected, below 20 K the heat is still primarily transported by phonons, and the electronic component becomes more important as the temperature increases. However, alloying reduces strongly the lattice thermal conductivity with regard to pure bismuth because of strong point defect scattering. The relaxation time τ for this process is expressed by (Klemens, 1955)

$$\tau^{-1} = A\Gamma\omega^4, \tag{13}$$

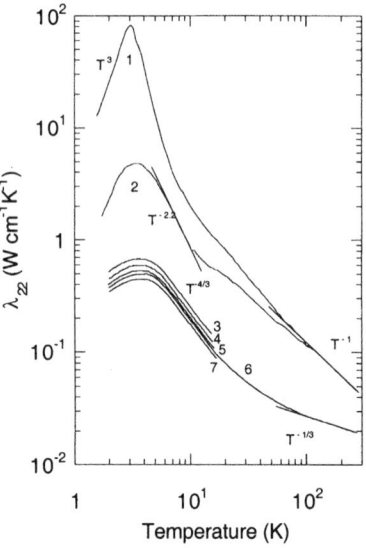

FIG. 18. Temperature dependence of the phonon thermal conductivity λ_{22} of single-crystal samples. (1) Pure bismuth; (2–7) $Bi_{1-x}Sb_x$ alloys of compositions $x = 0.001$ (2), 0.085 (3), 0.1 (4), 0.12 (5), 0.135 (6), and 0.15 (7). After Kagan and Red'ko (1995). Dimensions of the sample were $4 \times 4 \times 40$ mm. As in Bi, a size effect was observed in $Bi_{1-x}Sb_x$ alloys.

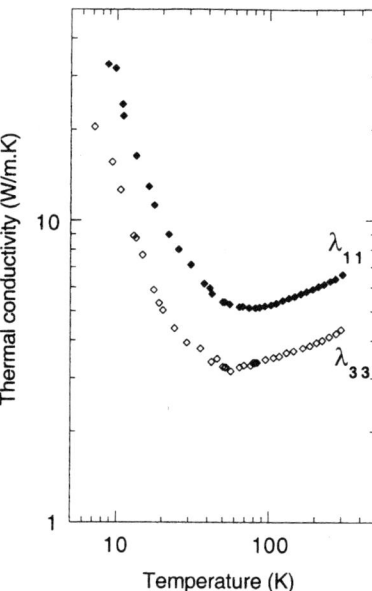

FIG. 19. Anisotropy of the thermal conductivity represented for a $Bi_{0.93}Sb_{0.07}$ alloy. After Lenoir et al. (1996).

where A is a constant, ω refers to the phonon frequency, and Γ, which is a measurement of the strength of the scattering, is given by

$$\Gamma = \sum_i f_i \left(1 - \frac{M_i}{\bar{M}}\right)^2, \qquad (14)$$

where M_i is the mass of the ith type of impurity, $\bar{M}$ the mean atomic mass, and f_i the fractional concentration of this impurity. Since the atomic masses of bismuth and antimony differ greatly ($M_{Bi} \cong 209$ and $M_{Sb} \cong 122$), point-defect scattering phenomena are relatively important. The decrease of the lattice thermal conductivity with respect to bismuth is spectacular near the dielectric maximum, where the defects have a far greater effect on the thermal conductivity than at higher temperatures (Fig. 19).

The thermal conductivity of alloys is anisotropic as in pure bismuth, λ_{11} being always greater than λ_{33} (Fig. 19). The $\lambda_{11}/\lambda_{33}$ ratio is around 1.7.

d. Thermoelectric Figure of Merit

The larger values of the Seebeck coefficient and the lower values of the lattice thermal conductivity in the alloys result in higher values of the figure

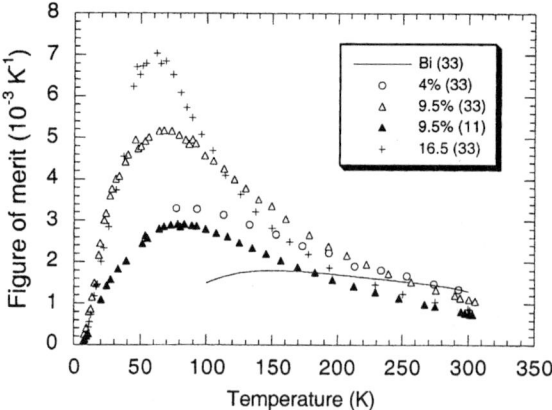

FIG. 20. Temperature dependence of the figure of merit of various $Bi_{1-x}Sb_x$ alloys. After Lenoir et al. (1996, 1998b).

of merit than for pure bismuth (Fig. 20). Values measured along the trigonal axis are higher than those measured in the basal plane (Fig. 20). The strong anisotropy of the thermal conductivity is mainly responsible for the anisotropy of the figure of merit. The temperature dependence of the figure of merit is independent of the composition. It slowly increases from 300 K up to a maximum around 70 K and then drops sharply at low temperatures (Smith and Wolfe, 1962; Lenoir et al., 1996, 1998b).

Lenoir et al. (1998b) have observed the presence of two maxima in the curve of the figure of merit versus Sb content around 70 K in undoped Bi–Sb alloys (Fig. 21) corresponding to the situation where the tops of the T and H bands are on the same level as the L bands, for $x \cong 0.09$ and $x \cong 0.16$. Such behavior was previously reported by Grabov et al. (1985) for $T = 82$ and $T = 95$ K. Among the physical reasons developed by Lenoir et al. (1998b) in order to explain the antimony dependence of the figure of merit observed at low temperatures, interband hole scattering occurring through transitions between bands of light and heavy holes could be an important factor affecting Z. This mechanism of hole scattering may essentially lower the mobility of light L-holes and increase the figure of merit of n-type semiconductors. Within the $0.09 < x < 0.16$ composition range, the interband scattering is maximal when the heavy hole extremum is on the same level as the light hole maximum. This feature is realized at the limits of the range. The figure of merit may be lowered outside this range because of the decrease of the thermal energy gap. Therefore, the interband hole scattering may explain the existence of the two maxima in the evolution of the figure of merit with Sb content at low temperatures. The same mechan-

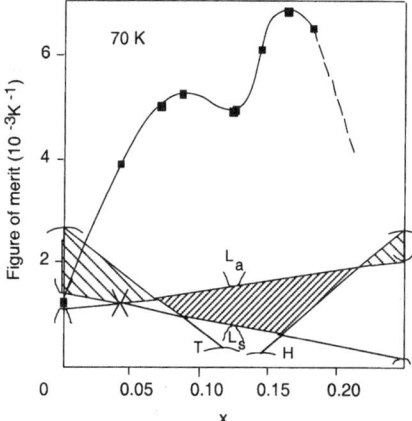

FIG. 21. Antimony content dependence of the thermoelectric figure of merit at 70 K of $Bi_{1-x}Sb_x$ alloys, as reported by Lenoir et al. (1998b).

ism of interband scattering can certainly also explain why the bismuth–antimony alloys cannot be good p-type thermoelectric materials.

Some recent results obtained on doped, by tin or tellurium, samples (Belaya et al., 1994) do not present any substantial improvement of the figure of merit for $T > 80$ K, as previously observed by other authors (Smith and Wolfe, 1962; Yim and Amith, 1972). At lower temperatures, it is possible to obtain a p-type thermoelement with acceptor impurities (Smith and Wolfe, 1962; Yim and Amith, 1972; Red'ko, 1995) over a small range of temperature of about 40 K. However, as was discussed by Red'ko (1995), the conditions are not favorable to obtain a high p-type figure of merit (the reported values never exceed $1 \times 10^{-3} \, K^{-1}$) because of the peculiarities of the complex valence band structure.

2. POLYCRYSTALLINE SAMPLES

As previously seen, a rigorous analysis of the transport properties in Bi–Sb single crystals is not an easy task due in part to the complex band structure. The complexity is still greater when we are interested in the transport properties of polycrystals, where the effects of anisotropy, microstructure, local inhomogeneities, grain boundaries, etc., have to be taken into account.

Powder metallurgy offers the advantage of improving the mechanical strength of alloys, but unfortunately degrades the thermoelectric performance as a result of random orientation. Particular attention has been devoted to the effect of grain size on the figure of merit in semiconducting alloys. Indeed, a substantial reduction of λ_L is expected in these solid solutions due to phonon scattering at grain boundaries at temperatures higher than the Debye temperature θ_D ($\theta_D = 120$ K in bismuth), as is known to occur at high temperatures for Si–Ge alloys. The physical reason is that the high-frequency phonons are strongly scattered by the point defects in the solid solution, leaving most of the heat to be carried by the low-frequency phonons that have very long free path lengths. These phonons are thus much more susceptible to boundary scattering (Goldsmid and Penn, 1968). Early work performed by Cochrane and Youdelis (1972) on $Bi_{0.88}Sb_{0.12}$ alloys synthesized by grinding zone-melted rods has, however, not shown preferential scattering by a particular powder size, the main effects affecting the figure of merit being the variations in electrical resistivity. Impurities such as oxygen insertion during powder elaboration could be an hindering factor for the improvement of the figure of merit, as well as the fact that the powder particles were too large (> 38 μm) for any effect to be seen. However, the study of Suse et al. (1993), performed on the same alloy composition but prepared by arc plasma, showed an enhancement of the figure of merit by a factor of 2 as compared to a single crystal (Fig. 22), for temperatures greater than 150 K, that was attributed to phonon scattering at grain boundaries (particle size 1–5 μm). The value of the thermal conductivity (1.2 $Wm^{-1}K^{-1}$) measured at room temperature

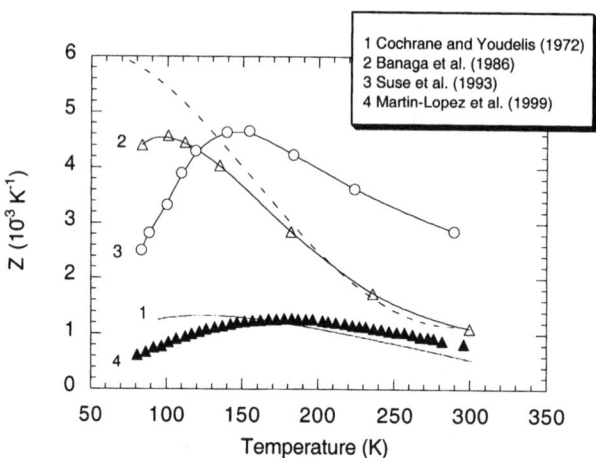

FIG. 22. Temperature dependence of the figure of merit of $Bi_{0.88}Sb_{0.12}$ polycrystalline samples from various references. The dashed line indicates Z in the trigonal direction (Z_{33}) of the $Bi_{0.88}Sb_{0.12}$ single crystal (from Yim and Amith, 1972).

was found to be two-thirds of that of a single crystal along the trigonal axis.

This huge increase of the figure of merit in this polycrystalline alloy has stimulated a great deal of experimental and theoretical work. Goldsmid *et al.* (1995) determined the conditions for high-temperature boundary scattering in these solid solutions by a simplified theory. They concluded that it is possible to obtain even larger reductions in the thermal conductivity of $Bi_{0.88}Sb_{0.12}$ alloys if the grain size is decreased to a fraction of a micrometer (reduction of 50% for a size of 0.13 μm). Experimentally, the lattice thermal conductivity of a polycrystalline sintered BiSb sample having mean grain sizes between 1 and 19 μm was investigated by Volckmann *et al.* (1996). As predicted by the theory, the lattice thermal conductivity was found to decrease as the grain size was reduced. Such a decrease occurred, however, at grain sizes somewhat larger than expected. Devaux *et al.* (1997b) also observed a large decrease of the thermal conductivity in $Bi_{0.88}Sb_{0.12}$ alloys prepared by arc plasma, having grain sizes between 0.1 and 2.5 μm. But, as previously reported by Volckmann *et al.* (1996), an enhanced figure of merit was not achieved in the samples because of a dominating concurrent increase in electrical resistivity. Recently, the transport properties of mechanically alloyed Bi–Sb alloys consolidated either by sintering or by extrusion was investigated by Martin-Lopez *et al.* (1999) in samples having grain sizes ranging from 2 to 10 μm. Despite favorable factors such as the increase (in absolute values) of the thermopower and a decrease of the thermal conductivity, no improvement of the figure of merit was observed (Fig. 22), again because of excessive electrical resistivity.

Experimental studies evidenced that scattering of phonons at grain boundaries is effective in Bi–Sb alloys at high temperatures. Merely decreasing the grain size reduces the thermal conductivity but also adversely increases the electrical resistivity. It is not impossible that small grain-size samples can degrade the high charge carrier mobility in alloys. Recent work performed by Brochin (2000) on bismuth silica nanocomposites having grain sizes of 1 μm and 0.1 μm clearly demonstrated that the carrier mobility is drastically reduced in these disordered materials.

As stated earlier, the presence of a strong texture in these layered materials is highly desirable. Banaga *et al.* (1986), by studying the hot extrusion of a $Bi_{0.88}Sb_{0.12}$ polycrystalline alloy with a rod of small diameter (1–5 mm) observed a strong texture with the trigonal axis of the alloy along the extrusion direction. They reported figures of merit slightly higher than those for the average values of the figure of merit of a single crystal of the same composition in the entire temperature range studied (80–300 K). They explained their results by the presence of both a favorable texture and potential barriers.

V. Concluding Remarks

The main features of Bi–Sb alloys have been discussed in this survey, including both preparation and transport properties. Although the band structure of alloys is well described at low temperatures, a rigorous discussion of the transport properties is rendered difficult by the effects of nonparabolicity and the large temperature dependence of the band parameters.

The unique peculiarities of the electronic L-band are mainly responsible for the great thermoelectric performance of semiconducting Bi–Sb single crystals around 80 K. Despite a great amount of experimental work, little or no improvement in the figure of merit of single crystals has been achieved since the first work of Smith and Wolfe (1962). Even doping attempts were unsuccessful. It is possible to suppose that further investigations of the band parameters and the scattering mechanism in Bi–Sb alloys will promote the improvement of thermoelectric characteristics of these unique semiconductors.

Some encouraging results were obtained on polycrystalline samples. However, it is not clear whether a real improvement in the figure of merit can be achieved in small grain size samples.

Acknowledgments

The authors apologize for any inadvertent omissions or misstatements of fact or opinion attributed to others.

References

A. A. Abricosov and L. A. Falkovskii (1963). Theory of the electron energy spectrum of metals with a bismuth type lattice. *Sov. Phys. JETP*, 16(3), pp. 769–777.

V. G. Alekseeva, N. F. Zaets, A. A. Kudryashov, and A. B. Ormont (1976). Dependence of the forbidden band width of Bi–Sb semiconducting solid solutions on antimony concentration. *Sov. Phys. Semicond.*, 10(12), 1332–1334.

M. P. Banaga, O. B. Sokolov, T. E. Benderskaya, L. D. Dudkin, A. B. Ivanova, and I. I. Fridman (1986). Peculiarities of the structure and thermoelectric properties of extruded samples of $Bi_{0.88}Sb_{0.12}$. *Neorg. Mater.*, 22(4), pp. 619–622.

A. D. Belaya, S. A. Zayakin, and V. S. Zemskov (1994). Single crystals of Bi–Sb solid solutions for magneto-thermoelectric and galvano-themomagnetic cooling. *J. Adv. Mater.*, 1(2), pp. 158–165.

A. D. Belaya, S. A. Zayakin, and V. S. Zemskov (1995). Mechanical properties of Bi–Sb single crystal solid solutions grown by Czochralski method. *Proceedings of the XIV International Conference on Thermoelectrics, St. Petersburg*, pp. 37–41.

A. Borschchevsky (1995). In *CRC Handbook of Thermoelectrics*, edited by D. M. Rowe, CRC Press, p. 83.

N. B. Brandt and S. M. Chudinov (1971). Oscillation effects in semimetallic $Bi_{1-x}Sb_x$ alloys under pressure. *Sov. Phys. JETP*, 32(5), pp. 815–822.

N. B. Brandt and E. A. Svistova (1970). Electron transitions in strong magnetic fields. *J. Low Temp. Phys.*, 2(1), pp. 1–35.

N. B. Brandt, E. A. Svistova, and R. G. Valeev (1969). Investigation of the semiconductor-metal transition in the bismuth–antimony system in a magnetic field. *Sov. Phys. JETP*, 28(2), pp. 245–254.

N. B. Brandt, E. A. Svistova, and M. V. Semenov (1971). Electron transitions in antimony-rich bismuth–antimony alloys in strong magnetic fields. *Sov. Phys JETP*, 32, p. 238.

N. B. Brandt, S. M. Chudinov, and V. G. Karavaev (1972a). Investigation of gapless states in bismuth–antimony alloys under pressure. *Sov. Phys. JETP*, 34(2), pp. 368–375.

N. B. Brandt, Ya. G. Ponomarev, and S. M. Chudinov (1972b). Investigation of the gapless state in bismuth–antimony alloys. *J. Low Temp. Phys.*, 8(5–6), pp. 369–420.

N. B. Brandt, Kh. Dittmann, and Ya. G. Ponomarev (1972c). Metal–semiconductor transitions in $Bi_{1-x}Sb_x$ alloys under pressure, *Sov. Phys. Solid State*, 13, 2408.

N. B. Brandt, S. M. Chudinov, and V. G. Karavaev (1976). Investigation of the zero-gap state induced by a magnetic field in bismuth–antimony alloys, *Sov. Phys. JETP*, 43, p. 1198.

N. B. Brandt, R. Hermann, G. I. Golysheva, L. I. Devyatkova, D. Kusnik, W. Kraak, and Ya. G. Ponomarev (1982). Electron Fermi surface of semimetallic alloys $Bi_{1-x}Sb_x$ ($0.23 \leqslant x < 0.56$), *Sov. Phys. JETP*, 56(6), pp. 1247–1256.

F. Brochin (2000). Ph.D. Thesis, INPL, Nancy.

D. M. Brown and F. K. Heumann (1964). Growth of bismuth–antimony single-crystal alloys, *J. Appl. Phys.*, 35(6), pp. 1947–1951.

D. M. Brown and S. J. Silverman (1964). Electrons in Bi–Sb alloys. *Phys. Rev.*, 136(1A), pp. 290–299.

P. W. Chao, H. T. Chu, and Y. H. Kao (1974). Nonlinear band-parameter variations in dilute bismuth–antimony alloys. *Phys. Rev. B*, 9(10), pp. 4030–4034.

K. D. Chaudhuri and T. K. Dey (1975). Heat conduction in bismuth–antimony alloy single crystals between 4.2 and 300 K. *J. Low Temp. Phys.*, 20(3/4), pp. 397–405.

G. Cochrane and W. V. Youdelis (1972). Transport and thermoelectric properties of bismuth and Bi-12 at. pct Sb alloy powder compacts. *Metal. Trans.*, 3, pp. 2843–2850.

M. Cook (1922). The antimony–bismuth system. *J. Inst. Met.*, 28, 421.

P. Cucka and C. S. Barrett (1962). The crystal structure of Bi and solid solutions of Pb, Sn, Sb, and Te in Bi. *Acta. Cryst.*, 15, pp. 865–872.

K. F. Cuff, R. B. Horst, J. L. Weaver, S. R. Hawkins, C. F. Kooi, and G. M. Enslow (1963). The thermomagnetic figure of merit and Ettingshausen cooling in Bi–Sb alloys. *Appl. Phys. Lett.*, 2(8), pp. 145–146.

Z. M. Dashevskii, N. A. Sidorenko, S. Ya. Skipidarov, N. A. Tsvetkova, and A. B. Mocolov (1991). Cryogenic thermoelectric coolers with passive high T_c superconductor legs. *Proceedings of the X International Conference on Thermoelectrics, Cardiff*, pp. 142–146.

A. Demouge, B. Lenoir, Yu. I. Ravich, H. Scherrer, and S. Scherrer (1995). Estimation of carrier mobilities and densities in $Bi_{0.96}Sb_{0.04}$ alloys from galvanomagnetic coefficients. *J. Phys. Chem. Solids*, 56, 1155.

X. Devaux, F. Brochin, A. Dauscher, B. Lenoir, R. Martin-Lopez, H. Scherrer, and S. Scherrer (1997a). Production of ultrafine powders of Bi–Sb solid solution. *Nanostruct. Mat.*, 8, p. 137.

X. Devaux, F. Brochin, A. Dauscher, B. Lenoir, R. Martin-Lopez, H. Scherrer, and S. Scherrer (1997b). Observation of the grain size influence on the thermoelectric properties of polycrystalline bismuth–antimony alloys. *Proceedings of the XVI International Conference on Thermoelectrics, Dresden*, p. 199.

M. S. Dresselhaus (1971). Electronic properties of group V semimetals. *J. Phys. Chem. Sol.*, 32, Suppl. 1, pp. 3–33.

M. Dugué (1965). Propriétés électriques des solutions solides bismuth–antimoine. *Phys. Stat. Sol.*, 11, pp. 149–158.

V. S. Edelman (1976). Electrons in bismuth. *Adv. Phys.*, 25(6), pp. 555–613.

M. R. Ellett, R. B. Horst, L. R. Williams, and K. F. Cuff (1996). Shubnikov–de Haas investigations of the $Bi_{1-x}Sb_x$ ($0 < x < 0.3$) system. *J. Phys. Soc. Jpn.*, 21(Suppl.), pp. 666–672.

M. G. Fee (1993). Peltier refrigerator using a high T_c superconductor. *Appl. Phys. Lett.*, 62(10), pp. 1161–1163.

Y. Feutelais, G. Morgant, J. R. Didry, and J. Schnitter (1992). Thermodynamic evaluation of the system bismuth–antimoine. CALPHAD, 16, p. 111.

C. F. Gallo, B. S. Chandrasekhar, and P. H. Sutter (1963). Transport properties of bismuth single crystals. *J. Appl. Phys.*, 34(1), pp. 144–152.

H. J. Goldsmid (1995). In *CRC Handbook of Thermoelectrics*, edited by D. M. Rowe. CRC Press, p. 75.

H. J. Goldsmid and A. W. Penn (1968). Boundary scattering of phonons in solid solutions, *Phys. Lett.*, 27A, p. 523.

H. J. Goldsmid, K. K. Gopinathan, D. N. Matthews, K. N. R. Taylor, and C. A. Baird (1988). High T_c superconductors as passive thermoelements. *J. Phys. D*, 21, p. 344.

H. J. Goldsmid, H. B. Lyon, and E. H. Volckmann (1995). A simplified theory of phonon boundary scattering in solid solutions. *Proceedings of the XIV International Conference on Thermoelectrics, St. Petersburg*, pp. 16–19.

S. Golin (1968). Band model for bismuth–antimony alloys. *Phys. Rev.*, 176(3), pp. 830–832.

V. M. Grabov, G. A. Ivanov, V. L. Naletov, and A. F. Panarin (1985). Interband scattering of charge carriers and thermoelectric figure of merit of bismuth–antimony alloys. *Soviet Seminar of Semiconductor Materials for Thermoelectric Conversion, Leningrad*, pp. 30–31 (in Russian).

V. M. Grabov, G. A. Ivanov, V. L. Naletov, M. G. Bondarenko, and O. N. Uryupin (1995). Thermoelectric figure of merit of horizontal zone-leveling prepared bismuth–antimony single crystals. *Proceedings of the XIV International Conference on Thermoelectrics, St. Petersburg*, pp. 115–118.

V. M. Grabov, O. N. Uryupin, and V. A. Komarov (1998). Thermoelectric properties of polycrystalline bismuth and bismuth antimony alloys. *Proceedings of the XVII International Conference on Thermoelectrics, Nagoya*, p. 138.

O. S. Gryaznov, G. A. Ivanov, B. Ya. Moizhes, V. N. Naumov, V. A. Nemchinskii, N. A. Rodionov, and N. A. Red'ko (1982). The effect of interband scattering on kinetic phenomena in p-type $Bi_{1-x}Sb_x$. *Sov. Phys. Solid State*, 24(8), pp. 1326–1330.

M. Hansen and K. Anderko (1958). In *Constitution of Binary Alloys*, p. 332. McGraw Hill, New York.

R. Hartman (1969). Temperature dependence of the low field galvanomagnetic coefficients of bismuth. *Phys. Rev.*, 181, p. 107.

K. Hiruma, G. Kido, K. Kawauchi, and N. Miura (1980). Shubnikov–de Haas effect and semimetal–semiconductor transition in bismuth–antimony alloys in high magnetic fields. *Solid State Comm.*, 33, pp. 257–260.

R. B. Horst and L. R. Williams (1980a). Application of solid state cooling to spaceborne infrared focal planes. *Proceedings of the III International Conference on Thermoelectric Conversion, Arlington*, pp. 183–199.

R. B. Horst and L. R. Williams (1980b). Potential figure of merit of the BiSb alloys. *Proceedings of the III International Conference on Thermoelectric Conversion, Arlington*, pp. 139–173.

K. Hüttner and G. Tamman (1905). Über die Ligierungen des Antimons und Wismuts. *Z. Anorg. Chem.*, 44, p. 131.

A. F. Ioffe (1957). Semiconductor Thermoelements and Thermoelectric Cooling. Infosearch, London, p. 39.

G. A. Ivanov and A. M. Popov (1964). Electrical properties of bismuth–antimony alloys. *Sov. Phys. Solid State*, 5(9), pp. 1754–1761.

G. A. Ivanov, V. A. Kulikov, V. L. Naletov, A. F. Panarin, and A. R. Regel (1973). Thermoelectric figure of merit of pure and doped bismuth–antimony alloys in magnetic fields. *Sov. Phys. Semicond.*, 6(7), pp. 1134–1137.

D. M. Jacobson (1973). Magnetoresistance anisotropy in bismuth and bismuth–antimony crystals. *Phys. Stat. Sol. (b)*, 58, pp. 243–250.

A. L. Jain (1959). Temperature dependence of the electrical properties of bismuth–antimony alloys. *Phys. Rev.*, 114(6), pp. 1518–1528.

P. Jandl and U. Birkholz (1994). Thermogalvanomagnetic properties of Sn-doped $Bi_{95}Sb_5$ and its application for solid state cooling. *J. Appl. Phys.*, 76, p. 7351.

H. J. Juretschke (1955). Symmetry of galvanomagnetic effects in antimony. *Acta. Cryst.*, 8, p. 716.

V. D. Kagan and N. A. Red'ko (1991). Phonon thermal conductivity of bismuth alloys. *Sov. Phys. JETP*, 73(4), pp. 664–671.

V. D. Kagan and N. A. Red'ko (1995). Phonon thermal conductivity of the thermoelectric Bi–Sb alloys. *Proceedings of the XIV International Conference on Thermoelectrics*, St. Petersburg, pp. 78–81.

Y. H. Kao (1963). Cyclotron resonance studies of the Fermi surfaces in bismuth. *Phys. Rev.*, 129, p. 1122.

P. G. Klemens (1955). The scattering of low frequency lattice waves by lattice imperfections. *Proc. Phys. Soc. A*, 68, 1113.

H. J. Koh, P. Rudolph, and T. Fukuda (1995). Growth of $Bi_{1-x}Sb_x$ mixed crystals by a new melt injection technique. *J. Cryst. Growth*, 154, 151–155.

N. V. Kolomoetz, S. Ya. Skipidarov, N. A. Sidorenko, and R.S. Erofeev (1990). Cryogenic temperature level thermoelectric coolers. In *Proceedings of the IX International Conference on Thermoelectrics*, Pasadena, pp. 128–135.

W. Kraak, G. Oelgart, G. Schneider, and R. Herrmann (1978). The semiconductor–semimetal transition in $Bi_{1-x}Sb_x$ alloys with $x \geq 0.22$. *Phys. Stat. Sol. (b)*, 88, pp. 105–110.

R. Kuhl, W. Kraak, H. Haefner, and R. Herrmann (1976). The semimetal–semiconductor transition in bismuth–antimony alloys. *Phys. Stat. Sol. (b)*, 77, K109–K111.

V. L. Kuznetsov, M. V. Vedernikov, P. Yandl, and U. Birkholz (1994). Exploring the limits of thermoelectric refrigeration at liquid nitrogen temperatures. *Tech. Phys. Lett.*, 20(9), 757–758.

B. Lax and J. G. Mavroides (1960). In *Advances in Solid State Physics* (F. Seitz and D. Turnbull, eds.), Vol. 11, p. 261. Academic Press, New York.

B. Lenoir, A. Demouge, D. Perrin, H. Scherrer, S. Scherrer, M. Cassart, and J.-P. Michenaud (1995). Growth of $Bi_{1-x}Sb_x$ alloys by the traveling heater method. *J. Phys. Chem. Solids*, 56, 99.

B. Lenoir, M. Cassart, J.-P. Michenaud, H. Scherrer, and S. Scherrer (1996). Transport properties of Bi-rich Bi–Sb alloys. *J. Phys. Chem. Solids*, 57, 89.

B. Lenoir, M. O. Selme, A. Demouge, H. Scherrer, Yu. V. Ivanov, and Yu. I. Ravich (1998a). Electron and hole transport in undoped $Bi_{0.96}Sb_{0.04}$ alloys. *Phys. Rev. B*, 57, 11242.

B. Lenoir, A. Dauscher, M. Cassart, Yu. I. Ravich, and H. Scherrer (1998b). Effect of antimony content on the thermoelectric figure of merit of $Bi_{1-x}Sb_x$ alloys. *J. Phys. Chem. Solids*, 59, 129.

R. Martin-Lopez, B. Lenoir, A. Dauscher, X. Devaux, H. Scherrer, S. Scherrer, and M. Zandona (1995). Bi–Sb semiconductor alloy synthesized by mechanical alloying. In *Proceedings of the Second European Workshop on Thermoelectrics*, Nancy, p. 34.

R. Martin-Lopez, B. Lenoir, A. Dauscher, X. Devaux, W. Dümmler, H. Scherrer, M. Remy, and M. Zandona (1997). BiSb semiconducting alloy synthesized by mechanical alloying. *Scripta Mat.* 37, 219.

R. Martin-Lopez, B. Lenoir, X. Devaux, A. Dauscher, and H. Scherrer (1998a). Mechanical alloying of BiSb semiconducting alloys. *Mat. Sci. Eng. A*, 248, 147.

R. Martin-Lopez, Z. Zayakin, B. Lenoir, F. Brochin, A. Dauscher, and H. Scherrer (1998b). Mechanical properties of extruded $Bi_{85}Sb_{15}$ alloy prepared by mechanical alloying. *Philos. Mag. Lett.*, 78, 283.

R. Martin-Lopez, A. Dauscher, H. Scherrer, J. Hejtmanek, H. Kenzari, and B. Lenoir (1999). Thermoelectric properties of mechanically alloyed Bi–Sb alloys. *Appl. Phys. A*, 68, 657.

J. W. McClure and K. H. Choi (1977). Energy band model and properties of electrons in bismuth. *Solid State Commun.* 21, 1015.

E. E. Mendez (1979). Ph.D. Thesis, MIT.

E. E. Mendez, A. Misu, and M. S. Dresselhaus (1981). Pressure-dependent magnetoreflection studies of Bi and $Bi_{1-x}Sb_x$ alloys. *Phys. Rev. B*, 24(2), 639–648.

J.-P. Michenaud and J.-P. Issi (1972). Electron and hole transport in bismuth. *J. Phys. C*, 5, 3061–3072.

V. A. Nemchinskii and Yu. I. Ravich (1991). Transport effects in bismuth at 77 K. *Sov. Phys. Solid State*, 33, 1165.

G. Oelgart and R. Herrmann (1976). Cyclotron masses in semiconducting $Bi_{1-x}Sb_x$ alloys. *Phys. Stat. Sol. (b)*, 75, pp. 189–196.

G. Oelgart, G. Schneider, W. Kraak, and R. Herrmann (1976). The semiconductor–semimetal transition in $Bi_{1-x}Sb_x$ alloys. *Phys. Stat. Sol. (b)*, 74, K75–K78.

T. Okumura, M. Yamashita, and Y. Kibayashi (1991). Ga doped n-type Bi–Sb alloys prepared by sintering process. In *Proceedings of the X International Conference on Thermoelectrics, Cardiff*, pp. 44–48.

D. A. Petrov and V. M. Glasov (1985). High-speed determination of the solidus line in the phase diagram of the system Bi–Sb for ultrarapidly cooled melts. *Dokl. Phys. Chem.*, 283, 863.

S. D. Probert and C. B. Thomas (1979). Transport properties of some bismuth–antimony alloys. *Applied Energy*, 5, 127–140.

Yu. I. Ravich and A. V. Rapoport (1992). Temperature dependences of the carrier mobilities and densities in bismuth. *Sov. Phys. Solid State*, 34(6), 960–963.

Yu. I. Ravich, Yu. I. Ivanov, and A. V. Rapoport (1995). Thermogalvanomagnetic effects in bismuth. *Semiconductors*, 29, 458.

N. A. Red'ko (1990). Thermal conductivity of bismuth alloys under conditions of combined phonon–impurity scattering of phonons. *Sov. Tech. Phys. Lett.*, 16(11), 868–869.

N. A. Red'ko (1995). Thermoelectric efficiency of semiconducting Bi–Sb alloys. In *Proceedings of the XIV International Conference on Thermoelectrics, St. Petersburg*, pp. 82–84.

N. A. Red'ko, V. I. Belitskii, V. V. Kosarev, N. A. Rodionov, and V. I. Pol'shin (1986). Bands of heavy holes and sign of the thermoelectric power of Bi–Sb alloys. *Sov. Phys. Solid State*, 28(12), pp. 2111–2112.

N. A. Rodionov, N. A. Red'ko, and G. A. Ivanov (1979). Transport effects in $Bi_{0.88}Sb_{0.12}$ alloys with low hole densities in the L_s band. *Sov. Phys. Solid State*, 21(9), 1473–1476.

N. A. Rodionov, G. A. Ivanov, and N. A. Red'ko (1981). Anomalous temperature dependence of the thermoelectric power of holes in semiconducting $Bi_{1-x}Sb_x$ alloys. *Sov. Phys. Solid State*, 23(7), 1231–1234.

N. A. Rodionov, G. A. Ivanov, and N. A. Red'ko (1982). Thermoelectric efficiency of $Bi_{1-x}Sb_x$ ($0.12 \leqslant x\, 0.14$) p-type alloys at low temperatures. *Sov. Phys. Solid State*, 24(6), 1074–1075.

B. Rönnlund, L. Ericsson, and O. Beckman (1965). Low field galvanomagnetic effects on Bi–Sb alloys at 4.2 K. *Arkiv Fysik*, 29, 237.

D. Schiferl and C. S. Barrett (1969). The crystal structure of arsenic at 4.2, 78 and 299°K, *J. Appl. Cryst.*, 2, 30.

M. A. Short and J. J. Schott (1965). Preparation of homogeneous single-crystal bismuth–antimony alloys. *J. Appl. Phys.*, 36, pp. 659–660.

N. A. Sidorenko (1994). Brittle thermoelectric semiconductors extrusion under high hydrostatic pressure. In *Proceedings of the XIII International Conference on Thermoelectrics, Kansas City*, pp. 260–266.

N. A. Sidorenko and A. B. Mosolov (1992). Cryogenic thermoelectric coolers with passive high-T_c superconducting legs. In *Proceedings of the XI International Conference on Thermoelectrics, Arlington*, pp. 289–298.

G. E. Smith and R. Wolfe (1962). Thermoelectric properties of bismuth–antimony alloys. *J. Appl. Phys.*, 33(3), 841–846.

Y. Suse, Y. H. Lee, H. Morimoto, T. Koyonagi, K. Matsubara, and A. Kawamoto (1993). Structure and thermoelectric properties of $Bi_{88}Sb_{12}$ ceramics using fine particles produced by hydrogen arc-plasma. In *Proceedings of the XII International Conference on Thermoelectrics, Yokohama*, pp. 248–251.

M. M. Tagiev, Z. F. Agaev, and D. Sh. Abdinov (1992). Thermoelectric properties of extruded specimens of $Bi_{0.85}Sb_{0.15}$ doped with lead. *Neorg. Mater.* 29(6), 868–869.

S. Tanuma (1959). Semiconducting properties of Bi–Sb alloys. *J. Phys. Soc. Japan*, 14, 1246.

C. B. Thomas and H. J. Goldsmid (1970). Thermogalvanomagnetic effects in tellurium-doped $Bi_{95}Sb_5$. *J. Phys. C: Solid State Phys.*, 3, 696.

E. J. Tichovolsky and J. G. Mavroides (1969). Magnetoreflection studies on the band structure of bismuth–antimony alloys. *Solid State Commun.*, 7, 927–931.

W. A. Tiller, K. A. Jackson, J. W. Rutter, and B. Chalmers (1953). The redistribution of solute atoms during the solidification of metals. *Acta Met.*, 1, 428–437.

C. Uher and H. J. Goldsmid (1974). Separation of the electronic and lattice thermal conductivities in bismuth crystals. *Phys. Stat. Sol.*, 65, 765–772.

M. P. Vecchi and M. S. Dresselhaus (1974). Temperature dependence of the band parameters of bismuth. *Phys. Rev. B*, 10(2), 771–774.

M. V. Vedernikov, V. L. Kuznetsov, A. V. Ditman, B. T. Melekh, and A. T. Burkov (1991). Efficient thermoelectric cooler with a thermoelectrically passive, high-T_c superconducting leg. In *Proceedings of the X International Conference on Thermoelectrics, Cardiff*, pp. 96–101.

V. N. Vigdorovich, G. A. Ukhlinov, and N. Yu. Dolinskaya (1973). Construction of the solidus in phase diagrams for alloys with low intercystalline liquation. *Ind. Lab.*, 39, 242.

E. H. Volckmann, H. J. Goldsmid, and J. Sharp (1996). Observation of the effect of grain size on the lattice thermal conductivity of polycrystalline bismuth antimoine. In *Proceedings of the XV International Conference on Thermoelectrics, Pasadena*, p. 22.

T. H. Weber and K. Cruse (1959). *Z. Anal. Chem.*, 166, 333.

R. Wolfe and G. E. Smith (1962). Effects of a magnetic field on the thermoelectric properties of a bismuth–antimony alloy. *Appl. Phys. Lett.*, 1(1), 5–7.

T. Yazaki (1968). Thermal conductivity of bismuth–antimony alloy single crystals. *J. Phys. Soc. Jpn.*, 25(4), 1054–1060.

T. Yazaki and Y. Abe (1968a). Galvanomagnetic investigations of the $Bi_{1-x}Sb_x$ ($0 < x \leqslant 0.15$) system at 77°K. *J. Phys. Soc. Jpn.*, 24(2), 290–295.

T. Yazaki and Y. Abe (1968b). Galvanomagnetic determinations of band parameters of $Bi_{97}Sb_3$ and $Bi_{90}Sb_{10}$ at 300°K. *J. Phys. Soc. Jpn.*, 25, 633.

W. M. Yim and A. Amith (1972). Bi-Sb alloys for magneto–thermoelectric and thermomagnetic cooling. *Solid-State Electron.* 15, pp. 1141–1165.

W. M. Yim and J. P. Dismukes (1966). Growth of homogeneous Bi-Sb alloy single crystals. In *Proceedings of the International Conference on Crystal Growth, Supplement to Physics and Chemistry of Solids, Boston*, edited by H. S. Speiser, pp. 187–195.

V. S. Zemskov, A. D. Belaya, and G. N. Kozhemyakin (1985). Growth of single crystals of bismuth–antimony alloys by Czochralski method. *J. Crystal Growth*, 71, pp. 243–245.

CHAPTER 5

Skutterudites: Prospective Novel Thermoelectrics

Ctirad Uher

DEPARTMENT OF PHYSICS
UNIVERSITY OF MICHIGAN
ANN ARBOR, MICHIGAN

I. INTRODUCTION	139
II. STRUCTURAL ASPECTS AND BONDING	141
1. Binary Skutterudites	141
2. Filled Skutterudites	149
III. BAND STRUCTURE	156
IV. VIBRATIONAL PROPERTIES	167
1. Phonon Modes and the Density of States	167
2. Evidence of "Rattling"	170
V. SAMPLE PREPARATION ASPECTS	174
VI. MAGNETIC PROPERTIES	180
1. Mössbauer Spectroscopy	181
2. Magnetic Susceptibility	184
VII. TRANSPORT PROPERTIES	196
1. Electronic Transport	196
2. Thermal Conductivity	214
VIII. CONCLUSIONS	246
REFERENCES	247

I. Introduction

As the search for novel, promising thermoelectrics intensifies, skutterudites have emerged as prospective candidates for achieving figures of merit well in excess of unity. Exceptionally high power factors, together with a realistic prospect of a significant reduction in the lattice thermal conductivity, have attracted great attention, and many laboratories worldwide are focusing efforts on optimizing the material in order to achieve high thermoelectric efficiency.

A question that is usually asked first is what the name "skutterudite" means and where it comes from. The name "skutterudite" is derived from a small mining town in Norway called Skutterud where a $CoAs_3$-based mineral had been mined extensively. The compounds, identical in structure to $CoAs_3$ — the structure first identified by Oftedal in 1928 — have since become known as skutterudites. Skutterudite structure offers interesting possibilities to alter both the electronic and lattice properties. A clever materials scientist can exploit this potential and synthesize a superior thermoelectric material.

The early interest in skutterudites dates back to the mid-1950s, when the researchers in the then Soviet Union screened a vast number of compounds and alloys in their search for the most promising thermoelectrics. At that time the only known forms of skutterudites were the binary compounds with the $CoAs_3$ structure. It soon became obvious that, in spite of their outstanding electronic properties, notably exceptionally high mobilities and modest Seebeck coefficients, skutterudites possessed too high a thermal conductivity to make them an interesting thermoelectric material. The fundamental requirement placed on a good thermoelectric — a large thermoelectric figure of merit defined as $Z = S^2\sigma/\kappa$, where S is the Seebeck coefficient, σ the electrical conductivity, and κ the thermal conductivity — was impossible to achieve with the binary skutterudites.

In the mid-1970s, Professor Jeitschko and his colleagues realized that the open structure of skutterudites, typified by the presence of two large voids in the unit cell, might be able to accept foreign ions that would effectively "fill" the structure. Experiments with the lanthanoid group of elements resulted in the synthesis of a large number of filled skutterudites.

Since the early 1990s, and following a major rejuvenation of interest in thermoelectric materials for both cooling and power generation applications, an extensive critical assessment of prospective materials conducted by a team of researchers at the Jet Propulsion Laboratory in Pasadena led to a rediscovery of skutterudites as prime candidates for novel thermoelectrics. In 1994, Dr. Slack proposed the concept of phonon glass electron crystal (PGEC) as one of the desirable features a material should possess to maximize the thermoelectric figure of merit. Realizing that filled skutterudites — via a very disruptive influence of the filler ions on the lattice thermal conductivity — are the canonical example of the PGEC concept, Morelli and Meisner demonstrated a dramatic suppression of the lattice thermal conductivity in $CeFe_4Sb_{12}$.

Theoretical and experimental effort worldwide is generating exciting new results, and it is time to summarize the current understanding of the physical properties of skutterudites and attempt to assess their prospect as a viable thermoelectric material. A brief summary of the research into skutterudites prepared by Drs. Nolas, Morelli, and Tritt has appeared recently in the Annual Review of Materials Science (Nolas *et al.*, 1999). In my review of the field, I hope both to provide an in-depth account of the relevant physical, chemical, and materials issues pertaining to skutterudites and to include the latest exciting developments. I trust this work will serve as a reference material to many researchers in the field.

II. Structural Aspects and Bonding

1. BINARY SKUTTERUDITES

Binary skutterudites are structures with the general formula MX_3 where M is one of the group 9 transition metals Co, Rh, or Ir and X stands for P, As, or Sb. Binary skutterudites form with all nine possible combinations of the M and X elements and crystallize with the body-centered-cubic structure in the space group $Im3$, first identified by Oftedal in 1928. The unit cell contains 32 atoms arranged in eight groups of MX_3 blocks. In terms of the crystallographic designation applicable to the $Im3$ space group, the metal atoms occupy the c-sites and the pnicogen[1] atoms occupy the g-sites of the structure (Fig. 1). From the structural and bonding perspective, a more informative picture of the skutterudite phase consists of an infinite three-dimensional array of trigonally distorted and tilted MX_6 octahedrons that share corners with six neighboring octahedra (Fig. 2). In this arrangement, the metal atom M lies in the center and is octahedrally coordinated by the pnicogen atoms. It should be noted that the metal atoms are far apart from each other and no atom M has another metal atom as its nearest neighbor. Tilting of the octahedra gives rise to a formation of planar rectangular

---

[1]The term pnicogen (and pnictides) is often used for the elements of the nitrogen family—N, P, As, Sb, and Bi. Its root is the Greek word meaning to choke, and the allusion to the Germanic name for nitrogen, Stickstoff, or choking gas. In the context of skutterudites, the pnicogens are limited to P, As, and Sb.

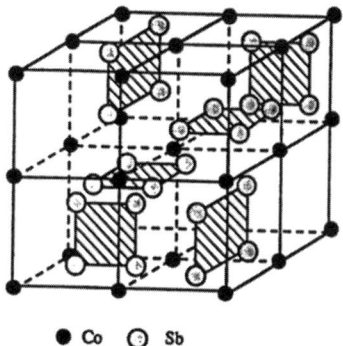

FIG. 1. The unit cell of the binary skutterudite structure shifted by one-quarter distance along the body diagonal. The metal atoms (solid circles) form a simple cubic sublattice. The pnicogen atoms (open circles) are arranged in planar rectangular four-membered rings that form linear arrays along the (100), (010), or (001) crystallographic directions. There are six such pnicogen rings in the unit cell. Two of the eight small cubes are empty, giving rise to two voids (cages) per unit cell.

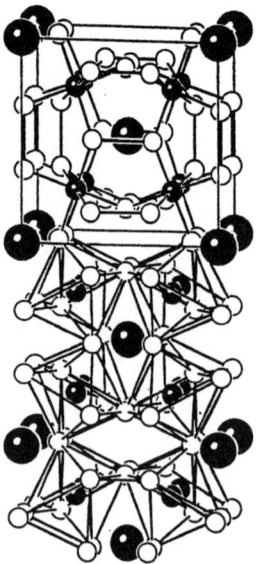

FIG. 2. Crystal structure of the filled skutterudite. Large black spheres represent the filler atoms, small black spheres are the metal atoms M, and open circles stand for the pnicogen atoms. The cubic cell is outlined in the upper part of the figure. The lower section of the figure emphasizes the octahedral pnicogen coordination of the M atoms. (From Danebrock et al., 1996.) Reprinted with permission from Elsevier Science.

four-membered rings of pnicogen atoms (X_4) that are a characteristic feature of the skutterudite structure. Since the pnicogen ring species are anions, one can classify skutterudites as polyanionic compounds. The rings form linear arrays along the (100), (010), or (001) crystallographic directions with adjacent rings being orthogonal to each other. The exact position of the pnicogen atoms is specified by two positional parameters y and z. Thus, the skutterudite structure is completely determined by giving the lattice parameter and the two positional parameters y and z. In terms of the pnicogen positional parameters and the lattice constant, the important interatomic distances are (Rundqvist and Ersson, 1968; Rosenqvist, 1953):

$$d(M - X) = a[(\tfrac{1}{4})^2 + (y - \tfrac{1}{4})^2 + (z - \tfrac{1}{4})^2]^{1/2} \qquad (1)$$

$$d_1(X - X) = 2az \qquad (2)$$

$$d_2(X - X) = a(1 - 2y). \qquad (3)$$

In his original identification of the skutterudite structure, Oftedal predicted a *square* planar configuration of the pnicogen rings, i.e., the two pnicogen distances d_1 and d_2 being equal. From Eqs. (2) and (3) this leads to what is often called the Oftedal relation,

$$2(y + z) = 1. \qquad (4)$$

Assuming regular octahedral coordination for the metal atom M, the positional parameters y and z must further satisfy the condition

$$y(2z - 1) = z - \tfrac{3}{8}. \qquad (5)$$

The simultaneous solution of Eqs. (4) and (5) yields $y = z = \tfrac{1}{4}$, the condition that signifies a structural change from the skutterudite structure to that of the more symmetric ReO_3-type structure with only one MX_3 group per unit cell and the space group *Pm3m*. Hence, to stay within the skutterudite phase, the constraints imposed by Eqs. (4) and (5) must not be viewed as rigid. As discussed by Rundqvist and Ersson (1968), violating either one of Eqs. (4) and (5) costs energy: A departure from the Oftedal relation is unfavorable to the X–X bonds, and a distortion of the ideal MX_6 octahedron environment is clearly detrimental to the M–X bonds. In fact, insisting on the Oftedal relation would make the square ring structure so large that virtually no bonding interaction would remain between the pnicogen atoms unless the M–X distance becomes unreasonably large. Consequently, rather than assessing a very large energy penalty for violation of only one of the constraints, it is less taxing to the structure to violate both constraints, each with only a modest energy penalty. This is exactly what happens — the octahedral MX_6 complex undergoes a slight trigonal antiprismatic distortion while the pnicogen rings assume a rectangular rather than square coordination.

The crystal structure determination of binary skutterudites of both synthetic and mineral forms was made by a number of research groups. Some of the studies assumed the validity of the Oftedal relation (Rosenqvist, 1953; Zhuravlev and Zhdanov, 1956), while the first hint of a distorted octahedron and the rectangular coordination of the pnicogen atoms were provided by Ventriglia (1957). Subsequent detailed studies (Rundqvist and Ersson, 1968; Rundqvist and Hede, 1960; Kjekshus and Pedersen, 1961; Mandel and Donohue, 1971; Kjekshus *et al.*, 1973; Kjekshus and Rakke, 1974; Schmidt *et al.*, 1987) have firmly established small deviations from the Oftedal relation for all binary skutterudites. Kjekshus *et al.* (1973) attributed rectangular distortions ($d_1 \neq d_2$) of the X_4 rings to their anisotropic environment — the presence of two cavities per MX_3 unit cell symmetrically arranged on opposite sides of the X_4 groups. Figure 3 provides graphic illustration of the Oftedal relation, the condition for ideal octahedral coordination, and the actual positional parameters of all known binary skutterudites. Table I collects the important structural parameters. It is interesting to point out a subtle trend in the positional parameters of Table I.

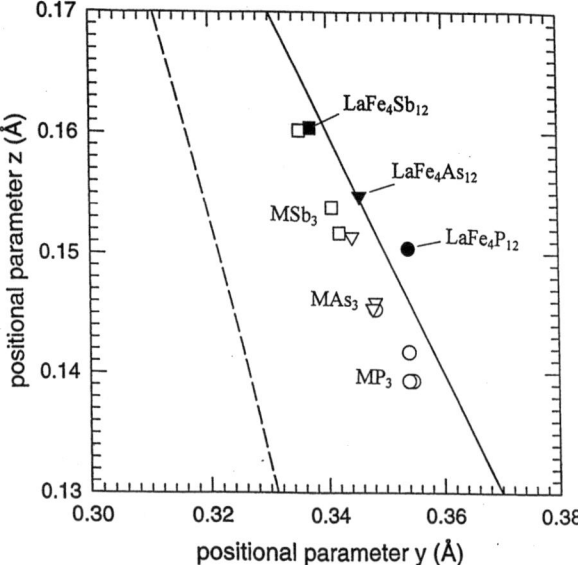

FIG. 3. Positional parameters for binary skutterudites (open circles for phosphides, open triangles for arsenides, and open squares for antimonides) and for La-based filled skutterudites. The solid line is the Oftedal relation, Eq. (4), and the dashed curve represents the ideal octahedral coordination, Eq. (5). The data are from Table I and Jeitschko and Brown (1977); Brown and Jeitschko (1980b); Evers et al. (1994).

TABLE I

STRUCTURAL PARAMETERS OF BINARY SKUTTERUDITES[a]

	Lattice constant (Å)	y (Å)	z (Å)	$d(M-X)$ (Å)	$d_1(X-X)$ (Å)	$d_2(X-X)$ (Å)	R(void) (Å)
CoP_3	7.7073	0.3482	0.1453	2.222	2.240	2.340	1.763
RhP_3	7.9951	0.3547	0.1393	2.341	2.227	2.323	1.909
IrP_3	8.0151	0.3540	0.1393	2.345	2.233	2.340	1.906
NiP_3	7.8192	0.3540	0.1417	2.280	2.216	2.283	—
$CoAs_3$	8.2055	0.3442	0.1514	2.337	2.478	2.560	1.825
$RhAs_3$	8.4507	0.3482	0.1459	2.434	2.468	2.569	1.934
$IrAs_3$	8.4673	0.3477	0.1454	2.441	2.456	2.574	1.931
$CoSb_3$	9.0385	0.3351	0.1602	2.520	2.891	2.982	1.892
$RhSb_3$	9.2322	0.3420	0.1517	2.621	2.807	2.917	2.024
$IrSb_3$	9.2503	0.3407	0.1538	2.617	2.850	2.943	2.040

[a]Data adapted from Kjekshus and Rakke (1974) and Nolas et al. (1996b).

Progressing from phosphides to arsenides to antimonides, one positional parameter (y) generally decreases while the other one (z) increases. This small gradual change leads to a higher coordination number of pnicogen atoms in the sequence MP_3, MAs_3, MSb_3 and, in turn, it results in a shift from more localized bonding in phosphides to more delocalized bonding in antimonides.

Addressing the issue of bonding, the key experimental input that has to be taken into account is the fact that the binary skutterudites are diamagnetic semiconductors, that is, the relevant bonding scheme must have no unpaired spins. It has already been noted that the distance between the metal atoms is large, and there are no nearest metal atoms to any given M atom. This implies that there is no significant M–M bonding in the skutterudite structure. Referring to Fig. 4, each pnicogen atom X has two other pnicogens as its nearest neighbors and it also bonds with two nearest metal atoms. The pnicogen ring structure X_4 holds together via σ bonds, i.e., each pentavalent pnicogen atom (ns^2np^3) contributes two valence electrons, one each to bond with its two nearest neighbor pnicogen atoms. The fact that the X–X–X bond angle is strictly 90° suggests that the dominant contribution comes from the p_x and p_y orbitals. The remaining three valence electrons participate in bonding with the two nearest metal atoms. Since there are six pnicogen atoms in octahedral coordination around each metal atom M, the pnicogens contribute the total of $(5 - 2) \times \frac{1}{2} \times 6 = 9$ electrons toward the MX_6 octahedral complex. This is just enough to engage nine valence electrons of the Co-like metal (d^7s^2) to form the 18-electron rare-gas configuration that favors diamagnetism and semiconducting behavior.

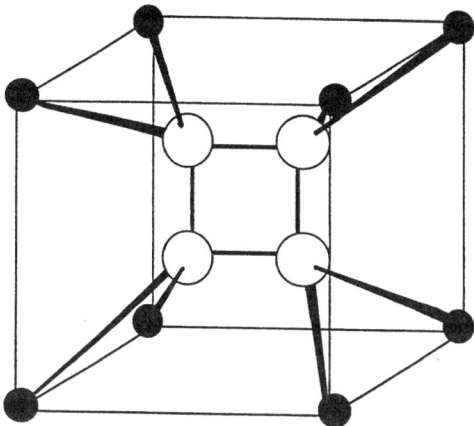

FIG. 4. Coordination of pnicogen atoms (white spheres) in the skutterudite structure. Pnicogen atoms form a planar rectangular cluster. Each pnicogen atom has four near neighbors: two pnicogen atoms and two metal atoms (dark spheres).

From the perspective of the metal atom M, it contributes $6 \times \frac{1}{2} = 3$ electrons for bonding with the 6 neighboring pnicogen atoms. These electrons occupy the octahedral d^2sp^3 hybrid orbitals that are the essence of the M–X bonding. The Co-like metal is thus left in the 3+ state with six nonbonding electrons that adopt the maximum spin-pairing configuration and therefore the low-spin d^6 state. It is also interesting to note that electronegativity of pnicogens is very close to that of the Co-like metals, and the M–X bond thus has only a small degree of ionic character.

The preceding bonding description is a modified version of one of the bonding models developed by Dudkin and his colleagues (Dudkin, 1958) in a series of papers published in the late 1950s. Although various aspects of this model have been criticized (Kuzmin, 1967), particularly its apparent omission to highlight the importance of the X_4 ring structure that some researchers view as pivotal for skutterudites, its basic premise remains valid and its predictive power served well during the past 40 years of research on these fascinating materials. Schematics of Dudkin's model are shown in Fig. 5. On the basis of this model, one can predict the character of the skutterudite compounds. As depicted in Fig. 5, there are no unpaired electrons in the cobalt family of skutterudites [Co(Rh,Ir)P$_3$, Co(Rh,Ir)As$_3$, and Co(Rh,Ir)Sb$_3$] and thus, indeed, these structures are diamagnetic semiconductors. The hypothetical iron family of binary skutterudites [Fe(Ru,Os)P$_3$, Fe(Ru,Os)As$_3$, and Fe(Ru,Os)Sb$_3$] with one less electron in their inner d-shells should be paramagnetic semiconductors. Unfortunately, this prediction is impossible to verify because such skutterudites have not yet been synthesized, nor do they exist in nature. In the case of the equally elusive nickel family of skutterudites [Ni(Pd,Pt)P$_3$, Ni(Pd,Pt)As$_3$, and Ni(Pd,Pt)Sb$_3$], the additional nonbonding electron has no choice but to be promoted into higher energy levels (possibly into the conduction band) leading to a likely scenario that such compounds would be paramagnetic

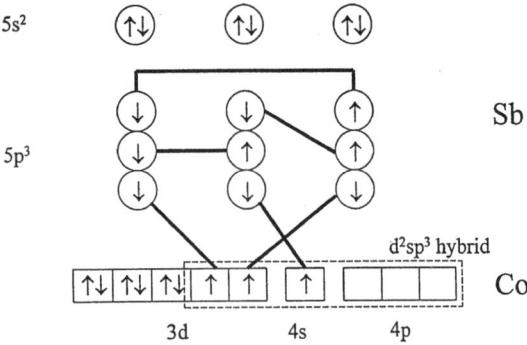

FIG. 5. Schematic illustration of Dudkin's bonding model. Adapted from Dudkin (1958).

metals. Again, with the exception of NiP_3 and possibly PdP_3, no binary skutterudites with the transition metals from the Ni column are known to exist. Although it is doubtful that PdP_3 is a stable compound (Rundqvist and Ersson, 1968), the existence of NiP_3 is firmly established (Jolibois, 1910; Biltz and Heimbrecht, 1938; Rundqvist and Larsson, 1959) and the delocalized seventh electron indeed leads to metallic conductivity and paramagnetism (Hulliger, 1961). It is interesting to note that Ni is somewhat "tolerated" in phosphide skutterudites, whereas it is distinctly less so in arsenides and even less in antimonides.

The inability to prepare pure binary skutterudites with transition metals other than those of the Co group does not mean that cobalt cannot be *partially* replaced by its immediate neighbors in the period table—iron and nickel. The skutterudite structure can accommodate such a partial replacement, and several studies have attempted to establish substitutional solid solution limits. It is important to realize that an elemental impurity will substitute for a component of the solid solution only if it is capable of forming the same bond as the component it replaces, and if the covalent radii of the two species are not too different. Apart from the number of d-electrons, iron and nickel have similar valence electron structure as cobalt. They can readily form octahedral bonds, and when replacing trivalent cobalt in the skutterudite structure, the valency of Fe and Ni will be such as to closely match the size of the covalent radius of Co^{3+} (1.22 Å). In the case of Ni it is clearly its tetravalent state Ni^{4+} (1.21 Å) that preserves the d^6 configuration and promotes a single electron into a conduction band, making nickel an electron donor. When iron substitutes for cobalt, there are two valence states of iron that have radii comparable to that of the trivalent cobalt: the divalent state Fe^{2+} (1.23 Å) and the trivalent state Fe^{3+} (1.22 Å). Quite apart from the fact that the divalent iron does not provide enough electrons for bonding, the trivalent ion exactly matches the size of the trivalent cobalt it replaces, and thus Fe^{3+} is the most likely configuration in the solid solution. This, however, leaves only five electrons in the d-shell and it is impossible to pair all spins. Substitution of Fe for Co is therefore likely to give rise to paramagnetism.

The actual solid solution limit in antimonide skutterudites was first established by Dudkin and Abrikosov (1959), who found that up to 10% of cobalt atoms can be replaced by nickel. In the case of iron this figure increases to near 25%. A somewhat wider boundary is possible if one allows for a slight deviation in pnicogen stoichiometry. In arsenide skutterudites the substitution limits are higher and the arsenide structure seems to prefer nickel to iron. Pleass and Heyding (1962) report 16% and 65% as the respective limits for Fe and Ni substitutions of Co. These authors confirmed the electron donor character of nickel and they detected paramagnetic signals in the Fe-substituted $CoAs_3$ compounds with the effective moment per iron atom of about $2\mu_B$, where μ_B is the Bohr magneton. This is in agreement with the presumed Fe^{3+} state and the presence of an unpaired electron in the nonbonding d-orbital. What is surprising, and contrary to the Dudkin's model that assumes "inactiveness" of iron atoms in $CoSb_3$, is the profound influence of iron on the conduction process. With as little as 1% of Fe substituted for Co, the structure attains metallic character.

An example of the solid solution field for arsenide skutterudites (Roseboom, 1962) is shown in Fig. 6. The phase diagram suggests a possibility of a coupled replacement of two cobalt atoms by one iron and one nickel atom:

$$2Co^{3+}(d^6) \rightleftarrows Fe^{2+}(d^6) + Ni^{4+}(d^6). \tag{6}$$

Structures with symmetrically substituted cobalt thus preserve the total number of electrons and all ions are in low-spin d^6 configuration. Magnetic measurements by Nickel (1969) confirmed that $(Fe_{0.45}Ni_{0.55})As_3$ is indeed a diamagnetic solid, and Pleass and Heyding (1962) observed semiconducting behavior in $(Fe_{0.46}Ni_{0.54})As_3$. This, again, does not conform to the Dudkin's model that would predict a strong paramagnetism due to the unpaired spin of Fe and metallic conduction on account of the conduction electron of Ni. It does appear as if nickel and iron form pairs (Fe^-Ni^+) equivalent to Co with the conduction electron of Ni occupying the nonbonding d-orbital of Fe.

In addition to giving rise to planar rectangular rings of pnicogen atoms, the tilt of the MX_6 octahedrons also creates large empty spaces (icosahedral voids) in the structure. Referring to the unit cell depicted in Fig. 1, we note that the X_4 rings are located in only six out of eight cubes, the two remaining cubes being empty. This is necessary to keep the ratio $Co^{3+}:[As_4]^{4-}$ equal to 4:3 and thus assure the overall charge neutrality of the structure. The voids occupy a body-centered position of the cubic lat-

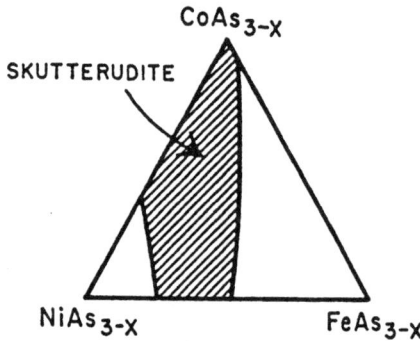

FIG. 6. Solid solution field for arsenide skutterudites at 800°C. The Ni and Fe end members have not been synthesized, nor do they exist as minerals. Adapted from Roseboom (1962).

tice, the so-called a-site. Thus, instead of designating the binary skutterudites as MX_3, one can equivalently write their chemical formula as $\square_2M_8X_{24} = 2\square M_4[X_4]_3$ where $\square$ highlights the presence of the void and $[X_4]$ reminds one of the presence of planar rectangular four-membered rings of pnicogen atoms. It is often convenient to consider only one-half of the unit cell including its single void and describe skutterudites as $\square M_4[X_4]_3$. In this scheme, each transition metal contributes nine electrons and each pnicogen contributes three electrons to the covalent bonding. The total valence electron count in the $\square M_4X_{12}$ complex is then 72 (or 144 for the unit cell). This valence electron count represents an alternative criterion for a semiconducting behavior of skutterudite structures. The voids in the structure are large enough to accommodate foreign species, and when occupied, such skutterudites are called filled skutterudites.

2. FILLED SKUTTERUDITES

The first filled skutterudite—$LaFe_4P_{12}$—was synthesized by Jeitschko and Brown in 1977. Subsequently, filled skutterudites were also prepared as arsenides (Brown and Jeitschko, 1980a) and antimonides (Brown and Jeitschko, 1980b). Since the mid-1990s, following their identification as promising novel thermoelectrics, interest in the physical properties of filled skutterudites has grown rapidly and there is now a large amount of data available in the literature. Filled skutterudites span a broad range of materials that include metals, semiconductors, and even heavy-fermion metals (Morelli and Meisner, 1995; Gajewski et al., 1998) and ferromagnets (Danebrock et al., 1996). It is also worth pointing out that among filled skutterudites are several superconductors (Meisner, 1981), and one of them, $LaRu_4As_{12}$, has a transition temperature as high as 10.3 K (Shirotani et al., 1997). Interestingly, one of the superconducting skutterudites is $PrRu_4As_{12}$, the structure filled with a magnetic rare earth. Table II lists

TABLE II

SUPERCONDUCTING FILLED SKUTTERUDITES AND THEIR TRANSITION TEMPERATURES

Skutterudite	$T_c(K)$	Reference
$LaFe_4P_{12}$	4.1	Meisner (1981)
$LaRu_4P_{12}$	7.2	Torikachvili et al. (1987)
$LaOs_4P_{12}$	1.8	Shirotani et al. (1996)
$LaRu_4As_{12}$	10.3	Shirotani et al. (1997)
$PrRu_4As_{12}$	2.4	Shirotani et al. (1997)
$LaRu_4Sb_{12}$	2.8	Shirotani et al. (1997)

some of the superconducting skutterudites and their respective transition temperatures.

Filled skutterudites are compounds of a general form RT_4X_{12}, where R stands for an electropositive filler that may be a lanthanide, actinide (Brown and Jeitschko, 1980c), or even alkaline-earth ion (Stetson et al., 1991; Evers et al., 1994), T represents the transition metal of the group 8 elements (T = Fe, Ru, Os), and X = P, As, or Sb. For the reader's convenience, a list of filled skutterudites including their lattice constants and cell volumes is given in Table III. It is interesting to note the trend in the lattice constant (or cell volume): It increases going from phosphides to arsenides to antimonides and, within each series, from iron to ruthenium to osmium. The divalent Eu and the alkaline earth filled skutterudites have generally larger lattice parameters. Most recently, even a monovalent thallium was successfully inserted into the voids of the skutterudite structure (Sales et al., 2000). Because of their large lattice constant, antimonides have the largest voids. However, because the positional parameters y and z in antimonides differ less than in arsenides and especially phosphides, the volume of the icosahedral void *per* lattice parameter is actually the smallest in antimonides. In Subsection 1 of Section II, I noted a gradual change in the positional parameters of the binary skutterudites toward a higher coordination number of the sequence MP_3, MAs_3, and MSb_3. This trend is also reflected in the positional parameters of the filled skutterudites. Figure 3 includes the data for $LaFe_4X_{12}$ (X = P, As, Sb), and other filled skutterudites conform to the behavior. As first pointed out by Brown and Jeitschko (1980b), the positional parameters of the antimony atoms in $LaFe_4Sb_{12}$ fall on the same side of Oftedahl's line as the parameters of the binary skutterudites. Therefore, the shortest X–X distances in $LaFe_4Sb_{12}$ and in all binary skutterudite compounds correspond to each other but do not correspond to the shortest P–P and As–As distances in $LaFe_4P_{12}$ and $LaFe_4As_{12}$, the positional parameters of which fall on the other side of Oftedahl's line.

Upon inspecting Table III, one notices a distinct lack of entries representing compounds filled with smaller (and heavier) rare earths. This should not be too surprising. Since the $[T_4X_{12}]^{4-}$ framework is rather rigid, small-radius ions cannot attain optimal bonding distances within the now much oversized cage, and competing phases may become more stable and thus prevent the nucleation of the desired skutterudite phase. In general, using conventional synthesis routes, rare-earth-filled skutterudites up to and including Eu can be prepared. By relying on the use of high pressure (several gigapascals) to aid the synthesis, it has been demonstrated (Sekine et al., 1998) that rare earths such as Gd and Tb can also be inserted into the voids of the skutterudite structure.

Whereas binary skutterudites with the group 8 transition metals do not exist, in the filled skutterudites the filler ion R is supposed to supply the missing electron of the iron group elements and provide enough electrons

TABLE III

LATTICE PARAMETERS AND CELL VOLUMES FOR FILLED SKUTTERUDITES

Compound	a_o(Å)	V(Å^3)	Reference[a]
LaFe$_4$P$_{12}$	7.8316	480.3	1
CeFe$_4$P$_{12}$	7.7920	473.1	1
PrFe$_4$P$_{12}$	7.8149	477.3	1
NdFe$_4$P$_{12}$	7.8079	476.0	1
SmFe$_4$P$_{12}$	7.8029	475.1	1
EuFe$_4$P$_{12}$	7.8055	475.6	1
ThFe$_4$P$_{12}$	7.7999	474.5	2
LaRu$_4$P$_{12}$	8.0561	522.9	1
CeRu$_4$P$_{12}$	8.0376	519.3	1
PrRu$_4$P$_{12}$	8.0420	520.1	1
NdRu$_4$P$_{12}$	8.0364	519.0	1
EuRu$_4$P$_{12}$	8.0406	519.8	1
ThRu$_4$P$_{12}$	8.0461	520.9	2
LaOs$_4$P$_{12}$	8.0844	528.4	1
CeOs$_4$P$_{12}$	8.0626	524.1	1
PrOs$_4$P$_{12}$	8.0710	525.7	1
NdOs$_4$P$_{12}$	8.0638	524.3	1
LaFe$_4$As$_{12}$	8.3252	577.0	3
CeFe$_4$As$_{12}$	8.2959	570.9	3
PrFe$_4$As$_{12}$	8.3125	574.4	3
LaRu$_4$As$_{12}$	8.5081	615.9	3
CeRu$_4$As$_{12}$	8.4908	612.1	3
PrRu$_4$As$_{12}$	8.4963	613.3	3
LaOs$_4$As$_{12}$	8.5437	623.6	3
CeOs$_4$As$_{12}$	8.5249	619.5	3
PrOs$_4$As$_{12}$	8.5311	620.9	3
NdOs$_4$As$_{12}$	8.5291	620.4	3
ThOs$_4$As$_{12}$	8.5183	618.1	2
LaFe$_4$Sb$_{12}$	9.1395	763.4	4
CeFe$_4$Sb$_{12}$	9.1350	762.3	4
PrFe$_4$Sb$_{12}$	9.1351	762.3	4
YbFe$_4$Sb$_{12}$	9.1580	768.1	5
CaFe$_4$Sb$_{12}$	9.1621	769.1	6
SrFe$_4$Sb$_{12}$	9.1782	773.1	6
BaFe$_4$Sb$_{12}$	9.2022	779.2	6
LaRu$_4$Sb$_{12}$	9.2700	796.6	4
CeRu$_4$Sb$_{12}$	9.2657	795.5	4
PrRu$_4$Sb$_{12}$	9.2648	795.3	4
NdRu$_4$Sb$_{12}$	9.2642	795.1	4
EuRu$_4$Sb$_{12}$	9.2824	799.8	4
SrRu$_4$Sb$_{12}$	9.2891	801.5	6
BaRu$_4$Sb$_{12}$	9.2022	779.2	6
LaOs$_4$Sb$_{12}$	9.3029	805.1	4
CeOs$_4$Sb$_{12}$	9.3011	804.6	4
PrOs$_4$Sb$_{12}$	9.2994	804.2	4
NdOs$_4$Sb$_{12}$	9.2989	804.1	4
SmOs$_4$Sb$_{12}$	9.3009	804.6	4
EuOs$_4$Sb$_{12}$	9.3187	809.2	4
SrOs$_4$Sb$_{12}$	9.3222	810.1	6
BaOs$_4$Sb$_{12}$	9.3401	814.8	6

[a] References: 1, Jeitschko and Brown (1977); 2, Brown and Jeitschko, 1980c; 3, Brown and Jeitschko, 1980a; 4, Brown and Jeitschko, 1980b; 5, Dilley et al. (1998); 6, Evers et al. (1994).

to saturate the bonds and stabilize the structure. Completely filled skutterudites (i.e., both voids are occupied) contain 34 atoms in the conventional unit cell.

From the perspective of thermoelectricity, the optimal void fillers would be neutral atoms with large atomic displacement parameters, perhaps something like Xe atoms. This would ensure a large reduction in the lattice thermal conductivity but only a minimal perturbation of electronic transport properties. Unfortunately, it is difficult to keep such rare-gas species "locked" in the voids of the skutterudite structure, and the attempts to do so have failed.

To discuss bonding in filled skutterudites, it is useful to consider the fundamental building block of the structure—the polyanionic complex $[T_4X_{12}]^{4-} \sim [TX_3]^{1-}$. This polyanion is electronically saturated just like the isostructural MX_3 block of the binary skutterudites, and the ironlike group 8 transition metal is in the low-spin d^6 configuration, just as is the group 9 cobaltlike element. Whereas the interactions between T–X and X–X are essentially covalent, the interaction of the R filler with the neighboring X atoms can be assumed as ionic. To be isoelectronic with the binary skutterudites, the polyanion $[T_4X_{12}]^{4-}$ should couple with a tetravalent R^{4+} ion, resulting in a diamagnetic and semiconducting filled skutterudite $R^{4+}[T_4X_{12}]^{4-}$ with a valence electron count of 72. This scenario is plausible only with Ce (provided Ce is in the 4+ state) among the rare-earth elements, and possibly Th (exclusively tetravalent) and U among the actinides. Cerium is substantially tetravalent in $CeFe_4P_{12}$ and $CeFe_4As_{12}$, but not in the corresponding antimonide $CeFe_4Sb_{12}$. The evidence for this is a large cell volume contraction in phosphide and arsenide skutterudites and the lack of it in antimonides (Grandjean et al., 1984). This is shown in Fig. 7, where the cell volume of $CeFe_4P_{12}$ is much smaller than the cell volume of its neighbors, $LaFe_4P_{12}$ and $PrFe_4P_{12}$. The same behavior is seen in arsenides. Furthermore, the changing valence of Ce as one goes from phosphides to antimonides is also reflected in the temperature dependence of resistivity. Thus, $CeFe_4P_{12}$ and $CeFe_4As_{12}$ have a distinctly semiconductorlike resistivity consistent with the predominantly tetravalent nature of Ce, valence electron count of 72, and fully saturated bonds. In contrast, $CeFe_4Sb_{12}$ shows the metallic (or semimetallic) conduction that is symptomatic of trivalent Ce. All other rare-earth ions favor a lower charge state than 4+, most often the 3+ state. Taking La as an example, all skutterudites of the form $La^{3+}[T_4X_{12}]^{4-}$ are somewhat electron deficient (valence electron count of 71) and should be paramagnets with metallic properties, as indeed they are. To achieve the desired electron count of 72 and thus to force the structure back into the semiconducting domain, one must charge-compensate for the void fillers. This can be done by partially replacing some of the pnicogen atoms or, alternatively, by replacing some of the metal atoms. An example of the former is

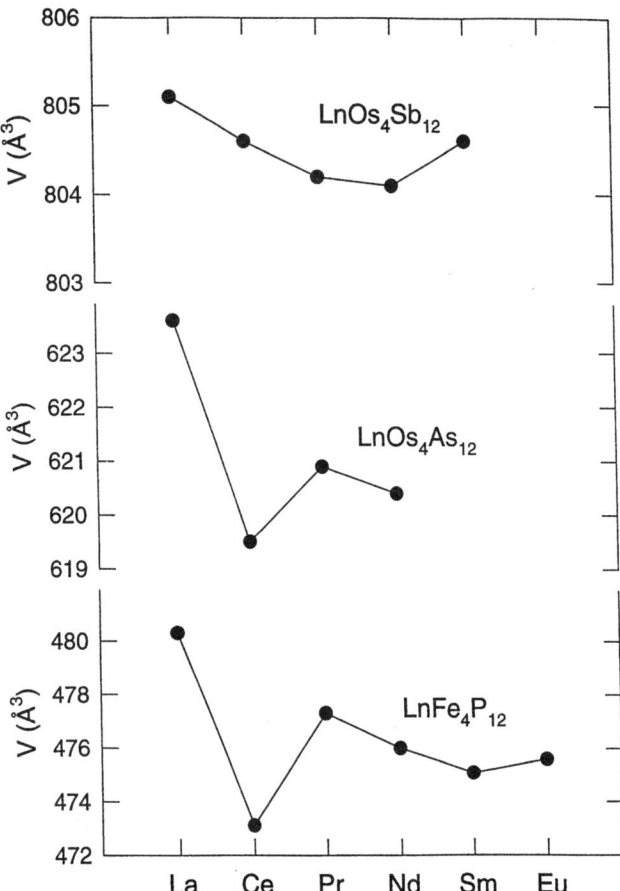

FIG. 7. Cell volumes of rare earth-filled skutterudites. Note a large volume contraction in Ce-filled phosphides and arsenides that is often interpreted as a sign of the 4+ charge state of Ce. The graph was constructed using the data in Table III.

$LaIr_4Ge_3Sb_9$ (Tritt et al., 1996), where Ge is added in an attempt to compensate for three electrons donated by La. Compensating on the metal site, a favored approach is to form compounds such as $CeFe_{4-x}Co_xSb_{12}$ (Fleurial et al., 1996a; Chen et al., 1997a), $LaFe_{4-x}Co_xSb_{12}$ (Sales et al., 1996), and even $CeFe_{4-x}Ni_xSb_{12}$, where Ni instead of Co substitutes for Fe (Chapon et al., 1999). By varying x one can, in principle, force the structure from the metallic to the semiconducting regime of conduction. In order to succeed, one must keep in mind that the atoms one compensates with must be compatible with the bonding environment and must have covalent radii that closely match the species they substitute for. When charge compensating on the metal site, one also must be aware of the possibility that the species might have intermediate valence and that the octahedral coordination with pnicogen atoms may stabilize more than one valence state. This is particularly important for elements such as Ru and Os that often assume both 2+ and 4+ charge states, and for Rh and Ir that can be in the 3+ and 5+ states.

In the early work and until about 1995, it was generally assumed that when preparing filled skutterudite alloys of correct stoichiometry, all voids of the structure will be occupied. This need not be the case. By a careful analysis of the variation of the lattice constant with the amount of the filler and in combination with the microprobe and chemical wet analysis, Chen et al. (1997a) observed that the amount of Ce the structure of $Ce_yFe_{4-x}Co_xSb_{12}$ can accommodate is controlled by the fraction of Co in the sample. While virtually all voids are occupied in $CeFe_4Sb_{12}$, the fractional occupancy of voids decreases with increasing Co content and, for pure $CoSb_3$, no more than 10% of voids are filled. Attempts to increase the Ce content failed as evidenced by minute amount of a second phase that nucleated and was detected by X-ray structural analysis and magnetic susceptibility. By painstakingly determining the critical Ce concentration across a wide range of Co compositions, the authors were able to establish the phase diagram shown in Fig. 8. Cerium is not the only rare earth for which fractional void occupancy is restricted; a similar constraint is imposed on La, except that here the fractional occupancy is higher, up to 23% of voids can be filled with La in pure $CoSb_3$ (Nolas et al., 1988). Constraints on the void occupancy exist also for species other than rare earths; for example, a monovalent thallium can fill up to 22% of the voids in $CoSb_3$ (Sales et al., 2000). Restrictions on the occupancy of voids apparently exist also in phosphide skutterudites. While attempting to prepare fully occupied $CeCo_4P_{12}$ and $LaCo_4P_{12}$, Zemni et al. (1986) instead prepared $Ce_{0.25}Co_4P_{12}$ and $La_{0.2}Co_4P_{12}$ as determined from refinements of their single crystal data.

In determining the structure of a new crystalline solid, an integral part of the task is to specify the so-called atomic displacement parameter (ADP) for each distinct atomic site. The ADP—in years past known as the thermal parameter—is a measure of the mean-square displacement amplitude of an atom about its equilibrium lattice site. Thus, the magnitude of the ADP depends on how "vigorously" the atoms are vibrating and possibly also on whether any static disorder is present on that site. In a crystalline solid, the ADPs reflect the underlying structural anisotropy. To come up with just a single parameter reflecting the displacement amplitudes in a solid, one converts the ADPs into an isotropic quantity U_{iso} (usually given in units of Å^2) that measures the mean-square displacement amplitude of the atom averaged over all directions. X-rays and, more recently, neutron diffraction have been the usual techniques of collecting the ADP data.

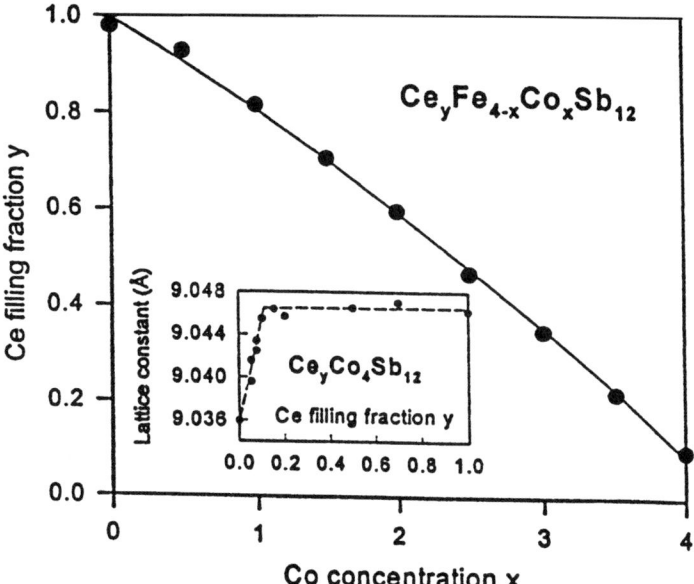

FIG. 8. Phase diagram showing the fractional void occupancy by Ce as a function of Co content in $Ce_yFe_{4-x}Co_xSb_{12}$. The inset shows the variation of the lattice constant of $Ce_yCo_4Sb_{12}$ with increasing cerium filling. A sharp break near $y = 0.1$ represents the maximum amount of cerium that can be accommodated in the cages for this compound. (Adapted from Chen *et al.* (1997a) and Uher *et al.* (1998). Copyright 1998 by IEEE.

From the time the first filled skutterudites were synthesized it was clear that the ADPs of the filler ions are unusually large in comparison to the other atoms of the structure. In fact, the term "rattling" was first used in the original paper of Brown and Jeitschko (1980b). It is well known that the ADPs serve as an estimate of the physical parameters such as the Debye temperature, lattice specific heat, and sound velocity. If we treat the "rattling" ion as a localized harmonic oscillator (Einstein oscillator), the mean square displacement amplitude $\langle u^2 \rangle$ is given by

$$U_{iso} \equiv \langle u^2 \rangle = \frac{h}{8\pi^2 mv} \coth \frac{hv}{2k_B T}, \qquad (7)$$

where m is the mass of the rattler and v is its frequency. At high temperatures Eq. (7) reduces to the classical formula $U_{iso} = k_B T/K$, where $K = m(2\pi v)^2$ is the force constant of the oscillator. Thus, from the slope of U_{iso} vs. T one can estimate the Einstein temperature $\theta_E = hv/k_B$. As an example of the ADPs, Fig. 9 depicts the U_{iso} for $La_{0.75}Fe_3CoSb_{12}$ that

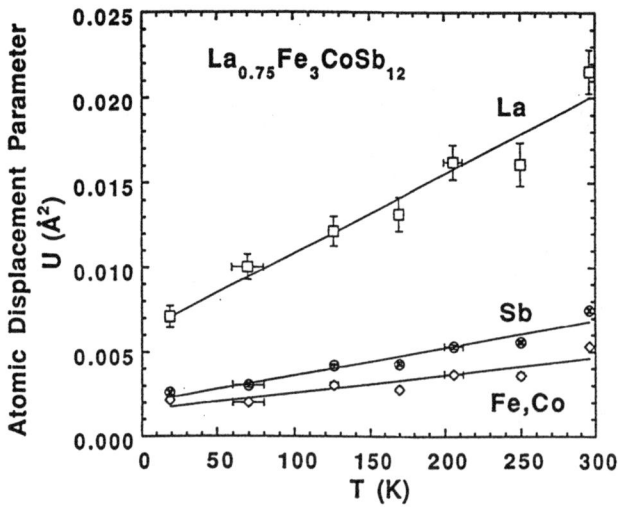

FIG. 9. Atomic displacement parameters of La, Fe, Co, and Sb in the filled skutterudite $La_{0.75}Fe_3CoSb_{12}$ (From Sales, 1998.)

clearly indicates an unusually large atomic displacement parameter for La (Sales, 1998). Recently, Dr. Sales and his colleagues at the Oak Ridge National Laboratory drew attention to the fact that the room temperature U_{iso} — the isotropic version of the ADPs — could be used to estimate thermal conductivity of solids such as skutterudites (Sales et al., 1999). The key point here is the use of the kinetic formula for thermal conductivity ($\kappa = \frac{1}{3}C_v v_s l$) and the assumption that the mean free path of phonons l is primarily limited by the separation of the local oscillators (rattlers). The authors showed that the room temperature thermal conductivities estimated using these isotropic ADPs are in very good agreement with the experimental values. The ADPs are thus a very useful tool for screening prospective novel thermoelectrics.

III. Band Structure

Considerable insight into the electronic properties of a material can be gained from considering its crystal structure and applying appropriate models for the chemical bonding topologies (King, 1989). However, the detailed picture of the electronic bands including the position of the band edges and the band dispersion, can only be obtained from computations of the electronic band structure. The availability of efficient computational algorithms coupled with the progress in computing technology has led to the development of increasingly sophisticated and reliable techniques of band structure calculations. Such theoretical approaches offer a powerful tool to check the relevance of the various two-electron bond models postulated for the skutterudite structure. Furthermore, the theoretical results provide guidance to experimentalists to focus their search in the material and compositional domain that appears most fruitful from the perspective of the desired material properties.

In the first of such calculations, Jung *et al.* (1990) used the tight-binding method to compute the dispersion relation and the density of states (DOS) for $[Fe_4P_{12}]^{3-}$, the primitive unit cell that is the fundamental structural block of $LaFe_4P_{12}$. The essential outcome of these calculations is that the t_{2g}- and e_g-block bands are separated by a rather large gap (~ 2.7 eV) with the Fermi level cutting the highest occupied valence band so that it is half-filled. As all skutterudites filled with the lanthanoid $3+$ species have the building lattice blocks $[T_4X_{12}]^{3-}$ that are isoelectronic and isostructural with the $[Fe_4P_{12}]^{3-}$ lattice, they all should be metals. On the other hand, the $[Fe_4P_{12}]^{4-}$ lattice and its isoelectronic M_4X_{12} lattices have the t_{2g}-block bands completely filled and the structure should be semiconducting. This is the case of a $4+$ filler ion such as U^{4+}, Th^{4+}, or a strictly tetravalent form of Ce^{4+}, and of course this also holds for the binary skutterudites.

Figure 10 indicates that the Fermi level for $[Fe_4P_{12}]^{3-}$ falls in the energy region where the DOS is small (~ 5 electrons/eV/formula unit) and where the Fe 3d-orbital character is negligible. Thus, the highest occupied band has mainly phosphorus character. This has an interesting implication — since electrons near the Fermi level are primarily responsible for superconductivity, the observed superconducting properties of $LaFe_4P_{12}$ (see Meisner, 1981) are governed by the electrons of its phosphorus sublattice. In spite of completely ignoring the contribution of the lanthanoid cations (their important role will emerge in the discussions that follow) this work substantially confirmed the phenomenological two-electron bond model outlined in the previous section.

Further insight into the electronic band structure has been gained by using first principles methods based on density functional theory. Such computations are usually carried out within the local density approximation (LDA) that makes use of the local exchange correlation energy and treats the potential as that of a uniform electron gas with the local density $\rho(\mathbf{r})$. There are different variants of this method depending on the choice of the specific basis set expansions of the wave function in the computations (plane waves, muffin-tin orbitals, local orbitals, etc.). Using the extended general-potential linearized-plane-wave (LAPW) method, Singh and Pickett (1994) calculated the electronic structure of the binary skutterudites $IrSb_3$, $CoSb_3$, and $CoAs_3$. In all three materials there is a clear separation (pseudogap) between the valence and conduction bands (Fig. 11a) that gives rise to the

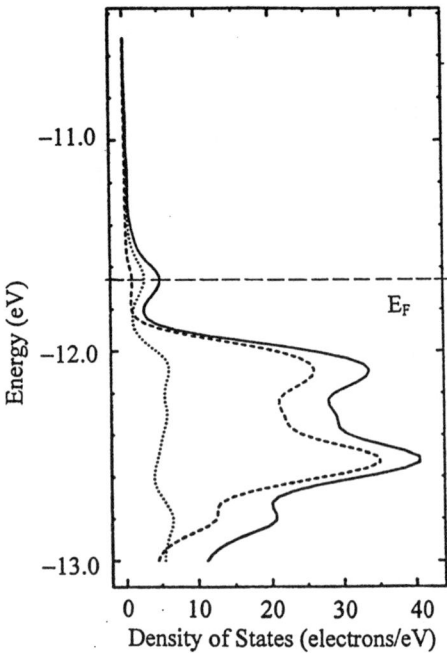

FIG. 10. Total density of states (solid curve), and the contributions of the Fe $3d$ (dashed curve) and the phosphorus orbitals (dotted curve) to the density of states calculated for the t_{2g}-block bands of $LaFe_4P_{12}$. Horizontal dashed line indicates the Fermi energy (from Jung et al., 1990). Reprinted with permission. Copyright by American Chemical Society.

gap in the DOS. This corresponds to the gap seen in optical experiments. The conduction as well as valence bands are derived from hybridized combinations of the transition metal $3d$-states and pnicogen p-states. All three materials have substantial indirect pseudogaps and the magnitudes of the direct pseudogaps at the zone center are 1.21 eV ($IrSb_3$), 0.80 eV ($CoSb_3$), and 0.95 eV ($CoAs_3$). An important result of the calculations is the occurrence of a single band that crosses the pseudogap and touches (or nearly so for $CoSb_3$) the conduction band minimum at the zone center. The theory thus predicts the structures to be very small gap (~ 50 meV for $CoSb_3$) or zero gap ($IrSb_3$ and $CoAs_3$) semiconductors. The band crossing the pseudogap arises from hybridization of the d-orbitals of the transition metal with the p-states of the pnicogen. Moving away from the zone center, the pnicogen p-character increases. Because this band is just a single one among the many bands forming the manifold, it leaves little mark on the DOS in the gap region and it is difficult to detect it by optical means. On the other hand, this band is of vital importance for transport properties and determines the nature of the conduction process in binary skutterudites.

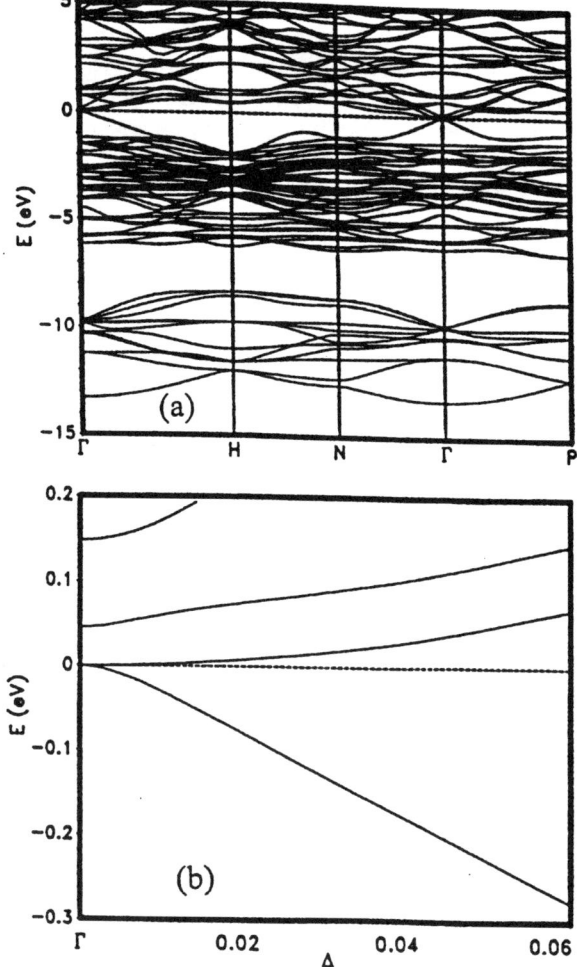

FIG. 11. (a) Band structure of IrSb$_3$. The dashed line denotes E_F. (b) *Expanded scale showing the region near the Fermi energy along the Γ–H line.* Note the dispersion of the gap-crossing band that, except for a very small region near the Γ point, is linear in energy. Similar linear dispersion is observed in CoSb$_3$ but not in CoAs$_3$, where the band has a normal parabolic shape. (From Singh and Pickett, 1994.) Copyright 1994 by the American Physical Society.

This band has a remarkable character—while it is parabolic in CoAs$_3$, in antimonides it has linear dispersion right up to near the zone center, Fig. 11b. The slopes of a linear region of dispersion are -3.45 eV Å for IrSb$_3$ and -3.10 eV Å for CoSb$_3$. The parabolic region very near the Γ point can only be accessed with doping levels below about 3×10^{16} cm^{-3}, some two orders of magnitude below the doping levels typical of the currently best crystals of CoSb$_3$. As this band governs transport properties, one expects the binary antimonide skutterudites to show unusual doping dependence in their Seebeck coefficient and electrical conductivity: in the constant scattering time and degenerate regime the former should vary as $n^{-1/3}$ instead of $n^{-2/3}$, while the latter should be proportional to $n^{2/3}$ instead of n. From the viewpoint of thermoelectricity, this also implies that the power factor $S^2\sigma$ will be less sensitive to the doping levels than is usual in semiconductors with a parabolic band dispersion.

The electronic structure of binary phosphide skutterudites, CoP$_3$ and NiP$_3$, was studied by Llunell et al. (1996) and by Zhukov (1996) using a related version of the LDA — the one based on the linear muffin-tin orbital method in the atomic sphere approximation (LMTO-ASA). Again, there is a clear separation between the valence and conduction bands in CoP$_3$ (the indirect pseudogap is 1.26 eV) with the pseudogap crossed by a completely filled single band with a linear dispersion, Fig. 12a. The indirect and direct bandgaps are 0.07 and 0.28 eV, respectively. The values are to be compared with 0.45 eV, the gap determined by Ackermann and Wold (1977) using optical means. Llunell et al. also computed the band structure for NiP$_3$, Fig. 12b. The atoms of nickel provide one extra electron per formula unit and the electrons reside at the bottom of the conduction band. The calculated pseudogap is here substantially less (~ 0.57 eV) than in CoP$_3$ and the pseudogap crossing band penetrates the bottom of the conduction bands. The projections of the DOS for the transition metal atoms show both CoP$_3$ and NiP$_3$ as having the main peak in the region below the pseudogap. Thus, the d-orbitals of the transition metal are almost completely filled and the band states that determine the conductivity of the compounds (CoP$_3$ a semiconductor and NiP$_3$ a metal) are constructed predominantly from the atomic states of phosphorus. As strong evidence for this statement, the authors note that the phosphorus atoms contribute 51.7 states/eV to the total DOS at the Fermi level of NiP$_3$, while the nickel atoms contribute only 13.7 states/eV. The results add weight to a statement in Jung et al. (1990) that the phosphorus atoms shape the nature of superconductivity in LaFe$_4$P$_{12}$.

Rather than viewing the band structure of skutterudites as unusual, Sofo and Mahan (1999) pointed out that the band structure of CoSb$_3$ is in fact typical of narrow-gap semiconductors and the bands in the proximity of the Fermi level can be described by the two-band Kane model. The approach they used is based on the full-potential linearized augmented plane wave method (FP-LAPW) that returned the lattice constant within 1% of the experimental value, and the direct bandgap of 0.22 eV when the positional parameters y and z were relaxed to their minimum energy value. The bandgap is, however, extremely sensitive to the Sb position. If the experimental positional parameters are used instead of the parameters that

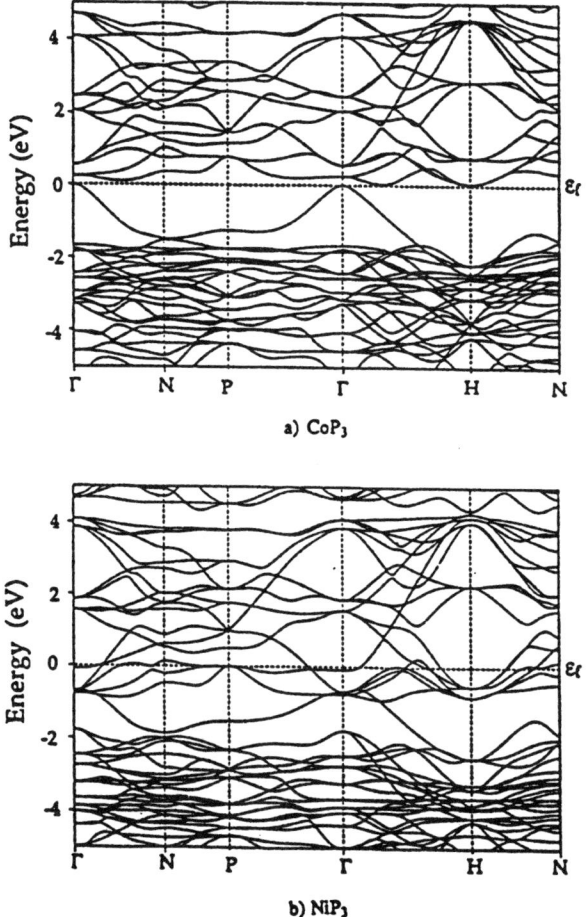

FIG. 12. (a) Band structure for CoP_3. The indirect bandgap (from Γ to a point on the $\Gamma-H$ line) is 0.07 eV; the direct gap at Γ is 0.28 eV. (b) Band structure of NiP_3. In this compound, the metal atoms provide one more electron per formula unit than in CoP_3. These extra electrons occupy the bottom of the conduction band, and the compound is a metal (from Llunell *et al.*, 1996). Copyright 1996 by the American Physical Society.

minimize the energy, the bandgap comes out as 0.05 eV, in very good agreement with the results of Singh and Pickett.

The two-band Kane model refers to the $\mathbf{k} \cdot \mathbf{p}$ perturbation that leads to the nonparabolic dispersion relation of the form

$$\frac{\hbar^2 k^2}{2m^*} = \varepsilon_k \left(1 + \frac{\varepsilon_k}{E_g}\right) \tag{8}$$

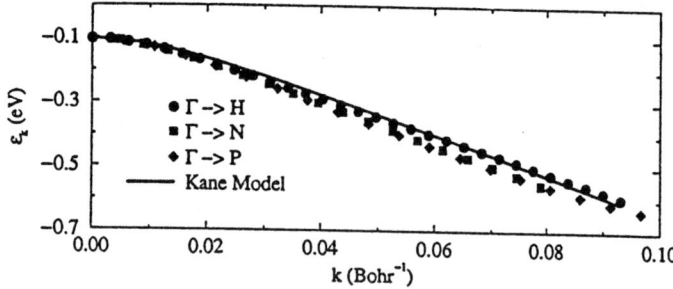

FIG. 13. Closeup of the nonparabolic valence band of CoSb$_3$ along three different high-symmetry directions of the Brillouin zone together with the fitted Kane model band (from Sofo and Mahan, 1999).

where E_g is the bandgap. The authors have shown (see Fig. 13) that the nonparabolic valence band of CoSb$_3$ can be fit very well with the functional form of Eq. (8). For the bandgap of $E_g = 0.22\,\text{eV}$, the fit returns, depending on the direction, the effective mass of $m^* = 0.069$–$0.071\,m_e$. This value is in good agreement with both the transport effective mass (Caillat et al. 1996a) and the effective mass obtained from the temperature dependence of the Shubnikov–de Haas oscillation amplitude (Rakoto et al., 1999).

I have already noted that, in order to simplify computations, the early band structure calculations for LaFe$_4$P$_{12}$ ignored the presence of the La-filler atoms. That this was a serious omission has been shown in the more recent treatment of the filled skutterudites by Nordstrom and Singh (1996), Singh and Mazin (1997), Singh et al. (1997), and Harima (1998). In a series of papers, Singh and his colleagues extended their self-consistent LAPW calculations to the filled skutterudites with the Ce filler and to the La-filled iron/cobalt alloyed antimonide skutterudites. They found both CeFe$_4$P$_{12}$ and CeFe$_4$Sb$_{12}$ to be narrow-gap semiconductors with bandgaps of 0.34 eV and 0.1 eV, respectively. The manifold of the bands, and more specifically its separation into the valence and conduction band blocks, is reminiscent of the binary skutterudites CoSb$_3$ and CoAs$_3$. A new and distinct feature of the electronic band structure is the presence of the spin-orbit split Ce 4f-bands, which dominate the lowest lying conduction bands, Fig. 14. The influence of the Ce 4f-states extends also to the valence bands. Although qualitatively similar overall, the CeFe$_4$P$_{12}$ and CeFe$_4$Sb$_{12}$ bands differ in detailed features, particularly near the Fermi level. Thus, while in the antimonide, the highest valence band is singly degenerate at the Γ point, in the phosphide, there are two parabolic bands at the valence band maximum at the Γ point. The band corresponding by the symmetry to the single highest band in CeFe$_4$Sb$_{12}$ is the second highest valence band of CeFe$_4$P$_{12}$. This band is primarily of Fe 3d character with no admixture of

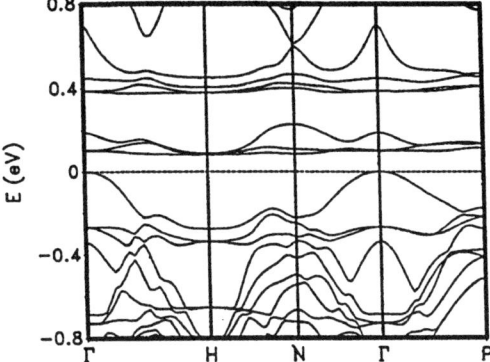

FIG. 14. Band structure of $CeFe_4Sb_{12}$ for energies close to the Fermi level. Note the presence of the flat, spin-orbit split Ce $4f$-bands at the bottom of the conduction band manifold (from Nordstrom and Singh, 1996). Copyright 1996 by the American Physical Society.

Ce $4f$ states. The highest valence band in $CeFe_4P_{12}$ relates to the fourth highest band in antimonide (located some 0.4 eV below the highest antimonide valence band), and both of these bands result from a strong hybridization of the Ce $4f$, Fe $3d$, and pnicogen p-states. I note these details here to drive home the point that Ce $4f$ states are important and they exert a strong influence, via hybridization, on both the conduction and valence band structure.

One of the most important consequences of the presence of the Ce $4f$ states is the very heavy band mass predicted for Ce-filled skutterudites. The flatness of the Ce $4f$ conduction bands implies values of $\sim 10-20$ m_e ($CeFe_4Sb_{12}$) and $6-8$ m_e ($CeFe_4P_{12}$), and even the valence band masses are large, 0.8 m_e for the phosphide and 2.2 m_e for the antimonide. An unusual combination of very small bandgaps with exceptionally high effective masses makes the Ce-filled skutterudites outstanding candidates for large thermoelectric power factors. The calculations also indicate that in order to place the Ce $4f$ bands in a reasonable position, the Ce configuration must be f^1 (and not f^0) corresponding to the trivalent state of cerium.

Brown and Jeitschko (1980b) and subsequently numerous studies have established that $CeFe_4Sb_{12}$ is a metal rather than a semiconductor. The discrepancy with the theoretical prediction of a finite gap for $CeFe_4Sb_{12}$ is likely the result of an underestimation of intra-atomic correlation effects and a consequent overestimation of the degree of hybridization, the inherent deficiency of the LDA treatment as applied to the $4f$ orbitals.

In their LAPW calculations on La-filled skutterudites, Singh and his colleagues (Singh and Mazin, 1997; Singh et al., 1997) have shown that while the electronic structure of $La(Fe,Co)_4Sb_{12}$ is strongly perturbed from that of the binary $CoSb_3$, a partial replacement of iron by cobalt, i.e., within the series $La(Fe_{1-x}Co_x)_4Sb_{12}$ for $0 < x < 0.75$, the calculated bands near the Fermi level are only weakly dependent on the presence of cobalt. The calculations yield two Γ-point valence bands in the proximity of the Fermi level—the highest band of mostly Sb p-character, and a second band about 0.1 eV below the top of the highest valence band. This lower lying band has a low dispersion and high effective mass ($\sim 3m_e$) and derives from the Fe/Co states. The authors argue that the close proximity of these two highest valence bands with very different masses is the origin of the high thermopowers observed in the material. A similar outcome, that is, the two highest valence bands (phosphorus p-derived band and a somewhat lower lying Fe $3d$-band) being intersected by the Fermi level, has been reported by Harima (1998) for $LaFe_4P_{12}$.

An important new finding of all these calculations points to the fact that the filler species such as lanthanoids play an active role in the formation of the electronic structure, and they exert strong influence on the detailed shape and form of the bands in the vicinity of the Fermi level. Thus, the original assumption that the "cage" fillers are good only for the reduction of the lattice thermal conductivity via their interference with the normal modes of vibration is far from the truth. The cage-filling atoms, under the right circumstances, may also aid in enhancing the electronic properties and thus have even more positive influence on the optimization of thermoelectric properties. Further support for this scenario is provided by the results of computational studies on the effect of defects and impurities in the $CoSb_3$ structure (Akai *et al.*, 1997, 1998). By placing different impurities in the voids of $CoSb_3$ and calculating (using either the LMTO-ASA method or the LAPW method) changes in the band structure, the authors were able to relate the resulting carrier type with a particular impurity occupying the voids. Moreover, their results indicate that the void center position is the most stable location for the impurity ion. The authors also pointed out a possible reason for the p-type character of the undoped $CoSb_3$—a few Sb atoms that occupy voids and contribute holes.

The recent theoretical efforts are starting to shed light on the fundamental band structure issues that underlie the promising thermoelectric properties of skutterudites. However, complete understanding of these materials, including the power to predict the properties with a high degree of reliability, will not be achieved until the theoretical results are corroborated by the detailed experimental studies of the band structure and the Fermi surface. At the present time, optical studies are sporadic and experiments probing the electronic structure and the Fermi surface have begun only recently. Even these rather preliminary results indicate the richness of the data that awaits a skilled experimenter. Shubnikov–de Haas studies seem particularly important as they address issues such as the cyclotron effective mass, nonparabolicity, and an energy bandgap. The data obtained by Rakoto *et al.* (1998, 1999) indicate that the Fermi surface of holes in CoSb$_3$ is a sphere located at the center of the Brillouin zone and that the number of equivalent valleys is just one. From the temperature dependence of the oscillation amplitude the authors evaluate the cyclotron effective mass that increases from 0.11m_e to 0.15m_e as the hole concentration increases from 1.45 to 4.06 × 10^{18} cm^{-3}. The experimental value of the energy gap of 35 ± 2 meV agrees well with the value predicted by theory. It is of considerable interest to have such studies done on filled skutterudites, provided samples with high enough mobility can be prepared.

As I have noted in the preceding paragraphs, an unequivocal assignment of the charge state of the filler ions as well as of the oxidation state of the transition metal is perhaps the most important task in order to be able to explain and rationalize the bonding and be in a position to predict physical properties of filled skutterudites. Experimental probes of the valence state of the rare earth species are well established and refined but, so far, only rarely have they been applied to skutterudites. The exception is a technique called X-ray absorption near-edge structure (XANES) that is a proven method of determining the valence state of rare earths. XANES measurements have enough sensitivity to clearly distinguish between the 3+ and 4+ charge states of the rare earth ions that in this experimental technique are manifested by a single-peak, respectively double-peak structure of the L-edge XANES spectra; see Fig. 15. Applying XANES to filled phosphide

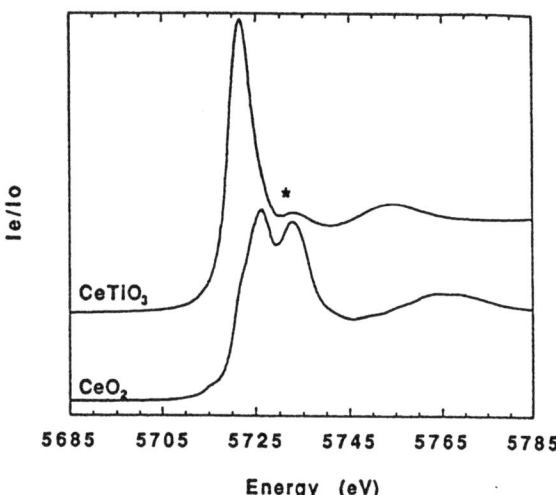

FIG. 15. An example of Ce L_3-edge XANES spectra for trivalent Ce such as found in CeTiO$_3$, and for tetravalent Ce characteristic of CeO$_2$. Note in particular a single-peak, respectively, double-peak structure that is used to assign a trivalent, respectively, tetravalent state of Ce. The asterisk denotes a very small peak in the CeTiO$_3$ data that arises from EXAFS and is not electronic in origin (from Xue *et al.*, 1994). Reprinted with permission from Elsevier Science.

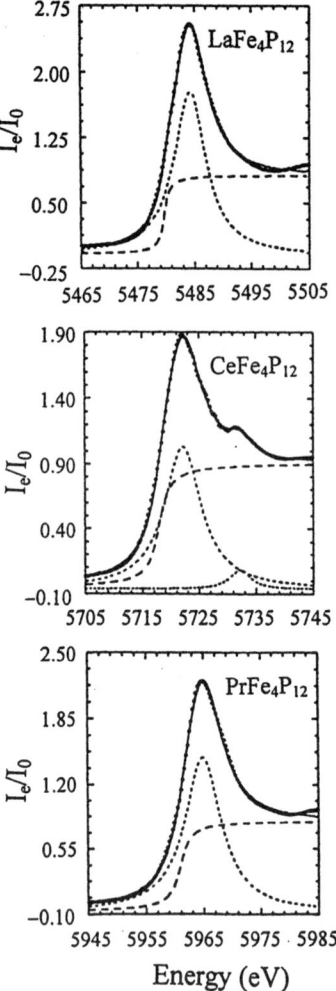

FIG. 16. The XANES data for $LaFe_4P_{12}$, $CeFe_4P_{12}$, and $PrFe_4P_{12}$. Note the presence of a small second peak of electronic origin in the spectra of $CeFe_4P_{12}$. This suggests a more complex bonding consistent with hybridization of Ce $4f$ electrons. The dotted and dashed lines are the fitted arctangent and Lorentzian contributions to the total, fitted intensity shown by a solid line (from Xue *et al.*, 1994). Reprinted with permission from Elsevier Science.

skutterudites, Xue *et al.* (1994) compared the L_3-edge of the Ce ion in $CeFe_4P_{12}$ with the corresponding L_3-edge of La and Pr (Fig. 16). The Ce L_3-edge is broader and shows an additional quite small peak at an energy expected for the Ce^{4+} second peak. However, from the edge energy of the main peak the authors concluded that Ce in $CeFe_4Sb_{12}$ is predominantly trivalent and the more complex bonding is the result of hybridization of Ce $4f$ electrons with the transition metal $3d$ and pnicogen $3p$ electrons. A similar conclusion has been reached in a recent study (Lee et al., 1999) of Ce- and Pr-filled ruthenium phosphides.

We can only hope that in the future more effort will be directed toward experimental studies of the band structure and bonding in the skutterudite structure. A plethora of fascinating phenomena that these materials display should be a strong incentive for such studies.

IV. Vibrational Properties

1. Phonon Modes and the Density of States

The lattice thermal conductivity is one of the key parameters determining the thermoelectric figure of merit. It is thus essential to understand the lattice dynamics in skutterudites and to have a good grasp of the phonon density of states. The crystal structure of skutterudites determines their symmetry ($Im3$-T_h^5) and, having this information, one can proceed to categorize lattice vibrations of the structure using group theory. The total of 45 zone center vibrations of the skutterudite-type compounds is decomposed into the irreducible representation

$$\Gamma = 2A_g + 2E_g + 4F_g + 2A_u + 2E_u + 7F_u. \tag{9}$$

Thus, there are 19 distinct zone center phonon modes. From the far-infrared reflection spectra of $CoAs_3$ and $CoSb_3$, Lutz and Kliche (1982) measured seven of these modes and compared them with the calculated values using their force-constant model. The respective frequencies of these modes are given in Table IV. The experimentally obtained modes in Table

TABLE IV

Phonon Frequencies Obtained from the Far-Infrared Reflection Spectra of $CoAs_3$ and $CoSb_3$ Compared with the Calculated Modes (After Feldman and Singh, 1996)[a]

Skutterudite	f_1	f_2	f_3	f_4	f_5	f_6	f_7 (cm^{-1})
$CoAs_3$: Observed	324	306	292	242	208	163	118
Calculated	327.9	308.6	285.1	243.6	209.1	162.0	119.9
$CoSb_3$: Observed	275	257	247	174	144	120	78
Calculated	275.2	256.6	244.1	173.9	142.4	118.7	79.6
$RhSb_3$: Observed	243	225	215			115	
$IrSb_3$: Observed	213	197	187	144	132	112	

[a]Also included are experimental values of infrared active modes of $RhSb_3$ (from Kliche and Bauhofer, 1987) and of $IrSb_3$ (from Slack and Tsoukala, 1994).

TABLE V

Zone Center Phonon Frequencies in Units of cm^{-1} for CoSb$_3$[a]

Mode	LK model	Singh et al. (1996)	Experiment (Lutz and Kliche, 1982)	LDA (Feldman and Singh, 1986)
A_g	162	149		150
	183	177		178
A_u	69	109		110
	250	242		241
F_u	79	78	78	
	119	120	120	
	143	145	144	
	174	176	174	
	242	242	247	
	258	260	257	
	277	275	275	
E_g	140	141		
	194	181		
F_g	71	84		
	103	96		
	162	158		
	188	176		
E_u	95	139		
	275	262		

[a]LK model refers to the calculations of Singh et al. (1996) using the force constant model of Lutz and Kliche (1982).

IV were useful in the calculations of the vibrational spectrum of CoSb$_3$ made by Feldman and Singh (1996) and by Singh et al. (1996). The strategy was to use first principle local density approximation (LDA) assuming frozen phonons on the A_g and A_u modes and refit the force constant model of Lutz and Kliche to these calculations and to the experimental frequencies. This strategy has not produced a good fit. A modified model, which included bond angle forces, was developed and a very good agreement was obtained. Phonon frequencies of all zone center modes were calculated using this new improved model and the results are given in Table V. The phonon densities of states obtained from this model are displayed in Fig. 17. A notable feature here is a gap near 100 cm^{-1} and a sudden departure from low-frequency parabolic behavior near 70 cm^{-1}. In addition to the dominant role of the acoustic modes, there are significant contributions of the optic branches in this frequency range (mostly due to rigid motions of the Sb$_4$ rings) and they are responsible for the break in the parabolic frequency dependence. In the order of increasing frequency, the Sb modes evolve from substantially rigid librational Sb$_4$ ring character to the twisting motions of the rings, rigid translational motions of the Sb$_4$ and, eventually, bond stretching motions in

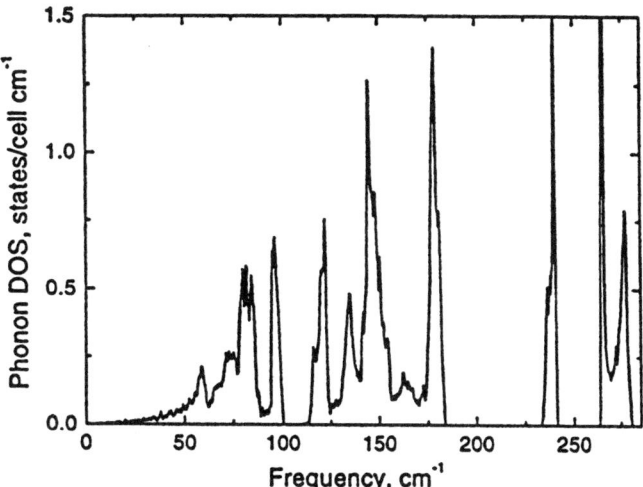

FIG. 17. Phonon density of states of CoSb$_3$ based on the model of Feldman and Singh (from Singh et al., 1999).

the rings. Separated by a wide gap from the essentially Sb modes, the transition metal modes dominate above 230 cm^{-1}. Singh et al. (1999) calculated phonon density of states D weighted by the squared group velocity $\langle v_g^2 \rangle D$. The effect of this function is to significantly enhance the importance of the low-frequency part of the spectrum ($\leqslant 100$ cm^{-1}) and virtually suppress any contribution from modes above 200 cm^{-1}. This function, multiplied by the relaxation time τ and called the phonon transport function, enters in the Boltzmann theory of thermal conductivity and it is thus a very good indicator of which modes represent the dominant contribution to lattice thermal conductivity; see Fig. 18. It should be clear that the transition metal modes are substantially irrelevant and that the bulk of heat conduction is due to the acoustic and, to a lesser extent, optic modes, both of which originate with Sb. This finding leads to the following dogma: To reduce the thermal conductivity of binary skutterudites, it is imperative to scatter Sb-related phonons!

How does the filler ion modify the host lattice vibrational spectrum? This important question was considered by Singh et al. (1999) who made LAPW frozen phonon calculations of the total energy for LaFe$_4$Sb$_{12}$ and CeFe$_4$Sb$_{12}$ as a function of the rare earth position, holding fixed all the other structural degrees of freedom. This approach yielded "bare" Einstein frequencies of 68 cm^{-1} for Ce and 74 cm^{-1} for La. The two frequencies fall straight into the range of Sb motions that strongly contribute to the lattice thermal conductivity. The next step was to derive the force constant model for LaFe$_4$Sb$_{12}$ and calculate the phonon density of states. This was done by

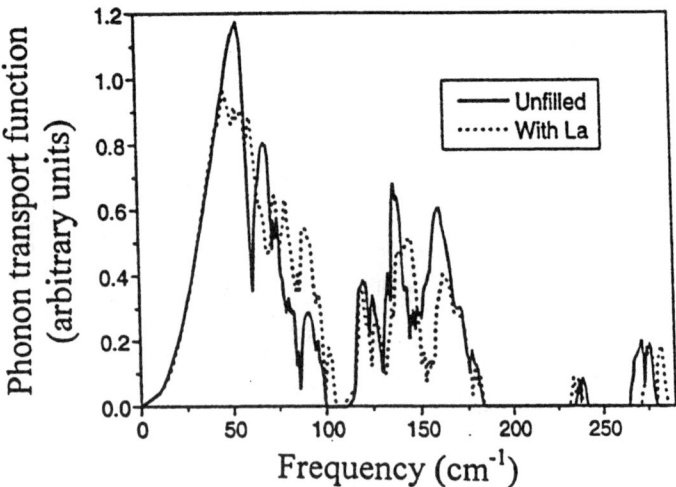

FIG. 18. Phonon transport function ($\langle v_g^2 \rangle D\tau$) for $CoSb_3$ (solid curve) and for $LaFe_4Sb_{12}$ (dashed curve) assuming constant relaxation time τ (from Singh et al., 1999).

assuming that $LaFe_4Sb_{12}$ is similar to $CoSb_3$ but with additional force constants. The authors identified strong force constants between the La and near neighbor Sb atoms and between the La and the near neighbor Fe atoms, the former being a restoring force while the latter is a nonrestoring force of about half magnitude. With such force constant model, they calculated phonon transport function for $LaFe_4Sb_{12}$, which is shown as a dotted curve in Fig. 18. One observes significant changes in the 50–175 cm^{-1} range dominated by Sb motions, but little difference in the acoustic part of the spectrum. Singh et al. also calculated the force constant model for $CeFe_4Sb_{12}$ using the result for $LaFe_4Sb_{12}$ but with somewhat softer force constants for the rare earth. The calculated difference spectrum between $LaFe_4Sb_{12}$ and $CeFe_4Sb_{12}$ is shown in Fig. 19. Strong coupling between the La and Sb motions is responsible for the observed two-peak structure and is believed to be the essential ingredient for the reduction of the lattice thermal conductivity in the filled skutterudites.

2. EVIDENCE OF "RATTLING"

Large ADPs of the filler ions discussed in Subsection 2 of Section II, together with a drastically reduced lattice thermal conductivity (see Subsection 2 of Section VII), are believed to be the consequence of the rattling motion of the filler ions in the voids of the skutterudite structure. But is there an independent verification that such rattlers exist? If rattling is the

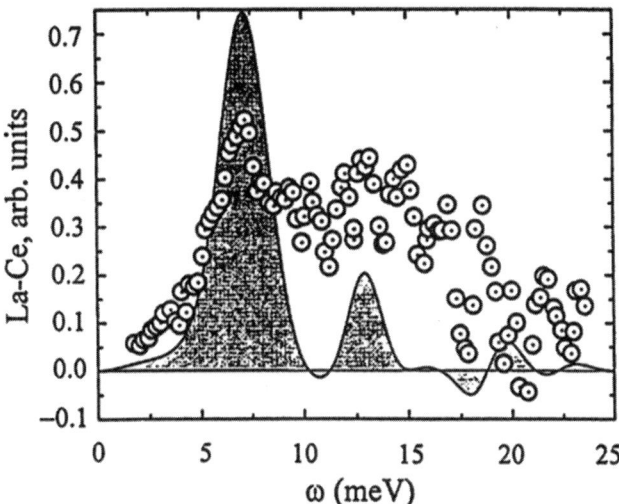

FIG. 19. Calculated difference spectrum for $LaFe_4Sb_{12}$ and $CeFe_4Sb_{12}$ (solid curve) and the difference in the vibrational spectra of the same two compounds obtained from inelastic neutron scattering studies. Since the scattering cross-section for Ce is much less than for La, the difference spectrum tends to emphasize the features of the vibrational spectrum due to La (from Singh et al., 1999, which uses data of Keppens et al., 1998). Reprinted with permission from Macmillan Magazines Ltd.

essential mechanism for reducing the thermal conductivity, it is imperative that one understand how such a mechanism actually impedes the phonon transport. Because systems displaying localized incoherent vibrational modes are rather rare and, among metallic systems, with their close-packed structure, virtually nonexistent, it is important to understand how the localized modes influence mechanical and thermodynamic properties of the skutterudite structure. In other words, what footprints are such localized vibrations likely to leave behind? These issues have been addressed.

In principle, one should be able to detect localized vibrational modes in the Raman spectra of filled skutterudite compounds, and in the behavior of their thermodynamic and mechanical properties. Regarding the Raman scattering studies on filled skutterudites, there are only three reports in the literature. Nolas et al. (1996a) measured Raman spectra of binary antimonide skutterudites and compared them with the spectra obtained on filled skutterudites. They observed significant broadening of the lines upon filling the voids, but no unique modes of the rattler were detected. The energy range covered in this study was too high to detect the presence of the localized modes. In the previous section I mentioned that Singh et al. (1999) estimated the "bare" Einstein frequency of the filler ions to be 68 cm^{-1} for Ce and 74 cm^{-1} for La. In interactions with the normal modes of Sb, the "bare" Einstein frequencies would likely be further downshifted and thus fall out of the window of the energies covered in this Raman scattering experiment. The mismatch between the energy regime where the localized modes should manifest their presence and the energy range covered by the experiment was even greater in the microprobe Raman studies of Sekine *et al.* (1998) made on rare earth-filled ruthenium phosphide skutterudites. As in Nolas *et al.* (1996a), the authors do observe line broadening and frequency shifts that, in this case, scale well with the cell volume. The notable exception is Ce-filled Ru_4P_{12} where the cell volume is anomalously small. Most recently, Nolas and Kendziora (1999) extended their earlier Raman measurements to lower temperatures, expanded the detector's range to below 50 cm^{-1}, and made use of polarization to identify the A_g (singly degenerate) Raman modes of the structure. These improvements in the experimental technique led to a better resolution of the line shifts and broadening (see Fig. 20), believed to be caused by the lanthanide ions

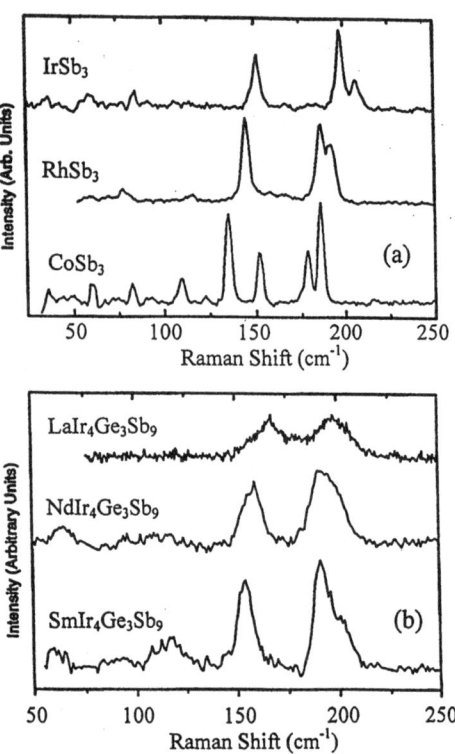

FIG. 20. Room temperature Raman spectrum of (a) binary antimonide skutterudites, (b) filled skutterudites taken with the scattered light polarized parallel to the laser line (from Nolas and Kendziora, 1999). Copyright 1999 by the American Physical Society.

vibrations, but have not resulted in the detection of the actual local vibration modes of the rattling rare earth ions.

Perhaps the most convincing evidence of rattling comes from inelastic neutron scattering studies, investigations of the elastic constants, and low temperature specific heat measurements. Keppens *et al.* (1998) correctly anticipated that any kind of a local oscillator ought to leave a mark on the specific heat and on the density of phonon states. Since binary skutterudites have no rattling species, it is convenient to compare the data on filled skutterudites with the data on the parent binary skutterudite. This is done in Fig. 21, where the experimental specific heat data for $CoSb_3$ and for $La_{0.9}Fe_3CoSb_{12}$ are displayed together with the curves representing two distinct Einstein oscillators. In addition, this figure also shows a dashed curve that represents the equation

$$C_p(La_{0.9}Fe_3CoSb_{12}) = C_p(CoSb_3) + \gamma T + AC_{E1} + BC_{E2} \qquad (10)$$

where γ, A, and B are the fitting parameters and C_{E1} and C_{E2} are specific heat contributions of the two Einstein oscillators, given by $C_E = (\theta/T)^2 e^{\theta/T}/(e^{\theta T} - 1)^2$. The two data sets are, indeed, very different. However, the very close match between the dashed line and the data points for $La_{0.9}Fe_3CoSb_{12}$ clearly shows that the specific heat of the filled skutterudite

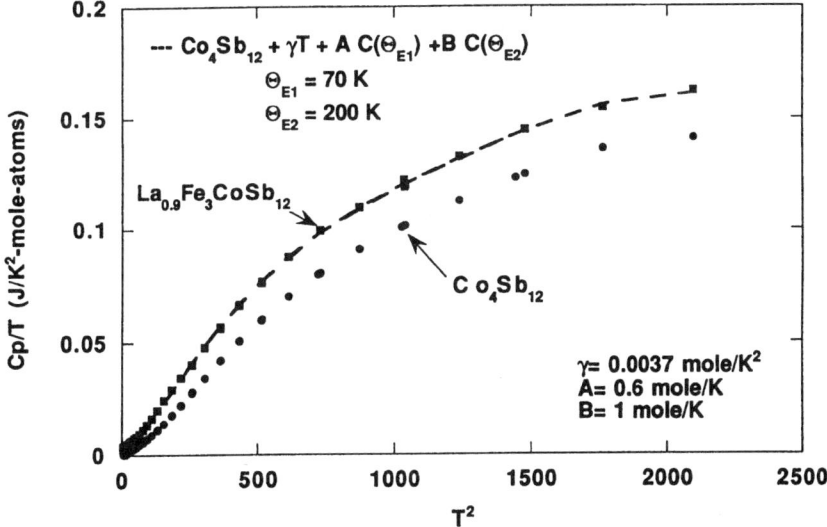

FIG. 21. Temperature dependence of specific heat of $CoSb_3$ and $La_{0.9}Fe_3CoSb_{12}$. The dashed line through the $La_{0.9}Fe_3CoSb_{12}$ data represents a fit to Eq. (10) (from Keppens *et al.*, 1998). Reprinted with permission from Macmillan Magazines Ltd.

can be viewed as the specific heat of the unfilled structure plus *two* additional terms representing Einstein oscillators with the Einstein temperatures of $\theta_{E1} = 70\,K$ and $\theta_{E2} = 200\,K$. The necessity of having two Einstein oscillators to fit the data implies that there are two different La ion motions. In addition, Keppens *et al.* also observed two well-defined peaks (at energies comparable to those of the two Einstein oscillators) in the vibrational spectrum of La obtained from the inelastic neutron studies, and significant changes in both shear and compression moduli. The results beg the following question: Why two Einstein temperatures and two peaks instead of just one? Keppens *et al.* speculated that perhaps the higher temperature mode is the one where the La^{3+} ion moves toward one of the Sb atoms while the lower temperature mode represent a situation where the ion moves toward a "void." A more specific explanation was provided by Singh *et al.* (1999) whose theoretical calculations mentioned in the previous section also yield a two-peak structure (see Fig. 19). They associated the lower peak with the La oscillations (slightly down-shifted because of interactions with the acoustic modes of Sb), while the higher temperature peak they view as being due to the La atoms driven by the natural oscillations of the Sb atoms. An excellent agreement between the predictions based on Eq. (10) and the experimental data is very strong evidence of a local vibrational mode (rattler) in the filled skutterudite structure. The peaks in the La vibrational density of states are not exactly δ-function sharp because of hybridization with acoustic phonons that broadens the peaks. Nevertheless, the results provide strong and convincing evidence for the existence of local, incoherent vibrations in the filled skutterudite structure.

V. Sample Preparation Aspects

To study physical properties of materials that might be useful thermoelectrics, one needs sufficiently large and homogeneous samples. Thus, the task is to identify appropriate growth conditions that result in at least millimeter-size single crystals or, if this is difficult to achieve, in homogeneous polycrystalline specimens that can be used in the as-grown form or subjected to further processing. Although single crystals are always an advantage, they are not an absolute necessity. A vast majority of the prospective novel thermoelectrics are materials with a cubic structure, and anisotropy is not a major concern. It is only when one needs to ascertain the role of grain boundaries that the availability of single crystals is essential.

In order to secure suitable samples for property measurements, scientists are quite willing to commit heroic effort with little regard for the economy of the process. Once the desired material properties are established, it is the task of the materials technologist to come up with a fabrication process that closely reproduces the optimal material characteristics yet, at the same time, is economically viable.

In general, novel thermoelectric materials are compounds or intermetallic alloys that have a large number of atoms in their unit cells and are formed from elements that often have significantly different melting points and vapor pressures, and the phase diagrams of these compounds are rather complicated. A simple binary solid–liquid equilibrium phase diagram with a full range of solid solutions such as the one describing Bi–Sb alloys or even Si–Ge alloys is certainly not the norm for the novel thermoelectric structures under consideration. Thus, one has to pay the utmost attention to the sample preparation conditions, and the skutterudites are a case in point.

The key to successful crystal growth is a thorough understanding of the respective phase diagram that describes thermodynamic equilibrium between various phases of the system. Skutterudites are compounds with the highest pnicogen concentration forming stable compounds with Co, Rh, and Ir at ambient conditions. There are typically other, lower pnicogen content compounds within any one of the phosphide, arsenide, or antimonide series. Thus, as already hinted at, the equilibrium phase diagrams are complex. To illustrate the point, Fig. 22 depicts the phase diagram for $CoSb_3$ (Feschotte and Lorin, 1989), one of the most actively studied skutterudites. Other skutterudites have substantially similar phase diagrams.

We note three stable compounds that can form in the higher pnicogen concentration range: CoSb (γ-phase), which grows congruently and is a metal, and the two semiconducting phases $CoSb_2$ (δ-phase) and $CoSb_3$ (ε-phase), each forming peritectically. The skutterudite structure—the ε-phase $CoSb_3$—is obtained in a peritectic reaction at 873°C from the δ-phase and the liquid phase. To remind the reader, a peritectic reaction involves formation of a homogeneous compound or solid solution from another (different) solid phase and a liquid at a specific fixed temperature called the peritectic temperature, T_p. In the case of $CoSb_3$ we can describe this reaction as

$$\delta\text{-phase} + \text{Liq} \rightleftarrows \varepsilon\text{-phase} \quad \text{at } T_p = 873°C. \tag{11}$$

Peritectic temperatures for other skutterudites are given in Table VI. Because of very slow kinetics, growth rarely proceeds under such conditions that the peritectic reaction is fully completed. With these facts in place, there are a couple of strategies one can use to prepare skutterudite samples of either polycrystalline or single crystal nature.

The early favored method of preparing polycrystalline samples (Zobrina and Dudkin, 1960) was to dissolve a stoichiometric quantity of cobalt in antimony at 1220–1250°C and cool the alloy to room temperature. To ensure completion of the peritectic reaction leading to the formation of the

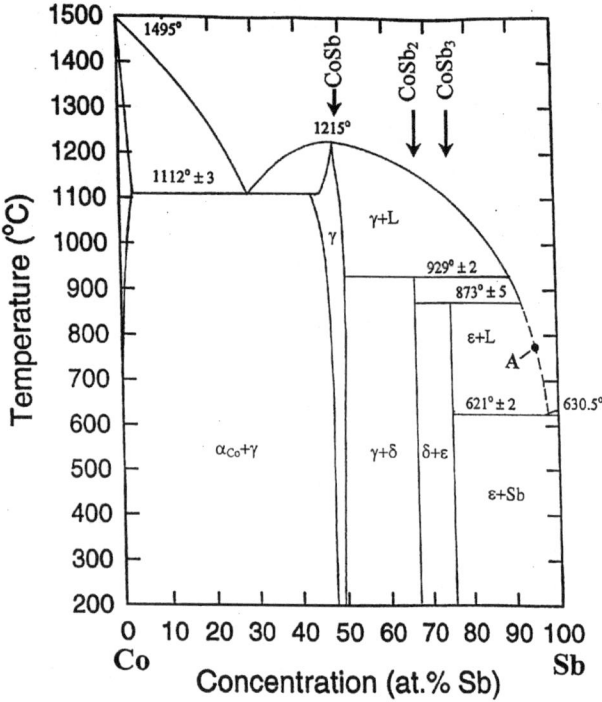

FIG. 22. Phase diagram of $CoSb_3$. Point A indicates a possible starting composition to grow a single crystal from melt (from Feschotte and Lorin, 1989). Reprinted with permission from Elsevier Science.

TABLE VI

PERITECTIC TEMPERATURES OF THE BINARY SKUTTERUDITES

Skutterudite	Peritectic temperature (°C)	Reference
CoP_3	>1000	Biltz and Heimbrecht (1939)
$CoAs_3$	960	Caillat et al. (1995b)
$CoSb_3$	873	Feschotte and Lorin (1989)
RhP_3	>1200	Odile et al. (1978)
$RhAs_3$	>1000	Caillat et al. (1995b)
$RhSb_3$	900	Caillat et al. (1996b)
IrP_3	>1200	Rundqvist and Ersson (1968)
$IrAs_3$	>1000	Kjekshus (1961)
$IrSb_3$	1141	Caillat et al. (1993a)
NiP_3	>850	Rundqvist and Larsson (1959)

$CoSb_3$ phase, the samples were annealed for a week or two at a somewhat lower temperature (550–600°C) than that corresponding to the fusion of the eutectic with antimony. To increase sample density and uniformity, the alloy was crushed and hot pressed at 500°C under pressure of 3000 kg/cm². Finally, the sample was annealed at 800°C (below the peritectic temperature) in argon for a couple of days. A more recent variant of this technique (Nakagawa et al., 1996) consisted of induction melting of Co and Sb under argon and quenching the melt by pouring it into a copper mold. Upon annealing the ingot at 580–800°C range for 20–100 hours under argon, a pure phase of n-type $CoSb_3$ was obtained. In contrast, when the cast ingot was crushed, mechanically alloyed, and hot-pressed, the resulting samples were p-type. The different nature of conduction is the result of different processing techniques. Unlike the annealed ingots, the mechanically alloyed powders contain high levels of impurities that originate from the milling process (especially Fe), and they also pick up very high amounts of oxygen. The hot-pressed samples also have much smaller grain size and this, together with the impurities they contain, contributes to more than a factor of two lower thermal conductivity.

Polycrystalline structures can also be synthesized by the solid–liquid sintering process. This consists of sealing stoichiometric amounts of the constituent elements in an evacuated quartz ampoule and heating the mixture for a prolonged time (a week or so) at temperatures just below the peritectic temperature. The end product is a polycrystalline compound that is typically 80–85% dense and that may contain small amounts of other phases. The phase purity can be improved by repeating grinding and reannealing until X-rays show complete absence of any impurity phases. Further densification of the structure, to about 95–97% of the theoretical density, can be achieved by hot pressing. With appropriate adjustments of the sintering temperature, duration of annealing, and number of intermediate grinding and reannealing stages, this growth technique works well for most of the skutterudites.

An inspection of Fig. 22 suggests that skutterudites can also be grown from melts. There is a narrow window of nonstoichiometric melt compositions (such as represented by point A in Fig. 22) where the melt is richer in pnicogen than the concentration corresponding to the intersection of the peritectic isotherm with the liquidus, but not as rich as the eutectic point near the pnicogen's endpoint. As the melt cools and its temperature reaches the liquidus, the skutterudite phase starts to emerge. Slowly lowering furnace temperature while maintaining a large temperature gradient (to aid phase separation), and using a quartz ampoule with a pointed bottom (to promote nucleation of a single crystal), it is possible to grow large single crystals of skutterudites. A pure skutterudite phase grows at the lower part of the ampoule until the melt cools close to the temperature of the eutectic point. Afterwards, the remaining melt solidifies as a eutectic mixture of the ε-phase and the pnicogen phase at the upper part of the ampoule. The two different phases — pure skutterudite and eutectic mixture — are easily discerned upon examination of the ingot. Caillat et al. (1996b) used this so-called gradient freeze technique with great success in preparing large single crystals of $CoSb_3$ and $RhSb_3$. Doping elements can be introduced directly into the melt and the crystals are robust with at least 99.5% theoretical density. The gradient freeze technique is less successful when the peritectic region extends too close to the eutectic point, as in the case of $IrSb_3$. This leaves only a very narrow range of compositions with steep temperature variation available for crystal growth, and the resulting ingot consists of a mixture of $IrSb_3$ and the eutectic with no clear delineation between the two phases.

Two other techniques are worth pointing out as possible means of preparing single crystal skutterudites: the flux-assisted growth, and the chemical vapor transport method. The former is a special technique, often reflecting the intuition and years of experience of a crystal grower in identifying a suitable flux from which the desired crystal could be grown at a considerably lower temperature and where the flux does not form phases that could impede nucleation of the intended compound. In the optimal case, the flux would be one of the constituents. This seems to work well for antimonide skutterudites (Mandrus et al., 1995), where a large excess of Sb (Co:Sb = 1:6.7) serves as a flux. The excess flux is spun off while still molten, leaving behind shiny crystals with dimensions of a few millimeters on a side. Skutterudites also grow well using tin flux, the method used first by Jolibois (1910) and employed extensively by Biltz and Heimbrecht in their preparation of CoP_3 (1939) and NiP_3 (1938). Tin is then easily removed by leaching in diluted HCl.

The chemical vapor transport method is a classical growth technique that has a potential to yield high-quality single crystals. Several attempts have been made to apply this technique to skutterudites. The most notable effort was that of Ackermann and Wold (1977), who grew CoP_3, $CoAs_3$, and $CoSb_3$. The transport agent in this case was chlorine, and they obtained single crystals with dimensions up to $3 \times 3 \times 1$ mm in a period of 3 weeks. Optical measurements indicated that while CoP_3 has a gap of 0.45 eV, $CoAs_3$ and $CoSb_3$ appeared, surprisingly, gapless and displayed a metallic character in their temperature dependence of resistivity. Thus, considerable nonstoichiometry of the constituents had to be present in the crystals despite having the appearance of a very high quality crystalline structure.

One should note that skutterudites ($RhSb_3$, $CoSb_3$, and $IrSb_3$) were also prepared as thin films by electron evaporation (Chen et al., 1997b), by rf sputtering (Anno et al., 1996), and by pulsed laser deposition (Caylor et al., 1999). Moreover, thin film multilayer deposition has led to a recent exciting development that concerns the preparation of metastable skutterudite structures from elementally modulated amorphous intermediate products

(Hornbostel et al., 1997a; Sellinschegg et al., 1998). The essence of this technique is the low-temperature deposition of short modulation wavelength (repeat distance $\sim 20\text{–}30\,\text{Å}$) multilayers consisting of the elemental layers. Provided the layers are thin enough, they will mix (interdiffuse) before nucleation sets in, and they will form an amorphous intermediate product. The composition of this amorphous intermediate is controlled by changing the ratio of the elemental layer thicknesses. The compound crystallizes upon annealing at relatively low temperatures ($\sim 100\text{–}200°\text{C}$), and the process is controlled by nucleation kinetics rather than being diffusion limited. The resulting compounds are thermodynamically metastable and decompose at temperatures around $500°\text{C}$ into a mixture of elemental constituents or more stable binary compounds. The technique works well for both binary and filled skutterudites. As an example, it was possible to prepare $FeSb_3$ and even skutterudites with a post-transition-metal filler such as Sn. Furthermore, the technique works well for the full range of lanthanide-filled skutterudites, including $LuCo_4Sb_{12}$. Lutetium is a cousin of lanthanum (both have an empty $4f$ electron shell), but is much smaller and heavier and thus has more space to rattle in the skutterudite cage. Indeed, the preliminary results indicate (Sellinschegg et al., 1998) that the thermal conductivity of $LuCo_4Sb_{12}$ is significantly lower than that of $LaCo_4Sb_{12}$ and approaches 1.4 W/m-K at 300 K. The ability to explore the properties of these metastable skutterudites that do not form via the usual synthetic routes opens new opportunities to optimize physical properties and may even point out the most profitable directions for improving the thermoelectric performance of skutterudites.

Various attempts have been made to speed up the processing of skutterudites, especially to cut down on long hours and days required for the solid–liquid sintering process. Spin casting under argon on the copper drum rotating with various surface speeds has been tried for $Co_{0.97}Cr_{0.03}Sb_3$ as well as for the filled skutterudite $LaFe_4Sb_{12}$ (Morimura et al., 1997; Kitagawa et al., 1998a). This technique produces ribbons of typically 20 μm thickness with grain sizes that decrease with the increasing surface speed of the copper drum. After several hours of annealing at $600\text{–}700°\text{C}$, the ribbons became a phase pure skutterudite and, in the case of $LaFe_4Sb_{12}$, they attained high power factors in the range $20\text{–}50$ $\mu\text{W/cm-K}^2$ at 550 K. Another approach that aims to cut down on the processing time is to start with powders of small diameter. One of the promising routes that also have a mass production potential is to atomize the melt by purified argon (Uchida et al., 1998). This fast processing method results in a fine-grained powder of less than 100 μm diameter that is not contaminated, as often happens during the ball milling process. The following hot-press sintering may be done by conventional means or with the aid of spark plasma sintering, which seems to require shorter sintering time.

Finally, I wish to mention that new high pressure-assisted synthesis techniques are being developed, and these promise to be fruitful in the preparation of filled skutterudites that are difficult or cannot be formed via more conventional approaches. For instance, Takizawa *et al.* (1999) have been able to insert Sn and Pb into the voids of $CoSb_3$.

From the perspective of practical applications of skutterudites, it will be important to understand their resistance to the environmental conditions and their "aging" characteristics. How skutterudites withstand long-term exposure to air at elevated temperature (Wilson and Mikhail, 1989) will be one of the important criteria determining their operational range, lifetime, and viability as thermoelectrics.

VI. Magnetic Properties

As I noted in Subsection of 2 of Section II, skutterudites are solids that encompass a rich variety of physical properties. Among them, one of the most fascinating and important characteristics is magnetism. Studies of the magnetic properties have greatly aided our understanding of the transport behavior in skutterudites, and the magnetic response of the structure provided pivotal input for rationalizing bonding in these materials. It is fair to say that without the insight gained from magnetic measurements our knowledge of skutterudites would have been very incomplete.

The fundamental point about the binary skutterudites is that they are diamagnetic solids. This means that they lack permanent magnetic dipoles, and the magnetic susceptibility is small and negative. This finding was a foundation stone for developing a picture how the pnicogen and transition metal atoms bond together and form the crystal structure. The fact that the binary skutterudites cannot have an unpaired electron in their structure has led directly to the conclusion that the transition metal atoms must be in the low-spin d^6 state. Since rare earths are the archetypal species with which one fills the voids of the structure, magnetism is an integral part of the world of filled skutterudites and one can expect a spectrum of interesting magnetic properties.

Studies of the magnetic properties of skutterudites have focused on the behavior of two important parameters: the magnetic susceptibility and the Mössbauer effect. Magnetic susceptibility provides information about the global magnetic character of a solid; its magnitude and sign determines what kind of magnetic behavior one deals with; and from its temperature dependence one can extract important magnetic parameters and any hints of a developing long-range order. The relative ease with which one can measure magnetic susceptibility, especially given a widespread use of very sensitive SQUID magnetometers, makes this parameter a favored tool of the magnetic assessment of a given solid. The Mössbauer effect, on the other hand, is a rather specialized spectroscopic technique that probes the magnitude of the magnetic field acting on a specific nucleus (Mössbauer nucleus) or the presence of the gradient of the electric field at the nucleus. A combination of the two techniques yields rich information on the magnetic state of a material. In the following two sections I illustrate and discuss the results obtained with these techniques as they relate to skutterudites.

1. Mössbauer Spectroscopy

The Mössbauer effect, or recoilless gamma-ray resonance spectroscopy, probes the charge distribution and magnetic field intensity in an immediate vicinity of a nucleus. As such it is a sharp spectroscopic tool that provides information about the local atomic environment. Mössbauer spectra are characterized by several parameters:

1. The position of the resonance maximum δ, often called the isomer or chemical shift, which is sensitive to the electron configuration of the atom including its oxidation state
2. The linewidth Γ
3. The magnitude of the resonance measured by the total area enclosed by the resonance curve
4. The quadrupole splitting $\Delta = e^2qQ/2$ (q is the gradient of the electrostatic field at the nucleus, and Q is the nuclear quadrupole moment) observed when the Mössbauer nuclide is in an environment where the charge distribution does not have cubic symmetry
5. The splitting due to the magnetic field acting on the nucleus via the magnetic hyperfine interaction

For our purposes, the most important parameters are the isomer shift δ and the quadrupole splitting Δ.

The first Mössbauer studies on skutterudites were made by Kjekshus and colleagues (Kjekshus et al., 1973; Kjekshus and Rakke, 1974), who studied the ^{121}Sb and ^{57}Fe Mössbauer spectra in the binary antimonide skutterudites $CoSb_3$, $RhSb_3$, $IrSb_3$, and $Fe_{0.5}Ni_{0.5}Sb_3$. The observed isomer shifts of the ^{121}Sb transition were comparable, and their magnitude (-9 to -10.1 mm/s) was consistent with covalent bonding. Unfortunately, the spectra lines were broadened to the extent that it was not possible to resolve the expected quadrupole splitting directly. Instead, the authors provided an order of magnitude estimate of the quadrupole interaction from the overall linewidth Γ. They were particularly interested in the relation between the magnitude of the quadrupole interaction and the difference in the bond lengths d_1 and d_2 (see Eqs. (2) and (3)). Since the difference between d_1 and d_2 is inevitably the result of an electron imbalance, this should be reflected in the magnitude of the quadrupole split. They found an approximately linear relationship between the quadrupole interaction and $(d_2 - d_1)$. From the Mössbauer split of ^{57}Fe, the authors concluded that the Fe in $Fe_{0.5}Ni_{0.5}Sb_3$ indeed has a formal $3d^6$ configuration, implying the transfer of one electron from the Ni atom to the nonbonding $3d$ orbital of Fe.

The first Mössbauer spectrum on filled skutterudites was taken by Shenoy et al. (1982), who were interested in the magnetic moment of Fe in $LaFe_4P_{12}$ using the ^{57}Fe resonance. The data (see Fig. 23a) show a well-defined quadrupole splitting that can be fit by superposing appropriate Lorentzians. Repeating the measurement in the presence of an external magnetic field (6.1 T) applied parallel to the direction of gamma ray resonance absorption, the spectrum changes to that shown in Fig. 23b. The data analysis of this spectrum (the various Lorentzians fitted to the data) returned a hyperfine magnetic field equal to the applied field with a maximum error of ± 0.1 T. This important result indicates that Fe in $LaFe_4P_{12}$ has a magnetic moment of less than $0.01\,\mu_B$. In other words, the polyanion $[Fe_4P_{12}]^{3-}$ does not carry a magnetic moment! On one hand, this result is in accord with the fact

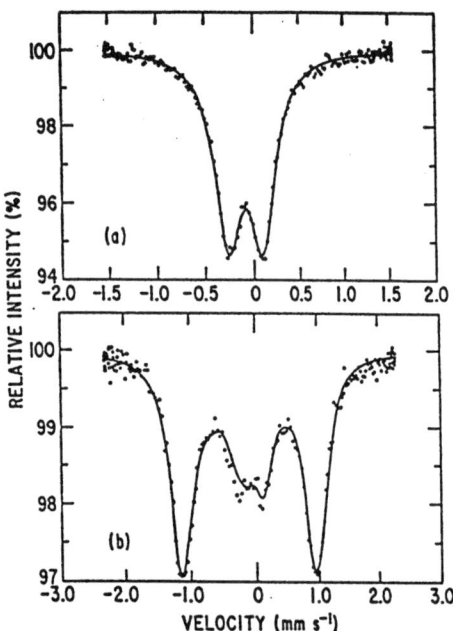

FIG. 23. Mössbauer spectrum of $LaFe_4P_{12}$ at 4.2 K in (a) $H_{ext} = 0$ and (b) $H_{ext} = 6.1$ T. Solid lines represent least-squares fit to the data describing the superposition of numerous Lorentzians (from Shenoy et al., 1982).

that $LaFe_4P_{12}$ is a superconductor. On the other hand, the result is surprising as one would expect the polyanion $[Fe_4P_{12}]^{3-}$ complex to have one of the four Fe atoms in the d^5 configuration that should have a spin-only moment of $1.73\,\mu_B$. The Mössbauer measurement shows no presence of such atoms.

Grandjean et al. (1983, 1984) reexamined the Mössbauer spectra of $LaFe_4P_{12}$ and extended the study to include $CeFe_4P_{12}$ and $EuFe_4P_{12}$. Regarding $LaFe_4P_{12}$, the authors confirmed the nonmagnetic state of Fe in this skutterudite and found zero magnetic moment also on the polyanionic framework of $CeFe_4P_{12}$ and $EuFe_4P_{12}$. They concluded that Fe must have a very similar electronic configuration in all three compounds and, whatever it is, its magnetic moment is zero. Based on this result they suggested that the nonbonding Fe $3d$ states ($\sim t_{2g}$ block bands) are filled in all three compounds and are located well below the Fermi level. According to the authors, these bands cannot be responsible for the magnetic properties, at least in filled phosphide skutterudites.

Obviously, it was of considerable interest to check the magnetic state of Fe in other skutterudites, and most notably in those with the iron-antimonide framework. This was accomplished by Kitagawa et al. (1998b), who measured the Mössbauer spectra on La- and Ce-filled structures. The spectra showed virtually the same isomer shift for the two compounds, but different quadrupole splittings, 0.374 mm/s for $LaFe_4Sb_{12}$, and 0.406 mm/s for $CeFe_4Sb_{12}$. The authors related the quadrupole splits to the density of states at the Fermi level and made the statement that the Fe atoms are in nonmagnetic state at room temperature. However, the crucial test to decide the magnetic state of Fe — to carry out the measurements in an external magnetic field — has not been done and thus the measurements are inconclusive.

The recently synthesized $YbFe_4Sb_{12}$ is a skutterudite filled with a rare earth that is next to the last in the lanthanide series. If Yb were trivalent, the structure would not hold together because the Yb^{3+} ion is too small to form adequate bonds. However, if this ion was divalent, it might be large enough to bond effectively. With the divalent Yb the structure resembles the alkaline earth-filled skutterudites and it is of interest to probe the magnetic state of the $[Fe_4Sb_{12}]^{2-}$ polyanion. Leithe-Jasper et al. (1999) did just that by taking the Mössbauer spectra over a wide range of temperatures and in an applied magnetic field of 10.5 and 13.5 T directed parallel to the incident gamma rays. By carefully measuring the hyperfine field they observed its value always to coincide with the external magnetic field. From this they concluded that the Fe atoms carry no magnetic moment. This is a surprising result, because the $[Fe_4Sb_{12}]^{2-}$ polyanion should have a net moment as there are two Fe atoms in the nominally d^5 configuration. In the latest and most extensive study of the $Ce_yFe_{4-x}Co_xSb_{12}$ solid solutions, Long et al. (1999) also noted that Fe in the structure is nonmagnetic. Again, however, the effect of the external magnetic field was not considered, and thus the argument regarding the magnetic state of Fe has not been substantiated.

Mössbauer spectroscopy is a truly microscopic probe with the ability to examine the magnetic and electrical properties of Fe (and several other Mössbauer nuclides) directly. Unfortunately, in its application to skutterudites the full potential of the technique has not always been utilized. The key issue — the magnetic state of Fe — still stands as the most challenging problem to be solved.

2. Magnetic Susceptibility

With the exception of NiP_3, all other binary skutterudites show a small, negative, and weakly temperature-dependent magnetic susceptibility χ. To illustrate the point, Fig. 24 presents the data for $CoSb_3$ that include both polycrystalline specimens as well as a single crystal. Table VII lists room

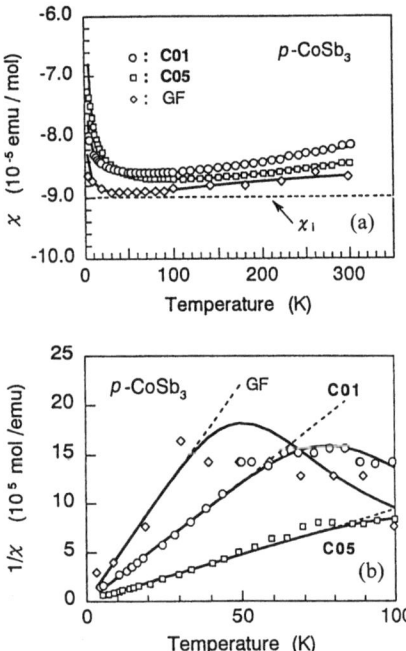

FIG. 24. (a) Temperature dependence of molar magnetic susceptibility of two polycrystalline samples and a single crystal of $CoSb_3$. The solid lines are the fits according to Eq. (12). The dashed line indicates the ion core contribution. The data for the single crystal are from Morelli et al. (1995). (b) Inverse molar magnetic susceptibility of the same samples as in (a). The diamagnetic contribution of the ion cores was subtracted from the data. Molar weight of $CoSb_3$ is 424.2 g (from Anno et al., 1998a).

TABLE VII
ROOM TEMPERATURE SUSCEPTIBILITIES OF BINARY SKUTTERUDITES

Skutterudite	Magnetic susceptibility χ (in units of 10^{-6} emu/g)	Reference
CoP_3	−0.092	Ackermann and Wold (1977)
$CoAs_3$	−0.158	Pleass and Heyding (1962)
	−0.106	Ackermann and Wold (1977)
$CoSb_3$	−0.165	Hulliger (1961)
	−0.111	Ackermann and Wold (1977)
	−0.189	Mandrus et al. (1995)
	−0.121	Morelli et al. (1997)
	−0.200	Morelli et al. (1995)
RhP_3	−0.281	Hulliger (1961)
$RhAs_3$	−0.214	Hulliger (1961)
	−0.219	Pleass and Heyding (1962)
$RhSb_3$	−0.203	Hulliger (1961)
$IrAs_3$	−0.251	Hulliger (1961)
	−0.253	Pleass and Heyding (1962)
	−0.210	Kjekshus and Pedersen (1961)
$IrSb_3$	−0.260	Hulliger (1961)
	−0.160	Kjekshus and Pedersen (1961)

temperature magnetic susceptibilities of most of the binary skutterudites. The data are uncorrected for the core diamagnetism of the constituent species that represents the dominant contribution to the diamagnetic signal. The core susceptibility is independent of the temperature, and for $CoSb_3$ it is approximately $\chi_{core} = -0.212 \times 10^{-6}$ emu/g. By subtracting the core diamagnetism, one would obtain a very small positive (paramagnetic) value that would be equivalent to the effective magnetic moment of much less than one Bohr magneton (1 μ_B = 9.27 × 10^{-24} J/T). Such small values testify to the essentially covalent bonding in the binary skutterudites. Had the bonds been of ionic character, just the "spin only" contribution to the magnetic susceptibility would yield a value of several μ_B.

Referring to Fig. 24a, χ shows weak temperature dependence, and one can distinguish two regions: a very gradual decrease of the diamagnetic signal as the temperature increases above 50 K, and a rapidly decreasing susceptibility below 50 K. In the former case, the decrease in χ reflects the semiconducting nature of $CoSb_3$—with the increasing temperature the carrier density increases and the paramagnetic contribution increases. At low temperatures, the rapid decrease in the measured (negative) susceptibility is due to the presence of a paramagnetic impurity. This follows from Fig. 24b, where the data are plotted in the form χ^{-1} vs. T (after subtracting the diamagnetic contribution of the cores) and the Curie law behavior at low temperatures is clearly evident. Polycrystalline samples of $CoSb_3$, as might be expected, show a stronger paramagnetic signal than the single crystal. Anno et al. (1998a) fitted the susceptibility data assuming three distinct contributions: the temperature-independent diamagnetic susceptibility of the ion cores χ_{core}, the paramagnetic susceptibility of charge carriers χ_e (assumed proportional to their number density), and the paramagnetic susceptibility of magnetic impurities χ_{imp} (given by the Curie law):

$$\chi = \chi_{core} + \chi_e + \chi_{imp} = \chi_{core} + N_o e^{-E_g/2k_B T} + \frac{C}{T}. \quad (12)$$

Here N_o is a constant, E_g is the bandgap, and C is the Curie constant. The fits using Eq. (12) are shown in Fig. 24a as solid lines through the data points. The bandgap value from the fit comes out as $E_g \approx 70\text{--}80$ meV, in good agreement with the theoretical estimate of 50 meV (Singh and Pickett, 1994). The fitted Curie constants were 9.66×10^{-8}, 2.36×10^{-7}, and $5.19 - 10^{-8}$ emuK/g for the samples labeled C01, C05, and GF, respectively. Had impurities been identified (i.e., had their effective magnetic moments been known) one could, given the values of the Curie constant, calculate the respective concentrations of the magnetic impurities in the samples.

It is interesting to ask what happens to the magnetic susceptibility when some of the Co atoms are replaced by atoms of the Ni and Fe. This was considered by Pleass and Heyding (1962), who studied the magnetic response of arsenide skutterudites. Because Ni is a donor impurity, its presence on the sites of Co should lead to an enhanced paramagnetic response. While at 3 at% substitution the room temperature susceptibility remained weakly diamagnetic, at higher concentrations of Ni the paramagnetic response prevailed and the straight lines in the plot of χ^{-1} vs T yielded effective magnetic moments per Ni atom of $0.80\,\mu_B$ at 20% substitution and $1.24\,\mu_B$ at 40% substitution. In arsenides, the structure can accommodate up to 65 at% Ni; hence both the 20% and 40% substitution levels are well within the solid solubility limit. When considering Fe substitution on the sites of Co, it should be remembered that the solubility limit of Fe in arsenide skutterudites is considerably lower ($\leqslant 16$ at% Fe) than the solubility limit for Ni. Consequently, one is restricted to alloys with a smaller amount of magnetic material in the structure and the paramagnetic signal may not dominate over the core diamagnetism. This seems to be the case at least for low Fe concentrations. To assess the paramagnetic response in this case, one must subtract the core diamagnetism. Rather than accounting for the individual diamagnetic contributions of As, Co, and Fe, Pleass and Heyding made an approximation by assuming that the core diamagnetism in $Fe_xCo_{1-x}As_3$ is the same as in pure $CoAs_3$. By plotting the resulting paramagnetic response as χ^{-1} vs. T, they extracted the effective magnetic moments per Fe atom of $1.88\,\mu_B$ (for 15 at% Fe), $2.71\,\mu_B$ (for 10.7 at% Fe),

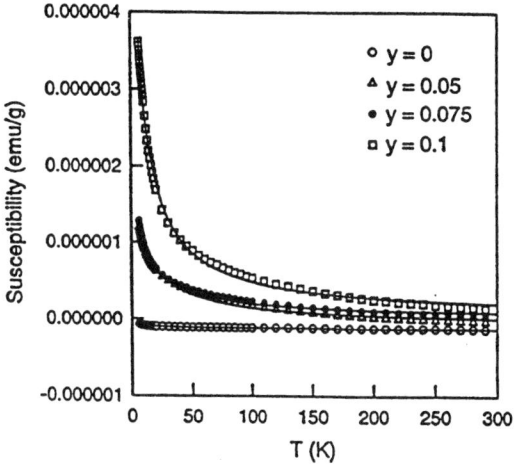

FIG. 25. Temperature dependence of magnetic susceptibility of Ce_yCoSb_3. Solid lines are fits using Eq. (13) with the following parameter values: $y = 0.05$, $C = 2.43 \times 10^{-5}$ emuK/g, $\theta = 30.2$ K, $\chi_o = -1.57 \times 10^{-7}$ emu/g; $y = 0.075$, $C = 1.85 \times 10^{-5}$ emuK/g, $\theta = 0.2$ K, $\chi_o = -3.64 \times 10^{-8}$ emu/g; $y = 0.10$, $C = 4.50 \times 10^{-5}$ emuK/g, $\theta = 10.6$ K, $\chi_o = -2.00 \times 10^{-8}$ emu/g (from Uher et al., 1997).

and 2.06 μ_B (for 5.9 at% Fe). The data are thus consistent with an unpaired electron in Fe, although the effective moment seems to decrease as the Fe concentration approaches the solubility limit.

A very interesting situation is encountered in the coupled replacement of Co with Fe and Ni, i.e., in the alloys $Fe_xNi_{1-x}As_3$ with $x \approx 0.5$ formed according to Eq. (6). The model of Dudkin predicts such structure to be a strong paramagnet because of the unpaired spin on Fe and, at the same time, a good metallic conductor because of the conduction electron on Ni. But this is not the case. Not only is such compound semiconducting, but the magnetic susceptibility is small. The magnetic moment per metal atom is much smaller than the values obtained with the $Fe_xCo_{1-x}As_3$ system and, in fact, independent measurements (Nickel, 1969) have indicated that $Fe_{0.5}Ni_{0.5}Sb_3$ is a diamagnetic solid. As already noted in Subsection 1 of Section II, it appears that electron transfer occurs between Fe and Ni resulting in the formation of (Fe^-Ni^+) pairs equivalent to Co.

By inserting filler species into the voids of the skutterudite structure, one effectively dopes the system. In the case of rare earth, the ions such as La^{3+} or Ce^{3+} donate three electrons, and this should be immediately evident in the behavior of magnetic susceptibility. From the data in Fig. 25, the reader will appreciate a rapid development of a paramagnetic signal in Ce_yCoSb_3 that becomes progressively stronger with the increasing amount of Ce (remember, $y \approx 0.1$ is a limiting Ce occupancy of voids in $CoSb_3$). From the fits of the susceptibility using the Curie–Weiss form (solid lines through the data points in Fig. 25),

$$\chi = \frac{C}{T - \theta} + \chi_o. \qquad (13)$$

Uher et al. (1997) determined the Curie constant for each compound, which, in turn, allowed them to deduce the effective magnetic moment per Ce atom of 2.58, 1.84, and 2.48 μ_B for the samples with $y = 0.05$, 0.075, and 0.1, respectively. The average effective moment of 2.30 μ_B is consistent with the theoretical value of 2.54 μ_B for trivalent cerium.

We now turn our attention to filled skutterudites. There have been several investigations of mostly phosphide- and antimonide-based compounds. In one of the first reports on the magnetic properties of filled skutterudites, Guertin et al. (1987) studied the magnetic response of UFe_4P_{12}, perhaps the most exotic of filled skutterudites. This compound is a ferromagnetic semiconductor with the Curie temperature $T_c = 3.15$ K, and its semiconducting character is believed to originate from strong hybridization of the f-electrons of U (Meisner et al., 1985). It is unlikely that the ferromagnetic order is mediated by conduction electrons (the structure is semiconducting). Rather, the magnetic coupling between the uranium ions is based on a superexchange mechanism.

In filled skutterudites that contain Fe, a question arises regarding the magnetic state of Fe and what fraction, if any, of the total magnetic moment it might account for. On one hand, there are strong arguments for the Fe atoms in phosphides to have a zero or vanishingly small magnetic moment; on the other hand, in antimonides, there are reasons to believe that the Fe–Sb polyanion carries a magnetic moment.

The evidence for a nonmagnetic state of Fe in phosphide skutterudites is quite compelling. Perhaps the most persuasive experimental evidence is the occurrence of superconductivity with a surprisingly high transition temperature in $LaFe_4P_{12}$; see Table II. Had Fe carried a magnetic moment, it would surely break the Cooper pairs via exchange coupling. Moreover, an essentially zero moment of Fe is also heralded by the Mössbauer measurements discussed in the preceding section. The bottom line is that in rare-earth-filled iron phosphide skutterudites, there is no evidence whatsoever for a contribution of Fe to the magnetic susceptibility of the structure. Thus, the paramagnetism observed in $LnFe_4P_{12}$, Ln = La, Ce, Pr, Nd, Sm, as well as the surprisingly strong ferromagnetism of $EuFe_4P_{12}$ ($T_c \sim 100$ K), both originate with the rare earth filler ions. Specifically, let us consider the case of magnetic susceptibility of $CeFe_4P_{12}$ and $LaFe_4P_{12}$.

Experimental results indicated a small and weakly temperature-dependent paramagnetic response from which one can obtain room temperature effective moments of $1.07\,\mu_B$ per formula unit for $CeFe_4P_{12}$ and $1.46\,\mu_B$ per formula unit for $LaFe_4P_{12}$. In the case of Ce-filled phosphide, a pronounced dip in its cell volume (see Fig. 7) gave a hint of the 4+ valence state of Ce that would make the polyanion $[Fe_4P_{12}]^{4-}$ saturated, and the compound should be a diamagnetic solid. To rationalize its weak paramagnetism, Grandjean et al. (1984) proposed that the paramagnetic behavior is the result of a mixed (or intermediate) valence of Ce. To achieve the overall magnetic moment of $1.07\,\mu_B$, one requires a little over 40% of Ce atoms to be in the 3+ state, assuming the free ion moment of Ce^{3+} is $2.54\,\mu_B$. Since $CeFe_4P_{12}$ shows a semiconducting temperature dependence, an alternative scenario is a strong hybridization of the Ce $4f$ level with a broad band of the iron-phosphorus polyanion, which leads to a gap opening in the density of states at the Fermi level. The idea of hybridization is supported by electron spin resonance experiments (Martins et al., 1994) that indicate the absence of g shift and thermal broadening line for all rare earths.

The situation of $LaFe_4P_{12}$ is somewhat different. Since the lanthanum ions invariably assume a 3+ state, the valence electron count is 71 (one electron missing), the polyanion $[Fe_4P_{12}]^{3-}$ cannot be completely saturated, and the paramagnetic behavior of $LaFe_4P_{12}$ is not surprising. The problem is that the Mössbauer measurements on the phosphide skutterudites leave no doubt about the Fe atoms being nonmagnetic. It is thus difficult to contemplate that the missing electron originates from the nonbonding localized Fe $3d$ states, because this would imply that one-quarter of the Fe atoms have a low-spin d^5 configuration with a magnetic moment of $1.73\,\mu_B$. Although this value is not too far from the experimental value of $1.46\,\mu_B$, the Mössbauer measurements clearly deny the possibility of Fe being magnetic. So where does the moment of $LaFe_4P_{12}$ come from? Perhaps a hint comes from the weak temperature dependence of the susceptibility that points towards some kind of strong Pauli paramagnetism.

$EuFe_4P_{12}$ presents an interesting case of its own. Eu usually occurs as Eu^{2+}, and since Fe is clearly nonmagnetic in this lattice, Eu^{2+} ions are the species responsible for the ferromagnetic order below 100 K. The experimental magnetic moment of $6.2\,\mu_B$ per formula unit is a bit lower than the value expected for the Eu^{2+} ion ($7.94\,\mu_B$ per formula unit), but the discrepancy could be rectified by assuming that some of Eu ions assume the Eu^{3+} configuration that nominally has zero magnetic moment. This would effectively dilute the overall magnetic moment of the structure. However, the situation may be more complicated because of the influence of crystal field effects.

Turning our attention to the filled antimonide skutterudites, we see a very different picture. Here the experimental evidence suggests that the polyanionic iron-antimony framework carries a magnetic moment and its contribution is a significant part of the magnetic state of the system. In their work on magnetic properties of alkaline earth-filled antimonide skutterudites, i.e.,

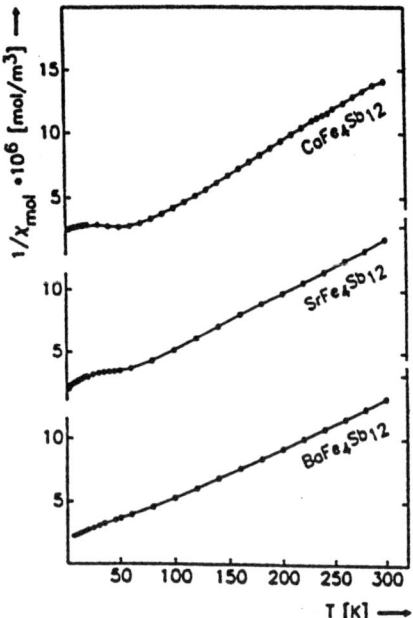

FIG. 26. Inverse magnetic susceptibility as a function of temperature for the alkaline earth antimonide skutterudites, RFe_4Sb_{12}, $R = Ca$, Sr, and Ba, measured in a magnetic field of 3 T (from Danebrock et al., 1996). Reprinted with permission from Elsevier Science.

in the compounds RFe_4Sb_{12} where $R = Ca$, Sr, and Ba, Danebrock et al. (1996) observed (see Fig. 26) strong Curie–Weiss behavior from which they extracted effective magnetic moments per formula unit between 3.7 and 4.0 μ_B. Since alkaline earths are nonmagnetic, the observed nonzero moments must come from the polyanionic framework, which in this case is $[Fe_4Sb_{12}]^{2-}$. Each alkaline earth atom transfers two electrons to the polyanionic complex, and thus two out of four Fe atoms are expected to carry uncompensated spins, i.e., they have the magnetic d^5 configuration. Expressed per Fe^{3+} per formula unit, the observed magnetic moments are then in the range 2.6–2.8 μ_B. Compared to the theoretical spin-only value of the moment, $\mu_{eff} = 2[s(s+1)]^{1/2}$ μ_B, the experimental magnetic moment is larger and the difference was ascribed to the effect of spin–orbit coupling.

The fact that the polyanionic framework has a significant moment in the case of alkaline-earth-filled antimonide skutterudites prompted the same authors to look closely into the possibility that there is also a magnetic moment associated with the $[Fe_4Sb_{12}]^{3-}$ polyanionic complex, i.e., when one fills the voids with the rare-earth ions. In this case one would have two magnetic contributions, one originating with the rare earth and the other

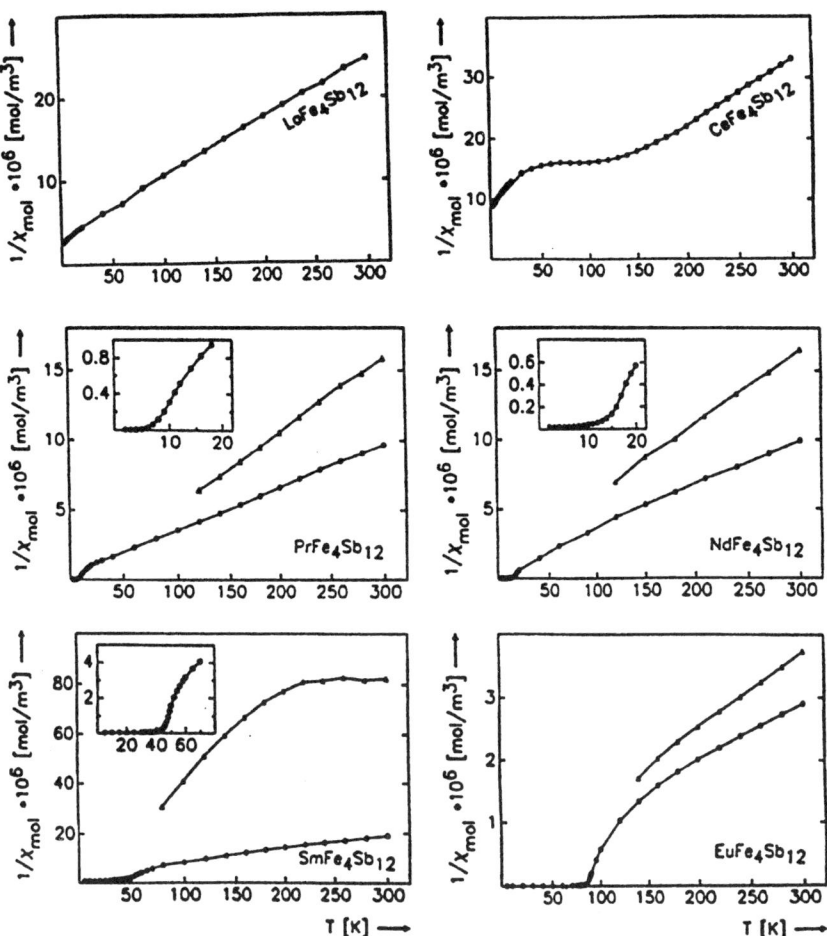

FIG. 27. Inverse magnetic susceptibility as a function of temperature for LnFe$_4$Sb$_{12}$, (Ln = La, Ce, Pr, Sm, Nd, and Eu) measured in a field of 3 T. The insets in the diagram for PrFe$_4$Sb$_{12}$, SmFe$_4$Sb$_{12}$, and NdFe$_4$Sb$_{12}$ show the low-temperature behavior in a magnetic field of 0.1 T. Of the two curves in the main diagrams, the upper ones indicate the inverse susceptibilities obtained from the rare earth component after subtracting the susceptibility of LaFe$_4$Sb$_{12}$ (for the compounds with Ln = Pr, Nd, Sm) and CaFe$_4$Sb$_{12}$ (Ln = Eu) from the total experimental values (adapted from Danebrock et al., 1996). Reprinted with permission from Elsevier Science.

one with the [Fe$_4$Sb$_{12}$]$^{3-}$ complex. This, of course, would stand in contrast to the case of phosphide skutterudites. Exploring a range of rare earth fillers, Danebrock et al. noted that they all display Curie–Weiss behavior at least at elevated temperatures (Fig. 27). The respective magnetic moments per formula unit obtained from the Curie constant came out as $\mu_{\text{exp}} = 2.4\,\mu_B$ for

$CeFe_4Sb_{12}$, $\mu_{exp} = 3.0\,\mu_B$ for $LaFe_4Sb_{12}$, $\mu_{exp} = 4.6\,\mu_B$ for $PrFe_4Sb_{12}$, $\mu_{exp} = 4.5\,\mu_B$ for $NdFe_4Sb_{12}$, and $\mu_{exp} = 8.4\,\mu_B$ for $EuFe_4Sb_{12}$. The low-temperature behavior of susceptibility of the last three listed compounds was indicative of ferromagnetism with Curie temperatures of $T_c = 5\,K$, 13 K, and 82 K, respectively.

Among the preceding skutterudites, $LaFe_4Sb_{12}$ represents a special case. Because the La^{3+} ion has no magnetic moment, the measured nonzero values of the magnetic moment must necessarily come from the magnetism of the $[Fe_4Sb_{12}]^{3-}$ polyanion. Thus, in this case, there is no doubt about the Fe being in a low-spin but magnetic d^5 state. It is of considerable interest to know and understand magnetic properties of $LaFe_4Sb_{12}$, as this compound can serve as a benchmark against which to compare the influence of the magnetic rare earths in the sister lanthanide-filled iron antimonide skutterudites. I should point out that the diamagnetic signal observed in $LaIr_4Ge_3Sb_9$ (Nolas et al., 1996b) is no surprise because the framework $[Ir_4Ge_3Sb_9]^{3-}$ is nonmagnetic.

The issue of whether the polyanionic complex contributes to the magnetic moment of $CeFe_4Sb_{12}$ is controversial. Assuming the Ce^{3+} valence and a contribution from the $[Fe_4Sb_{12}]^{3-}$ polyanion comparable to that in the case of $LaFe_4Sb_{12}$, the magnetic moment of $CeFe_4Sb_{12}$ should be considerably larger ($2.54\,\mu_B$ for Ce^{3+} plus a contribution from the polyanionic complex) than the experimental value determined by Danebrock et al. (1996), $\mu_{exp} = 2.4\,\mu_B$. To come close to the measured moment, the authors proposed a roughly 40/60 split for the Ce^{3+}/Ce^{4+} ions. This is not exactly a rigorous estimate, because they did not have a good handle on the actual value of the magnetic moment of the polyanion in the $CeFe_4Sb_{12}$ compound. Nevertheless, Chen et al. (1997a) also concluded from their measurements that Ce in $CeFe_4Sb_{12}$ likely exhibits intermediate valence behavior at low temperatures. Furthermore, they studied the effect of magnetic dilution by alloying Co on the site of Fe in compounds of the series $Ce_yFe_{4-x}Co_xSb_{12}$. With the increasing Co concentration, fewer Fe atoms contribute to the overall moment, and since Co is expected to be in the nonmagnetic d^6 state, the overall magnetic moment rapidly decreases.

A different perspective on the problem was offered by Sales et al. (1997), who assumed that cerium is trivalent at all temperatures. To explain low-temperature anomalies, they invoked hybridization that arises from the proximity of the Ce $4f$ level to the Fermi energy. They also pointed out that in a cubic environment, the ground state of the cerium ion with a degeneracy of 6 is split into a doublet and a quartet. The best fit of the data is achieved when one assumes that the doublet lies lower and is separated by $(350 \pm 50)\,K$ from the quartet, and the effective moment is $2.9\,\mu_B$ rather than the ionic value of $2.54\,\mu_B$. At temperatures below 15 K, the Curie law yields an effective moment of $1.5\,\mu_B$ that agrees well with the value of $1.4\,\mu_B$, the magnetic moment expected from the doublet. Yet another viewpoint has been advanced by Gajewski et al. (1998). Based on the specific heat coefficient γ and magnetic susceptibility of $CeFe_4Sb_{12}$, they argued that the low temperature properties are due to the itinerant electrons and the susceptibility corresponds to Pauli paramagnetism of the conduction electrons. They justified their argument by pointing out that the Wilson–Sommerfeld ratio $R = (\chi/\gamma)(\pi^2 k_B^2/\mu_{eff}^2)$ is of the order of unity, an indication of the itinerant nature of carriers.

A prominent "bulge" on the curve of inverse susceptibility of $CeFe_4Sb_{12}$, similar to the one shown in Fig. 27, was seen previously by Morelli and Meisner (1995) who characterized it as a signature of a heavy-fermion-like behavior. The authors noted that as the temperature approaches absolute zero, the susceptibility becomes very large, with values exceeding most of the established heavy fermion systems (Stewart, 1984). Based on the correlation among f-electron spacing (Ce–Ce distance ~ 6.5 Å), susceptibility, and electronic specific heat, Morelli and Meisner predicted the linear coefficient of the specific heat to reach values of $\gamma \sim 1000$ mJmol^{-1}K^{-1}! The heavy fermion character of the Ce-filled antimony skutterudites was explored further in Gajewski et al., (1998).

At this stage, there is no universally agreed-upon mechanism that causes anomalies in the low-temperature magnetic behavior of $CeFe_4Sb_{12}$ nor is it clear how large a magnetic contribution comes from the $[Fe_4Sb_{12}]^{3-}$ polyanion. If anything, the latest Mössbauer study (Long et al., 1999) on $CeFe_4Sb_{12}$ indicates a zero moment associated with the $[Fe_4Sb_{12}]^{3-}$ polyanionic framework. It should also be noted that the reported effective magnetic moments in this compound vary between 2.6 and 4.1 μ_B, and this rather wide margin allows for various alternative interpretations. More experimental work is needed; especially, electron spectroscopy studies and X-ray-absorption measurements near the Ce L_{III} absorption edge would be most useful, in order to shed light on the valence state of Ce and the electronic configuration in $CeFe_4Sb_{12}$.

Correlated electron effects, of which heavy fermion behavior is one of the most interesting examples, are often observed in systems with elements that are close to the f-electron valence instability. Such systems can be described as having a characteristic energy (or temperature) above which the structure displays magnetic (Curie law) behavior, while at low temperatures the material shows Pauli-like susceptibility. The structure thus exhibits elements of localized behavior as well as characteristics one associates with itinerant electrons. Once the $CeFe_4Sb_{12}$ was recognized as having characteristics of a heavy fermion system, considerable interest developed in searching for the correlated electron behavior in other filled skutterudites. One of the primary targets was $YbFe_4Sb_{12}$ (Dilley et al., 1998; Leithe-Jasper et al., 1999), a skutterudite filled with Yb ions that are known to exhibit an intermediate

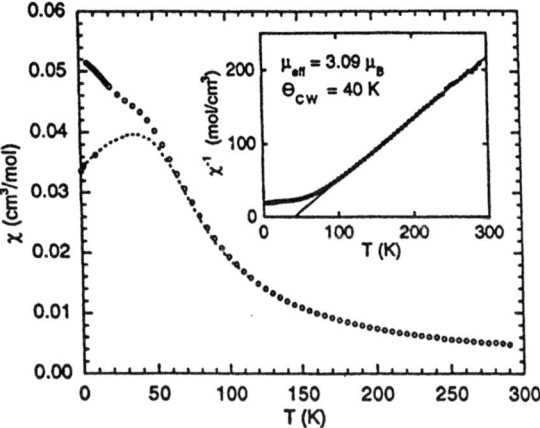

FIG. 28. Magnetic susceptibility χ (open circles) of YbFe$_4$Sb$_{12}$ measured at 5.5 T and the estimated high-field susceptibility (filled circles) as a function of temperature. The dashed line is the conjectured behavior of the high-field susceptibility. The inset shows inverse magnetic susceptibility χ^{-1} vs. temperature. The line through the data shows a fit to a Curie–Weiss law above 100 K (from Dilley et al., 1998). Copyright 1998 by the American Physical Society.

valence. Moreover, Yb is also interesting for its location within the rare earth series — it is the $4f$-hole analog of Ce. Being placed far down in the rare earth series, Yb would be expected to have too small a radius to bond adequately with Sb had its ions assumed an exclusively 3+ valence state. However, Yb sometimes occurs in a divalent state, in which case its ionic radius is larger and bonding is possible. Dilley et al. (1998) measured the lattice constant of YbFe$_4$Sb$_{12}$ and found it comparable to that of EuFe$_4$Sb$_{12}$ (recall Eu having valence 2+) and considerably larger than the lattice parameter of the trivalent fillers such as Sm. Magnetic measurements performed by the authors delineated two distinct regimes of magnetism. At temperatures above 100 K they observed a Curie–Weiss behavior with an effective magnetic moment of 3.09 μ_B per formula unit, whereas at low temperatures (below 50 K) the magnetic behavior consists of a saturable magnetization of localized magnetic moments combined with a susceptibility evident at high fields that reflects Pauli paramagnetism of the conduction electrons (Fig. 28). Thus, the structure displays the features that have both localized and itinerant character. Specific heat measurements made by the authors indicated an enhanced linear coefficient of specific heat, and the Wilson–Sommerfeld ratio of 2.62 suggested that the enhancement in the susceptibility and the linear term in the specific heat are due to heavy electrons. The authors attributed the heavy carrier mass to intermediate valent Yb ions and to a Kondo lattice of screened Yb^{3+} moments. Although plausible, this conclusion is only tentative. The problem is that one does not know the magnetic moment of the iron–antimony polyanionic framework. If the magnetic moment of $[Fe_4Sb_{12}]^{2-}$ was anywhere near the value of $2.6\,\mu_B$ per formula unit observed in $EuFe_4Sb_{12}$ (recall Eu being divalent), the Yb ion would be required to supply only a very small moment ($\sim 0.5\,\mu_B \approx (3.09-2.6)\,\mu_B/\text{f.u.}$). On the other hand, if for some reason the iron–antimony framework had zero magnetic moment, then the experimental value of $3.09\,\mu_B$ per formula unit would require a significant fraction of the Yb ions to be in the 3+ state ($\mu_{eff} = 4.5\,\mu_B/\text{f.u.}$), since the magnetic moment of the divalent Yb is zero. Clearly, it is essential to know the magnetic contribution of the polyanionic framework before one can make judgment about the presence and abundance of the Yb^{3+} ions in this skutterudite. In the background, of course, there also looms the question of how large fraction of the Yb^{3+} ions the skutterudite structure can tolerate and still form adequate bonds.

The puzzle concerning the total charge balance in $YbFe_4Sb_{12}$ has not been resolved even in subsequent investigations by Leithe-Jasper et al. (1999). Although the susceptibility behavior they observed was similar to that of Dilley et al. (1998), the resulting effective magnetic moment was considerably larger ($\mu_{eff} = 4.49\,\mu_B/\text{f.u.}$). The authors made both X-ray absorption edge measurements and Mössbauer studies, but the results are very puzzling. Although they resolved two distinct peaks in the Yb L_{III} edge indicating the presence of both Yb^{2+} and Yb^{3+} ions for an average valence of 2.68, their Mössbauer studies suggested that the Fe atoms carry no magnetic moment! If true, the only magnetic species in the structure are the Yb^{3+} ions. It so happens that the expected effective moment of Yb^{3+} is $4.54\,\mu_B$, a value that is comparable to the measured moment of $4.49\,\mu_B$ per formula unit. However, this is likely nothing more than a coincidence, since one has to find an answer to what happened to the magnetic moment of the $[Fe_4Sb_{12}]^{2-}$ and $[Fe_4Sb_{12}]^{3-}$ polyanions expected to contribute moments of 2.45 and $1.73\,\mu_B/\text{f.u.}$, respectively.

Surveying the field of magnetic properties in the filled skutterudites, on one hand one witnesses the richness of the phenomena that result from altering the filler ions and one appreciates the important data the magnetic studies provide regarding the charge state and bonding in the skutterudite compounds. On the other hand, one also experiences frustration with less than complete results and their conflicting interpretation. Consequently, detailed understanding of the magnetic phenomena and an accurate assessment of how much various magnetic entities contribute to the overall magnetic moment are still lacking. This is one area where a major effort would pay dividends, and it is something that is needed to be resolved before one can say that we understand the physics of skutterudites.

VII. Transport Properties

1. ELECTRONIC TRANSPORT

a. Binary Skutterudites

As part of their extensive survey of semiconducting materials aimed at identifying promising thermoelectrics, Dudkin and his colleagues at the Baikov Institute of Metallurgy in Moscow were first to carry out an in-depth study of the electronic transport properties of binary skutterudites. Their primary focus was $CoSb_3$ and how its transport behavior is affected by replacing some of the Co atoms with Ni and Fe, and a related question concerning the role of doping on the Sb sites. The results of this study were published in a series of papers between 1956 and 1960 and, until recently, formed one of the primary sources of information about the transport processes in skutterudites.

Measurements on pure $CoSb_3$ — the benchmark for the study — revealed (Dudkin and Abrikosov, 1956) an activation behavior of transport with the energy gap of 0.5 eV and a room temperature value of the resistivity within the range 33–100 mΩ-cm. Stoichiometric samples displayed n-type conduction, with the room temperature thermopower attaining large values on the order of $-200\,\mu V/K$. The mobility of electrons in this polycrystalline sample was quoted (Dudkin and Abrikosov, 1957) as 290 cm^2/V-s. In samples departing from stoichiometry in the direction of higher Sb content, the thermopower was smaller in magnitude. In samples with large Sb excess, the thermopower changed sign and attained small positive values. Positive thermopower was also observed in stoichiometric $CoSb_3$ in the regime of intrinsic conduction (above about 500–600 K), suggesting that the mobility of holes is much higher than the mobility of electrons.

In attempting to alter transport properties of $CoSb_3$, Dudkin and Abrikosov (1959) considered replacing some of the Co atoms with six different metals (Cu, Zr, Al, Ti, Fe, and Ni), and searched for a suitable doping agent for the Sb site among the elements likely to form tetrahedral bonds similar to those formed by antimony in the skutterudite structure: Si, Ge, Sn, Pb, Bi, Se, and Te. Of all these elements, only four proved to have a strong influence on the transport properties: Ni and Fe on the Co site, and Sn and Te as antimony dopants.

The valence state of Ni that closely matches the octahedral covalent radius of the trivalent Co is Ni^{4+}. Upon substitution of Ni for Co, each Ni atom introduces an additional electron that cannot be accommodated in the skutterudite structure, and this electron is thermally excited into the conduction band. One thus expects the Ni atoms to act as electron donors. This is borne out by the experimental data. However, there are subtleties involved. One of them is the observation that a small amount of Ni (up to about 1 at%), while causing an expected increase in the electrical conductivity, also leads to an increased thermopower in comparison to the pure $CoSb_3$. Remarkably, the higher conductivity does not appear to be solely due to the expected increase in the carrier density. An important contribution comes from the carrier mobility, which more than doubled relative to the value measured on pure cobalt triantimonide. In conjunction with the apparent improvement in the crystalline character of the samples containing $\lesssim 1$ at% Ni, Dudkin and Abrikosov (1957) also noted an increase in the microhardness of the structure. As an explanation they proposed that a small amount of Ni impurity "corrects" the defects and distortions in the crystal lattice of the skutterudite structure. This correction, however, applies only to samples with low Ni content ($\lesssim 1$ at%). With the concentration of Ni above 1 at%, the trend rapidly reverses. Although the electrical conductivity continues to increase on account of the steeply rising free carrier density, the mobility nosedives, the thermopower diminishes, and the microhardness decreases. A plausible account of this behavior is the formation of Ni impurity levels of low activation energy that release electrons into the conduction band at relatively low temperatures. As a consequence, one expects a continuously increasing carrier density that will have an adverse impact on the magnitude of the thermopower. Excess of Ni will also affect interatomic bonding, and the readily ionized Ni impurities will scatter free charge carriers, leading to a degradation of the carrier mobility. As far as the temperature dependence of the transport parameters is concerned, samples with low Ni concentration ($\lesssim 1$ at% Ni) display a weak (nearly constant) temperature dependence of resistivity. As the Ni content increases, the resistivity attains a metallic character. At elevated temperatures (above about 600 K), the intrinsic conduction takes over, the resistivity decreases with increasing temperature, and the Hall effect changes sign from negative to positive, reflecting the higher mobility of holes.

When Fe substitutes for Co, its d-shell has one electron less than the d-shell of cobalt. Since there are two valence states of iron, Fe^{2+} (1.23 Å) and Fe^{3+} (1.22 Å), that closely match the octahedral covalent radius of Co^{3+} (1.22 Å), it is not clear *a priori* which Fe ion substitutes for the trivalent cobalt. Making a plausible assumption that it is the Fe^{3+} state that matches perfectly with Co^{3+}, iron and cobalt then have the same valence electron structure (they differ in how the inner, nonvalence $3d$-shells are filled, and the unpaired electron of Fe will lead to paramagnetism). According to Dudkin, iron should then be inactive in $CoSb_3$. In reality, experimental data show great sensitivity to the presence of iron just as they did in the case of Ni. Comparing the results for cobalt triantimonide containing 1 at% Ni and 1 at% Fe (see Fig. 29a), we notice a similar trend in the thermopower: the initial increase with increasing temperature followed by a break near 600 K and a decreasing thermopower afterward. The thermopower of the Fe-substituted sample is, however, only about one-half that

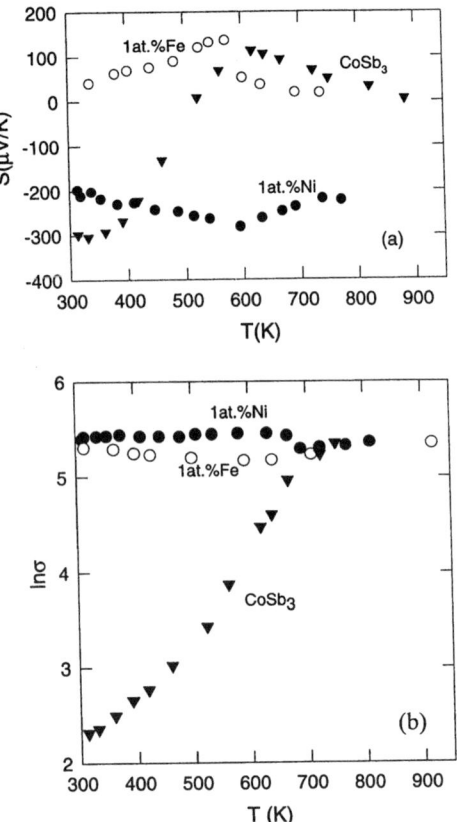

FIG. 29. (a) Temperature dependence of thermopower for polycrystalline samples of $CoSb_3$ containing 1 at% Ni (solid circles) and 1 at% Fe (open circles). Solid triangles represent the data for nominally undoped $CoSb_3$. (b) Electrical conductivity as a function of temperature for the same three samples. (Adapted from the data of Dudkin and Abrikosov (1956, 1957, 1959).

of the Ni-substituted sample. The electrical conductivity of the two samples (Fig. 29b) also displays a distinct break near 600 K, but at lower temperatures the 1 at% Ni sample shows a weak temperature dependence while the conductivity of the 1 at% Fe sample increases as one cools toward room temperature. To explain the transport behavior of Fe-substituted cobalt triantimonide, Dudkin and Abrikosov (1959) argued that it is not the direct effect of Fe atoms in the lattice, but rather the shift in the skutterudite phase stoichiometry in the direction of increasing antimony content that causes the observed features. This, in their view, leads to the formation of lattice defects at concentrations comparable to those of Fe substituting for Co.

Of several prospective doping agents on the Sb site, only two—tellurium and tin—proved to be electrically active (Zobrina and Dudkin, 1960). Although the solubility limits of Te (~ 0.25 at%) and Sn (~ 0.15 at%) are not large, the influence of the two species on the transport properties of $CoSb_3$ is dramatic. Tellurium acts as a donor, and as little as 0.1 at% Te increases the electrical conductivity of cobalt triantimonide by an order of magnitude. The thermopower, negative in sign, initially increases with increasing Te content, reaches a peak value of $\sim -300\,\mu V/K$ at a concentration of 0.05 at% Te, and then gradually decreases. Temperature dependence of the thermopower is similar to that of Ni-substituted $CoSb_3$. It increases in magnitude up to about 600 K, where the intrinsic conduction starts to dominate. From then on, the thermopower decreases rapidly as the mobile holes start to "short" the thermoelectric voltage. Meanwhile, the electrical conductivity decreases with increasing temperature until the onset of intrinsic conduction sets in. The Hall coefficient data confirm that the crossover from the impurity-dominated regime to the domain of intrinsic conduction is accompanied by a change in the sign of the dominant carriers in favor of holes. Room temperature carrier densities (electrons) and mobilities of samples with the Te content between 0.05 and 0.2 at% are in the range of 5.9×10^{18} to 2.1×10^{19} cm^{-3}, and 95 to 70 cm^2/V-s, respectively.

Although tin impurity is an acceptor, its influence on the transport properties is equally dramatic. Just 0.1 at% Sn yields a carrier density (holes) of 1.6×10^{19} cm^{-3}, a room temperature conductivity of 1600 Ω^{-1} cm^{-1}, and a high mobility ~ 590 cm^2/V-s. Electrical conductivity is a weak function of temperature until the regime of intrinsic conduction is reached, and then it decreases rather rapidly. Thermopower is considerably smaller (only about 60 $\mu V/K$ for 0.1 at% Sn), in accord with the much enhanced electrical conductivity. The respective influence of Sn and Te impurities is given by their ability to substitute for antimony. Of all the potential dopants, Sn and Te have the tetrahedral covalent radii that most closely match that of Sb. Furthermore, the donor (acceptor) character of Te (Sn) follows from their symmetrical position with respect to the location of Sb in the periodic table. Regardless of the carrier sign, the data suggest that the energy spectrum of the doped $CoSb_3$ includes a narrow impurity band that dominates the transport behavior until the intrinsic conduction takes over at temperatures near 600 K.

Transport studies of comparable significance, this time on arsenide skutterudites, were made by Pleass and Heyding (1962). Samples were prepared by the direct interaction of arsenic and metal powders in sealed, evacuated tubes followed by prolonged annealing at 750°C. All three pure arsenides—$CoAs_3$, $RhAs_3$, and $IrAs_3$—displayed p-type semiconducting behavior with room temperature resistivities on the order of 8 mΩ-cm for $CoAs_3$ and 5 mΩ-cm for $RhAs_3$. The intrinsic regime of conduction seems to set in at lower temperatures (not much above 200 K) than for the binary

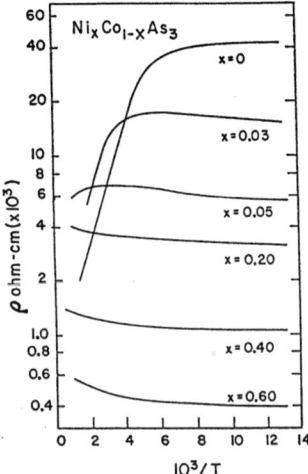

FIG. 30. Temperature dependence of the electrical resistivity of $Ni_xCo_{1-x}As_3$ (adapted from Pleass and Heyding, 1962). Reprinted with permission from NRC Research Press.

antimonide skutterudites. Thermopower was quoted only for $CoAs_3$; at 320 K it was about 140 μV/K and decreased to 85 μV/K at 600 K.

Studying the influence of Ni as it replaces some of the Co atoms, Pleass and Heyding observed a similar trend as for the antimonide skutterudites — Ni acts as a donor and merely 1 at% of nickel is enough to drive the conduction n-type. With Ni concentrations above 10 at%, the transport behavior is essentially metallic (see Fig. 30). Magnetic susceptibility measurements suggested an effective moment per nickel atom roughly equivalent to one unpaired electron, consistent with the view that one of the nickel electrons is promoted into the conduction band. As noted before, according to Dudkin's model, there should be no significant change in the semiconducting character of $CoAs_3$ when some of the Co atoms are replaced with Fe. The data, however, indicate a profound influence of Fe on the transport properties (see Fig. 31). Samples with 1 at% Fe display metallic conduction, and at higher concentrations of iron the material turns n-type. Pleass and Heyding have not addressed the issue of "shifted stoichiometry," although they pointed out that there were significant deviations (deficiency of As), especially for the Fe-substituted $CoAs_3$. The structure should properly be designated as $Fe_xCo_{1-x}As_{3-y}$, where $3-y$ varies between 2.66 and 2.94. Somewhat surprisingly, the deviation from stoichiometry did not appear to be a function of the degree of Fe substitution, nor did it seem to influence electrical resistivity. The effective moment per iron atom was equivalent to between one and two unpaired electrons.

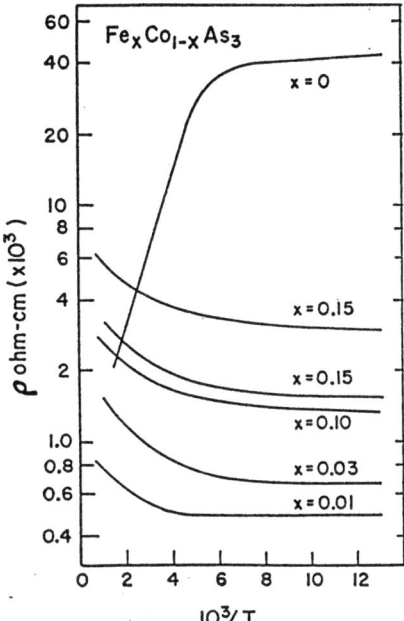

FIG. 31. Temperature dependence of the electrical resistivity of $Fe_xCo_{1-x}As_3$ (adapted from Pleass and Heydring, 1962). Reprinted with permission from NRC Research Press.

Among the early reports on skutterudites (prior to 1990) that mentioned at least some of the transport parameters or that attempted to use transport properties to characterize the structure, one should not omit the work done at Brown University and the studies conducted at the Max-Planck-Institute in Stuttgart. At Brown University the effort concentrated on the study of single crystals of skutterudites prepared mostly via the chemical vapor transport technique (CoP_3, $CoAs_3$, $CoSb_3$) (Ackermann and Wold, 1977) or from a tin flux (RhP_3) (Odile et al., 1978). Although all three Co skutterudites were diamagnets showing p-type transport, only CoP_3 displayed a semiconductorlike character and optical absorption measurements indicated the band edge at 0.45eV. The other crystals, including also RhP_3, had a positive temperature coefficient of resistivity over the entire range 10–300 K, and no band edge was detected. Room temperature electrical conductivity was quite high and roughly doubled in the series CoP_3 ($1.79 \times 10^3 \, \Omega^{-1} \, cm^{-1}$), $CoAs_3$ ($3.26 \times 10^3 \, \Omega^{-1} \, cm^{-1}$), $CoSb_3$ ($7.14 \times 10^3 \, \Omega^{-1} \, cm^{-1}$). Conductivity of RhP_3 ($7.13 \times 10^3 \, \Omega^{-1} \, cm^{-1}$) was comparable to that of $CoSb_3$ and a factor of four higher than that of CoP_3. Room temperature values of the thermopower were very small and positive (50 $\mu V/K$ for CoP_3, 1 $\mu V/K$ for $CoAs_3$, 1 $\mu V/K$ for $CoSb_3$, and 32 $\mu V/K$ for

RhP$_3$). The authors stated that the values were only approximate because they encountered difficulties in establishing a thermal gradient on their small crystals. It should be noted that the preceding transport results, most notably the behavior of resistivity of CoAs$_3$, CoSb$_3$, and RhP$_3$, and the values of the thermopower are inconsistent with the more recent studies on single crystal skutterudites. Although one would expect the vapor-grown crystals to be of the highest purity and structural integrity, the transport data clearly suggest the dominance of an extrinsic conduction mechanism, perhaps associated with self-doping. Thus, the transport data on these crystals may not be fully representative of the skutterudite structure.

The group in Stuttgart combined far-infrared reflection studies with transport measurements in order to obtain information about the carrier dynamics in CoAs$_3$ (Kliche and Bauhofer, 1988) and RhSb$_3$ (Kliche and Bauhofer, 1987). They worked with polycrystalline, hot-pressed samples as well as single crystals grown by the chemical vapor transport method. However, the resulting crystals were too small for direct use and were thus powdered and then hot-pressed. The carrier transport in CoAs$_3$ was activated, typical of a semiconducting material, with a decrease in the resistivity of three orders of magnitude between 10 and 300 K. The authors noted that they observed enhanced free carrier contribution on the powdered and then hot-pressed single crystals. This is akin to the metallic conductivities observed on single crystals by Ackermann and Wold (1977), and it supports the notion that the vapor-grown crystals may not have the right stoichiometry. The Hall constant was positive and strongly temperature dependent. There were two distinct regions where the Hall coefficient decreased with increasing temperature, suggesting activation across the intrinsic gap ($E_g \sim 0.18$ eV) and, at lower temperatures, activation of extrinsic holes ($E_a \sim 0.024$ eV). Comparing free carrier density obtained from reflectivity measurements with the one based on transport studies (Fig. 32), we see a factor of 10 discrepancy between the two. This can be rectified if one assumes an effective carrier mass $m^* = 0.1 m_e$ in the expression linking the carrier density and plasma frequency. Above 200 K, there was good agreement between the temperature dependencies of the optical and transport data.

In the early 1990s, as a result of various stimuli, including concerns about the undesirable effect of certain liquid coolants on the ozone layer and a generally increased environmental awareness, thermoelectricity reemerged as a potential savior and new enthusiasm was injected into this long-hibernating field. Realizing that the existing materials base of thermoelectricity (Bi$_2$Te$_3$ alloys, PbTe-based alloys, and Si–Ge alloys) is unlikely to make a major impact and that the maximum values of the figure of merit achieved with these materials will not exceed $ZT = 1$, the researchers at the Jet Propulsion Laboratory (JPL) undertook the task of identifying and evaluating other materials as prospective thermoelectrics. The attention

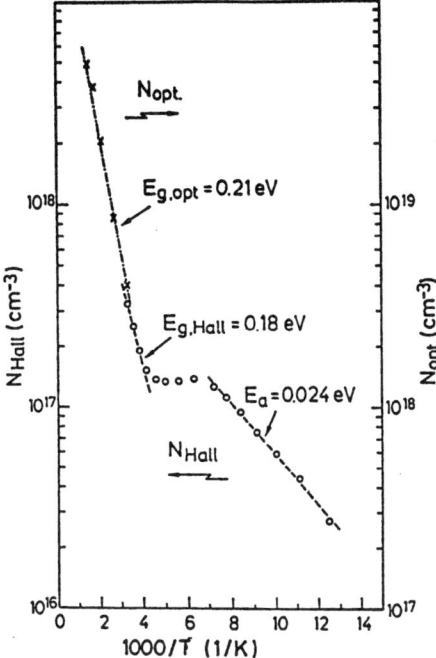

FIG. 32. Temperature dependence of the carrier concentration in $CoAs_3$ determined from the Hall effect and far-infrared reflectivity measurements. Note the factor of 10 difference between the carrier concentrations obtained using the two techniques (from Kliche and Bauhofer, 1988). Reprinted with permission from Elsevier Science.

focused on small gap semiconductors with very high mobilities and substantial Seebeck coefficients, and skutterudites turned out to be among the best prospects. Since 1992, skutterudites have become one of the primary targets of extensive studies worldwide.

In their examination of skutterudites, the JPL team did not just focus on the transport behavior, but also tackled related problems such as thermal expansion (Caillat et al., 1994a), (see Table VIII), and the issues pertaining to phase diagrams of skutterudites (Caillat et al., 1993a, b) and the growth of large crystals (Caillat et al., 1996b). The experimental studies commenced with $IrSb_3$ (Caillat et al., 1992a, b), which was prepared from Sb-rich melts by a vertical gradient freezing technique that yielded large single crystals. The work demonstrated that skutterudites can support very high mobility (~ 1200 cm^2/V-s), a consequence of the predominantly covalent bonding and reasonable Seebeck coefficients (~ 70 μV/K at 300 K). The resistivity increased with increasing temperature, but even at 900°C the mobility remained rather large at 220 cm^2/V-s. Near 400°C the power factor reached

TABLE VIII
THERMAL EXPANSION COEFFICIENTS OF SEVERAL SKUTTERUDITES OVER THE RANGE 25–450°C

Skutterudite	Thermal expansion coefficient	Reference
$CoSb_3$	$13.5 \times 10^{-6}/°C$	Caillat et al. (1994a)
	$6.36 \times 10^{-6}/°C$	Caillat et al. (1996a)
$RhSb_3$	$12.7 \times 10^{-6}/°C$	Caillat et al. (1994a)
$IrSb_3$	$7.96 \times 10^{-6}/°C$	Caillat et al. (1994a)
	$8.0 \times 10^{-6}/°C$	Kjekshus (1961)
	$6.6 \times 10^{-6}/°C$	Slack and Tsoukala (1994)
$Tl_{0.22}Co_4Sb_{12}$	$7.4 \times 10^{-6}/°C$	Sales et al. (2000)

values of 20 μW/cm-K^2, mostly on account of the Seebeck coefficient that peaked between 400 and 600°C at a value of 150 μV/K. It should be noted however, that a report of $ZT = 2$ obtained on $IrSb_3$ (Caillat et al., 1994b) is invalid because of false measurements and as such should be disregarded. Further studies on $IrSb_3$ were reported at the 13th International Conference on Thermoelectrics (Caillat et al., 1995a). This time, the samples were of polycrystalline form prepared via the solid–liquid phase sintering process. As-prepared structures were invariably p-type with room temperature resistivities comparable to those of single crystals and mobilities even marginally higher (1380 cm^2/V-s at $7.1 \times 10^{18} cm^{-3}$). Doping with Pt (solubility limit in $IrSb_3$ is about 0.15 at% Pt) resulted in n-type transport with two orders of magnitude higher resistivity and large thermopowers of -120 to $-500\,\mu$V/K, depending on the amount of Pt impurity (Fig. 33). Such large

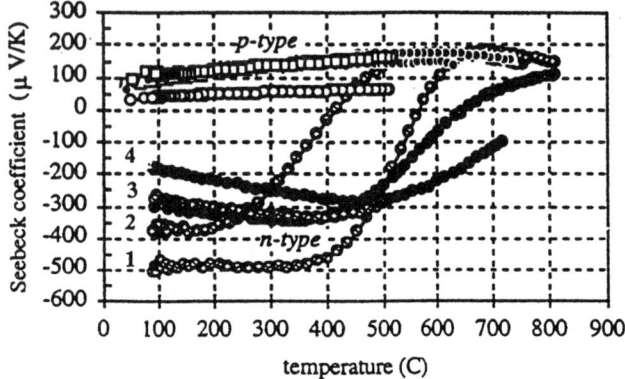

FIG. 33. Seebeck coefficient as a function of temperature for p- and n-type $IrSb_3$. Doping levels of the n-type samples are 1, 0.1 at% Pt; 2, 0.05 at% Pt; 3, 0.13 at% Pt; and 4, 0.15 at% Pt (from Caillat et al., 1995a).

values of thermopower are the consequence of a heavy effective mass of electrons. While p-type samples showed a slow rise in thermopower with the increasing temperature up to about 600°C, the thermopower of the n-type specimens remained high in the range of 400–500°C and then rapidly decreased as the mobile holes shorted the Seebeck voltage upon the onset of intrinsic conduction. The power factor of $IrSb_3$ appeared very promising. Unfortunately, the thermal conductivity was too high (8–11 W/mK at room temperature) and even though it decreased down to values on the order of 4 W/mK at 600°C, it had an adverse effect on the figure of merit. Thus, the best value of ZT achieved on p-type samples was only 0.35 at 550°C, about half as large for the n-type samples. $IrSb_3$ samples of comparable quality were also prepared by Slack and Tsoukala (1994) and by Tritt et al. (1996). The first team concentrated on studies at temperatures above 300 K while the other group focused on low-temperature measurements. At their common starting point at 300 K, the two data sets were in good agreement in spite of significantly different mass densities of their respective samples, 98% (Slack and Tsoukala, 1994) of the theoretical density versus 82% (Tritt et al., 1996). When the high-temperature results were compared with those of Caillat et al. (1995a), good accord was found and the highest figure of merit of 0.21 was obtained at 700 K. The low-temperature transport was notable for its metal-like resistivity down to about 60 K that gave way to an activated transport at the lowest temperatures. The bandgap extracted from the data was very small, ~10 meV. Another interesting feature was a prominent peak on the thermopower near 20 K that the authors ascribed to phonon drag. A strong phonon drag effect was seen previously in $CoSb_3$ (Morelli et al., 1995) and was explained as the consequence of a large phonon mean free path at low temperatures.

Measurements on polycrystalline samples of $CoAs_3$ and $RhAs_3$ (Caillat et al., 1995b) and on CoP_3 (Watcharapasorn et al., 1999) confirmed the uniformly high hole mobilities across the entire spectrum of binary skutterudites. In general, hole mobilities of binary skutterudites, for a given carrier density, are higher than the mobilities of state-of-the-art bulk semiconductors such as silicon, germanium, or gallium arsenide. The record belongs to $RhSb_3$ with a value of 8000 cm^2/V-s measured on a sample with a carrier density of 3.5×10^{18} cm^{-3} (Caillat et al., 1996b; Fleurial et al., 1995a). In Table IX are collected values of room temperature mobility measured on the binary, ternary, and filled skutterudites.

To gain further insight into the nature of conduction mechanisms, it was essential to extend transport studies to liquid helium temperatures. The first such studies were made by Morelli et al. (1995) and they focused on the behavior of the mobility in single crystals of $CoSb_3$ prepared by the gradient freeze technique from Sb-rich melts. The room temperature hole concentrations of the crystals ranged between 2.6×10^{17} and 4.1×10^{18} cm^{-3} and on cooling to 4 K decreased by no more than 20%. The temperature depend-

TABLE IX
Room Temperature Mobility of the Binary, Ternary, and Filled Skutterudites

Skutterudite	Mobility (cm^2/V-s)	Carrier density (cm^{-3})	Reference
CoP$_3$	748	$p = 3.3 \times 10^{19}$	Watcharapasorn et al. (1999)
CoAs$_3$	320	$p = 8.0 \times 10^{17}$	Sharp et al. (1995)
	3378	$p = 6.1 \times 10^{17}$	Caillat et al. (1995b)
CoSb$_3$	175	$n = 1.0 \times 10^{18}$	Sharp et al. (1995)
	200	$n = 1.0 \times 10^{18}$	Caillat et al. (1996a)
	270	$p = 1.9 \times 10^{18}$	Nolas et al. (1998)
	290	—	Dudkin and Abrikosov (1956)
	310	$p = 5.0 \times 10^{18}$	Sharp et al. (1995)
	1070	$p = 1.0 \times 10^{17}$	Sales et al. (1997)
	1800	$p = 7.0 \times 10^{16}$	Mandrus et al. (1995)
	1800	$p = 2.5 \times 10^{18}$	Anno et al. (1998a)
	1944	$p = 4.4 \times 10^{18}$	Fleurial et al. (1996b)
	2060	$p = 3.6 \times 10^{18}$	Borshchevsky et al. (1995a)
	2775	$p = 2.6 \times 10^{17}$	Morelli et al. (1995)
	2835	$p = 1.2 \times 10^{18}$	Caillat et al. (1996a)
	3445	$p = 4.0 \times 10^{17}$	Caillat et al. (1995c)
	6000	$p = \sim 10^{17}$	Arushanov et al. (1997)
	6100	$p = 1.5 \times 10^{17}$	Fess et al. (1997)
RhP$_3$	745	$p = 6.0 \times 10^{19}$	Odile et al. (1978)
RhAs$_3$	2368	$p = 1.0 \times 10^{19}$	Caillat et al. (1995b)
RhSb$_3$	1000	$p = 1.0 \times 10^{19}$	Sharp et al. (1995)
	8000	$p = 3.5 \times 10^{18}$	Zobrina and Dudkin (1960); Fleurial et al. (1995a)
IrSb$_3$	10	$n = 7.2 \times 10^{17}$	Caillat et al. (1995a)
	460	$p = 1.0 \times 10^{19}$	Sharp et al. (1995)
	1046	$p = 2.4 \times 10^{19}$	Borshchevsky et al. (1995a)
	1150	$p = 1.9 \times 10^{19}$	Caillat et al. (1992a)
	1150	$n = 1.2 \times 10^{19}$	Nolas et al. (1996b)
	1200	$p = 1.8 \times 10^{19}$	Caillat et al. (1992c)
	1320	$p = 1.1 \times 10^{19}$	Slack and Tsoukala (1994)
CeFe$_4$P$_{12}$	24.9	$p = 1.4 \times 10^{19}$	Watcharapasorn et al. (1999)
CeFe$_4$Sb$_{12}$	1.5	$p = 5.5 \times 10^{21}$	Fleurial et al. (1996b)
Ir$_{0.5}$Rh$_{0.5}$Sb$_3$	2120	$p = 6.7 \times 10^{18}$	Slack and Tsoukala (1994)
Ru$_{0.5}$Pd$_{0.5}$Sb$_3$	35	$p = 1.2 \times 10^{20}$	Caillat et al. (1995d)
	53	$p = 1.0 \times 10^{20}$	Nolas et al. (1996c)
	74	$p = 1.8 \times 10^{20}$	Caillat et al. (1996c)

ence of the hole mobility was suggestive of a combination of scattering by phonons (above 100 K) with its characteristic $T^{-3/2}$ variation, and scattering by neutral impurities resulting in a saturation below 20 K (Fig. 34). The authors specifically noted the absence of ionized impurity scattering that, if present, should have resulted in a $T^{3/2}$ dependence of mobility at low temperatures. In view of the rather large carrier density and the undoubtedly

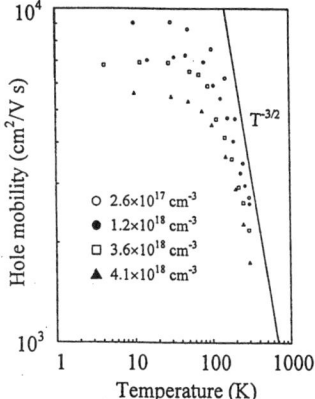

FIG. 34. Temperature dependence of the hole mobility of four p-type $CoSb_3$ single crystals below 300 K (adapted from Morelli et al., 1995).

extrinsic origin of the holes, the lack of ionized impurity scattering was surprising. The authors surmised that this might be due to a strong screening effect because the dielectric constant of $CoSb_3$ is large, $\varepsilon_\infty \sim 25$ (Kliche and Lutz, 1984). Plotting the saturation value of the mobility as a function of hole concentration, the authors obtained an approximately $p^{-1/3}$ variation. This is the variation predicted by Singh and Pickett (1994) for the energy independent scattering of holes that occupy the highly nonparabolic valence band, a prominent feature of the band structure of $CoSb_3$. Caillat et al. (Caillat et al., 1996a; Borshchevsky et al., 1995a) extended measurements on these crystals (and included additional p- and n-type crystals of $CoSb_3$) to temperatures up to 900 K. The two sets of data — one obtained from low-temperature measurements and the other one from high-temperature studies — matched very well at 300 K. The hole mobility followed the $T^{-3/2}$ variation (acoustic phonon scattering) up to 500 K and then started to decrease (Fig. 35). The Hall coefficient and the electrical resistivity of lightly doped p-type $CoSb_3$ were activated with energy of 0.63 and 0.75 eV, respectively. Above 650 K, the Hall coefficient decreased rapidly, and this was the main reason for an apparent sharp drop in the hole mobility at the highest temperatures. n-Type samples (0.1–0.2 at% Te or Pd impurity) had 25 times higher resistivity for a given carrier concentration than the p-type crystals and, consequently, much lower mobility ($\lesssim 100$ cm^2/V-s at 300 K). Their thermopower, however, was considerably higher, approaching values of -500 μV/K for the 0.08 at% Te sample at 300 K. In the extrinsic regime of conduction, the thermopower remained high for n-type samples and was increasing for p-type crystals. Once intrinsic conduction set in, the thermopower of both n- and p-type crystals decreased rapidly. In the former case

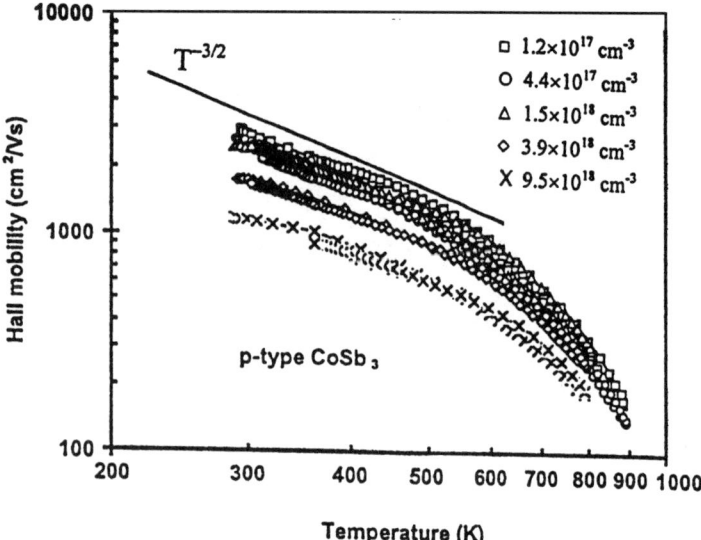

FIG. 35. Hole mobility of p-type $CoSb_3$ single crystals at high temperatures (adapted from Caillat et al., 1996a).

the minority holes started to short the Seebeck voltage, and in the latter case the light holes seemed to be compensated by heavy electrons. The very different nature of the p- and n-type transport is due to vastly different effective masses of holes ($m^*_{hole} \sim 0.16 m_e$) and electrons ($m^*_{el} \sim 1.65 m_e$). The power factor of both n- and p-type samples reached $30 \mu W/cm-K^2$. This value, however, required different carrier concentrations: For n-type samples the optimal electron concentration was 2×10^{20} cm^{-3}, while for p-type crystals the value was 1×10^{19} cm^{-3}, the highest hole concentration used in the study. It is likely that even higher power factors could be achieved with more heavily p-type doped $CoSb_3$. The best figure of merit, $ZT = 0.52$, was obtained at 600 K on n-type crystals with a carrier density of 1.4×10^{20} cm^{-3}. The primary limitation on the magnitude of the figure of merit was high lattice thermal conductivity.

The issue concerning the dominant scattering process that controls low-temperature transport properties of $CoSb_3$ has yet to be settled. Mandrus et al. (1995) measured the Hall effect on flux-grown single crystals of $CoSb_3$ and obtained the results displayed in Fig. 36. The data divide into three regions: a steeply rising Hall coefficient between 5 and 50 K, a slow increase from 60 to 150 K, and a sharp fall above 150 K. The authors explained this trend as follows: At the lowest temperatures the behavior of

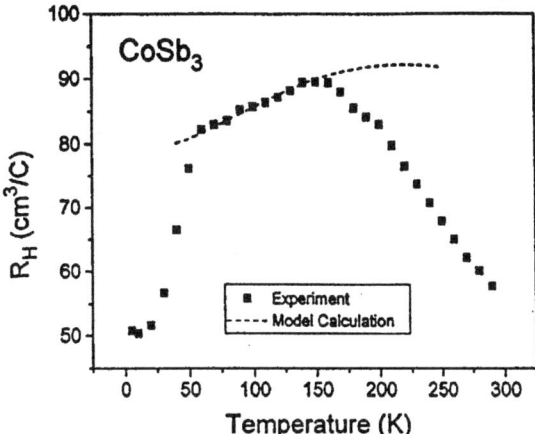

FIG. 36. Temperature dependence of the Hall coefficient of a flux-grown single crystal of $CoSb_3$. The dashed line represents a model calculation assuming a single type carrier, parabolic bands, and scattering from ionized impurities (from Mandrus et al., 1995). Copyright 1995 by the American Physical Society.

the Hall coefficient is consistent with the onset of impurity band conduction. Between 60 and 150 K, the Hall coefficient is relatively constant, as would be expected in the saturation, or exhaustion, regime in which conduction takes place via ionized impurities with a density that is essentially constant. The region above 150 K represents the onset of intrinsic conduction with a rapidly rising density of thermally excited carriers. The key point in this picture is the influence of ionized impurities. Combining the Hall effect data with the behavior of the electrical resistivity, the authors obtained the temperature-dependent mobility depicted in Fig. 37. The mobility clearly decreases below 200 K and can be fitted with a reasonable accuracy by a $T^{3/2}$ temperature dependence, the functional form appropriate for scattering by ionized impurities. Decreasing mobility at low temperatures was also observed by Sharp et al. (1995) on their hot- and cold-pressed polycrystalline $CoSb_3$, and by Arushanov et al. (1997) and Fess et al. (1997) on single crystals of $CoSb_3$ prepared by chemical vapor transport using chlorine and iodine, respectively. All these studies contrast sharply with the data of Fig. 34, and the source of the discrepancy is not immediately clear. Mandrus et al. (1995) and Sharp et al. (1995) mention that their samples are not perfect: Powder X-ray scans detected Sb impurity in the crystals grown from Sb flux, and the polycrystalline samples were less than 90% of the theoretical density. At this stage, lacking a detailed analytical and structural assessment of the samples, one can conclude only that the low-temperature transport

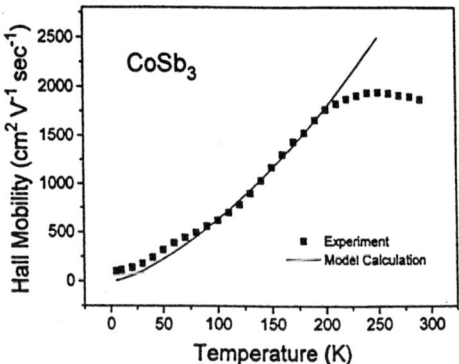

FIG. 37. Hall mobility of a flux-grown crystal of $CoSb_3$. The solid line represents a $T^{3/2}$ dependence appropriate for ionized impurity scattering (from Mandrus et al., 1995). Copyright 1995 by the American Physical Society.

of binary skutterudites appears very sensitive to the sample preparation process, to the purity of the constituent elements[2], and to the sample microstructure. The last point — the influence of the microstructure — has been well documented by measurements made in the laboratory of Prof. Matsubara (Anno et al., 1998a; Matsubara et al., 1996). The results demonstrated that the segregation of phases such as $CoSb_2$, CoSb, Co, and Sb is one of the important factors that governs the carrier scattering mechanism. Of particular interest was the distinct behavior of two p-type samples made from the highest purity Co (99.998%) and Sb (99.9999%), one sample hot-pressed at 650°C and the other one hot-pressed at 750°C. The initial powder mixture was Sb-rich to prevent the segregation of Co- and Sb-poor phases. By hot-pressing both samples at temperatures above the melting point of antimony, excess Sb was melted and removed during the process. Hot-pressing the samples at two different temperatures resulted in their vastly different microstructures. The sample pressed at 650°C had a grain size of 3 μm; the sample pressed at 750°C had an average grain size of 300 μm, some two orders of magnitude larger. Although the carrier densities were comparable (3.1×10^{18} cm^{-3} as opposed to 2.5×10^{18} cm^{-3}), the transport behavior of the two samples was entirely different (Fig. 38a, b). The small-grain sample with its semiconductorlike character strongly resembled the flux-grown crystals of Mandrus et al. (1995). On the other hand, the large-grain sample possessed characteristics of the melt-grown crystals

---

[2] It is worth pointing out that using a slightly less pure cobalt, 99.95% instead of cobalt of 99.998% purity, but otherwise the same technique, Mandrus et al. (1995) observed metallic rather than semiconducting behavior in their crystals.

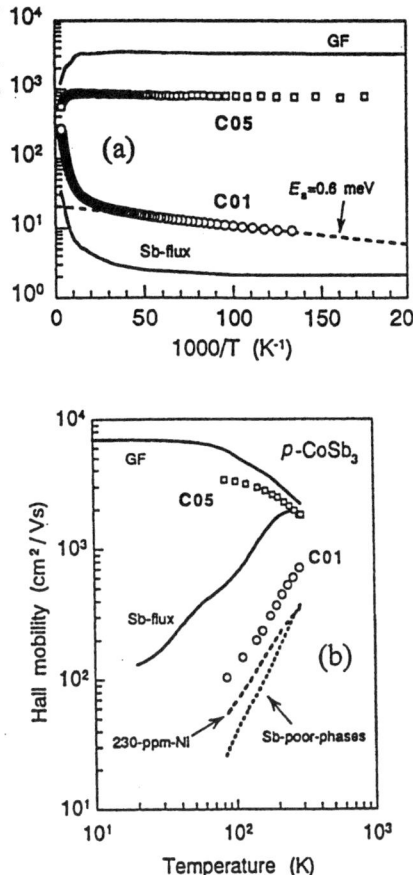

FIG. 38. (a) Temperature dependence of electrical conductivity of polycrystalline $CoSb_3$. Open circles (sample C01) are the data for the sample hot pressed at 650°C, which has an average grain size of 3 μm; open squares (sample C05) represent the sample hot pressed at 750°C with an average grain size of 300 μm. The solid line marked GF are the data of Morelli et al. (1995); the curve marked Sb-flux are the data of Mandrus et al. (1995). (b) Temperature dependence of the Hall mobility for the same samples as in (a) (from Anno et al., 1998a).

of Morelli et al. (1995). Moreover, temperature dependence of the mobility of this large-grain sample was very similar to that of the crystals of Morelli et al. (1995) and, specifically, it lacked any sign of the presence of ionized impurity scattering. This stands in contrast to the mobility of the small-grain sample, which mimicked the behavior of the Sb-flux grown crystal of Mandrus et al. (1995). The microstructure clearly appears to have a strong bearing on the scattering mechanism at low temperatures, and the influence of ionized impurity scattering rises with the decreasing grain size. Additional data on the influence of grain size on the carrier mobility of $CoSb_3$ are given in Nakagawa et al. (1997).

Outstanding electronic properties of the binary skutterudites have stimulated much interest and, among other things, gave impetus to revisiting the role of doping and its influence on transport behavior. Although the original work of Dudkin and Abrikosov and of Pleass and Heyding provided a firm foundation and guidance, the crystal growth techniques developed recently coupled with the availability of chemicals of exceptionally high purity have made such an inquiry a worthwhile undertaking. The spectrum of dopants that has been explored encompassed "traditional" impurities such as Ni, Fe, Te, and Sn (Koyanagi et al., 1996; Stokes et al., 1999; Yang et al., 2000), as well as elements that have not been looked at before. Surprising results abound in either case. For instance, Schujman et al. (1999) observed that $IrSb_3$ and $RhSb_3$ are able to incorporate very large amounts of Sn that substitutes for Sb (up to 22% in the case of $IrSb_3$ and up to 14% in $RhSb_3$). Remarkably, in spite of the nominally large excess of holes associated with Sn occupancy of the Sb sites, the Seebeck coefficient remained negative. This can be sustained only if the presence of Sn induces mixed valence in some fraction of Rh (or Ir) ions and/or if the excess of Sn is accommodated by the vacancies on the transition metal sites. Assuming both mechanisms at play, the structure can be described as $(Rh_a^{5+}Rh_b^{3+}\square_y)Sb_{12-x}Sn_x$, where Rh_a^{5+} represents a significant fraction ($\frac{1}{3}$ to $\frac{1}{2}$) of Rh atoms attaining the charge state 5+, and the number of vacancies adjusts itself to keep the overall cation stoichiometry at $a + b + y = 4$. It is interesting to compare these findings with the data of Koyanagi et al. (1996), who found that in $CoSb_3$ only 2.4% of Sb can be replaced by Sn. The much suppressed ability of $CoSb_3$ to incorporate Sn is likely due to the fact that Co does not favor the 5+ charge state, whereas in Rh and Ir this high charge state is easier to generate. Of course, electron spectroscopy techniques such as ultraviolet photoelectron spectroscopy should be used to substantiate these more or less hand-waving arguments.

Large doping levels are apparently achievable also with the donor impurity. I already mentioned that Zobrina and Dudkin (1960) reported on $CoSb_3$ being able to accommodate only about 0.25 at% of Te, a rather low solubility limit. In their recent work, using a spark plasma sintering method, Nagamoto et al. (1998) were able to replace up to 5.6% of Sb atoms with Te. Such heavy doping resulted in an electron concentration reaching up to 6×10^{20} cm^{-3}, yet the thermopower remained in excess of $-100\,\mu V/K$ at 300 K and increased in magnitude with the increasing temperature. In fact, the dimensionless figure of merit attained a surprisingly large value of $ZT \sim 0.8$ at 700 K, primarily on account of the large thermopower. The authors proposed the existence of a second, heavy electron band ($m_2 = 8m_e$) separated from the lighter ($m_1 = 1.7m_e$) electron band by about 0.13 eV. It is this heavy electron band that seems to enhance the thermopower and lead to an exceptionally large value of the figure of merit (see Fig. 39).

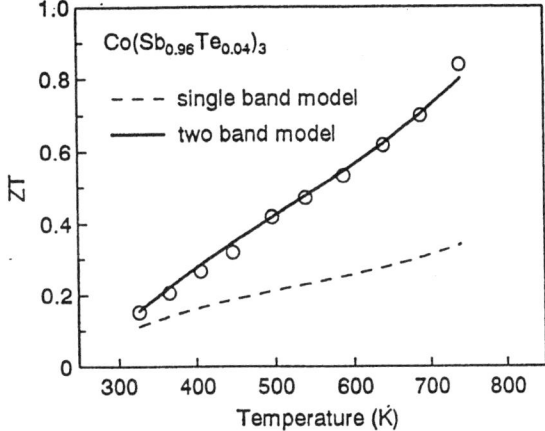

FIG. 39. Temperature dependence of the dimensionless figure of merit ZT for a heavily Te-doped $CoSb_3$ (from Nagamoto et al., 1998). Copyright 1998 by IEEE.

Of the elements previously unexplored, the work on $CoSb_3$ with Pd and Pt dopants (Tashiro et al., 1997; Anno et al., 1999a) is of particular interest. Just like Ni, both Pd and Pt are electron donors and Co atoms can be replaced with up to 5 at% Pd or 5 at% Pt. Remarkably, the solubility limit can be doubled to 10 at% if both Pd and Pt are introduced simultaneously in equal amounts. Such Pd + Pt double-doping expands the range of possible electron concentrations up to $\sim 10^{21}$ cm^{-3}. Although the electron mobility decreases with increasing carrier concentration, the carrier effective mass increases, and the large effective mass supports a reasonably high Seebeck coefficient (in excess of $-100\,\mu V/K$) even at carrier densities $\sim 10^{21}$ cm^{-3}. Benefiting from a large reduction in lattice thermal conductivity due to the presence of a large carrier density (at 300 K the lattice thermal conductivity is reported to drop down to $\sim 4\,W/m\text{-}K$ for carrier concentrations $\sim 10^{20}$ cm^{-3}), the dimensionless figure of merit of the double-doped $CoSb_3$ with Pd + Pt can be high—the authors report $ZT = 0.85$ at 800 K. In their most recent work on doping with Ni and Fe, Anno et al. (1999b) make use of a detailed analysis of the magnetic susceptibility to extract information on the valence state of Ni and Fe in $CoSb_3$. They arrive at a surprising result that nickel is in the Ni^{3+} state with a d-shell configuration of $3d^7$ and spin $S = \frac{1}{2}$. This is contrary to predictions of Dudkin's model, which favors nickel to be in the Ni^{4+} state. Detailed electron spectroscopy studies are needed to ascertain the valence state of both Ni and Fe.

An interesting attempt to influence the transport properties of skutterudites is to admix and disperse a judiciously chosen second phase into the structure. If the second phase is such that it has lower resistivity than the skutterudite matrix, there is a chance that the resulting composite structure will have lower electrical resistivity. At the same time, the dispersed second phase will be a source of phonon scattering, and thus such a composite should have a lower lattice thermal conductivity. This approach has been tried by Katsuyama *et al.* (1998a), who prepared a composite based on $CoSb_3$–$FeSb_2$. The second phase here is $FeSb_2$, which has a marcasite structure and an electrical resistivity that at elevated temperatures is significantly lower than the resistivity of $CoSb_3$. By mechanically grinding the composite powder of the molar ratio $CoSb_3$:$FeSb_2 = 0.8:0.2$ to disperse $FeSb_2$ in the $CoSb_3$ matrix, and by hot-pressing the powder, the authors observed a resulting composite having lower resistivity, lower thermal conductivity, and thermopower that approaches (above 700 K actually exceeds) that of $CoSb_3$. The dimensionless figure of merit of the composite material was 0.42 at 750 K, significantly higher than what can be achieved with pure $CoSb_3$.

2. THERMAL CONDUCTIVITY

a. Binary Skutterudites

From the discussion and data of the preceding section, one comes to the conclusion that the binary skutterudites possess outstanding electronic properties, but that their potential as promising thermoelectrics is seriously devalued by their high thermal conductivity relative to traditional thermoelectric materials. Since thermal conductivity, and more specifically its lattice component, is an important parameter that directly determines the thermoelectric figure of merit, it makes sense to devote close attention to its behavior.

First of all, it is useful to estimate the relative contribution of charge carriers and phonons to the transport. Using the Wiedemann–Franz law and the experimental values of electrical resistivity and thermal conductivity, we conclude that phonons are by far the dominant heat conduction channel. Even in heavily doped $CoSb_3$, the carrier thermal conductivity does not exceed more than 20% of the total thermal conductivity. This means that there is at least a theoretical chance to reduce the thermal conductivity to values more compatible with those of the state-of-the-art thermoelectrics ($\lesssim 3$ W/mK). Whether this is practically feasible, without destroying the excellent electronic properties of the structure, is another matter that will be considered in subsequent sections.

Table X collects room temperature values of the thermal conductivity of binary skutterudites reported in the literature. Although the data on arsenides and phosphides are sparse, it seems that thermal conductivity increases on going from antimonide to arsenide to phosphide. This is consistent with the general trend that heavy-element compounds typically

TABLE X

ROOM TEMPERATURE VALUES (IN W/mK) OF THERMAL CONDUCTIVITY OF BINARY SKUTTERUDITES

Skutterudite	κ(W/mK)	Form of the material	Reference
$CoSb_3$	5.2	Polycrystal	Dudkin and Abrikosov (1959)
	5.5	Cold-pressed, 70% density	Sharp et al. (1995)
	4.8	Submicron grain size	Nakagawa et al. (1996)
	8.4	Hot-pressed	Nolas et al. (1998)
	8.5	Hot-pressed, 90% density	Sharp et al. (1995)
	9.3	Hot-pressed	Stokes et al. (1999)
	9.2	Hot-pressed, >95% density	Sales et al. (2000)
	10.5	Hot-pressed, annealed	Nakagawa et al. (1996)
	11.5	Polycrystal, >95% density	Uher et al. (1997)
	10.5	Hot-pressed, 97% density	Katsuyama et al. (1998b)
	10.6	Polycrystal, >95% density	Yang et al. (2000)
	11.8	Hot-pressed, 98% density	Fleurial et al. (1996a)
	10.5–12	Single crystals	Morelli et al. (1995)
	~10	Single crystal	Caillat et al. (1996a)
	10.5–11	Single crystals	Caillat et al. (1995c)
$IrSb_3$	8–9.5	Polycrystal, 85% density	Caillat et al. (1995a)
	17.7	Hot-pressed	Slack and Tsoukala (1994)
	20	Hot-pressed	Tritt et al. (1996)
	16	Hot-pressed, 98% density	Slack and Tsoukala (1994)
	12	Cold-pressed, 70% density	Sharp et al. (1995)
$RhSb_3$	13	Hot-pressed	Caillat et al. (1995b)
	10.7	Hot-pressed, 90% density	Sharp et al. (1995)
$CoAs_3$	14.5	Polycrystal	T. Caillat, personal communication
CoP_3	18.5	Hot-pressed, 87% density	Watcharapasorn et al. (1999)

have lower thermal conductivity than the same compounds made with lighter elements. However, this rule does not seem to apply within the antimonide series itself. $IrSb_3$ has higher thermal conductivity than $RhSb_3$, which in turn has higher thermal conductivity than $CoSb_3$. One contributing factor to this "anomalous" trend may be the fact that the melting point of antimonides increases significantly with increasing mass of the compound. Looking at the data for $CoSb_3$, we note that the values of the thermal conductivity cluster near 5.5 W/mK and near 10.5 W/mK. The lower value of the thermal conductivity is associated with poorly compacted or small grain size polycrystalline samples that are typically of no more than 70% theoretical density. In contrast, higher values of the conductivity reflect well compacted (usually hot-pressed) polycrystalline samples with relatively large grain size and single crystals, all materials with a density well in excess of 90%. The effect of grain size on thermal conductivity can be gleaned from Fig. 40, which shows thermal conductivity of mechanically alloyed raw

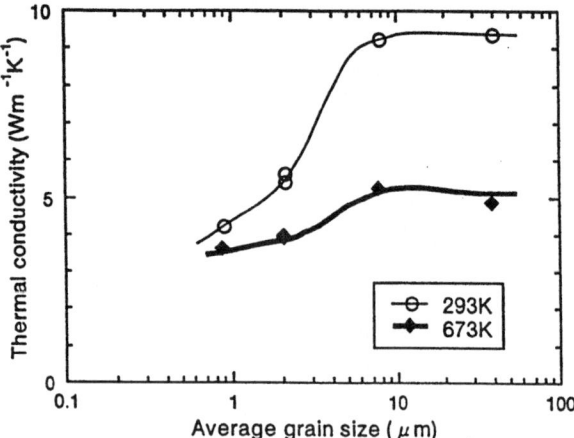

FIG. 40. Thermal conductivity of hot-pressed CoSb$_3$ as a function of the average grain size (from Nakagawa et al., 1997). Copyright 1997 by IEEE.

material powders that are subsequently hot-pressed (Nakagawa et al., 1997). This technique produces fine-texture samples, the grain size of which can be controlled by annealing. However, it should be mentioned that the micron or so size powders generated by mechanical alloying contain oxygen impurity and trace amounts of Fe that originate from contamination during crushing and mechanical alloying. Thus, the source of the reduction in thermal conductivity may not be just the boundary scattering of phonons; perhaps there is also a contribution from phonon-carrier scattering and point-defect scattering. The data in Fig. 40 suggest that polycrystalline samples with a grain size larger than 10 μm have a thermal conductivity similar to that of single crystals. In other words, the phonon mean free path at room temperature is not larger than 10 μm.

Figure 41 shows thermal conductivity of three single crystals of CoSb$_3$ measured from 4 to 300 K (Morelli et al., 1995). Also shown are two lines that represent fits to the data assuming the Debye model with Umklapp and boundary scattering of phonons (curve A), and the fit that includes scattering by vacancies with the density of 1×10^{18} cm^{-3} (curve B). The charge carrier contribution to the total thermal conductivity is less than 10% for all samples and temperatures. Thus, the measured thermal conductivity in Fig. 41 is essentially the lattice thermal conductivity. The very steep rise with decreasing temperature is characteristic of the dominance of phonon-phonon Umklapp scattering. It is unusual to observe such a rapid rise in the thermal conductivity in materials that are quite good electrical conductors. The different peak heights (near 12 K) are most likely due to the presence of vacancies on the pnictide sites. For comparison, Fig. 41 also

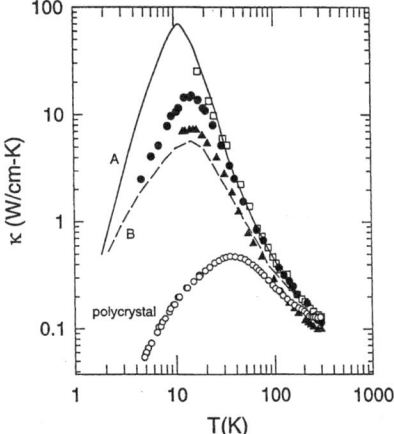

FIG. 41. Thermal conductivity of $CoSb_3$ single crystals below 300 K. Curve A is a fit to the Debye model assuming Umklapp and boundary scattering of phonons; curve B includes the effect of 1×10^{18} cm^{-3} vacancies. Open circles represent thermal conductivity of a polycrystalline $CoSb_3$ from Uher *et al.* (1997). (Adapted from Morelli *et al.*, 1995.)

includes the data on a well-compacted polycrystalline $CoSb_3$ (Uher *et al.*, 1997). The effect of the boundary scattering is obvious from a strong reduction of the peak height, while there is little difference in room temperature conductivities. Figure 42 illustrates the behavior of the thermal conductivity of $CoSb_3$ single crystals at temperatures above 300 K (Caillat

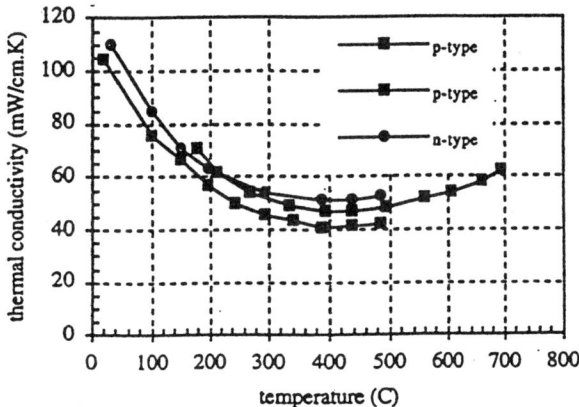

FIG. 42. High-temperature thermal conductivity of *p*- and *n*-type single crystals of $CoSb_3$ (from Caillat *et al.*, 1995c).

et al., 1995c). The conductivity decreases from its room temperature value of 10.8 W/mK and reaches a minimum of 4 W/mK near 400°C. At still higher temperatures the thermal conductivity increases because of the increasing carrier contribution. A similar trend was observed for $RhAs_3$ (Caillat *et al.*, 1995b), and a decreasing thermal conductivity with the increasing temperature was measured also in $IrSb_3$ (Slack and Tsoukala, 1994).

The value of the thermal conductivity (or more precisely, the value of the lattice part of the thermal conductivity) of the binary skutterudites at 300 K (taking $CoSb_3$ as an example) is too high by a factor of about 7 in comparison to the Bi_2Te_3-based alloys. Regardless of how outstanding the electronic properties of these materials are, they cannot fully compensate for the high thermal conductivity, and the figure of merit is not competitive. The dimensionless figure of merit of the undoped binary skutterudites is no more than $ZT = 0.15$ even in the temperature range of 500–600 K where it reaches its maximum. To lower the lattice thermal conductivity, one must increase the collision frequency of phonons. There are several ways to achieve this: One can control sample texture, i.e., rely on boundary scattering; one can make use of point-defect scattering by introducing impurities or prepare solid solutions; in heavily doped compounds electron–phonon scattering might be beneficial; one can enhance phonon scattering due to electron charge transfer in certain mixed-valence systems; and one can fill voids in the skutterudite structure with loosely bound rare-earth atoms. Different mechanisms usually operate on different segments of the phonon spectrum. One would like to benefit from all (or several) scattering mechanisms and affect phonons across a broad spectrum of frequencies. The trick is how to do it without seriously degrading electronic properties. In the following sections I review the options to lower the lattice thermal conductivity and illustrate the effectiveness of each method with data from the literature. Before doing so, it is instructive to consider how large a reduction in thermal conductivity is theoretically possible, i.e., what is the approximate value of the minimum lattice thermal conductivity in a typical skutterudite such as $IrSb_3$.

The concept of minimum thermal conductivity (Slack, 1979) is based on a perfectly rational notion that each solid has a finite thermal conductivity that cannot be smaller than a value corresponding to the shortest mean-free path of phonons that can be achieved at any given temperature for the structure under consideration. Thus, the question is, how short can the mean free path be?

In his original work, Slack assumed that the minimum mean free path of a phonon in a crystal must be of the order of one phonon wavelength. Shorter mean free paths are difficult to rationalize because lattice vibrations are waves. More recently, based on extensive studies of heat transport in amorphous and highly disordered solids, Cahill and Pohl (1989) developed an alternative approach to calculate the minimum thermal conductivity. In this model the thermal conductivity is viewed as a random walk of the thermal energy between neighboring atoms vibrating with random phases. With the lifetime of each oscillator (the time for the energy to diffuse from oscillator to oscillator) taken as one-half the period of vibrations, the shortest mean free path of heat conducting phonons that emerges from this interpretation corresponds to one-half of the phonon wavelength. The authors assumed the oscillating entities to be somewhat larger than single atoms. They divided the sample into regions of size $\lambda/2$ (λ being the wavelength), each with a frequency of vibrations given by $f = v/\lambda$, where v is the speed of sound. Taking again the time elapsed for the energy to diffuse from one region of localized oscillation to its neighboring region equal to one half-period of oscillations, they arrived at the minimum thermal conductivity

$$\kappa_{\min} = \left(\frac{\pi}{6}\right)^{1/3} k_B n^{2/3} \sum_i v_i \left(\frac{T}{\Theta_i}\right)^2 \int_0^{\Theta_i/T} \frac{x^3 e^x}{(e^x - 1)^2} dx. \tag{14}$$

Here $\Theta_i = v_i(\hbar/k_B)(6\pi^2 n)^{1/3}$ is the cutoff frequency for each polarization, the sum is taken over the three sound modes (two transverse and one longitudinal) with speeds of sound v_i, and n is the number density of atoms. The minimum thermal conductivity resulting from Eq. (14) is equal to 0.31 W/m-K, a value smaller than the one derived by Slack, ~ 0.4 W/m-K.

b. Effect of Doping

Low concentrations of impurities act as point defects that are strong scatterers of phonons. This so-called Rayleigh scattering does not affect all phonons equally. Because of its strong frequency dependence ($\sim \omega^4$), the point-defect scattering acts as a low-pass filter and scatters most effectively high-frequency phonons. The strength of the scattering depends on the mass difference between the impurity and the host atom, on the difference in their binding energies, and on elastic strains due to their different ionic radii. A convenient parameter to quantify these differences is the disorder parameter Γ defined as

$$\Gamma = c(1 - c)\left[\left(\frac{\Delta M}{M}\right)^2 + \varepsilon_b \left(\frac{\Delta \delta}{\delta}\right)^2\right]. \tag{15}$$

Here c is the atomic fraction of impurity atoms, ΔM is the mass difference between the impurity and the host atom, $\Delta \delta$ is the difference in average interatomic spacing, and ε_b is an adjustable parameter to account for the bond strength.

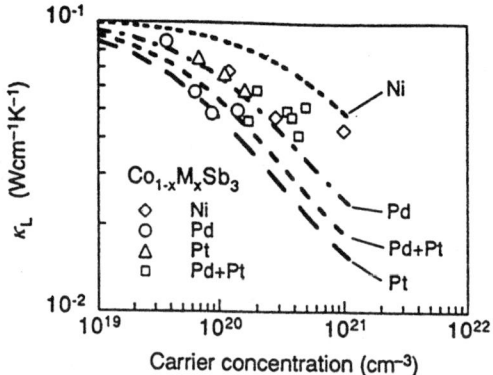

FIG. 43. Room temperature lattice thermal conductivity of $CoSb_3$ as a function of doping with the following n-type impurities: Ni, Pd, Pt, and Pd + Pt. The dashed lines are calculations for each compound based on the Debye model (from Anno et al., 1999a).

A strong influence of doping on the thermal conductivity of skutterudites was first reported by Dudkin and Abrikosov (1957), who observed a reduction of about 50% in the lattice thermal conductivity at 300 K when $CoSb_3$ was doped with 10 at% Ni, 10 at% Fe, or 0.25 at% Te. Although left without a comment, this is a surprisingly large effect on the lattice thermal conductivity, particularly in the case of Fe-substituted Co where the mass and size difference between Fe and Co is not more than 5%. A large reduction in the thermal conductivity with the increasing n-type doping (either Ni or Te) was also noted in high-temperature measurements of $CoSb_3$ (Caillat et al., 1996a), in Fe-doping studies of $CoSb_3$ (Stokes et al., 1999), and in extensive investigations of $CoSb_3$ doped with Fe, Ni, Pd, Pt, and PdPt (Tashiro et al., 1997; Anno et al., 1999a, b). The influence of the different dopants on the lattice thermal conductivity of $CoSb_3$ is shown in Fig. 43. The data are plotted as a function of carrier concentration for different n-type impurities (Anno et al., 1999a). Various dashed lines in Fig. 43 represent the Debye model calculations for each respective impurity and include (1) three-phonon scattering (both N and U processes), (2) point-defect scattering, and (3) electron–phonon scattering.

From a comparison of the experimental results with the calculated dependence of the lattice thermal conductivity on the carrier concentration (the dashed lines), Anno et al. conclude that the large reduction in the lattice thermal conductivity of n-type $CoSb_3$ is due to both point-defect scattering and electron–phonon scattering. The relative strength of the two scattering processes, however, depends on a particular impurity. For Pd, Pt, and PdPt the point-defect scattering dominates because of large differences in ΔM and $\Delta \delta$. On the other hand, for Ni the point-defect scattering is weak and the

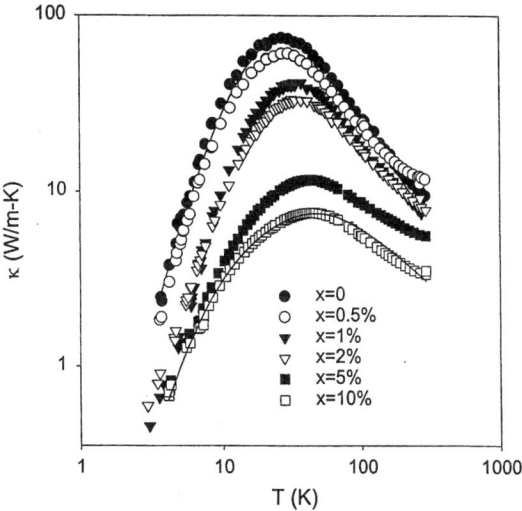

FIG. 44. Lattice thermal conductivity of $Co_{1-x}Fe_xSb_3$ below 300 K. The lines through the data points for samples with $x = 0$ and $x = 10\%$ are fits to Eq. (16) assuming the relaxation time given by Eq. (17). The fitting parameters for all samples are given in Table XI (from Yang et al., 2000). Copyright © 2000 by Technomic Publishing Company, Inc.

electron–phonon interaction plays a major role in reducing the lattice thermal conductivity on account of the relatively large effective mass associated with heavily Ni-doped $CoSb_3$. It is interesting to note that the model calculation for Ni seems to underestimate the experimental data. This would mean that there is an additional scattering process at play, a process that is not captured in the model calculations. The authors speculate that phonon scattering on paramagnetic Ni ions might be such a process. In this picture it is possible that a passing phonon gets absorbed through a dynamic exchange of electrons between Ni^{4+} and Co^{3+} ions. Alternatively, in the case that the phonon energy equals the energy separation of the crystal field-split or spin-orbit-split d-shell levels of Ni ions, phonons can be absorbed as they participate in phonon-assisted transitions between various low-lying states of the d-shell electrons. This kind of scattering was used with success to explain thermal conductivity of Fe-doped CdTe (Slack and Galginaitis, 1964).

Low-temperature behavior (3–300 K) of Fe-doped $CoSb_3$ (Yang et al., 2000) is shown in Fig. 44. Reduction in the room temperature thermal conductivity upon doping with 10 at% Fe is 58%, comparable to that seen in other Fe-doped skutterudites. An even more spectacular effect of the Fe impurity is seen near the dielectric peak (~25 K), where thermal conductivity is suppressed by an order of magnitude when 10% of Fe is added to the structure. After the electronic thermal conductivity was subtracted, the data were fit using the Debye model

$$\kappa_L = \frac{k_B}{2\pi^2 v}\left(\frac{k_B}{\hbar}\right)^3 T^3 \int_0^{\theta_{D/T}} \frac{\tau x^4 e^x}{(e^x - 1)^2}\, dx, \tag{16}$$

with the overall relaxation rate τ^{-1} determined by the contribution of the scattering processes that include boundary scattering, defect scattering and Umklapp processes:

$$\tau^{-1} = \tau_B^{-1} + \tau_D^{-1} + \tau_U^{-1} = v/L + A\omega^4 + B\omega^2 T \exp(-\theta_D/3T). \tag{17}$$

Here the Umklapp term is that due to Glassbrenner and Slack (1964). The curves through the data points indicate fits with the parameters given in Table XI. An interesting point is that the fits are very good even though they do not include a term representing electron–phonon interaction. The parameter that is strongly dependent on the presence of Fe impurity is the prefactor A in the point defect scattering. It increases by a factor of more than 25 if we compare the data for $CoSb_3$ and the $Co_{0.9}Fe_{0.1}Sb_3$. According to Klemens (1959), this prefactor is proportional to $c(1 - c)$ where c is the relative concentration of point defects. This implies that the point defect concentration increases with increasing Fe doping level. The authors propose that these point defects are vacancies on the Co site. With increasing Fe doping the density of these vacancies increases and they significantly enhance phonon scattering. The increasing density of vacancies with severed atomic bonds due to the presence of Fe is reasonable in light of the eventual instability of the skutterudite structure at higher Fe concentration and the lack of the existence of a $FeSb_3$ phase. Thermal conductivity of hot-pressed Fe-doped $CoSb_3$ was determined also at elevated temperature (Katsuyama

TABLE XI

FITTING PARAMETERS FOR THE LATTICE THERMAL CONDUCTIVITY OF $Co_{1-x}Fe_xSb_3$ PLOTTED IN FIG. 44 (ADAPTED FROM YANG ET AL., 2000)

x	L (μm)	A (10^{-43} s^3)	B (10^{-18} s/K)
0	10.54	2.79	5.38
0.005	10.86	4.52	5.38
0.010	3.18	4.36	5.38
0.020	3.11	6.73	5.38
0.050	3.23	37.10	5.38
0.100	2.36	77.27	5.38

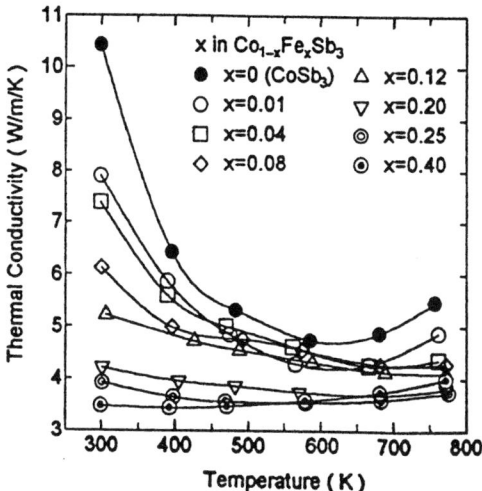

FIG. 45. Temperature dependence of the thermal conductivity for $Co_{1-x}Fe_xSb_3$ at high temperatures. Samples with a low concentration of Fe display a strongly T-dependent behavior, while samples with a heavy concentration of Fe (well beyond the solid solubility limit) are only weakly temperature dependent (from Katsuyama et al., 1998c).

et al., 1998b) by measuring thermal diffusivity, specific heat, and density and calculating the resulting thermal conductivity. Figure 45 shows the total thermal conductivity for a wide range of Fe concentrations measured from 300 K to nearly 800 K. For samples with small concentration of Fe, thermal conductivity decreases with increasing temperature, while the more heavily doped (alloyed) samples display nearly temperature-independent behavior. The carrier contribution amounts to about 10% of the total thermal conductivity for the sample with the smallest concentration of Fe, but this contribution increases and becomes 30% for $x \geqslant 0.12$ and about one-half of the total conductivity at $x = 0.40$. It should be mentioned that samples with $x \geqslant 0.06$ contained trace amounts of Sb, and in those with $x \geqslant 0.25$ X-rays detected the presence of $FeSb_2$. This should not be surprising in view of the 10% solubility limit of Fe in $CoSb_3$.

One of the promising approaches to reduce lattice thermal conductivity and, at the same time, create minimal damage to the electronic properties is to use doping or alloying with isoelectronic impurities. The impurity or the alloyed element in this case acts on the phonon spectrum (via the Rayleigh scattering), but this electrically neutral impurity does not disturb the charge carrier distribution to the same extent as do ionized impurities. Thus, there is a good chance that the overall effect will be an improvement in the figure of merit. An example is $CoSb_3$ doped isoelectronically on the cation as well as anion sites to form $Co_{0.97}Ir_{0.03}Sb_{2.85}As_{0.15}$ (Sharp et al., 1995). This material, with the additional Te-doping on the anion sites, resulted in high and rising values of $ZT \approx 0.6$ near 700 K.

A natural extension of the preceding concept is to admix and uniformly disperse a second component into the skutterudite structure. Although this is not absolutely essential, it helps if this second component has thermal conductivity lower than the skutterudite matrix. A finely dispersed second component will be a strong phonon scattering center. Provided it is reasonably inert to the carrier transport, that is, if the ratio σ/κ of the two-component skutterudite is larger than the same ratio for the pure skutterudite, there is a possibility that one can improve the thermoelectric figure of merit. This approach was found useful in Si–Ge alloys (Vandersande et al., 1992), and Anno et al. (1998b) have tried to replicate it in skutterudites. As the second component they chose PbS, which was added as a fine powder to the heavily n-doped $CoSb_3$ (Pd/Pt dopants) in a 5–20% molar ratio. Following a standard recipe of milling and hot pressing/sintering, the authors demonstrated that the thermal conductivity of well compacted samples containing PbS (using spark plasma sintering, the sample density was near 100% of the theoretical density) is indeed lower than the conductivity of the skutterudite matrix. There is a significant reduction ($\sim 30\%$) in thermal conductivity within the first 5 mol% of PbS and only minor changes thereafter. Moreover, the ratio σ/κ was larger than that of the host material. Since the Seebeck coefficient was almost independent of the PbS concentration, the authors found that the thermoelectric figure of merit for PbS concentration in the range 5–10 mol% increased and reached values in excess of 1×10^{-3} K^{-1}. General criteria for an enhancement of the thermoelectric power factor by making a composite mixture of two materials are discussed by Bergman and Fel (1999).

c. Solid Solutions

The formation of solid solutions from isostructural compounds is a time-honored technique of reducing lattice thermal conductivity that has been used in the development of virtually all state-of-the-art thermoelectrics. Solid solutions are archetypal examples of point defect scattering where both mass and strain fluctuations tend to effectively scatter high-frequency phonons. A particular solid solution will be of benefit to thermoelectricity provided the ratio of its mobility to its lattice thermal conductivity is larger than for the individual compounds constituting the solid solution.

A large number of solid solutions can be prepared by alloying binary or ternary skutterudites. Some of these, such as CoP_3–$CoAs_3$ (Lutz and Kliche, 1981), $RhSb_3$–$IrSb_3$ (Slack and Tsoukala, 1994), or $Fe_{0.5}Ni_{0.5}Sb_3$–$Ru_{0.5}Pd_{0.5}Sb_3$ (Borshchevsky et al., 1996), form a complete series of solid solutions obeying Vegard's law. Other combinations might be more restrict-

TABLE XII

LATTICE THERMAL CONDUCTIVITY OF SELECTED SKUTTERUDITE SOLID SOLUTIONS INCLUDING
APPROXIMATE VALUES OF THE CONDUCTIVITIES OF THE TWO END MEMBERS
(DATA FROM BORSHCHEVSKY et al., 1996)

	Lattice thermal conductivity in units of W/mK for:		
Solid solution	Solid solution	First end-member	Second end-member
$(CoSb_3)_{0.96}-(IrSb_3)_{0.04}$	4.5	10	10
$(CoSb_3)_{0.90}-(IrSb_3)_{0.10}$	3.2	10	10
$(CoSb_3)_{0.88}-(IrSb_3)_{0.12}$	2.9	10	10
$(CoAs_3)_{0.98}-(CoSb_3)_{0.02}$	5.7	14	10
$(CoSb_3)_{0.90}-(IrAs_3)_{0.10}$	2.5	10	14.5
$(CoSb_3)_{0.79}-(Fe_{0.5}Ni_{0.5}Sb_3)_{0.21}$	3.5	10	3
$(Ru_{0.5}Pd_{0.5}Sb_3)_{0.5}-(Fe_{0.5}Ni_{0.5}Sb_3)_{0.5}$	2.4	1.4	3

ive, and this gives rise to a miscibility gap for a certain range of compositions. Such partial solid solutions are quite numerous and include among others $CoAs_3-CoSb_3$ (Lutz and Kliche, 1981), $CoSb_3-IrSb_3$ (Borshchevsky et al., 1996), $CoAs_3-IrAs_3$ (Borshchevsky et al., 1996), and many binary-ternary skutterudite combinations. Effectiveness of the skutterudite solid solutions in reducing the lattice thermal conductivity was first demonstrated by Slack and Tsoukala (1994), who compared the results obtained on pure $IrSb_3$ and on $Ir_{0.5}Rh_{0.5}Sb_3$. The room temperature lattice thermal conductivity of $IrSb_3$ (~ 16 W/mK) was brought down to 9 W/mK upon the formation of the alloy. At the same time, there was only a small effect on the electronic properties, and the resistivity and mobility of $Ir_{0.5}Rh_{0.5}Sb_3$ actually improved over the values measured on $IrSb_3$. In spite of the reduction in the lattice thermal conductivity, its magnitude was still higher by a factor of more than 20 than the estimated minimum thermal conductivity.

The JPL researchers (Borshchevsky et al., 1995b, 1996) measured the lattice thermal conductivity of a number of skutterudite solid solutions and compared the results with those obtained on the individual compounds. Selected data are presented in Table XII. A reduction of about a factor of 3 in the lattice thermal conductivity is typical. In general, the lowest thermal conductivity is achieved in solid solutions that maximize mass and volume fluctuations. Temperature dependence of thermal conductivity for two solid solution samples based on $CoSb_3-IrSb_3$ is shown in Fig. 46. The thermal conductivity is nearly temperature independent and has a value of about 3 W/mK over a broad range of temperatures around 500 K. This is to be contrasted with the behavior of the end members ($CoSb_3$ and $IrSb_3$) of this partial solid solution system.

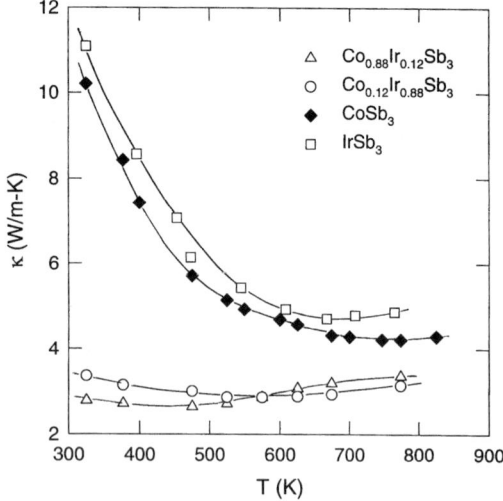

FIG. 46. Thermal conductivity of two solid solutions, $Co_{0.12}Ir_{0.88}Sb_3$ and $Co_{0.88}Ir_{0.12}Sb_3$. For comparison the data for the end-members, $CoSb_3$ and $IrSb_3$, are also included (adapted from the data of Borshchevsky et al., 1995b).

Knowing thermal conductivities of the individual constituent compounds one can, with a reasonable degree of reliability, predict the thermal conductivity of the resulting solid solution. For this purpose a convenient starting point is the theoretical work of Callaway and von Baeyer (1960) that relates the lattice thermal conductivity of an alloy to that of the pure compound. Detailed treatment, including a prescription of how to calculate the scattering parameter for ternary or quaternary skutterudites, can be found elsewhere (Slack, 1962; Abeles, 1963). Using this approach, Borshchevsky et al. (1996) calculated the lattice thermal conductivity of several solid solutions based on skutterudites. As expected, solid solutions with small mass and volume fluctuations are not very effective in reducing thermal conductivity. To achieve maximum reduction, one needs to select compounds where the fluctuations of atomic masses and volumes are large. The authors estimated that with properly chosen solid solutions one might be able to reduce lattice thermal conductivity of skutterudites down to about 1.5 W/mK.

d. Ternary Skutterudites

Ternary skutterudites should not be confused with heavily doped binary compounds. The point is that the half-unit cell complex $\square M_4X_{12}$ keeps the total valence electron count at 72, i.e., the resulting phases are isoelectronic with the binary compounds. To form ternary skutterudites, one can substitute on the cation site, for instance replacing trivalent Co^{3+} with divalent Fe^{2+} and tetravalent Ni^{4+} to form $\square Fe_2Ni_2Sb_{12}$, i.e., $Fe_{0.5}Ni_{0.5}Sb_3$. Alternatively, one can substitute on the anion site, for example, taking tetravalent Sn and hexavalent Te and replacing pentavalent Sb, resulting in $\square Co_4Sn_6Te_6$, which can be written as $CoSn_{1.5}Te_{1.5}$. Ultimately, one can substitute simultaneously on both the cation and anion sites and form ternaries such as $\square Fe_4Sb_8Te_4$, which one can write as $FeSb_2Te$. This scheme clearly expands the number of skutterudite compounds, and many new structures have been reported in the literature (Lutz and Kliche, 1981; Korenstein et al., 1978; Lyons et al., 1978; Bahn et al., 1969; Fleurial et al., 1997; Partik et al., 1996). Depending on where the substitution is made, the structure expands, typically by 0.6 to 1%. Although there are exceptions, the bandgaps of the ternary compounds are smaller than the bandgaps of the skutterudites from which the respective ternaries are derived.

In general, ternary skutterudites have low thermal conductivity. A ternary compound that generated considerable interest is $Ru_{0.5}Pd_{0.5}Sb_3$, which is derived from $RhSb_3$. Early reports (Caillat et al., 1995d; Fleurial et al., 1995b) indicated an extremely low value of thermal conductivity (well below 1 W/mK) over a broad range of temperatures up to 700 K. This would have been some 12–16 times smaller conductivity than the value for the parent $RhSb_3$. Furthermore, the temperature dependence had features akin to those of amorphous solids. Subsequent reports from the same group (Caillat et al., 1996c; Fleurial et al., 1997) revised the value upward to 2.5–3 W/mK. Comparable magnitude of thermal conductivity for this compound was reported also by Nolas et al. (1996c). High-temperature thermal conductivity of several ternary skutterudites is displayed in Fig. 47. Subtracting the electronic component that amounts to about one-third of the total thermal conductivity, the room temperature lattice thermal conductivity of $Ru_{0.5}Pd_{0.5}Sb_3$ is in the range 1.6–2 W/mK, a reduction by a factor of more than 5 in comparison to $RhSb_3$. Such low values of thermal conductivity cannot be explained as being due to the mass or volume fluctuations, as these are simply not large enough to cause such a dramatic effect. Clearly, if these exceptionally low values of thermal conductivity are real, they must have their origin in a different scattering mechanism. One such possible mechanism might be scattering due to valence fluctuations. Some transition metals, among them Fe, Ru, and Os, are known to form ions with different charge states, e.g., Ru^{2+} and Ru^{4+}, or Fe^{2+} and Fe^{3+}. The presence of these mixed charge states in a compound is often promoted by deviations from the expected stoichiometry. For instance, detailed microprobe analysis indicates that $Ru_{0.5}Pd_{0.5}Sb_3$ never ends up with this exact stoichiometry but is Pd deficient, more like $Ru_{0.53}Pd_{0.39}Sb_3$. To compensate for the charge imbalance (and any vacancies on the cation sites and excess of pnicogen that might be present), some of Ru will be in a higher charge state,

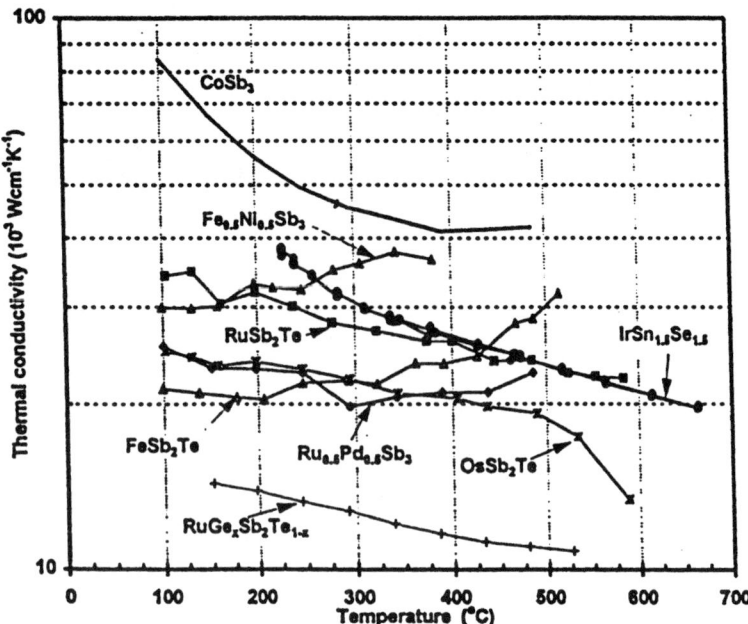

FIG. 47. High-temperature thermal conductivity of several ternary skutterudites (from Fleurial et al., 1995b).

Ru^{4+}. X-ray absorption near-edge structure studies (XANES) made on this ternary skutterudite (Nolas et al., 1996c) indicated the relative proportion of the expected Ru^{2+} and its higher charge state cousin Ru^{4+} as 48% vs. 52%. In the presence of both Ru^{2+} and Ru^{4+}, a phonon may induce a transition of a pair of $4d$ electrons from the Ru^{2+} ion ($4d^6 \to 4d^4$) to the neighboring Ru^{4+} ion ($4d^4 \to 4d^6$). From the perspective of phonons this is a scattering event the strength of which depends on the concentration of both Ru^{2+} and Ru^{4+} ions. Slack et al. (1996) and Nolas et al. (1996c) calculated the thermal resistivity associated with this process and arrived at values as high as 1.4 and 1.18 mK/W, respectively. These values are about three times too small to bring down the lattice thermal conductivity to a value reflecting the minimum thermal conductivity.

Low values of lattice thermal conductivity together with rather low electrical resistivity (on account of very high carrier concentrations and modest mobilities) are positive features as far as thermoelectricity is concerned. Unfortunately, high carrier densities inevitably mean low Seebeck coefficients and, indeed, $Ru_{0.5}Pd_{0.5}Sb_3$ has very low and weakly temperature dependent Seebeck coefficient. At 300 K it is only 20 μV/K, and even at 700 K it is no more than 40 μV/K. Although there are ternary skutterudites (e.g., $RuSb_2Te$) that show semiconducting behavior with Seebeck coefficient of 120–250 μV/K, in general it is difficult to alter the electronic properties of ternary skutterudites because of their high carrier density, and because the dopant might be compensated by fluctuations in the charge state of the transition metal.

e. Filled Skutterudites

Based on the discussion in the preceding section, it should be clear to the reader that in spite of a determined effort to reduce the lattice thermal conductivity of the binary skutterudites, success has been limited and the values of the conductivity remain too high to make the binary skutterudites competitive thermoelectric materials. If all that one could achieve with the binary skutterudites was $ZT \sim 0.3$–0.5, then the research effort was nothing more than an interesting exercise in materials science but of no consequence for thermoelectricity. Fortunately, the open crystal structure of skutterudites, typified by the presence of large voids, allows for a modification of the structure by filling the voids with foreign species such as rare-earth ions. The presence of such ions in the voids, and thus the formation of filled skutterudites, has a dramatic effect on transport properties. It is thanks to the influence of the void fillers that skutterudites continue to generate excitement in the thermoelectric community and that this class of materials has emerged as one of the most novel prospective thermoelectrics.

In Subsection 2 of Section II, I described how the void fillers modify the skutterudite structure, and their influence on the band structure was discussed in Section III. In this section I describe the influence of the fillers on the electronic and lattice properties. It will be evident that it is primarily the effect of the fillers on phonon propagation that makes the filled skutterudites promising thermoelectrics.

A dramatic reduction in the lattice thermal conductivity achieved upon filling the empty voids is strong evidence that an interesting concept formulated by Slack (1995) and known as phonon glass electron crystal (PGEC) is viable and that it can be realized in practice. The essence of this concept is decoupling of the electronic and vibrational degrees of freedom so that a PGEC material looks like a good crystalline solid with a large mobility, while its thermal properties resemble those of a glassy or amorphous solid with a very low thermal conductivity. The PGEC concept is the most influential new approach to engineering a prospective thermoelectric that has been developed in recent years. The challenge is to find material systems where this approach can be applied and used. Skutterudites with their voids and, in general, materials with open crystal structure that contain cavities, cages, and empty sublattices are favorable candidates (Sales, 1998; Uher et al., 1999) for a successful realization of the PGEC concept.

Key features of the PGEC concept as it is applied to filled skutterudites are loose bonding of the void filler species and their large thermal parameter as they "rattle" in the oversized cage of the skutterudite structure. The rattling motion of the fillers, that is, the localized vibrational modes modeled as Einstein oscillators (see Subsection 2 of Section IV), interfere with the normal modes of the structure, and this leads to a drastic reduction in the lattice thermal conductivity. At the same time, because of weak coupling of the filler with the rest of the structure and the fact that the electrical conduction is due mainly to the pnicogen orbitals that are "shielded" from the influence of the rattling ions, the motion of the charge carriers is relatively weakly perturbed. Consequently, the carrier mobility remains reasonably large, and, as far as the electronic properties are concerned, the structure preserves its substantially crystalline character.

The effectiveness of local vibrations of atoms or molecules in reducing the lattice thermal conductivity has been seen in several systems. One of the best examples is the study on $(KBr)_{1-x}(KCN)_x$ mixed crystals where the rotation of the CN molecule about its lattice site results in exceptionally low, amorphouslike thermal conductivity in spite of the crystalline nature of the structure (Cahill et al., 1992). In the case of filled skutterudites, the decisive influence of the rattling filler ion was first demonstrated by Morelli and Meisner (1995) on a polycrystalline sample of $CeFe_4Sb_{12}$. The measured thermal conductivity was smaller by a factor of nearly 10 than the thermal conductivity of the unfilled $CoSb_3$. Similar results were subsequently obtained on a number of filled skutterudites based on $IrSb_3$ (Grandjean et al., 1994; Nolas et al., 1996b) and $CoSb_3$ (Uher et al., 1997).

Referring to Fig. 48, several features are noteworthy regarding the influence of different rare-earth species in the voids of the polycrystalline $IrSb_3$. All three, La, Sm, and Nd, reduce thermal conductivity by at least an order of magnitude at 300 K and even more at lower temperatures. The dielectric peak in $IrSb_3$, a characteristic feature of the conductivity of crystalline solids, is virtually washed out in the filled skutterudites and the thermal conductivity is only weakly temperature dependent. The greatest reduction in thermal conductivity is achieved with smaller and heavier fillers such as Nd, as these may rattle more freely inside the voids and thus have a greater effect on the phonon spectrum. An additional point about Nd^{3+} and Sm^{3+} as opposed to La^{3+} is the occupancy of low-lying $4f$ electronic states into which the ground state of Nd^{3+} and Sm^{3+} is split because of the octahedral crystal field of the pnicogen atoms. These levels may be accessed by phonons, resulting in further phonon scattering. It is also possible that the small ions such as Nd^{3+} and Sm^{3+} do not occupy the center of the voids but are somewhat displaced (Hornbostel et al., 1997b). If so, the R^{3+} site symmetry would be lower and more level splitting would ensue. Because Nd^{3+} splits into more levels of smaller energy separation than does Sm^{3+}, the neodymium ion is likely to be more effective in scattering

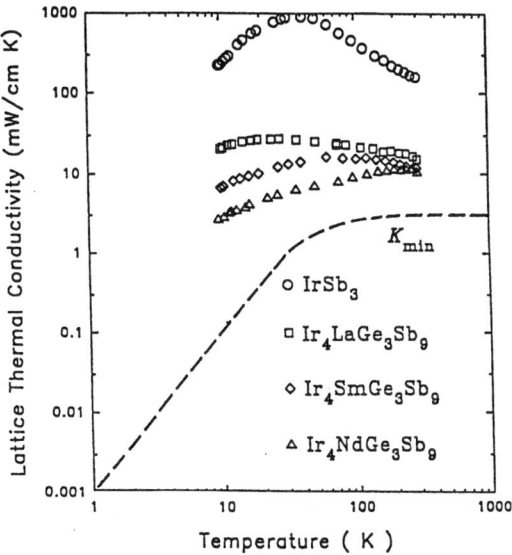

FIG. 48. Lattice thermal conductivity of IrSb$_3$ and of three filled skutterudites based on IrSb$_3$. The dashed line indicates the minimum thermal conductivity estimated for IrSb$_3$ (from Nolas et al., 1996b).

long-wavelength phonons. A progressively lower thermal conductivity of Ir$_4$NdGe$_3$Sb$_9$ in comparison to Ir$_4$SmGe$_3$Sb$_9$ as the temperature decreases is in support of this hypothesis. In all three filled skutterudites in Fig. 48, an attempt was made to partially substitute Ge for Sb in order to enhance Rayleigh scattering and thus suppress the high-frequency part of phonon spectrum. Unfortunately, in the form presented, the data do not allow one to discern the relative importance of the point-defect (Rayleigh) scattering against the influence of the void fillers and other possible phonon scattering mechanisms. What is important is that thermal conductivity has been drastically suppressed and that its magnitude at 300 K is no more than three times larger than the minimum thermal conductivity $\kappa_{min} \sim 0.3$ W/m-K estimated for IrSb$_3$.

Although the foregoing IrSb$_3$-based filled skutterudites have very low lattice thermal conductivity, their electronic transport properties, notably the resistivity and especially thermopower, are very poor. Electrical resistivity here is higher than for the parent IrSb$_3$ and the thermopower of less than 10 μV/K at 300 K is totally unacceptable for thermoelectric applications (Nolas et al., 1996b). In this particular case, the void fillers degrade both phonon and electron transport and the PGEC solid is obviously not realized.

In Subsection 2 of Section II, I noted that there are restrictions on the occupancy of the voids and that the filling fraction of Ce and La (and

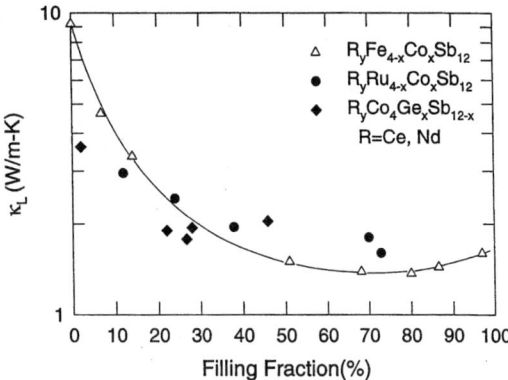

FIG. 49. Room temperature lattice thermal conductivity of antimonide skutterudites as a function of the filling fraction. Note a minimum on the curve for $Ce_yFe_{4-x}Co_xSb_{12}$ near the 70% filling fraction (courtesy of Dr. J.-P. Fleurial).

undoubtedly of other void fillers as well, once a detailed structural analysis is made) depends on the degree of compensation on the transition metal sites. From the point of view of thermal transport, one does not need all voids occupied in order to achieve maximum reduction in the thermal conductivity. It is actually more preferable to have randomness in the occupancy of voids, as this should maximize the frequency of phonon scattering in comparison to that of a perfectly ordered structure. The dependence of thermal conductivity on the filling fraction is shown in Fig. 49. The data support the premise that randomness in the void occupancy is desirable. This certainly seems to be the case with all rare-earth fillers. However, the most recent data of Sales *et al.* (2000) indicate that the minimum thermal conductivity of Tl-filled skutterudites occurs near complete Tl filling of the voids. In this case the excess thermal resistivity produced by the presence of monovalent thallium in the voids of $CoSb_3$ is qualitatively described by the $x^{1/3}$ dependence where x is the fraction of voids filled with Tl; see Fig. 50.

What is remarkable is that a very small amount of the filler (a few percent) has a strong influence on the heat transport. Figure 51 shows thermal conductivity of $CoSb_3$ where less than 10% of the voids are occupied (Uher *et al.*, 1997). There is no significant change after 5% of voids are filled. Similar outcome is observed with La fillers (Nolas *et al.*, 1998).

It should be understood that by introducing rare earth ions into the voids, one dopes the structure and alters the carrier density. Being electron donors, the filler ions tend to drive the structure *n*-type with very high concentration of free charge carriers, typically in the range 10^{20}–10^{22} cm^{-3}. Even a small amount of the filler raises the carrier density rapidly as shown

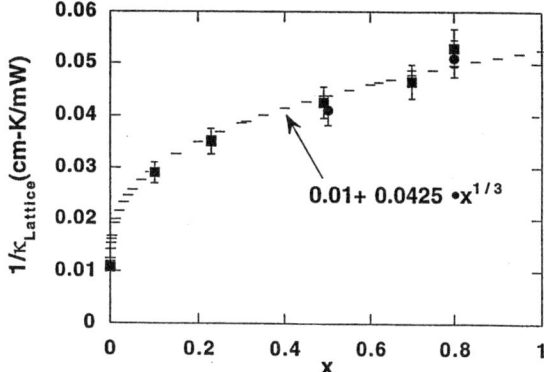

FIG. 50. Room temperature lattice thermal resistivity of $Tl_xCo_4Sb_{12}$ as a function of Tl filling x. The filled circles correspond to Fe compensated alloys (adapted from Sales *et al.*, 2000). Copyright 2000 by the American Physical Society.

in Fig. 52. It is truly amazing that in spite of the metal-like carrier densities, the Seebeck coefficient of many filled skutterudites remains high and magnitudes on the order of 150 μV/K at 300 K are not unusual. This can only be achieved provided the effective mass of the carriers is very high. Indeed, by assuming a single parabolic band in the regime of predominantly acoustic-phonon scattering, one can estimate the effective mass from the room temperature values of the carrier density and the Seebeck coefficient

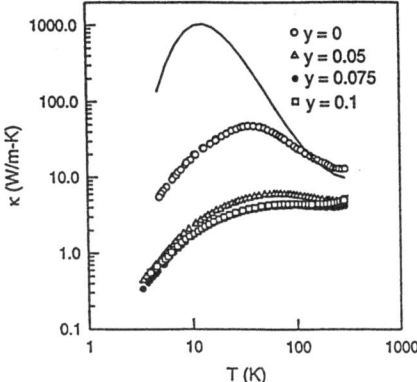

FIG. 51. Temperature dependence of thermal conductivity of $Ce_yCo_4Sb_{12}$. The solid line represents the data for a single crystal of $CoSb_3$ from Morelli *et al.* (1995). (From Uher *et al.*, 1997).

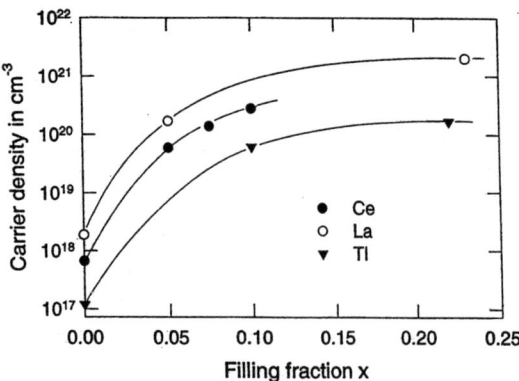

FIG. 52. Carrier density (electrons) as a function of the filling fraction for antimonide skutterudites with Ce, La, and Tl fillers.

using the equations

$$S = \pm \frac{k_B}{e}\left(\frac{2F_1(\eta)}{F_0(\eta)} - \eta\right) \quad (18)$$

$$n = \frac{4}{\sqrt{\pi}}\left(\frac{2\pi m^* k_B T}{h^2}\right)^{3/2} F_{1/2}(\eta). \quad (19)$$

Here $\eta = E_F/k_B T$ and $F_x(\eta)$ is a Fermi integral of order x. The resulting effective masses are large. An example of how the effective mass varies with even a small amount of the filler ion is seen from the data in Table XIII. A plot of the electron mass as a function of electron concentration is given in Fig. 53. The data displayed here are those of the Ce-filled samples in Table XIII as well as the effective masses of samples doped with Te and Pd (Caillat et al., 1996a). The results attest to the fact that Ce (and other void filler species) are strong n-type dopands, and the large values of m^* are consistent

TABLE XIII

ROOM TEMPERATURE CARRIER DENSITY, HALL MOBILITY, AND EFFECTIVE MASS FOR LOW FILLING FRACTIONS OF Ce IN CoSb$_3$ (ADAPTED FROM MORELLI ET AL., 1997)

Composition	Carrier density (10^{20} cm^{-3})	Hall mobility (cm^2/V-s)	Effective mass (in units of m_e)
Ce$_{0.050}$Co$_4$Sb$_{12}$	0.6	55	1.8
Ce$_{0.075}$Co$_4$Sb$_{12}$	1.4	41	2.4
Ce$_{0.100}$Co$_4$Sb$_{12}$	2.8	25	4.7

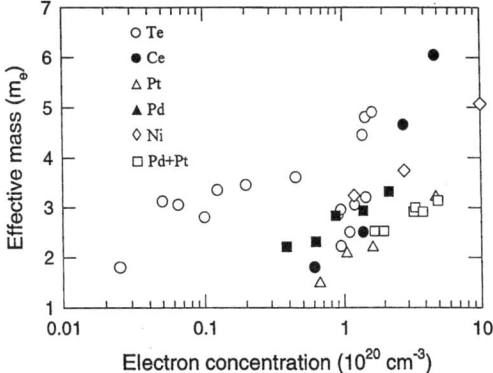

FIG. 53. Electron effective mass as a function of electron concentration in n-type $CoSb_3$. Various doping species are identified by different symbols. Note a general trend characterized by increasing effective mass with the increasing electron concentration. The data are from Caillat *et al.* (1996a), Morelli *et al.* (1997), and Anno *et al.* (1999a).

with band structure calculations that predict a heavy conduction-band mass in these structures (Singh and Pickett, 1994). Even filled skutterudites possessing large *hole* densities tend to have large Seebeck coefficients, implying a hole effective mass on the order of the free electron mass. The functional dependence of the hole effective mass on the carrier concentration is given approximately as $m_e p^{1/3}$ (Singh and Pickett, 1994; Caillat *et al.*, 1996a). These results underscore perhaps the most important difference between the filled skutterudites and the current state-of-the-art thermoelectrics: In filled skutterudites the optimum thermoelectric properties are realized at carrier densities at least two orders of magnitude larger than the densities in the conventional thermoelectrics such as Bi_2Te_3-based alloys.

Because La and Ce are the most frequently used fillers, it is useful to comment on their influence and compare their effectiveness in scattering phonons. Both La and Ce are the early lanthanides with close-packed structures at 300 K and a coordination number 12. The main difference between the two comes from the occupancy of the $4f$ shell in Ce while the $4f$ shells in La are empty. I have already noted one consequence of the occupied $4f$ shell in Ce—the intermediate valence that often complicates the properties of Ce compounds while La seems to prefer a $3+$ charge state. A direct comparison of the effect of Ce and La on thermal transport is shown in Fig. 54. Here the room temperature thermal resistivity W is plotted as a function of the void occupancy x for the Ce^{3+} and La^{3+} fillers within the range of their solid solubility in $CoSb_3$. We note both Ce^{3+} and La^{3+} as being effective in impeding phonon transport, and large enhancements in the lattice thermal resistivity are achieved with small amounts of

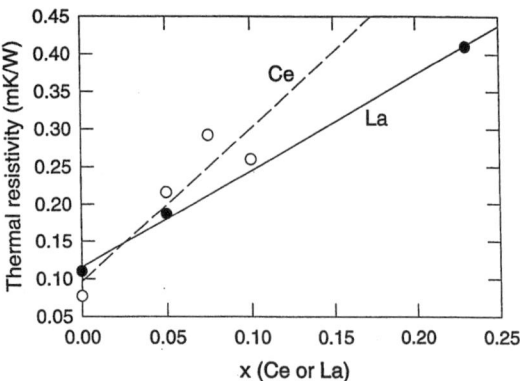

FIG. 54. Room temperature lattice thermal resistivity of $CoSb_3$ as a function of the fraction x of Ce and La in the voids. The filling fraction x is restricted to within the solid solubility limit of the respective fillers. The data for Ce-filled samples are from Morelli et al. (1997), and the data for La-filled samples are from Nolas et al. (1998).

the filler. Although the vibration frequencies (74 cm^{-1} for La and 68 cm^{-1} for Ce (Singh et al., 1999)) are comparable, Ce seems to have a somewhat stronger influence than La. The slope dW/dx (2.02 mK/W for the Ce filler and 1.28 mK/W for La) is some 20 to 30 times larger than the initial slope (0.066 mK/W; Nolas et al., 1998) one would expect assuming that the fillers act as point-defect scatterers. This again suggests that the void fillers behave as "rattlers" rather than the ordinary Rayleigh-type scattering centers. Optimally, one would like to fill the voids with different ions (size, mass) so as to affect a broad spectrum of phonon vibrations and thus maximize the impact of the rattlers. Since a small amount of the filler seems to be sufficient to cause a large reduction in thermal conductivity, there might be an opportunity to fill the structure with several different species. Although a specific study of the effect on the thermal conductivity of two or more fillers introduced simultaneously into the voids of the skutterudite structure has not yet been done, Rowe et al. (1998) attempted to partially substitute on the Ce sublattice with other rare earth species. Forming structures such as $Ce_{0.8}Ln_{0.2}Fe_4Sb_{12}$ where Ln = La, Yb, Sm, Er, and Eu, they observed a significant enhancement in the thermopower of $Ce_{0.8}Yb_{0.2}Fe_4Sb_{12}$. This translated into a nearly 50% larger power factor of this compound in comparison to $CeFe_4Sb_{12}$ at 700 K; see Fig. 55. Unfortunately, the authors have not measured thermal conductivity, and thus data concerning the figure of merit are not available. Nevertheless, one would expect lower thermal conductivity for the partially substituted compounds, and this should lead to the improvement in the thermoelectric properties.

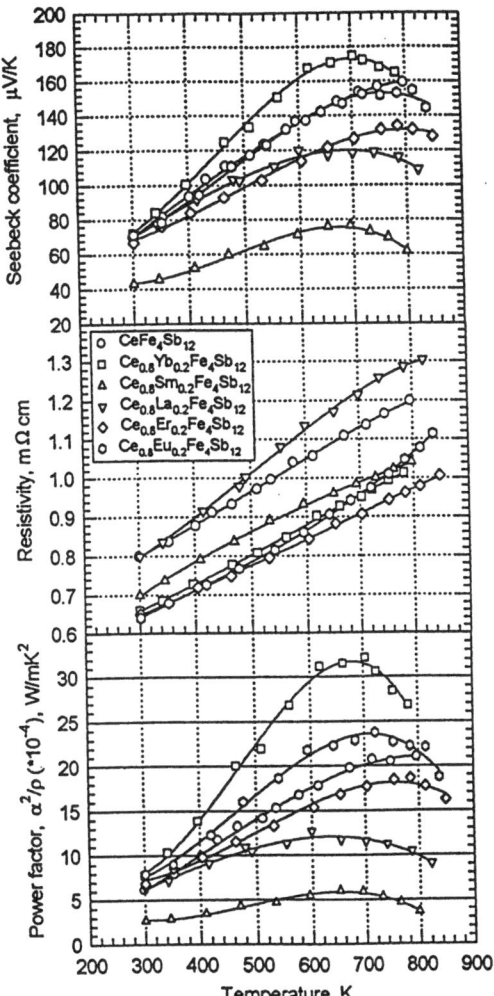

FIG. 55. Temperature dependence of Seebeck coefficient, electrical resistivity, and power factor for various partially substituted skutterudites. Note a large Seebeck coefficient for $Ce_{0.8}Yb_{0.2}Fe_4Sb_{12}$ that results in a nearly 50% enhancement in the power factor of this compound over that of $CeFe_4Sb_{12}$ (from Rowe et al., 1998). Copyright 1998 by IEEE.

Overall, there is little difference in the electrical and thermal transport properties of Ce- and La-filled skutterudites at and above 300 K. At low temperatures, because of the proximity of the Ce $4f$ level to the Fermi energy, the transport behavior of the Ce-filled skutterudites starts to reveal the influence of hybridization. This is notable in the electrical resistivity of some of the samples (see Fig. 56), and it is also reflected in the behavior of

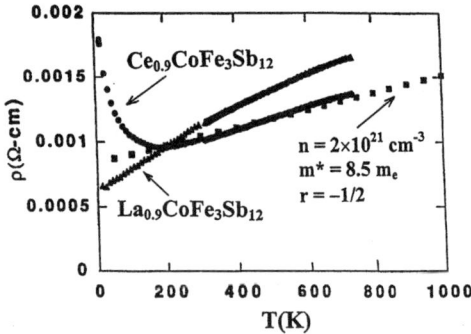

FIG. 56. Temperature dependence of resistivity for $Ce_{0.9}CoFe_3Sb_{12}$ and $La_{0.9}CoFe_3Sb_{12}$. Solid squares represent the model calculation based on a single parabolic band with the parameters indicated in the figure. Low temperature data of $Ce_{0.9}CoFe_3Sb_{12}$ are affected by the proximity of the Ce $4f$ level to the Fermi energy, (from Sales et al., 1997). Copyright 1997 by the American Physical Society.

the thermal conductivity. Although the thermal conductivity of both $LaFe_3CoSb_{12}$ and $CeFe_3CoSb_{12}$ is nearly the same at 300 K, as the temperature decreases, the conductivity of the Ce-filled skutterudite becomes progressively smaller in comparison to the La-filled compound. This has been explained as due to hybridization (Sales et al., 1996). In general, hybridization involving the Ce $4f$ level is considerably stronger in arsenides and especially phosphides, where it gives rise to a small gap and the resistivity increases to very high values $\sim 10^3$ Ω cm at 10 K (Shirotani et al., 1996; Meisner et al., 1985).

Very high carrier concentrations observed in filled skutterudites raise a question regarding the influence of the carrier–phonon scattering on the transport properties. In general, charge carriers are effective scatterers of low frequency phonons. Their influence shows as a significant reduction in the low-temperature (dielectric) peak of thermal conductivity. In binary skutterudites, both p- and n-type doping cause a significant reduction in the low-temperature thermal conductivity, the donors being more effective than the acceptors. At the carrier density of 10^{21}–10^{22} cm^{-3} that is representative of the filled skutterudites, it should not be surprising to see a significant influence of the carrier–phonon interaction on the thermal transport. The problem is that the resonant scattering by the void fillers (rattlers) is so strong that it masks contributions of the other scattering processes, including the one due to the charge carriers. In the absence of detailed calculations and fits of the thermal conductivity of filled skutterudites, one can offer only a rough estimate — up to perhaps 30% of the thermal resistivity arising from charge carrier scattering.

So far, the most promising filled skutterudites are structures such as $Ce_yFe_{4-x}Co_xSb_{12}$ and $La_yFe_{4-x}Co_xSb_{12}$ that include Fe in the formula unit. These compositions provide an opportunity to alter the nature of the conduction process from that of a substantially metallic ($CeFe_4Sb_{12}$) to one that has a semiconducting character (Co-rich compounds). Furthermore, these materials reveal a subtle interplay between the electron doping via the filler species (Ce or La) and the hole doping by Fe on the Co sites. Based on crystal chemistry arguments, Chen et al. (1997a) suggested that the carrier concentration per formula unit is given approximately by $(p,n) = 4 - x - 3y$. When this quantity is positive the samples will be p-type, and when it is negative they will be n-type. The phase diagram dividing the n- and p-type regions of conduction is shown in Fig. 57, and the predicted trend is indeed followed in practice. This scenario assumes that Fe is in the Fe^{2+} state (inferred from the susceptibility) and the concentration of the trivalent Ce is restricted by its solid solubility limit. In the case that Ni is used instead of Co, that is, for compounds $Ce_yFe_{4-x}Ni_xSb_{12}$, the equation that divides the phase diagram into the p- and n-type regimes of conduction is $(p,n) = 4 - 2x - 3y$. The factor of 2 represents two electrons donated by Ni when substituting for Fe.

An immediate consequence of the presence of Fe is a very low mobility that drops to values on the order of unity in units of $cm^2/V\text{-}s$. Thus, there is strong carrier scattering on the Fe site. The presence of Fe has also a strong and unanticipated influence on thermal conductivity. Figure 58

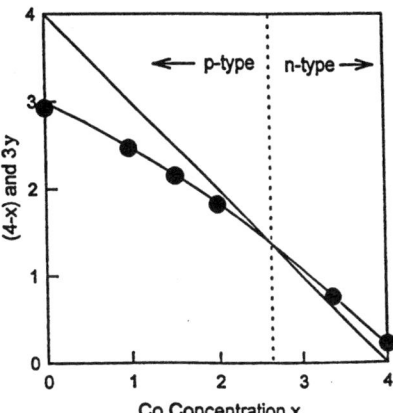

FIG. 57. Phase diagram delineating the p- and n-type regions of conduction in $Ce_yFe_{4-x}Co_xSb_{12}$. Solid line represents available holes per formula unit $(4-x)$, and the solid circles stand for available electrons that can fill these holes $(3y)$. Above the point where the curves cross, the samples will be n-type. The concentration of the trivalent Ce is restricted by its solid solubility limit (adapted from Chen et al., 1997a).

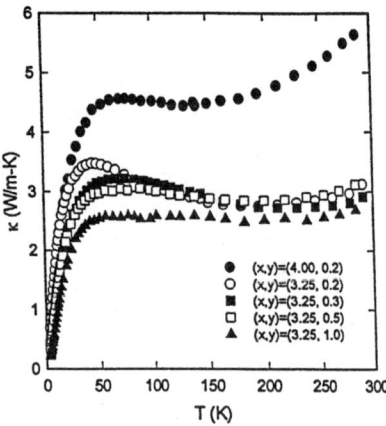

FIG. 58. Thermal conductivity of $Ce_yFe_{4-x}Co_xSb_{12}$. Note a dramatic effect resulting from the presence of a small fraction of Fe on the Co site that is in addition to the effect of Ce in the cages of the structure (from Uher et al., 1998). Copyright 1998 by IEEE.

shows the thermal conductivity of $Ce_yFe_{4-x}Co_xSb_{12}$ for various Ce filling fractions and two values of $x = 4$ (no Fe present) and $x = 3.25$. We note a substantial reduction in the thermal conductivity ($\sim 50\%$) in the Fe-containing samples compared with the sample that has no iron. This reduction is on top of the reduction caused by Ce rattling. In this series of samples, as the Ce concentration is increased, what remains of the low-temperature peak is depressed further and eventually the peak is washed out, while the room temperature thermal conductivity is almost independent of Ce concentration. Above about 30 K, Fe appears to be an additional strong phonon scatterer. This is quite surprising since Fe and Co have nearly the same mass and atom size and thus there should be no more than about 5% reduction in thermal conductivity arising from the mass-fluctuation scattering. To clarify this point, Meisner et al. (1998) investigated thermal conductivity on a series of "optimally" filled $Ce_yFe_{4-x}Co_xSb_{12}$ compounds. Such compounds had Ce content equal to the maximum filling consistent with the solid solubility limit of Ce for a given concentration of Co (see Fig. 8). The conclusion drawn from these results was that these fractionally filled compounds can be viewed as solid solutions of fully filled ($CeFe_4Sb_{12}$) and completely empty ($\square_2Co_4Sb_{12}$) skutterudites. The upshot is that the large reduction in thermal conductivity upon the presence of Fe in the structure is not due to the mass difference between Fe and Co but due to the mass difference between a site occupied by Ce and a vacancy $\square$. This of course represents a mass difference of 100%. Figure 59 is a plot of the lattice thermal resistivity at 300 K across the solid solution series from $\square_2Co_4Sb_{12}$ to $CeFe_4Sb_{12}$. The dashed line in this figure represents variation in the

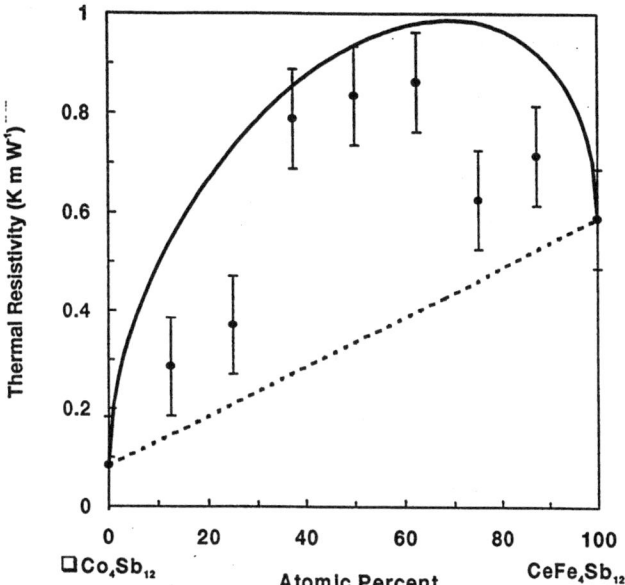

FIG. 59. Variation of the thermal resistivity of $(CeFe_4Sb_{12})_\alpha(\square Co_4Sb_{12})_{1-\alpha}$ solid solutions at 300 K. Dashed line represents variation from the rule of mixtures. Solid line includes additional thermal resistivity due to solid solution formation calculated from the theory of Callaway and von Baeyer (1960); from Meisner et al. (1998). Copyright 1998 by the American Physical Society.

thermal resistivity arising from the rule of mixtures. The solid line includes additional thermal resistivity due to solid solution formation calculated from the theory of Callaway and von Baeyer (1960). It accounts quite well for the increase in thermal resistivity due to the formation of the solid solution.

To illustrate a spectrum of transport parameters of Ce-filled skutterudites based on $CeFe_{4-x}Co_xSb_{12}$, Table XIV collects their room temperature values. These polycrystalline samples were of 98% theoretical density. High-temperature behavior of the filled samples is illustrated in Figs. 60–62. Particularly noteworthy are the data in Fig. 62 — they show that high values of ZT in excess of unity are possible with these Ce-filled skutterudites. ZT values approaching unity were also obtained with $La_{0.9}Fe_3CoSb_{12}$ (Sales et al., 1997). In fact, above about 450°C, the rare earth-filled Fe-rich skutterudites are superior to the best state-of-the-art thermoelectrics.

Figure 63 depicts representative data of high-temperature thermal conductivity for several filled antimonide skutterudites, including Fe-based as well as Ru- and Co-based structures. The lowest thermal conductivity is

TABLE XIV

ROOM TEMPERATURE PARAMETERS OF SEVERAL Ce-FILLED SKUTTERUDITES BASED ON $CeFe_{4-x}Co_xSb_{12}$[a]

Composition	a_o (Å)	n (10^{21} cm^{-3})	μ_H (cm^2/V-s)	ρ (mΩ-cm)	S (μV/K)	κ (W/mK)
$CeFe_4Sb_{12}$	9.1460	5.5	1.5	0.75	59	1.4
$CeFe_{3.5}Co_{0.5}Sb_{12}$	9.1349	5.7	1.5	0.71	76	2.2
$CeFe_3Co_1Sb_{12}$	9.1145	4.4	1.7	0.84	106	1.9
$CeFe_2Co_2Sb_{12}$	9.0909	0.65	2.9	3.89	125	2.8
Co_4Sb_{12}	9.0385	0.0044	1944	0.74	138	11.8

[a] The data were collected at both the Jet Propulsion Laboratory and the University of Michigan. a_o stands for the lattice constant, n is the carrier density, μ_H is the Hall mobility, ρ is the electrical resistivity, S is the Seebeck coefficient, and κ is the total thermal conductivity. (Adapted from Fleurial et al., 1996.)

measured for the $Ce_{0.29}Co_4Ge_{0.7}Sb_{11.3}$ sample in which the charge compensation is achieved by substituting on the pnicogen site. Above 600 K, a large carrier thermal conductivity contribution leads to a rapidly rising thermal conductivity. Beyond antimonide skutterudites, there is little information available on the thermal conductivity or for that matter any properties of filled skutterudites. Most recently, Watcharapasorn et al. (1999) prepared hot-pressed $CeFe_4P_{12}$ with 99% theoretical density. The samples were

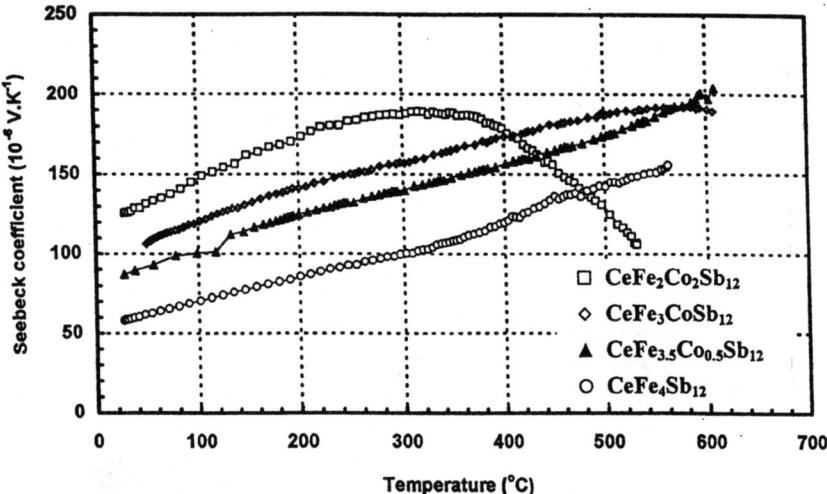

FIG. 60. High-temperature thermopower of several samples of $CeFe_{4-x}Co_xSb_{12}$ (from Fleurial et al., 1996). Copyright 1996 by IEEE.

5 SKUTTERUDITES: PROSPECTIVE NOVEL THERMOELECTRICS   243

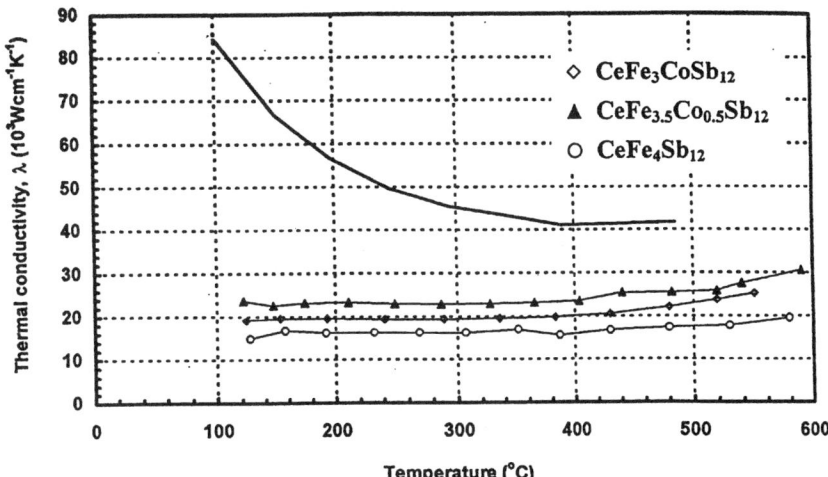

FIG. 61. High temperature thermal conductivity of $CeFe_{4-x}Co_xSb_{12}$. The solid curve represents the data for $CoSb_3$ (from Fleurial et al., 1996). Copyright 1996 by IEEE.

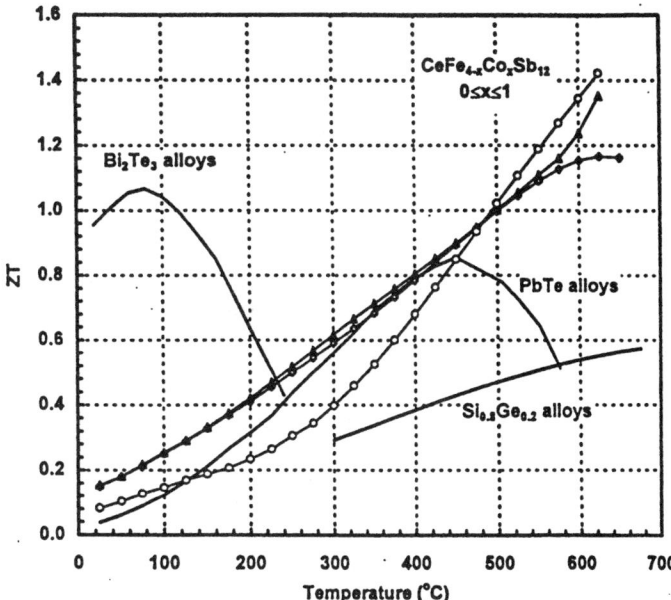

FIG. 62. Dimensionless figure of merit ZT as a function of temperature for $CeFe_{4-x}Co_xSb_{12}$. ▲, sample with $x = 1$; ○, sample with $x = 0.5$; and ◇, sample with $x = 0$. ZT values for the state-of-the-art traditional thermoelectric materials are shown for comparison (from Fleurial et al., 1997). Copyright 1997 by IEEE.

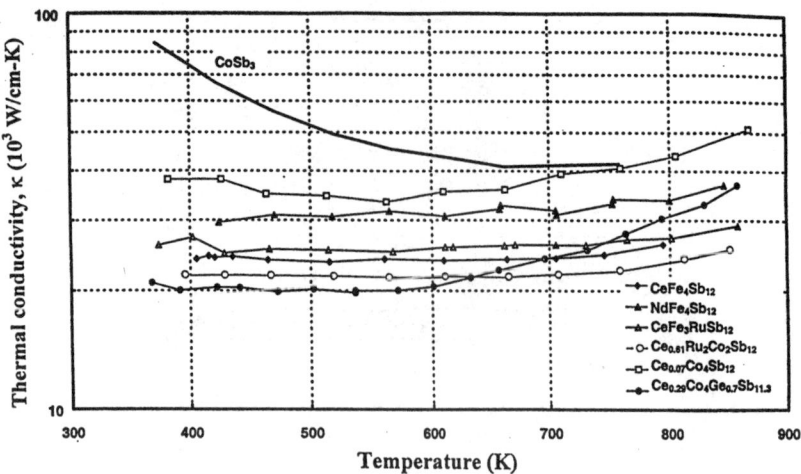

FIG. 63. High-temperature thermal conductivity of several filled skutterudites based on Fe, Ru, and Co antimonide. Thermal conductivity of $CoSb_3$ is shown for comparison (courtesy of Dr. J.-P. Fleurial).

p-type semiconductors with a carrier density of 1.4×10^{19} cm^{-3}, that is, more than two orders of magnitude lower carrier density than in $CeFe_4Sb_{12}$. Although the much lower carrier concentration was conducive to higher hole mobility (~ 25 cm^2/V-s), it was disappointing that the Seebeck coefficient only matched that of $CeFe_4Sb_{12}$. Furthermore, the samples showed very high thermal conductivity of 14 W/m-K at 300 K, 10 times the value of $CeFe_4Sb_{12}$. This large value of thermal conductivity is primarily due to a much smaller cage of $CeFe_4P_{12}$ (a cage radius of 1.8 Å as opposed to 1.936 Å for $CeFe_4Sb_{12}$) and thus very restricted motion of the Ce ions. Perhaps more telling is a comparison of the atomic displacement parameter, which for $CeFe_4P_{12}$ is 0.42 Å^2, while for $CeFe_4Sb_{12}$ it is 1.9 Å^2. The much reduced carrier density in the filled phosphides, and thus far less significant carrier–phonon scattering, may be an additional contributing factor to their very high thermal conductivity. It is unlikely that filled arsenides and phosphides, on their own, are going to emerge as useful thermoelectrics. However, if one succeeds with forming solid solutions between filled antimonides and phosphides, one might benefit from their lower carrier density, enhanced mass, and volume fluctuation scattering and, if such structures turn out to be semiconducting, one might be able to control their properties by doping.

To illustrate the overriding influence of the cage size on the thermal transport of filled skutterudites, Fig. 64 shows the lattice thermal conduc-

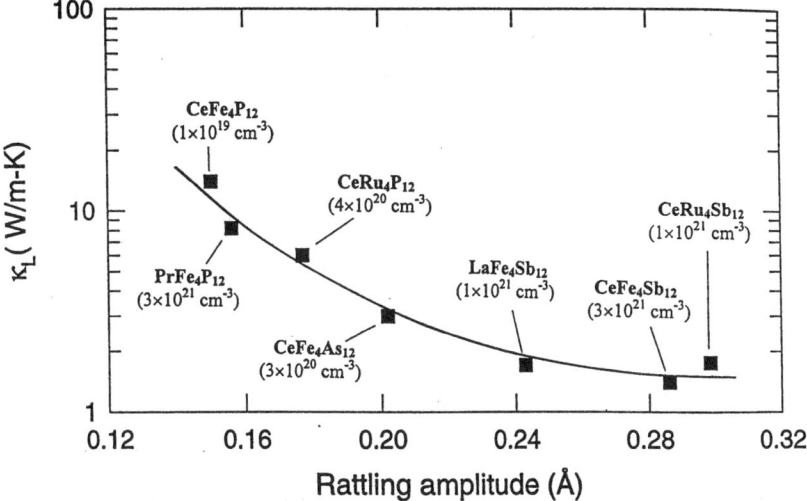

FIG. 64. Room temperature lattice thermal conductivity as a function of rattling amplitude for several fully filled skutterudites (data courtesy of Dr. J.-P. Fleurial).

tivity at 300 K as a function of the rattling amplitude. The trend is very clear and it is difficult to argue against the notion that a more freely vibrating ion, i.e., the larger the radius of the cage in relation to the ionic radius of the rattler, the greater the reduction in the lattice thermal conductivity.

What can we conclude about the effect of the rattling ions in skutterudites? Is the thermal conductivity truly that of a glasslike structure? From the data accumulated in the literature it should be clear that the resonant scattering of phonons is the dominant process that results in a drastically reduced thermal conductivity, and, within a factor of about 3, the lattice thermal conductivity approaches values representative of the minimum thermal conductivity. However, except perhaps for the strongly scattering structure of Nd-filled $Ir_4Ge_3Sb_9$, which relies on a combined effect of several different scattering mechanisms (and which has electronic properties hopelessly inadequate for thermoelectricity), the thermal conductivity of all other filled skutterudites has not yet reached the regime normally associated with that of a truly glasslike or amorphous solid. The remnants of the crystalline behavior are revealed in the flat or slowly rising thermal conductivity with decreasing temperature below 300 K, and the filled skutterudites still have a long way to go before one could call their temperature dependence akin to that of, say, vitreous silica.

VIII. Conclusions

Skutterudites represent an interesting material system with the real potential to make an impact in the field of thermoelectricity. The open nature of the skutterudite crystal structure offers exciting opportunities to tailor it in the direction of optimal transport properties that could maximize the thermoelectric figure of merit. Of particular importance is the existence of oversized voids or cages that can accommodate such foreign species as rare earth ions whose localized vibrations can greatly depress the heat conducting ability of the material. One useful indicator and predictor of how well the structure will conduct heat is the atomic displacement parameter (ADP). In skutterudites, the filler species show unusually large values of ADP, reflecting a loose bonding of the filler ion to its neighboring atoms; consequently, very low thermal conductivity can be achieved upon filling the structure. Another important aspect of the filled skutterudite compounds relates to the observation that the void fillers generally do not completely destroy the otherwise very favorable electronic transport properties. Although the exceptionally high values of the carrier mobility (especially that of holes) typical of the binary skutterudites are significantly degraded upon filling the voids, the structure nevertheless preserves useful electronic properties. As a result, the skutterudites can be viewed as an example of the recently developed PGEC concept — a new road map for developing novel, high-efficiency thermoelectric materials.

In addition to filling the void, the skutterudite structure is amenable to doping and the formation of solid solutions. One can also synthesize interesting ternary compounds. There is, therefore, a wide range of options to modify the structure and explore the parameter space of the material. Coupled with serious theoretical efforts that inform and guide the search, one has good reason to believe that, among the skutterudites, there will emerge a magic compound with superior thermoelectric properties.

Research efforts over the past few years have indicated the potential of skutterudites. Indeed, among skutterudites there are compounds that not only have exceeded the benchmark ZT value of unity, but have done so by a significant margin. In their current configuration, the skutterudites attain the maximum figure of merit at elevated temperatures, typically near 500–700 K. As a result, they are of considerable interest for such thermoelectric power generation applications as harvesting waste heat from various industrial operations. In comparison to the conventional state-of-the-art thermoelectrics, skutterudites are remarkable in that they achieve optimum performance in a carrier concentration range more reminiscent of a metal or semimetal than a moderately degenerate semiconductor. This is the consequence of the flat bands that arise from hybridization of the rare earth $4f$ level and the transition metal d-bands. The presence of the rare earth species in the finely tuned environment of the skutterudite structure also leads to a plethora of interesting magnetic, superconducting, transport, and optical phenomena. Indeed, many of the currently hotly pursued topics of many-body interactions in condensed matter can be observed in one of the compounds of the skutterudite family.

So far, the research effort has focused mostly on the properties of Co/Fe alloyed antimonide skutterudites—and especially those filled with either Ce or La. There are indications that useful thermoelectrics can also be found among Ru and Rh alloyed antimonide skutterudites. Fornari and Singh (1999) have suggested, based on their band structure calculations, that the n-type $La(Ru_{1-x}Rh_x)_4Sb_{12}$ compounds might be exceptionally good thermoelectrics and, furthermore, that their optimal performance might in fact be near room temperature. Experimental verification of this prediction would represent a major breakthrough in the field of thermoelectricity. As this volume is being published, new and exciting results are emerging on $CoSb_3$ partially filled with either divalent Ba or with trivalent Yb. These are n-type structures with the reported values of ZT approaching or even exceeding the value of unity near 700 K (Chen et al., 2000; Nolas et al., 2000). Having available efficient, n-type thermoelectric elements made of a substantially similar material as the already promising p-type filled skutterudites would be a great advantage from the perspective of fabrication of thermoelectric devices. How far one will be able to go with fine tuning the skutterudite structure to achieve the maximum figure of merit is to be seen in the near future. It is my hope that this review will serve as a useful reference for researchers who strive both to improve the properties of skutterudites and advance the field of thermoelectricity.

Acknowledgments

It has been a great pleasure for me to have the opportunity to return to the field of thermoelectrics during the past half-dozen years. I am indebted to my thesis mentor, Professor Julian Goldsmid, who, more than 25 years ago, ignited my interest in thermoelectrics. I also view with satisfaction my Ph.D. students who have made important contributions to the field of thermoelectricity. In particular, I wish to recognize Dr. Donald Morelli, with whom I collaborated on many exciting thermoelectric projects. I also acknowledge the hard work of my current graduate students, Jihui Yang, Wei Chen, and Siqing Hu, my postdoc, Dr. Jeffrey Dyck, and my collaborator, Dr. Gregory Meisner. I thank all colleagues who allowed me to use their data and figures in this review. Finally, I must note that the research into novel thermoelectrics would not have been possible, at least not in the United States, without the generous support provided by the Office of the Naval Research and the DARPA agency. As one of the beneficiaries, I am grateful for their support and encouragement.

References

A. Abeles, *Phys. Rev.* 131, 1906 (1963).
J. Ackermann and A. Wold, *J. Phys. Chem. Solids* 38, 1013 (1977).
K. Akai, H. Kurisu, T. Shimura, and M. Matsuura, *Proc. 16th International Conference on Thermoelectrics*, IEEE Catalog Number 97TH8291, Piscataway, NJ, p. 334 (1997).
K. Akai, H. Kurisu, T. Moriyama, S. Tamamoto, and M. Matsuura, *Proc. 17th International Conference on Thermoelectrics*, IEEE Catalog Number 98TH8365, Piscataway, NJ, p. 105 (1998).
H. Anno, K. Matsubara, Y. Notohara, T. Sakakibara, K. Kishimoto, and T. Koyanagi, *Proc. 15th International Conference on Thermoelectrics*, IEEE Catalog Number 96TH8169, Piscataway, NJ, p. 435 (1996).
H. Anno, K. Hatada, H. Shimizu, K. Matsubara, Y. Notohara, T. Sakakibara, H. Tashiro, and K. Motoya, *J. Appl. Phys.* 83, 5270 (1998a).
H. Anno, H. Tashiro, H. Kaneko, and K. Matsubara, *Proc. 17th International Conference on Thermoelectrics*. IEEE Catalog Number 98TH8365, Piscataway, NJ, p. 326 (1998b).
H. Anno, K. Matsubara, Y. Notohara, T. Sakakibara, and H. Tashiro, *J. Appl. Phys.* 86, 3780 (1999a).
H. Anno, H. Tashiro, and K. Matsubara, *Proc. 18th International Conference on Thermoelectrics*, IEEE Catalog Number 99TH8407, Piscataway, NJ, p. 169 (1999b).
E. Arushanov, K. Fess, W. Kaefer, Ch. Kloc, and E. Bucher, *Phys. Rev.* B56, 1911 (1997).
S. Bahn, T. Gödecke, and K. Schubert, *J. Less-Common Met.* 19, 121 (1969).
D. J. Bergman and L. G. Fel, *J. Appl. Phys.* 85, 8205 (1999).
W. Biltz and M. Heimbrecht, *Z. Anorg. Allg. Chem.* 237, 132 (1938).
W. Biltz and H. Heimbrecht, *Z. Anorg. Allg. Chem.* 241, 349 (1939).
A. Borshchevsky, J.-P. Fleurial, E. Allevato, and T. Caillat, *Proc. 13th International Conference on Thermoelectrics*, American Institute of Physics, New York, p. 3 (1995a).
A. Borshchevsky, J.-P Fleurial, E. Allevato, and T. Caillat, *Proc. 13th International Conference on Thermoelectrics* (B. Mathiprakasam, ed.), p. 3. American Institute of Physics, New York (1995b).
A. Borshchevsky, T. Caillat, and J.-P. Fleurial, *Proc. 15th International Conference on Thermoelectrics*, IEEE Catalog Number 96TH8169, p. 112 (1996).
D. J. Brown and W. Jeitschko, *J. Solid State Chem.* 32, 357 (1980a).
D. J. Brown and W. Jeitschko, *J. Less-Common Metals* 72, 147 (1980b).
D. J. Brown and W. Jeitschko, *J. Less-Common Met.* 76, 33 (1980c).
D. C. Cahill and R. O. Pohl, *Solid State Commun.* 70, 927 (1989).
D. G. Cahill, S. K. Watson, and R. O. Pohl, *Phys. Rev.* B46, 6131 (1992).
T. Caillat, A. Borshchevsky, and J.-P. Fleurial, *Proc. 27th Intersociety Energy Conversion Engineering Conference, San Diego, CA*, August (1992a).
T. Caillat, A. Borshchevsky, and J.-P. Fleurial, *Proc. 11th International Conference on Thermoelectrics, Univ. of Arlington, TX*, (K. R. Rao, ed.), p. 276 (1992b).
T. Caillat, A. Borshchevsky, and J.-P. Fleurial, *Proc. 11th International Conference on Thermoelectrics, Univ. of Arlington, TX* (K. R. Rao, ed. p. 98 (1992c).
T. Caillat, A. Borshchevsky, and J.-P. Fleurial, *J. Alloys and Compounds* 199, 207 (1993a).
T. Caillat, A. Borshchevsky, and J.-P. Fleurial, *J. Phase Equilibria* 14, 576 (1993b).
T. Caillat, A. Borshchevsky, and J.-P. Fleurial, *Proceedings of the XIth Symposium on Space Nuclear Power and Propulsion, Albuquerque, New Mexico*. American Institute of Physics, AIP Conference No. 301, p. 517 (1994a).
T. Caillat, A. Borshchevsky, and J.-P. Fleurial, *Proc. 12th International Conference on Thermoelectrics* (K. Matsuura, ed.), p. 132. Institute of Electrical Engineers of Japan, Tokyo (1994b).

T. Caillat, A. Borshchevsky, and J.-P. Fleurial, *Proc. 13th International Conference on Thermoelectrics* (B. Mathiprakasam, ed.), p.31 American Institute of Physics, New York (1995a).

T. Caillat, J.-P. Fleurial, and A. Borshchevsky, *Proc. 30th Intersociety Energy Conversion Engineering Conference, Amer. Soc. Mechan. Eng.,* (D. Y. Goswami, L. D. Kannberg, T. R. Mancini, S. Somasundaram, eds.), Vol. 3, p. 83 (1995b).

T. Caillat, A. Borshchevsky, and J.-P. Fleurial, *Proc. 13th International Conference on Thermoelectrics.* American Institute of Physics, NY, p. 58 (1995c).

T. Caillat, A. Borshchevsky, and J.-P. Fleurial, *Proc. 13th International Conference on Thermoelectricity.* (B. Mathiprakasam, ed.), p. 209. American Institute of Physics, New York (1995d).

T. Caillat, A. Borshchevsky, and J.-P. Fleurial, *J. Appl. Phys.* 80, 4442 (1996a).

T. Caillat, J.-P. Fleurial, and A. Borshchevsky, *J. Crystal Growth* 166, 722 (1996b).

T. Caillat, J. Kulleck, A. Borshchevsky, and J.-P. Fleurial, *J. Appl. Phys.* 79, 8419 (1996c).

J. Callaway and H. C. von Baeyer, *Phys. Rev.* 120, 1149 (1960).

J. C. Caylor, A. M. Stacy, T. Sands, and G. Gronsky, *Mat. Res. Soc. Symp. Proc.* 545, 327 (1999).

L. Chapon, D. Ravot, and J. C. Tedenac, *Mat. Res. Soc. Symp. Proc.* 545, 321 (1999).

B. Chen, J. H. Xu, C. Uher, D. T. Morelli, G. P. Meisner, J.-P. Fleurial, T. Caillat, and A. Borshchevsky, *Phys. Rev.* B55, 1476 (1997a).

B. Chen, J.-H. Xu, S. Hu, and C. Uher, *Mat. Res. Soc. Symp. Proc.*, 452, 1037 (1997b).

L. D. Chen, X. F. Tang, T. Goto, T. Hirai, J. S. Dyck, W. Chen, and C. Uher, *Proc. 19th International Conference on Thermoelectrics*, Cardiff, Great Britain, August 2000, submitted.

M. E. Danebrock, C. B. H. Evers, and W. Jeitschko, *J. Phys. Chem. Solids* 57, 381 (1996).

N. R. Dilley, E. J. Freedman, E. D. Bauer, and M. B. Maple, *Phys. Rev.* B58, 6287 (1998).

L. D. Dudkin, *Sov. Phys.-Tech. Physics* 3, 216 (1958).

L. D. Dudkin and N. Kh. Abrikosov, *Zh. Neorg. Khim.* 1, 2096 (1956).

L. D. Dudkin and N. Kh. Abrikosov, *Zh. Neorg. Khim.* 2, 212 (1957).

L. D. Dudkin and N. Kh. Abrikosov, *Sov. Phys.—Solid State* 1, 126 (1959).

C. B. H. Evers, L. Boonk, and W. Jeitschko, *Z. Anorg. Allg. Chem.* 620, 1028 (1994).

J. L. Feldman and D. J. Singh, *Phys. Rev.* B53, 6273 (1996).

P. Feschotte and D. Lorin, *J. Less-Common Metals* 155, 255 (1989).

K. Fess, E. Arushanov, W. Kaefer, Ch. Kloc, K. Friemelt, and E. Bucher, *Proc. 16th International Conference on Thermoelectrics.* IEEE Catalog Number 97TH8291, Piscataway, NJ, p. 347 (1997).

J.-P. Fleurial, T. Caillat, and A. Borshchevsky, *Proc. 13th International Conference on Thermoelectrics.* (B. Mathiprakasam, ed.), p. 40. *American Institute of Physics*, New York (1995a).

J.-P. Fleurial, T. Caillat, and A. Borshchevsky, *Proc. 14th International Conference on Thermoelectrics* (A. F. Ioffe Physical-Technical Institute, St. Petersburg, Russia, p. 231 (1995b).

J.-P. Fleurial, A. Borshchevsky, T. Caillat, D. T. Morelli, and G. P. Meisner, *Proc. 15th International Conference on Thermoelectrics.* IEEE Catalog 96TH8169, Piscataway, NJ, p. 91 (1996).

J.-P. Fleurial, T. Caillat, and A. Borshchevsky, *Proc. 16th International Conference on Thermoelectrics.* IEEE Catalog Number 97TH8291, Piscataway, NJ, p. 1 (1997).

M. Fornari and D. J. Singh, *Appl. Phys. Lett.* 74, 3666 (1999).

D. A. Gajewski, N. R. Dilley, E. D. Bauer, E. F. Freeman, R. Chau, M. B. Maple, D. Mandrus, B. C. Sales, and A. H. Lacerda. *J. Phys.: Condens. Matter* 10, 6973 (1998).

G. A. Glassbrenner and G. A. Slack, *Phys. Rev.* 134, A1058 (1964).

A. G. Grandjean, J. A. Hodges, D. J. Brown, and W. Jeitschko, *J. Phys. C: Solid State Phys.* 16, 2797 (1983).

F. Grandjean, A. Gerard, D. J. Braun, and W. Jeitschko, *J. Phys. Chem. Solids* 45, 877 (1984).
R. P. Guertin, C. Rossel, M. S. Torikachvili, M. W. McElfresh, M. B. Maple, S. H. Bloom, Y S. Yao, M. V. Kuric, and G. P. Meisner, *Phys. Rev.* B36, 8665 (1987).
H. Harima, *J. Magn. Mag. Mater.* 177-181, 321 (1998).
M. D. Hornbostel, E. J. Hyer, J. Thiel, and D. C. Johnson, *J. Am. Chem. Soc.* 119, 2665 (1997a).
M. D. Hornbostel, E. J. Hyer, J. H. Edvalson, and D. C. Johnson, *Inorg. Chem.* 36, 4270 (1997b).
F. Hulliger, *Helv. Phys. Acta* 34, 782 (1961).
W. Jeitschko and D. J. Brown, *Acta Crystallog.* B33, 3401 (1977).
P. Jolibois, *C. R. Acad. Sci.* 150, 106 (1910).
D. Jung, M.-H. Whangbo, and S. Alvarez, *Inorg. Chem.* 29, 2252 (1990).
S. Katsuyama, Y. Kanayama, M. Ito, K. Majima, and H. Nagai, *Proc. 17th International Conference on Thermoelectrics.* IEEE Catalog Number 98TH8365, Piscataway, NJ, p. 342 (1998a).
S. Katsuyama, Y. Shichijo, M. Ito, K. Majima, and H. Hagai, *J. Appl. Phys.* 84, 6708 (1998b).
V. Keppens, D. Mandrus, B. C. Sales, B. C. Chakoumakos, P. Dai, R. Coldea, M. B. Maple, D. A. Gajewski, E. J. Freeman, and S. Bennington, *Nature* 395, 876 (1998).
R. B. King, *Inorg. Chem.* 28, 3048 (1989).
H. Kitagawa, M. Hasaka, T. Morimura, and S. Kondo, *Proc. 17th International Conference on Thermoelectrics.* IEEE Catalog Number 98TH8365, Piscataway, NJ, p. 338 (1998a).
H. Kitagawa, S. Kondo, K. Oda, M. Hasaka, and T. Morimura, *Proc. 17th International Conference on Thermoelectrics.* IEEE Catalog Number 98TH8365, Piscataway, NJ, p. 319 (1998b).
A. Kjekshus, *Acta Chem. Scand.* 15, 678 (1961).
A. Kjekshus and G. Pedersen, *Acta Cryst.* 14, 1065 (1961).
A. Kjekshus, D. G. Nicholson, and T. Rakke, *Acta Chem. Scand.* 27, 1307 (1973).
A. Kjekshus and T. Rakke, *Acta Chem. Scand.* A28, 99 (1974).
P. G. Klemens, *Proc. Phys. Soc. (London)* A68, 1113 (1959).
G. Kliche and W. Bauhofer, *Mater. Res. Bull.* 22, 551 (1987).
G. Kliche and W. Bauhofer, *J. Phys. Chem. Solids* 49, 267 (1988).
G. Kliche and H. D. Lutz, *Infrared Phys.* 24, 171 (1984).
R. Korenstein, S. Soled, A. Wold, G. Collin, *Inorg. Chem.* 16, 2344 (1978).
T. Koyanagi, T. Tsubouchi, M. Ohtani, K. Kishimoto, H. Anno, and K. Matsubara, *Proc. 15th International Conference on Thermoelectrics.* IEEE Catalog Number 96TH8169, Piscataway, NJ, p. 107 (1996).
R. N. Kuzmin, in *Chemical Bonds in Semiconductors and Solids* (N. N. Sirota, ed.). Consultants Bureau, New York (1967).
C. H. Lee, H. Oyanagi, C. Sekine, I. Shirotani, and M. Ishii, *Phys. Rev.* B60, 13253 (1999).
A. Leithe-Jasper, D. Kaczorowski, P. Rogl, J. Bogner, M. Reissner, W. Steiner, G. Wiesinger, and C. Godart, *Solid State Commun.* 109, 395 (1999).
M. Llunell, P. Alemany, S. Alvarez, and V. P. Zhukov, *Phys. Rev.* B53, 10605 (1996).
G. J. Long, D. Hautot, F. Grandjean, D. T. Morelli, and G. P. Meisner, *Phys. Rev.* B60, 7410 (1999).
H. D. Lutz and G. Kliche, *J. Solid State Chem.* 40, 64 (1981).
H. D. Lutz and G. Kliche, *Phys. Stat. Solidi* (b) 112, 549 (1982).
A. Lyons, R. P. Druska, C. Case, S. N. Subbarao, and A. Wold, *Mat. Res. Bull.* 13, 125 (1978).
N. Mandel and J. Donohue, *Acta Cryst.* B27, 2288 (1971).
D. Mandrus, A. Migliori, T. W. Darling, M. F. Hundley, E. J. Peterson, and J. D. Thompson, *Phys. Rev.* B52, 4926 (1995).
G. B. Martins, M. A. Pires, G. E. Barberis, C. Rettori, and M. S. Torikachvili, *Phys. Rev.* B50, 14822 (1994).

K. Matsubara, T. Sakakibara, Y. Notohara, H. Anno, H. Shimizu, and T. Koyanagi, *Proc. 15th International Conference on Thermoelectrics* (T. Caillat, ed.), p. 96. IEEE 96TH8169, Piscataway, NJ, p. 96 (1996).

G. P. Meisner, *Physica B* 108, 763 (1981).

G. P. Meisner, M. S. Torikachvili, K. N. Yang, M. B. Maple, and R. P. Guertin, *J. Appl. Phys.* 57, 3073 (1985).

G. P. Meisner, D. T. Morelli, S. Hu, J. Yang, and C. Uher, *Phys. Rev. Lett.* 80, 351 (1998).

D. T. Morelli and G. P. Meisner, *J. Appl. Phys.* 77, 3777 (1995).

D. T. Morelli, T. Caillat, J.-P. Fleurial, A. Borshchevsky, J. Vandersande, B. Chen, and C. Uher, *Phys. Rev.* B51, 9622 (1995).

D. T. Morelli, G. P. Meisner, B. Chen, S. Hu, and C. Uher, *Phys. Rev.* B56, 7376 (1997).

T. Morimura, H. Kitagawa, M. Hasaka, and S. Kondo, *Proc. 16th International Conference on Thermoelectrics.* IEEE Catalog Number 97TH8291, Piscataway, NJ, p. 356 (1997).

Y. Nagamoto, K. Tanaka, and T. Koyanagi, *Proc. 17th International Conference on Thermoelectrics.* IEEE Catalog Number 98TH8365, Piscataway, NJ, p. 302 (1998).

H. Nakagawa, H. Tanaka, A. Kasama, K. Miyamura, H. Masumoto, and K. Matsubara, *Proc. 15th International Conference on Thermoelectrics.* IEEE Catalog Number 96TH8169, Piscataway, NJ, p. 117 (1996).

H. Nakagawa, H. Tanaka, A. Kasama, H. Anno, and K. Matsubara, *Proc. 16th International Conference on Thermoelectrics.* IEEE Catalog Number 97TH8291, Piscataway, NJ, p.351 (1997).

E. H. Nickel, *Chem. Geol.* 5, 233 (1969).

G. S. Nolas and C. A. Kendziora, *Phys. Rev.* B59, 6189 (1999).

G. S. Nolas, G. A. Slack, T. Caillat, and G. P. Meisner, *J. Appl. Phys.* 79, 2622 (1996a).

G. S. Nolas, G. A. Slack, D. T. Morelli, T. M. Tritt, and A. C. Ehrlich, *J. Appl. Phys.* 79, 4002 (1996b).

G. S. Nolas, V. G. Harris, T. M. Tritt, and G. A. Slack, *J. Appl. Phys.* 80, 6304 (1996c).

G. S. Nolas, J. L. Cohn, and G. A. Slack, *Phys. Rev.* B58, 164 (1998).

G. S. Nolas, D. T. Morelli, and T. M. Tritt, *Ann. Rev. Mater. Sci.* 29, 89 (1999).

G. S. Nolas, M. Kaeser, R. T. Littleton IV, T. M. Tritt, H. Sellinschegg, D. C. Johnson, and E. Nelson, *Mater. Res. Soc. Symp. Proc.*, San Francisco, April 2000, in press.

L. Nordstrom and D. J. Singh, *Phys. Rev.* B53, 1103 (1996).

J. P. Odile, S. Soled, C. A. Castro, and A. Wold, *Inorg. Chem.* 17, 283 (1978).

I. Oftedal, *Z. Kristallogr.* A66, 517 (1928).

M. Partik, C. Kringe, and H. D. Lutz, *Z. Kristall.* 211, 304 (1996).

C. M. Pleass and R. D. Heyding, *Canad. J. Chem.* 40, 590 (1962).

H. Rakoto, E. Arushanov, M. Respaud, K. M. Broto, J. Leotin, Ch. Kloc, E. Bucher, and S. Askenazy, *Physica* B246–247, 528 (1998).

H. Rakoto, M. Respaud, J. M. Broto, E. Arushanov, and T. Caillat, *Physica* B269, 13 (1999).

E. H. Roseboom, Jr., *Am. Mineralogist* 47, 310 (1962).

T. Rosenqvist, *Acta Met.* 1, 761 (1953).

D. M. Rowe, V. L. Kuznetsov, and L. A. Kuznetsova, *Proc. 17th International Conference on Thermoelectrics.* IEEE Catalog Number 98TH8365, Piscataway, NJ, p. 323 (1998).

S. Rundqvist and N.-O. Ersson, *Ark. Kemi* 30, 103 (1968).

S. Rundqvist and A. Hede, *Acta Chem. Scand.* 14, 893 (1960).

S. Rundqvist and E. Larsson, *Acta Chem. Scand.* 13, 551 (1959).

B. C. Sales, *MRS Bull.* 23, 15 (1998).

B. C. Sales, D. Mandrus, and R. K. Williams, *Science* 272, 1325 (1996).

B. C. Sales, D. Mandrus, B. C. Chakoumakos, V. Keppens, and J. R. Thompson, *Phys. Rev.* B56, 15081 (1997).

B. C. Sales, B. C. Chakoumakos, D. Mandrus, and J. W. Sharp, *J. Solid State Chem.* 146, 528 (1999).

B. C. Sales, B. C. Chakoumakos, and D. Mandrus, *Phys. Rev. B* 61, 2475 (2000).
Th. Schmidt, G. Kliche, and H. D. Lutz, *Acta Cryst.* C43, 1678 (1987).
S. B. Schujman, G. A. Slack, H. C. Nguyen, G. S. Nolas, R. A. Young, F. Mohammed, and T. Tritt, *Mat. Res. Soc. Symp. Proc.* 545, 45, (1999).
C. Sekine, H. Saito, T. Uchiumi, A. Sakai, and I. Shirotani, *Solid State Commun.* 106, 441 (1998).
H. Sellinschegg, D. C. Johnson, G. S. Nolas, G. A. Slack, S. B. Schujman, F. Mohammed, T. Tritt, and E. Nelson, *Proc. 17th International Conference on Thermoelectrics*, IEEE Catalog Number 98TH8365, Piscataway, NJ, p. 338 (1998).
J. W. Sharp, E. C. Jones, R. K. Williams, P. M. Martin, and B. C. Sales, *J. Appl. Phys.* 78, 1013 (1995).
G. K. Shenoy, D. P. Noakes, and G. P. Meisner, *J. Appl. Phys.* 53, 2628 (1982).
I. Shirotani, T. Adachi, K. Tachi, S. Todo, K. Nozawa, T. Yagi, and M. Kinoshita, *J. Phys. Chem. Solids* 57, 211 (1996).
I. Shirotani, T. Uchiumi, K. Ohno, C. Sekine, Y. Nakazawa, K. Kanoda, S. Todo, and T. Yagi, *Phys. Rev.* B56, 7866 (1997).
D. J. Singh and W. E. Pickett, *Phys. Rev.* B50, 11235 (1994).
D. J. Singh and I. I. Mazin, *Phys. Rev.* B56, R1650 (1997).
D. J. Singh, L. Nordstrom, W. E. Pickett, and J. L. Feldman, *Proc. 15th International Conference on Thermoelectrics*. IEEE Catalog 96TH8169, Piscataway, NJ, p. 84 (1996).
D. J. Singh, I. I. Mazin, S. G. Kim, and L. Nordström, *Mat. Res. Soc. Symp. Proc.*, 478, 187 (1997).
D. J. Singh, I. I. Mazin, J. L. Feldman, and M. Fornari, *Mat. Res. Soc. Symp. Proc.* 545, 3 (1999).
G. A. Slack, *Phys. Rev.* 126, 427 (1962).
G. A. Slack, in *Solid State Physics* (F. Seitz and D. Turnbull, eds.), Vol. 34, p. 1. Academic Press, New York (1979).
G. A. Slack, in *CRC Handbook of Thermoelectrics* (D. M. Rowe, ed.), p. 407. CRC Press, Boca Raton, FL (1995).
G. A. Slack and S. Galginaitis, *Phys. Rev.* 133, A253 (1964).
G. A. Slack and V. G. Tsoukala, *J. Appl. Phys.* 76, 1665 (1994).
G. A. Slack, J.-P. Fleurial, and T. Caillat, *Naval Research News* 18, 23 (1996).
J. O. Sofo and G. D. Mahan, *Mat. Res. Soc. Symp. Proc.* 545, 315 (1999).
N. T. Stetson, S. M. Kauzlarich, and H. Hope, *J. Solid State Chem.* 91, 140 (1991).
G. R. Stewart, *Rev. Mod. Phys.* 56, 755 (1984).
K. L. Stokes, A. C. Ehrlich, and G. S. Nolas, *Mat. Res. Soc. Symp. Proc.* 545, 339 (1999).
H. Takizawa, K. Miura, M. Ito, T. Suzuki, and T. Endo, *J. Alloys Compounds* 282, 79 (1999).
H. Tashiro, Y. Notohara, T. Sakakibara, H. Anno, and K. Matsubara, *Proc. 16th International Conference on Thermoelectrics*. IEEE Catalog Number 97TH8291, Piscataway, NJ, p. 326 (1997).
M. S. Torikachvili, J. W. Chen, Y. Dalichaouch, R. P. Guertin, M. W. McElfresh, C. Rossel, M. B. Maple, and G. P. Meisner, *Phys. Rev.* B36, 8660 (1987).
T. M. Tritt, G. S. Nolas, G. A. Slack, A. C. Ehrlich, D. J. Gillespie, and J. L. Cohn, *J. Appl. Phys.* 79, 8412 (1996).
H. Uchida, V. Crnko, H. Tanaka, A. Kasama, and K. Matsubara, *Proc. 17th International Conference on Thermoelectrics*. IEEE Catalog Number 98TH8365, Piscataway, NJ, p. 330 (1998).
C. Uher, B. Chen, S. Hu, D. T. Morelli, and G. P. Meisner, *Mat. Res. Soc. Symp. Proc.* 478, 315 (1997).
C. Uher, S. Hu, and J. Yang, *Proc. 17th International Conference on Thermoelectrics*. IEEE Catalog Number 98TH8365, Piscataway, NJ, p. 306 (1998).
C. Uher, J. Yang, and S. Hu, *Mat. Res. Soc. Symp. Proc.* 545, p. 247 (1999).

J. W. Vandersande, J.-P. Fleurial, J. S. Beaty, and J. L. Rolfe, *Proc. 11th International Conference on Thermoelectrics* (K. R. Rao, ed.), p. 21. The University of Texas at Arlington, TX (1992).

U. Ventriglia, *Periodico Mineral (Rome)* 26, 345 (1957).

A. Watcharapasorn, R. C. DeMattei, R. S. Feigelson, T. Caillat, A. Borshchevsky, G. J. Snyder, and J.-P. Fleurial, *J. Appl. Phys.* 86, 6213 (1999).

L. J. Wilson and S. A. Mikhail, *Thermochim. Acta* 156, 107 (1989).

J. S. Xue, M. R. Antonio, W. T. White, and L. Soderholm, *J. Alloys and Compounds* 207–208, 161 (1994).

J. Yang, D. T. Morelli, G. P. Meisner, and C. Uher, in *Thermal Conductivity 25/Thermal Expansion 13* (C. Uher and D. T. Morelli, eds.), p. 130. Technomic Publishing, Lancaster, PA (2000).

S. Zemni, D. Tranqui, P. Chaudouet, R. Madar, and J. P. Senateur, *J. Solid State Chem.* 65, 1 (1986).

B. N. Zobrina and L. D. Dudkin, *Sov. Phys.-Solid State* 1, 1668 (1960).

V. P. Zhukov, *Phys. Solid State* 38, 90 (1996).

N. N. Zhuravlev and G. S. Zhdanov, *Kristallografiya* 1, 509 (1956).

CHAPTER 6

Semiconductor Clathrates: A Phonon Glass Electron Crystal Material with Potential for Thermoelectric Applications

George S. Nolas

R&D DIVISION
MARLOW INDUSTRIES, INC.
DALLAS, TEXAS

Glen A. Slack and Sandra B. Schujman

DEPARTMENT OF PHYSICS
RENSSELAER POLYTECHNIC INSTITUTE
TROY, NEW YORK

I. INTRODUCTION	255
1. *Historic Perspective*	255
2. *Background*	257
3. *Clathrates, Crypto-Clathrates, and Local Modes*	259
II. SYNTHESIS METHODS	265
III. STRUCTURAL PROPERTIES	267
1. *Crystal Structure and Bonding*	267
2. *Atomic Rattlers*	273
3. *Vibration Frequencies*	278
4. *Band Structure*	282
IV. THERMAL CONDUCTIVITY	284
1. *Glasslike Conduction*	284
2. *Connection with Structural Properties*	285
V. ELECTRONIC PROPERTIES	290
1. *Doping Variation*	290
2. *Metals or Semiconductors*	293
VI. SUMMARY AND CONCLUSIONS	294
REFERENCES	295

I. Introduction

1. HISTORIC PERSPECTIVES

According to the *Merriam-Webster Dictionary*, 10th Collegiate Edition, clathrates are compounds formed by the inclusion of atoms or molecules of one kind into cavities in the crystal lattice of another. The word "clathrate" derives from the Latin "clathratus" meaning "furnished with a lattice." It is presently used in chemistry to describe a particular type of compound, usually a polyatomic compound, in which one component forms a cage structure imprisoning the other.

The crystalline complexes of water, H_2O, with simple molecules such as chlorine, Cl_2, have been known to form clathrate compounds for more than a century. In 1811 Davy (1811) mentioned the formation of a chlorine–water molecule; Faraday (1823) proposed the formula Cl_2-10H_2O in 1823. Von Stackelberg et al. (1947), Von Stackelberg (1949), Claussen (1951), and Pauling and Marsh (1952) determined the structure of these compounds, which had become known as "gas hydrates" because they entrap gas molecules in crystalline H_2O. Clathrate hydrates have been formed with various materials such as noble gases, halogen molecules, some hydrocarbons, and low molecular weight compounds such as sulfur dioxide, methyl iodide, and chloroform, to name a few. In these clathrate hydrates, or ice clathrates, the water molecules form a hydrogen-bonded framework where each water molecule is tetrahedrally bonded to four H_2O neighbors, as in normal ice, but with a more open structure, forming different types of cavities that can enclose atoms or molecules.

There are two common forms of ice clathrates:

1. Type I has a cubic cell constant of about 12 Å and holds 46 water molecules in the unit cell. There are two types of cavities in the unit cell: 2 smaller pentagonal dodecahedra, where 20 water molecules are arranged to form 12 pentagonal faces, and 6 tetrakaidecahedra, where 24 water molecules are arranged by forming 12 pentagons and 2 hexagons. There are eight cavities, or atomic "cages" (M) in total in the cubic unit cell, and the general formula is $8M-46H_2O$.

2. Type II clathrate also has a cubic cell with a lattice constant of about 17 Å, and there are 136 water molecules per unit cell. The water molecules are arranged in 16 pentagonal dodecahedra and 8 hexakaidecahedra, where the cage walls are 12 pentagons and 4 hexagons formed from 28 water molecules. There are 24 cavities in total per unit cell, and the general formula is $24M-136H_2O$.

The formation of ice clathrates occurs when the water is cooled and agitated in the presence of a sufficient concentration of the guest atoms. The crystallization takes place at the phase boundaries within the water, and the clathrates are only observed when the dimensions of the guest molecules or atoms are as close as possible to the free diameters of the two kinds of cavities. Ice clathrates encapsulating methane, CH_4, exist naturally under the sea in polar regions. There is currently much interest in these ice clathrates because of the huge amount of methane that is estimated to be trapped in this way. If successfully harvested this may represent a huge source of energy, potentially replacing all of the world's fossil fuel reserves combined (for a recent review, see Suess et al., 1999). Although much of the research on ice clathrates involves the recovery and transportation of the encapsulated methane gas, other research involves more scientific interests.

Measurements of the thermal conductivity of the ice clathrates with different molecules or atoms entrapped in their "cages" showed smaller values and anomalous temperature dependencies (Ross et al., 1981; Ross and Anderson, 1982; Cook and Laubitz, 1982; Dharma-Wardana, 1983; Tse and White, 1988; Tse, 1994) when compared to that of pure Ih ice (Slack, 1980). The thermal conductivity of pure, nonmetallic crystalline solids is expected to vary as $1/T$, because of umklapp scattering (see, for example, Ashcroft and Mermin, 1976), whereas in many different clathrates the thermal conductivity is much lower and increases slowly with increasing temperature. This is typical behavior for amorphous or glassy solids as well (see, for example, Cahill et al., 1992, and references therein). The very low thermal conductivity measured in ice clathrates is due to guest–host interactions whereby the localized guest vibrations interact strongly with the propagating host acoustic modes (Dharma-Wardana, 1983; Tse and White, 1988; Tse, 1994). This is also the case for many semiconducting clathrates, as discussed later in this review.

In 1965, Kasper et al. reported the existence of a clathrate phase, Na_8Si_{46}, isomorphic with that of the chlorine hydrate type. Silicon, germanium (Cros et al., 1968), and tin (Gallmeier et al., 1969) were found to form both type I and II clathrate structures in which the guest atoms are alkaline atoms. In these materials, the host lattice is formed by atoms of one kind bonded by strong covalent forces with bond lengths similar to those in diamond-structured Si, Ge, or Sn. In 1986, Einsenmann et al. reported the formation of ternary compounds of the form $A_8B_{16}E_{30}$ (where A is an alkaline-earth metal, and B is a group III element such as gallium or aluminum) isomorphic with type I hydrate clathrates. Nolas et al. (1998, 1999a) reported the first thermal conductivity measurements carried out on semiconducting germanium clathrate compounds. The amplitude and temperature dependence of the thermal conductivity of polycrystalline $Sr_8Ga_{16}Ge_{30}$ showed a behavior similar to that of amorphous materials and was found to be very near in value to the thermal conductivity of pure amorphous germanium (Nolas et al., 1998, 1999a). Since then Nolas and co-workers have intensified their investigation of these materials in searching for suitable candidates for thermoelectric applications as more ternary and quaternary compounds, prepared by filling the cavities with elements as diverse as rare-earth and halogen atoms, are being tested. Much of this work, as well as that of other research groups involved in this rapidly growing field, is discussed in this review.

2. Background

Besides the possibility of having a very low thermal conductivity, these clathrate systems have many other interesting properties that might lead to an entirely different range of applications. Cros et al. (1970) showed that the electronic properties of the silicon clathrates depended upon the fraction of filled cages. Type-I silicon and germanium clathrates (Me_xE_{46}, where Me represents an alkaline metal) are metallic, whereas type-II clathrates maintain the semiconducting properties of Si and Ge for low alkaline metal concentrations. The semiconductivity diminishes as the metal concentration increases. Ternary type-I clathrate compounds, where the addition of extra electrons provided by the atoms inside the cages is compensated by substituting tetravalent network atoms by trivalent atoms such as gallium, are semiconductors (Nolas et al., 1998). In 1994, Adams et al. (1994) made theoretical band-structure calculations showing that the optical bandgap of the ideal empty Si_{136} is 2.4 eV, compared to 1.107 eV calculated for diamond-structure silicon (this value is larger than the experimental bandgap of 1.17 eV at 0 K), which we can call Si_8. This larger bandgap makes this type of clathrate a very interesting material for optical applications. Ramachandran et al. (1999a) prepared for the first time an empty Si_{136} clathrate and measured a bandgap of ~ 2 eV. They also measured the bulk modulus for this empty clathrate, finding a value of about 90% of that of diamond-Si. An expansion of the energy bandgap, as compared to the energy gap of the original diamond-structured material, was measured by Chu et al. (1982) in $Ge_{38}As_8I_8$ clathrates, where a value of 0.9 eV was determined from electrical measurements compared to 0.67 eV for Ge_8.

The clathrate $Ba_2Na_6Si_{46}$ has a metal–superconductor transition at 3.5 K, as shown by Kawaji et al. (1995). Bryan et al. (1999) also found superconducting behavior in $Ba_8Ga_{16}Ge_{30}$, with a critical temperature $T_c = (7.5 \pm 0.2)$ K. In the case of the Ba/Na silicide, the s and d states of the barium atom are mixed with the ones of silicon, thereby enhancing the density of states at the Fermi level. There is no metal–superconductor transition in the pure alkaline-filled clathrates. In the case of $Ba_8Ga_{16}Ge_{30}$ there is no clear explanation for the superconducting behavior. According to Bryan et al. (1999), when stoichiometric the framework is electron-deficient because of the Ga substitution. The barium atoms would therefore donate their electrons to the valence band rather than to the conduction band. Since $Ba_8Ga_{16}Ge_{30}$ shows almost metallic conductivity, it is possible that the Ba is transferring only part of its charge, leaving the valence band only partially filled.

Until recently, there has been less knowledge about the structural properties of Ge and Sn clathrates than about those of Si clathrates, although all clathrates behave in rather similar ways. One difference, however, is the presence of tin lattice vacancies in the hexagonal rings of type I tin clathrates. This is not the case for the Si analog Na_8Si_{46} (Ramachandran et al., 1999b). The formula unit for the alkali–tin clathrates is Me_8Sn_{44} instead of Sn_{46} (Zhao and Corbett, 1994; Nolas et al., 1999b; Cohn et al., 1999). In the case of doping for charge compensation, as in $Cs_8Zn_4Sn_{42}$ (Nolas et al.,

1999b), ordering can occur between the compensating atom and the tin in the framework. In this example the positions for the compensating atoms of zinc are the same as the locations of the vacancies in the pure-tin compound (Nolas et al., 1999b). Another difference between tin and silicon or germanium clathrates is the formation of a third crystalline structure. $Ba_8Ga_{16}Sn_{30}$ forms a cubic, cagelike structure with symmetry $I\bar{4}3m$ and a lattice parameter of 11.5945 Å (Einsenmann et al., 1986; von Schnering et al., 1998a). Nolas (1999a) showed this compound to be a semiconductor with low thermal conductivity.

As discussed in detail later, the fact that clathrate compounds can possess "glasslike" thermal conductivity, the ability to vary the electronic properties with doping level, and the relatively good electronic properties obtained in these semiconductor materials indicates that this is a phonon glass electron crystal (PGEC) system. The ideal PGEC material would possess a thermal conductivity normally associated with amorphous materials (a "phonon glass") and electronic properties normally associated with good semiconductor single crystals (an "electron crystal"). Slack (1995) introduced the PGEC concept as a guide in searching for new thermoelectric materials with properties superior to those of materials presently used in devices. The first material system that demonstrated PGEC properties was the skutterudite system (Nolas et al., 1999c). The skutterudites have therefore received enormous attention as potential thermoelectric materials (Nolas et al., 1999c; see also Chapter 5 in this volume). Although "glasslike" thermal transport has been observed in some disordered crystalline dielectric materials (Cahill et al., 1992), the Ge clathrates are not the only compounds with "rattling" constituents resulting in "glasslike" behavior.

3. CLATHRATES, CRYPTO-CLATHRATES, AND LOCAL MODES

There are a fairly large number of molecules and compounds that entrap other atoms or molecules within their structures. A clathrate material is one in which the voids in the host lattice are present in the absence of a guest atom or molecule. These are called true clathrates. One good example in the field of thermoelectrics are the semiconducting skutterudites based on $CoAs_3$. Many other examples can be found in the oxide zeolites based on Si–Al–O networks, which are electrical insulators. In the second category are the crypto-clathrates, or hidden clathrates, which form cavities around guest particles only in the presence of the guests. These crypto-clathrate hosts are more numerous than the true clathrates and are the main new development in the field of thermoelectrics.

The cage structure in clathrates can vary over a large range of sizes. The simplest cage possible consists of a tetrahedron of four touching atoms of, say, carbon. This tetrahedron has a cavity in its center whose diameter is

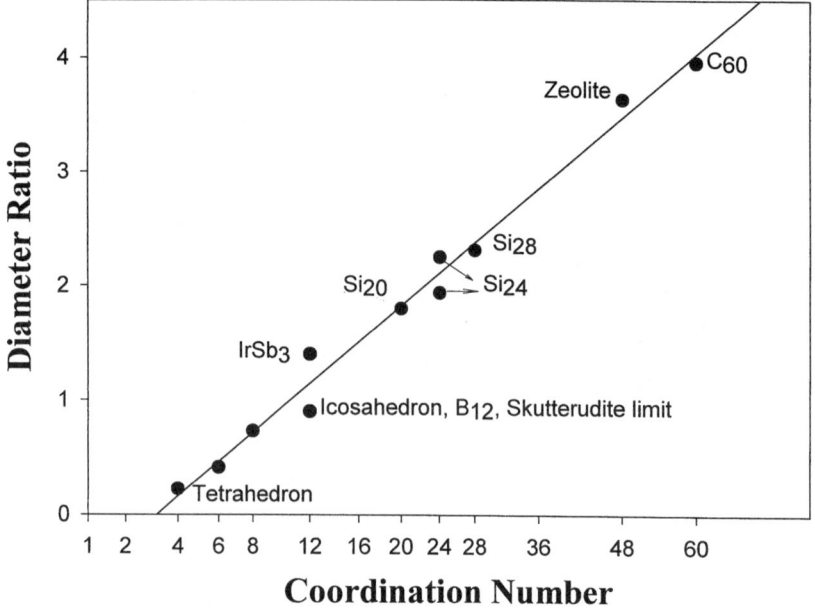

FIG. 1. The free cavity diameter to wall-atom diameter ratio as a function of the total number of atoms in the wall of the cage, i.e., the coordination number or CN. The lowest point is for the tetrahedron followed by the octahedron, cube, icosahedron, pentagonal dodecahedron, etc. The largest cage is for buckminsterfullerene or C_{60}. Still larger cages are possible.

0.2247 times the diameter of the four identical atoms of the host. This is the smallest size cavity with coordination number, CN, 4 (Fig. 1). If the tetrahedron is composed of four cesium atoms, the cavity diameter is 1.18 Å. A proton has a Bohr diameter of 1.06 Å. Thus, it could fit inside such a cavity formed by Cs.

The tetrahedral cavity in diamond is equal to the diameter of a carbon atom, which is 1.544 Å. It can trap hydrogen atoms with diameter 1.067 Å, but is too small for helium atoms with a diameter of 2.30 Å. Van Wieringen and Warmoltz (1956) showed that H and He atoms are soluble in and can diffuse through single crystals of both Si and Ge. Thus, these crystals are true clathrates for He and neutral hydrogen atoms.

In Fig. 1 we have plotted the interior void diameter, D, of cavities as a function of the coordination number, CN, which is the number of atoms in the cavity wall. These atoms, by definition, are bonded to their three nearest neighbor's wall atoms, and the bond length is equal to the atom diameter, d. We see that a straight line fits the data quite well for $4 \leqslant CN \leqslant 60$. Most of the cavities are spherical. In the case of $CN = 24$ for the tetrakaidecahedron, the cavity is an oblate ellipsoid with two slightly different diameters, both of which are plotted. Figure 1 allows us to estimate which atoms or molecules might be entrapped in which host. The equation of the straight line in Fig. 1 is

$$\frac{D}{d} = 0.54[\sqrt{CN} - \sqrt{3}]. \tag{1}$$

There are many different elemental and compound crystals that can accommodate interstitial light elements such as hydrogen, carbon, nitrogen, and oxygen. For example palladium is a well-known absorber for hydrogen, while titanium, zirconium, and hafnium can interstitially absorb a great deal of oxygen. These interstitial solid solutions can be thought of as clathrates. The host crystal structure is not disrupted by the presence of the guest atoms. These light atoms do not appreciably alter either the electronic or lattice thermal conductivity of the host, so they are not of importance for thermoelectrics. It is only when the guest atoms or molecules have masses comparable to or greater than that of the host that they become useful tools for controlling the lattice thermal conductivity of the host. This is especially interesting when the host is a semiconductor and the thermal conductivity is due mostly to phonons (quantized lattice vibrations).

There are two general methods of forming clathrates. The Type-I and Type-II ice clathrates and their Si, Ge, and Sn analogs are network structures or three-dimensional arrays of tetrahedrally bonded atoms or molecules built around various guests. They are also crypto-clathrates, because the structures do not form in the absence of the guests. There are no discernible clusters of the hosts in these structures. In NaSi, KSi, RbSi, and CsSi there are discernible Si_4 clusters (Busman, 1961), which are not bonded to each other. Thus, these silicides and their Ge, Sn, and Pb analogs are true cluster compounds, but are not clathrates for the alkali ions. Therefore, Si, Ge, and α-Sn in the diamond structure are true clathrates and network lattices for hydrogen and helium; they are crypto-clathrates and network lattices for the alkali, alkaline-earth, and rare-earth ions.

A similar list of properties can be made for boron and boron compounds. The low-temperature or α-boron is a cluster lattice formed from joined B_{12} units and is probably a clathrate for hydrogen and helium. The high temperature or β-rhombohedral boron (Slack et al., 1988a) is also a cluster compound formed from B_{84} and smaller clusters. It is a true clathrate for many different transition elements such as Cu, Fe, V, Zr, and Hf (Slack et al., 1988b). The LnB_{68} cubic form is a crypto-clathrate and cluster compound for numerous rare-earth metals, Ln, such as Gd, Y, and Yb (Slack et al., 1977). There are also cubic XB_6 and XB_{12} cluster compounds that are crypto-clathrates for (X=) K, Ca, Sr, Ba, Eu, and other rare-earth metals (Muetterties, 1967). The main reason to discuss these compounds here is that they are both clathrates and semiconductors, and the guest atoms have been shown to drastically lower their lattice thermal conductivities in a

TABLE I

Einstein Mode Wave Numbers in Some Boron and Beryllium Compounds

Compound	Cage shape	Coordination number of cage	Mode wave number (cm^{-1})	Reference[a]
LaB_6	Truncated cube	24	98	1
CeB_6	Truncated cube	24	90	1
PrB_6	Truncated cube	24	87	1
NdB_6	Truncated cube	24	83	1, 2
SmB_6	Truncated cube	24	77, 84	1–3
EuB_6	Truncated cube	24	85	1
YB_6	Truncated cube	24	69	4
YbB_6	Truncated cube	24	64	1
LuB_{12}	Truncated octahedron	24	Unknown	
YB_{68}	Dumbbell	Complex	140	5
$ThBe_{13}$	Snub cub	24	121	6
UBe_{13}	Snub cub	24	112	6

[a] References: 1, Korsukova et al. (1993); 2, Trunov et al. (1994); 3, Alekseev et al., 1991; 4, Schneider et al., 1987; 5, Werheit et al., 1991; 6, Tranquada et al., 1986.

manner similar to that of the Si, Ge, and Sn clathrates. The wave numbers of the Einstein modes of the heavy atoms in the cages have been measured for many of these compounds (Korsukova, 1993), as shown in Table I.

In the simpler structure boron compounds such as SmB_6 and LuB_{12} there are only 7 or 13 atoms, respectively, in the unit cell. Thus, the Einstein frequencies of the local rare-earth vibrational mode are well below the optic mode frequencies of the boron vibrations. Hence, each compound has only one Einstein frequency to interact with the acoustic phonons that transmit heat. In the case of YB_{68} there are 1652 atoms in the cubic unit cell and 413 atoms in the primitive unit cell (Slack et al., 1977). In this compound the optic modes extend over the range (Werheit et al., 1991; Slack et al., 1971) from 80 to 1100 cm^{-1}, and the yttrium vibration (Werheit et al., 1991) at 140 cm^{-1} does very little to scatter the acoustic phonons that lie between 0 and 80 cm^{-1}. The low thermal conductivity (Slack et al., 1971; Cahill et al., 1989a) of YB_{68}, shown in Fig. 2, is like that of a glass, but is mainly a result of the very large number of atoms in the unit cell. The Hf-doped β-boron in Fig. 2 shows a resonance dip in the lattice thermal conductivity, κ_g, versus temperature at about 100 K. The Hf resonance frequency is estimated to be about 150 cm^{-1} and can scatter some of the acoustic phonons in β-boron, which has 107 atoms in the rhombohedral unit cell (Slack et al., 1988a).

Of all the semiconducting hexaborides, the only ones that have been studied from the thermal conductivity point of view have been SmB_6 and

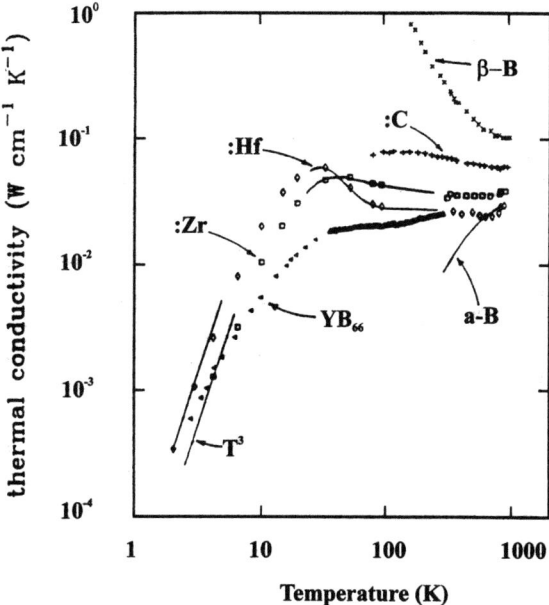

FIG. 2. The thermal conductivity versus temperature of YB_{68} and β-boron single crystals, polycrystalline β-boron containing various transition metal impurities, and amorphous boron (a-B) as a function of temperature. From Cahill et al. (1999a), Fig. 3.

EuB_6 (Sera et al., 1996; Kekelidze et al., 1991; L'vov et al., 1963). At temperatures below 100 K, samarium hexaboride is a semiconductor and its thermal conductivity is primarily due to phonons. At higher temperature, part of the samarium switches from Sm^{2+} to Sm^{3+}, and the samples become metallic. In EuB_6 the compound is metallic below 13 K, whereas it is a low-bandgap semiconductor with Eu^{2+} ions above 100 K. This changing behavior has been discussed by Etourneau et al. (1977). At low temperatures SmB_6 has an energy gap of 0.005 eV; at high temperatures EuB_6 has an energy gap of 0.38 eV. The lattice thermal conductivity of both in their semiconducting ranges is shown in Fig. 3. The data for SmB_6 is for single crystals of 0.7 mm in diameter (Sera et al., 1996). The data for EuB_6 is for polycrystalline samples (Kekelidze et al., 1991; L'vov et al., 1963; Dünner et al., 1984; Aivasov et al., 1979) corrected to theoretical density. The electrical conductivity used for subtracting the electronic thermal conductivity has been taken from data for high purity samples (Mercurio et al., 1974; Paderno et al., 1969). Some high-temperature data on the thermal conductivity of polycrystalline SmB_6 exists at temperatures to 1200 K (Markov et al., 1978), but no lattice component has been extracted.

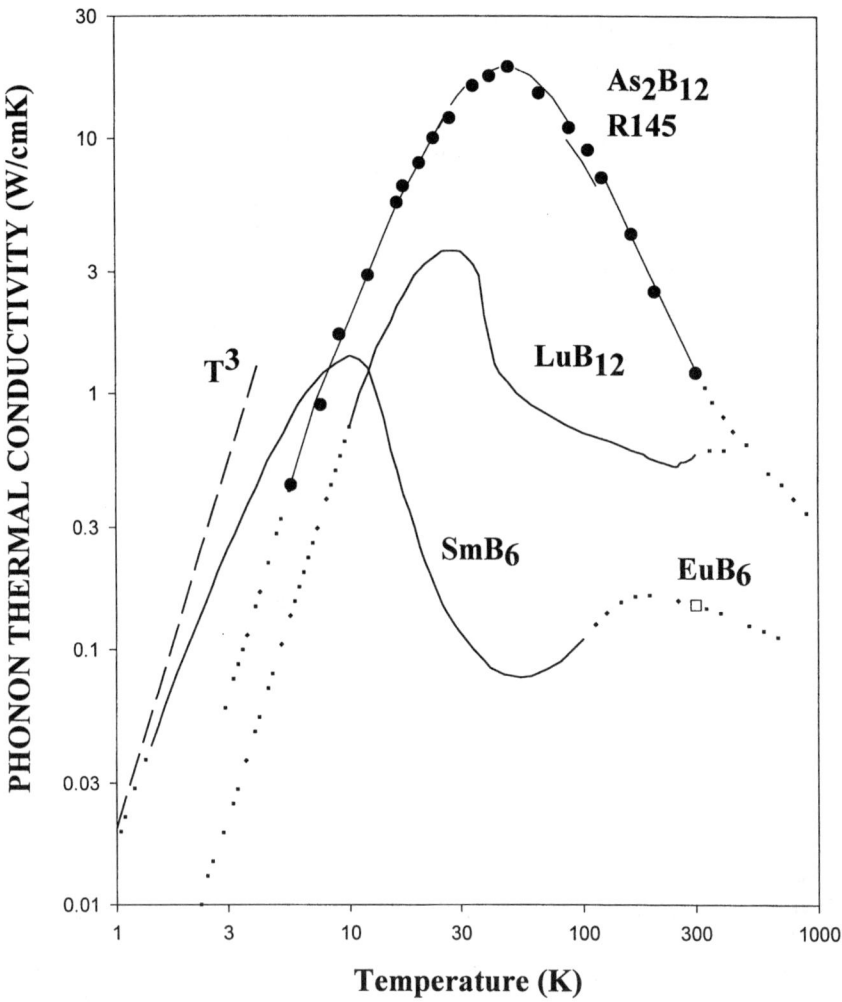

FIG. 3. The lattice thermal conductivity versus temperature of single crystal As_2B_{12} (or B_6As), LuB_{12}, SmB_6, and polycrystalline EuB_6. The data for SmB_6 and EuB_6 are combined to make a composite curve. From Slack (2000), Fig. 1. Reprinted with permission from Technomic Publishing Co., Inc., copyright 2000.

We note that Sm and Eu occur as neighbors in the periodic table, and that their vibration wave numbers in Table I are almost identical at 84 to 85 cm^{-1}, respectively. Thus we construct a composite curve of κ_g versus temperature for a semiconducting hexaboride. It exhibits a broad minimum at 55 K, which is caused by the vibrational Einstein mode at 84.5 cm^{-1}. The effect on the thermal conductivity extends from 10 to 300 K. If we had κ_g versus temperature data for CaB_6, SrB_6, or BaB_6, we would expect to see a curve similar to that for our composite crystal. We are looking for a model clathrate in which there is a very simple crystal structure, and a trapped "rattling" atom with a single vibration frequency in the acoustic phonon range. The SmB_6–EuB_6 curve in Fig. 3 is the closest approach to that which we have so far.

The thermal conductivity of a simple boride crystal with no low-frequency resonance is that for As_2B_{12} (or B_6As) taken from Slack et al. (1971). It is also shown in Fig. 3. Note that at 55 K the Sm resonance in SmB_6 has decreased κ_g by a factor of 200 compared to B_6As. This is a large and useful effect. The κ_g data for single crystal SmB_6 is decreasing as T^3 near a temperature of 1 K. From the sample size of 0.7 mm and the temperature of 373 K we can calculate the Casimir boundary scattering limit for κ_g (Smith et al., 1985). The dashed T^3 line in Fig. 3 is for a mean free path of 0.3 mm. This means the single crystal of SmB_6 is nearly perfect and the thermal conductivity below 1 K is limited only by boundary scattering of the phonons.

Other borides such as LuB_{12} and other compounds such as $ThBe_{13}$ and UBe_{13} possess entrapped atoms or ions that have low frequency Einstein modes. The lattice thermal conductivity of LuB_{12} (Misiorek et al., 1995), which behaves much like a metal with a large electronic thermal conductivity, is also shown in Fig. 3. The vibration frequency of the Lu in this structure is unknown. The cage size for $CN = 24$ in LuB_{12} should be similar to the $CN = 24$ cage size in YbB_{12} (see Fig. 1). Thus, we expect a Lu vibration frequency of about 69 cm^{-1}, since Yb^{3+} and Lu^{3+} have similar masses and ionic sizes. The large depression in κ_g of LuB_{12} between 40 and 200 K seen in Fig. 3 may be caused, in part, by "rattle" scattering of phonons. Just how this combines with the large electron–phonon interaction is uncertain. No thermal conductivity data on the Be_{13} compounds exists.

As outlined earlier, there is also a growing collection of thermal conductivity data on ice clathrates. The general effects are similar to those discussed here, but moved further down the temperature scale since the Debye temperature of ice is rather low (Ross et al., 1981; Ross and Anderson, 1982; Cook and Laubitz, 1982; Dharma-Wardana, 1983; Tse and White, 1988; Tse, 1994).

II. Synthesis Methods

There are several methods employed in forming Si, Ge, and Sn clathrates, as well as mixed-crystal clathrates containing two or more of these elements. In this section we outline a few different methods. Others can be found in the references at the end of this chapter.

When Cros (1970) first prepared silicon and germanium clathrates, he did so by thermal decomposition of the precursor MeSi or MeGe Zintl phases. Later Ramachandran et al. (1999c) did so following his method. For Si clathrates, for example, the precursor NaSi is prepared by mixing sodium metal and powdered silicon in an inert atmosphere, placing them in a tantalum ampoule sealed inside a steel bomb, and heating for 24 hours at 650°C. To obtain the stoichiometric NaSi compound the excess Na is removed by heating at 275°C for 8 hours under a vacuum of 10^{-6} torr. In order to get the clathrate structure, the NaSi is placed on a Ta boat and heated to 375°C under a 10^{-6} torr vacuum. Part of the sodium then evaporates; thus the clathrate composition is controlled by varying the heating time between 30 minutes to several hours. The fractional filling of the cages diminished for longer heating times. Reny et al. (1998) followed approximately the same procedure by heating under vacuum at temperatures between 340 and 440°C in order to get type-II clathrates with fewer than 14 sodium atoms per formula unit. For higher sodium concentrations they used a closed steel reactor where they annealed Na_xSi_{136} ($x < 14$) and excess Na in a temperature range between 370 and 440°C. The sodium concentration was varied by controlling the temperature. In these synthesis approaches the product of the reaction was sometimes a mixture of the two different clathrate structures: the stoichiometric type-I (Na_8Si_{46}) and the variable sodium content type-II Na_xSi_{136}. Washing with ethanol and water then reacts with and removes all the undecomposed NaSi. The clathrate itself does not decompose in water because the sodium is protected by its silicon cage. Ramachandran et al. (1999c) obtained pure phases of the type II clathrate (that is, only grains with a certain prefixed Na filling) by density separation of powder samples in a solution of dibromomethane (with a density of 2.477 g/cm^3) and tetrachloroethylene (1.614 g/cm^3). The density of Na_xSi_{136} varies between 2.05 and 2.33 g/cm^3 and depends on x.

It is also possible to produce clathrate compounds based on silicon, germanium or tin networks by more standard powder metallurgy techniques. Einsenmann et al. (1986) prepared ternary type-I compounds with composition $A_8B_{16}C_{30}$ (where A is Sr or Ba, B is Al or Ga, and C is Si, Ge, or Sn in this formula) by mixing stoichiometric quantities of the pure elements in alumina crucibles. They slowly heated the mixtures under argon at a rate of 2°C/min up to a maximum temperature of 1150°C for silicon, 1050°C for germanium, and 850°C for tin mixtures. The mixture was kept at maximum temperature for an hour, then cooled down to room temperature at the same rate of 2°C/min. They found that at the maximum temperatures the mixtures were completely liquid.

Nolas et al. (2000a) grew single crystals of $Sr_8Ga_{16}Ge_{30}$ and $Eu_8Ga_{16}Ge_{30}$ by mixing and reacting stoichiometric quantities of the high-purity elements and holding them for 3 days at 950°C inside a pyrolytic boron nitride crucible that was itself sealed inside a fused quartz ampoule that was evacuated and backfilled with 0.068 MPa of argon. The ampoule was then slowly cooled to 700°C, where it was left for 4 days. Then it was slowly cooled to room temperature. The resulting aggregate consisted of single crystal grains of the order of 1–3 mm in length that were stable in air and water. Chakoumakos et al. (1999) grew single crystals of $Sr_8Ga_{16}Ge_{30}$ by arc-melting high-purity Sr and Ge together in an argon atmosphere to form $SrGe_2$. Then they loaded stoichiometric amounts of $SrGe_2$, Ga shot, and Ge in a helium dry box into a carbonized silica tube. The constituents were then heated to 1050°C at 2°C/min, held at 1050°C for 20 hours, and then slowly cooled (0.02–0.1°C/hour) to 650°C, where they were held for several days, and finally cooled down to room temperature. Single crystals 5–10 mm in length of $Sr_8Ga_{16}Ge_{30}$ were obtained. $Sr_8Ga_{16}Ge_{30}$ melts congruently at 760°C (Schujman et al., 2000).

Single crystals of halogen-filled clathrates were grown by Chu et al. (1982) by vapor transport in a two-zone furnace. They enclosed Ge and As in stoichiometric quantities and varying amounts of iodine in a sealed fused silica tube. They kept the Ge at a temperature between 720 and 730°C and the As at a temperature between 600 and 610°C. In the case of $Ge_{38}As_8I_8$, the reaction involves the formation of GeI_2 in the high-temperature zone and the reaction of this compound with As_4 or with As_4 and I_2 in the lower temperature zone. The clathrate crystals were grown at high iodine pressures, for example, 5 atm, whereas for low iodine pressures, such as 1 atm, GeAs and $GeAs_2$ single crystals were obtained. Nolas et al. (1999b, 2000b) grew small crystals of different Sn clathrates by mixing and reacting the constituent elements for 2 weeks at 550°C inside a tungsten crucible that was itself sealed inside a stainless steel canister after the canister was evacuated and backfilled with high-purity argon. The resulting Sn clathrates consisted of small octahedrally shaped crystals with a shiny, somewhat blackish, metallic luster. These millimeter-sized crystals were generally not very reactive in air or moisture.

III. Structural Properties

1. CRYSTAL STRUCTURE AND BONDING

Clathrate compounds form in a variety of different structure types. The majority of work thus far on clathrates for thermoelectric research has been on two structure types that are isotypic with the clathrate hydrate crystal structures of type I and II (see, for example, Franks, 1973). The type I structure can be represented by the general formula X_8E_{46}, and that of the type II structure with the formula $X_8Y_{16}E_{136}$, where X and Y are typically alkali-metal or alkali-earth atoms and E represents a group 4 element, Si,

Ge, or Sn, although Zn, Cd, Al, Ga, In, As, Sb, or Bi can also be substituted for these elements to some degree. The key characteristic of both structure types is that the framework is formed by covalent, tetrahedrally bonded E atoms making up two different face-sharing polyhedra that are connected to each other by shared faces. As first pointed out by Slack (1995), if atoms that are trapped inside these polyhedra, or atomic "cages," are smaller than these "cages," they may "rattle" about and interact randomly with the lattice phonons, resulting in substantial phonon scattering. This is one of the most conspicuous aspects of these compounds and directly determines many of their interesting and unique properties, including their thermoelectric properties, as described in detail later. These compounds display an exceedingly rich number of physical properties, including semiconducting behavior, superconductivity, and thermal conductivity that is similar to amorphous materials in magnitude as well as in temperature dependence. All of these properties are a direct result of the nature of the structure and bonding in these materials. Figure 4 illustrates the different polyhedral "building

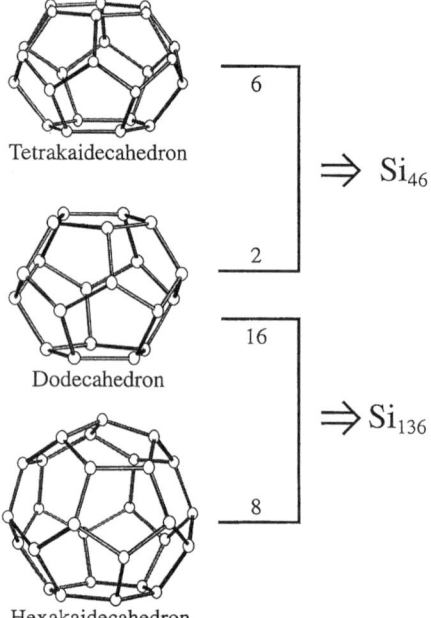

FIG. 4. The "building blocks" of the type I and II clathrates. The type I structure is formed by two pentagonal dodecahedra (top) and six lower-symmetry tetrakaidecahedra (middle) in the cubic unit cell connected by shared faces. The type II structure is formed by 16 dodecahedra (top) and eight hexakaidecahedra (bottom), also connected by shared faces in its larger cubic unit cell. In this way two "building blocks" displace all space. The "voids" in these structures are inside these 20-, 24-, or 28-atom cages. From Ramachandran et al. (1999c), Fig. 1.

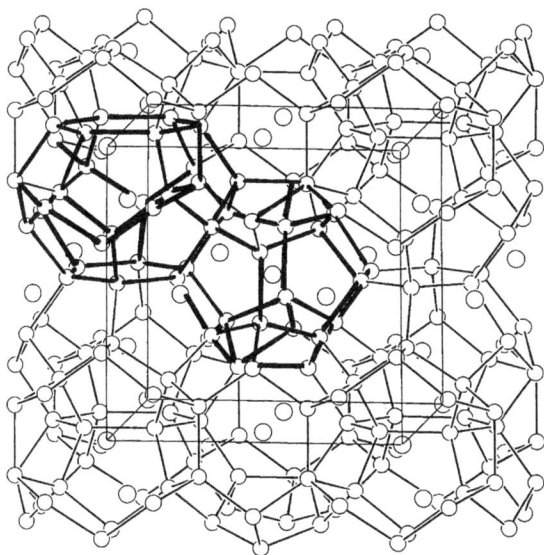

FIG. 5. The type I clathrate crystal structure. Only the group-IV elements are shown. Outlined are the two different polyhedra that form the unit cell, with the dodecahedron in the center and one tetrakaidecahedron on the upper left. From Nolas et al. (2000a), Fig. 2.

blocks" that form these two crystal structures. The type I structure (Fig. 5) comprises 2 pentagonal dodecahedra, E_{20}, each creating a void with $\bar{3}m$ symmetry, and 6 tetrakaidecahedra formed by 12 pentagonal and 2 hexagonal faces, E_{24}, each creating a void with $4m2$ symmetry. The corresponding unit cell is cubic with space group $Pm\bar{3}n$. The type II structure (Fig. 6) comprises 16 pentagonal dodecahedra and 8 hexakaidecahedra formed by 12 pentagonal and 4 hexagonal faces, E_{28}, creating a center with $\bar{4}3m$ symmetry. The cubic unit cell has a space group of $Fd\bar{3}m$. The type II compound can be formed nonstoichiometrically, that is, without all of the voids filled.

Extensive structural analyses have been reported on Si (Cros et al., 1962, 1968, 1971; Gallmeier et al., 1969; Einsenmann et al., 1986; Cros, 1970; Ramachandran et al., 1999c; Reny et al., 1998; Kröner et al., 1998a; von Schnering et al., 1998b), Ge (Gallmeier et al., 1969; Einsenmann et al., 1986; Nolas et al., 2000a; Chakoumakos et al., 1999; Schujman et al., 2000; von Schnering et al., 1998b; Westerhaus and Schuster, 1977; Czybulka et al., 1991; Kuhl et al., 1995; von Schnering and Menke, 1978; Kröner et al., 1998b,c,d; von Schnering et al., 1998c) and Sn[1] (Nolas et al., 1999b) type I and II clathrate compounds with varying compositions. In general the

[1] References to all work reported in the literature to date on Sn clathrates has been tabulated in Nolas et al. (1999b).

FIG. 6. A perspective of the type II clathrate crystal structure along the [100] direction. From Ramachandran et al. (1999c), Fig. 3.

average interatomic distances are slightly larger than that of the analogous diamond structured compounds. The average IV–IV–IV bond angles range from 105° to 126°, averaging close to the 109.5° that is characteristic of the tetrahedral angle found in the diamond structure. The volume per fourth column atom of these two clathrate compounds, however, is larger than those of their analogous diamond structured compounds ($\sim 15\%$). This is a good indication of the openness of these "open structures."

Type I germanium clathrates are perhaps the most studied compounds for potential thermoelectric applications because of their unique transport properties (Nolas et al., 1998, 1999a, 2000a; Cohn et al., 1999; Chakoumakos et al., 1999; Schujman et al., 2000; Iverson et al., 1999; Blake et al., 1999; Nolas, 1999b). From the extensive structural analyses on these compounds an estimate of the size of the polyhedra as a function of the entrapped "guest" can be obtained. Schujman et al. (2000) made a comparison between the structural properties of type I Ge clathrates and those of an "ideal" Ge_{46} clathrate. They defined the ideal clathrate structure as one where all the atomic bonds have the same length, that is, the average atomic diameter of the atoms forming the clathrate framework. The unique set of values for the atomic positions that will comply with the condition that all the bond lengths, d, are equal is $x = 0.182899$, $y = 0.309911$, and $z = 0.1162$, resulting in $a_o = 4.30211d$. A similar calculation for type-II clathrate gives a relationship between the lattice parameter and the average atomic diameter of the framework of $a_o = 6.20540d$. It is interesting to note that using the diameter

TABLE II

CALCULATION OF INTERATOMIC DISTANCES IN THE UNIT CELL OF THREE REPRESENTATIVE Ge CLATHRATES[a]

Bond	Ideal (Å)	$Sr_8Ga_{16}Ge_{30}$[b]	$Ba_8Ga_{16}Ge_{30}$[c]	$K_8Ga_{16}Ge_{30}$[d]
d_1	2.4496	2.498(3)	2.506	2.509
d_2	2.4496	2.506(2)	2.495	2.485
d_3	2.4496	2.489(6)	2.544	2.533
d_4	2.4496	2.429(7)	2.454	2.416
$\bar{d}$	2.4496	2.495()	2.5007	2.494
Me(1)-atom(2)	3.339	3.434(4)	3.435	3.439
Me(1)-atom(3)	3.488	3.550(2)	3.554	3.530
Me(2)-atom(1) (perpendicular to $\langle 100 \rangle$)	3.726	3.79508(7)	3.801	3.794
Me(2)-atom(3) (perpendicular to $\langle 100 \rangle$)	3.529	3.594(2)	3.623	3.618
Me(2)-atom(2) (along $\langle 100 \rangle$)	3.922	3.985(1)	3.999	3.982
Me(2)-atom(3) (along $\langle 100 \rangle$)	4.093	4.174(3)	4.159	4.145

[a] The ideal structure is empty and has no metal atoms in the polyhedra. The nomenclature follows from Fig. 7. The average interatomic distance $\bar{d}$ is given by $13\bar{d} = 6d_1 + 5d_2 + d_3 + d_4$. The values for $K_8Ga_{23}Ge_{23}$ have been altered to correspond to a K-filled $Ga_{16}Ge_{30}$ network. Values are for room temperature.
[b] From Schujman et al. (2000).
[c] From Eisenmann et al. (1986).
[d] From Westerhaus and Schuster (1977).

of a silicon atom ($d = 2.3516$ Å), the value we get for the lattice parameter is 14.9526 Å, which is strikingly similar to the lattice parameter for an empty silicon clathrate, as can be extrapolated from Fig. 10 of Ramachandran et al. (1999c). This agreement suggests that the extra donated electrons produce a lattice expansion, rather than the size of the guest atom. The values for the atomic positions are $x = -0.09304$ for the 32e sites and $x = -0.05697$ and $z = -0.2459$ for the 96g sites. These values are very similar to the ones given by Cros et al. (1971) for Na_3Si_{136}, which are -0.094, -0.058, and -0.246, respectively. It is not possible to adjust the values of x, y, and z so as to make all the bond angles equal.

In Table II a comparison between bond lengths and cage sizes for the ideal structure and the different compounds is shown (Schujman et al., 2000). The interatomic distances were calculated from refined atom positions, where the metal guest atoms were assumed to be at the center of the cages. The ideal structure was chosen to make all the nearest-neighbor interatomic distances equal; the real structures deviate from these values by $\pm 4\%$ or less. A comparison of the values for an ideal structure where d is

TABLE III

ANGLES (IN DEGREES AND DECIMAL FRACTIONS OF A DEGREE) BETWEEN BONDS FOR THE REFINED AND IDEAL STRUCTURES OF TYPE I CLATHRATES[a]

Angle	Ideal	$Sr_8Ga_{16}Ge_{30}$	$Ba_8Ga_{16}Ge_{30}$	$K_8Ga_{23}Ge_{23}$
α	106.67	107.15(8)	106.55	106.80
β	111.16	110.42(9)	110.55	109.99
γ	109.73	109.6(1)	110.97	111.28
δ	125.14	124.90(6)	124.49	124.37
ε	109.34	109.41(7)	108.72	108.61
φ	106.37	105.6(1)	106.21	105.76
θ	107.72	108.5(1)	108.32	108.92
κ	103.79	104.6(1)	105.28	106.06
Σ_1	360.01	360.0(2)	359.95	360.02
Σ_2	539.45	539.5(4)	539.48	539.64
Σ_3	537.52	538.0(5)	537.78	537.97

[a] The nomenclature follows from Fig. 7. Values are for room temperature. $\Sigma_1 = \delta + 2\gamma$ (hexagon); $\Sigma_2 = 2\alpha + 2\beta + \kappa$ (for **3-2-3-2-3** pentagon); $\Sigma_3 = 2\varphi + 2\theta + \varepsilon$ (for **2-3-1-3-2** pentagon). If these polygons are perfectly planar $\Sigma_1 = 360°$, $\Sigma_2 = \Sigma_3 = 540°$. Smaller values indicate slightly nonplanar polygons.

the average atomic diameter of the framework with several real clathrates with the same framework composition and different filler atoms (Sr, Ba, and K) showed a variation for the x, y, and z values of less than $\pm 2\%$. There are eight different interbond angles in the type I clathrate structure.

It is interesting to note that in both the ideal and the actual structures, the pentagons are not planar, but the hexagons are. This can be seen by calculating the angle sums Σ_1, Σ_2, and Σ_3 in Table III. For a planar hexagon, $\Sigma_1 = 360°$. The pentagons are of two different types (formed by atoms **3-2-3-2-3** and **2-3-1-3-2**), as shown in Fig. 7. For planar pentagons, Σ_2 and Σ_3 must be 540°. Σ_2 is close to 540° for all the clathrates in Table III, whereas Σ_3 is noticeably less than 540°. The dihedral angle between the individual triangles making up the **2-3-1-3-2** pentagons are between 12° and 13° of arc. In the **3-2-3-2-3** pentagons, this angle is 3° to 4° of arc. The calculation of the dihedral angles in the case of the hexagons shows that they are planar from symmetry considerations.

From a comparison of the distances between the filler and network atoms (Table II), it follows that the introduction of the filler atoms into the structure expands it slightly with respect to the ideal empty structure. However, once the cages are filled, the size of the filler atom does not influence the size of the tetrakaidecahedral cages. The pentagonal dodecahedral cages expand slightly upon filling. This is quite interesting and a positive result for thermoelectric applications, since it suggests that appropriate "guest" atoms incorporated into the polyhedra might further

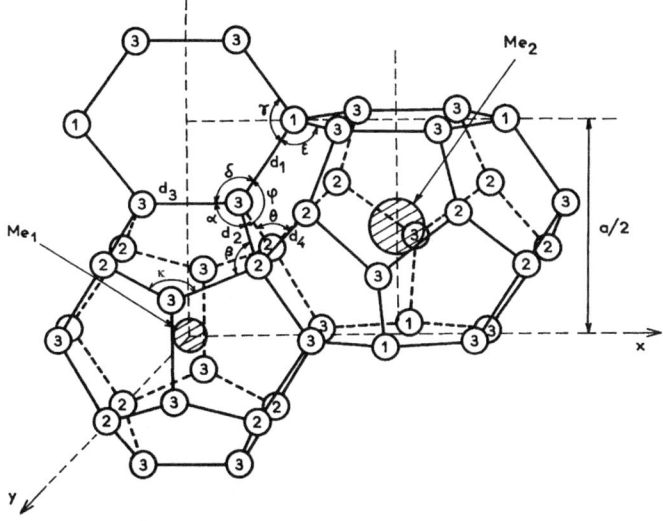

FIG. 7. The two polyhedra in the type I crystal structure illustrating the different crystallographic sites in this structure. The numbers 1, 2, and 3 represent the $6c$, $16i$, and $24k$ sites, respectively, of the group-IV framework atoms, while Me(1) and Me(2) represent the $2a$ and $6d$ sites, respectively, of the "guest" atoms. From Schujman et al. (2000), Fig. 3.

reduce κ_g toward κ_{min} (Nolas et al., 2000a; Schujman et al., 2000; Slack, 1997), the theoretical minimum thermal conductivity (Slack, 1979).

2. ATOMIC RATTLERS

As mentioned earlier, the specific framework and thermal parameters associated with the atoms in these compounds are an important aspect of this structure and have an effect on the transport properties. One of the more unique and interesting structural features is the thermal motion of the "guest" atoms inside their atomic "cages." We illustrate this by focusing on three type-I compounds (Nolas et al., 1999b, 2000a,b; Schujman et al., 2000; Chakoumakos et al., 1999), since most of the work to date in understanding the transport properties of these compounds has focused on this structure type.

As illustrated in Figs. 4, 5, and 7, in the type-I structure the "guests" occupy two distinct sites while the framework is made up of three distinct sites. From room temperature structural refinements the anisotropic atomic displacement parameters (ADPs) for Sr(2), obtained from single crystal and powder $Sr_8Ga_{16}Ge_{30}$, are enormous in comparison with those of the other atom positions (Einsenmann et al., 1986; Nolas et al., 2000a; Chakoumakos

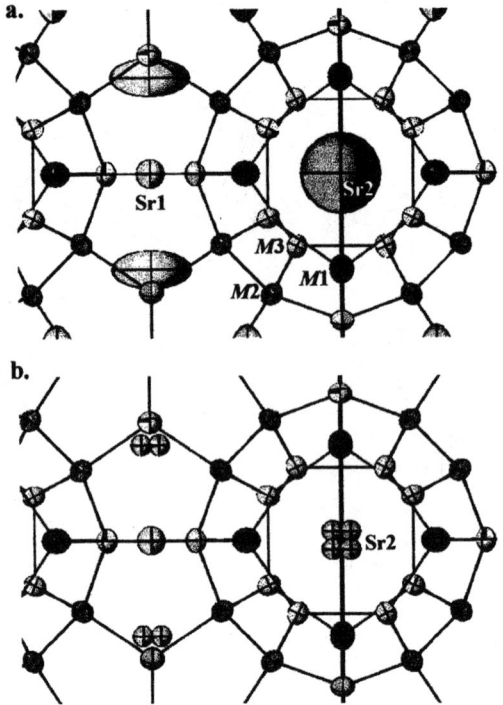

FIG. 8. Crystal structure projection on a (100) plane of $Sr_8Ga_{16}Ge_{30}$ illustrating the large anisotropic atomic displacement parameters for the single-site model (a), and the split-site model (b) where the combined static and dynamic disorder of the Sr(2) crystallographic site with isotropic atomic displacement parameters is indicated. From Chakoumakos et al. (1999), Fig. 2. Reprinted with permission from Elsevier Science.

et al., 1999). The difference in the ADPs between the Sr(2) atoms and the Sr(1) atoms is illustrated in the schematic shown in Fig. 8a. The Sr(1) site has a more symmetric ADP and is slightly larger in magnitude than that of the (Ga, Ge) framework atoms. The Sr(2) site exhibits an anisotropic ADP that is almost an order of magnitude larger than those of the other constituent atoms. Relatively large and anisotropic ADPs are typically observed for relatively small ions in the tetrakaidecahedra in this crystal structure. The enormous ADP for the Sr(2) site in $Sr_8Ga_{16}Ge_{30}$ implies the possibility of a static disorder in addition to the dynamic or "rattling" motion. The electrostatic potential within the polyhedra is not everywhere the same, and different points may be energetically preferred. Moreover, anisotropic refinement of the displacement parameters indicates a much smaller displacement amplitude in the $\langle 100 \rangle$ directions than in the perpendicular directions (Nolas et al., 2000a; Chakoumakos et al., 1999), as seen

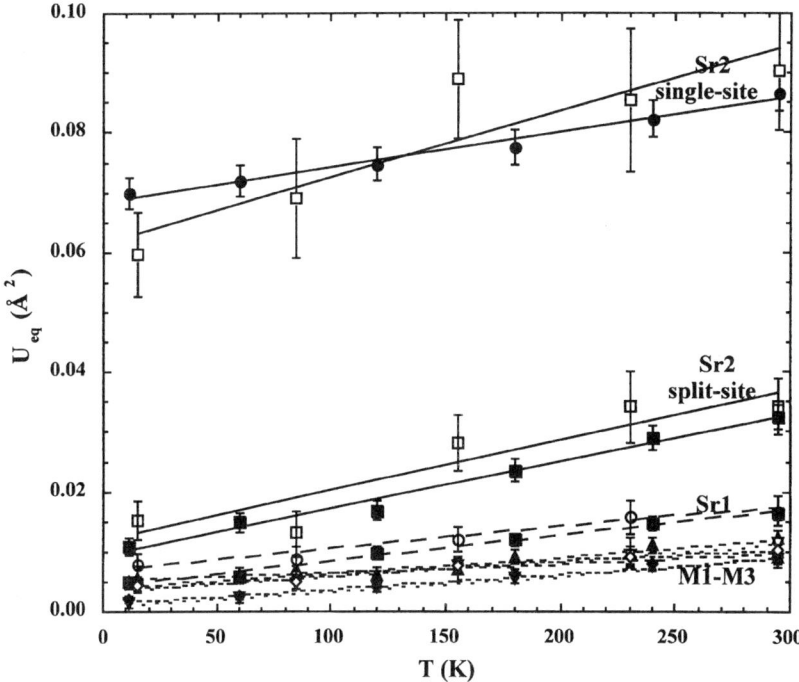

FIG. 9. Temperature dependence of the isotropic displacement parameters measured on single crystal (open symbols) and powder (closed symbols) $Sr_8Ga_{16}Ge_{30}$. The apparent $U(T)$ is enormous for Sr(2) in the single-site model. The large value of the low-temperature intercept suggests a large static disorder. The (Ga, Ge) sites are labeled M. From Chakoumakos et al. (1999), Fig. 4. Reprinted with permission from Elsevier Science.

in Fig. 8a. This suggests that the Sr(2) position at $\frac{1}{4}$, $\frac{1}{2}$, 0 could also be described by splitting it into four positions within the {100} planes, as shown in Fig. 8b. This would suggest that the atoms in the Sr(2) site can tunnel between the different energetically preferred positions. The possibility of a "freeze-out" of the "rattling" motion of Sr^{2+} in $Sr_8Ga_{16}Ge_{30}$ was originally indicated by low-temperature κ_g data.[2] The implication is that static disorder is associated with a spatial distribution of the Sr(2) positions inside the polyhedra. It is plausible that the large measured ADPs contain both a static as well as a dynamic component. This was best illustrated by Chakoumakos et al. (1999) from refinements of neutron diffraction data on single crystal and powder $Sr_8Ga_{16}Ge_{30}$, as shown in Fig. 9.

[2]In fact, thermal conductivity data (Cohn et al., 1999) was employed prior to the temperature dependent neutron scattering (Nolas et al., 2000a) data in providing reasonable estimates of the Sr-rattle frequencies along with explaining the static disorder associated with the positional disorder of the Sr in the tetrakaidecahedral cages.

Figure 9 shows the temperature dependence of the isotropic ADP, $U(T)$, for the atomic positions in $Sr_8Ga_{16}Ge_{30}$ in the split-site (static + dynamic disorder) model as well as the single site (dynamic disorder) model (Chakoumakos et al., 1999). At room temperature $U(T)$ is large for Sr(2) in the single-site model, while the low temperature intercept implies a large static positional disorder at absolute zero. With a model in which the Sr(2) fractionally occupies a multiply split site, the $U(T)$ values are dramatically reduced, but still remain the largest of all the atoms in the crystal. The split-site model implies less thermal disorder. Unconstrained refinement of the Sr(2) site occupancy indicates no deficiency in occupancy of the Sr(2) position for either model. The ADP for the Sr(1) site is also large, but more on par with the (Ga,Ge) framework sites (indicated by M in Fig. 9) because of its smaller cavity.

Nolas et al. (2000b) employed Rietveld-refined neutron diffraction in studying $Cs_8Zn_4Sn_{42}$ and Cs_8Sn_{44}. These data are shown in Figs. 10 and 11, respectively. In the case of $Cs_8Zn_4Sn_{42}$ the temperature dependent

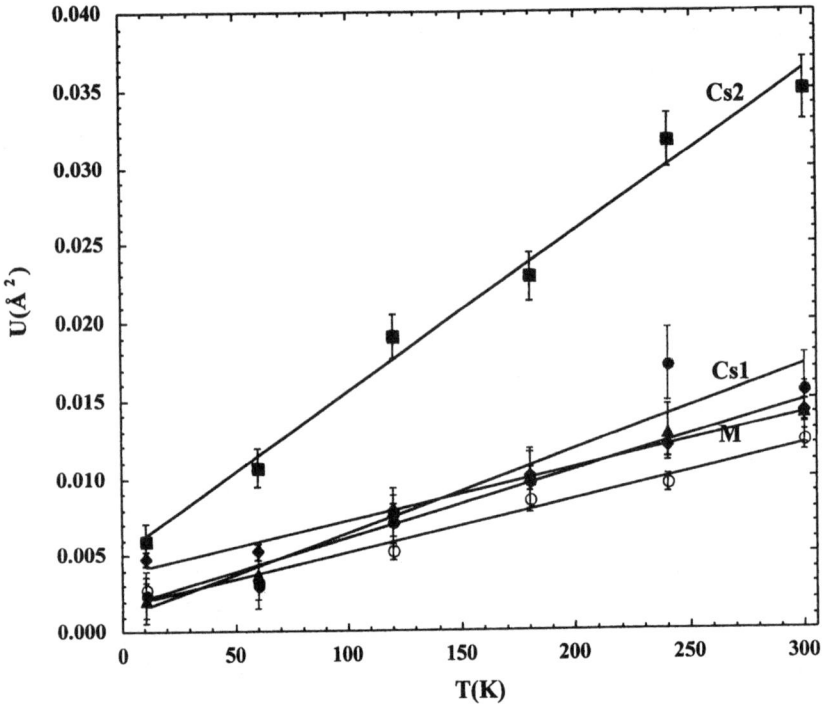

FIG. 10. Temperature dependence of the isotropic atomic displacement parameter for $Cs_8Zn_4Sn_{42}$. The (Zn, Sn) sites are labeled M in this figure. The apparent $U(T)$ is large for Cs(2) as compared to the other constituents. From Nolas et al. (2000b), Fig. 2.

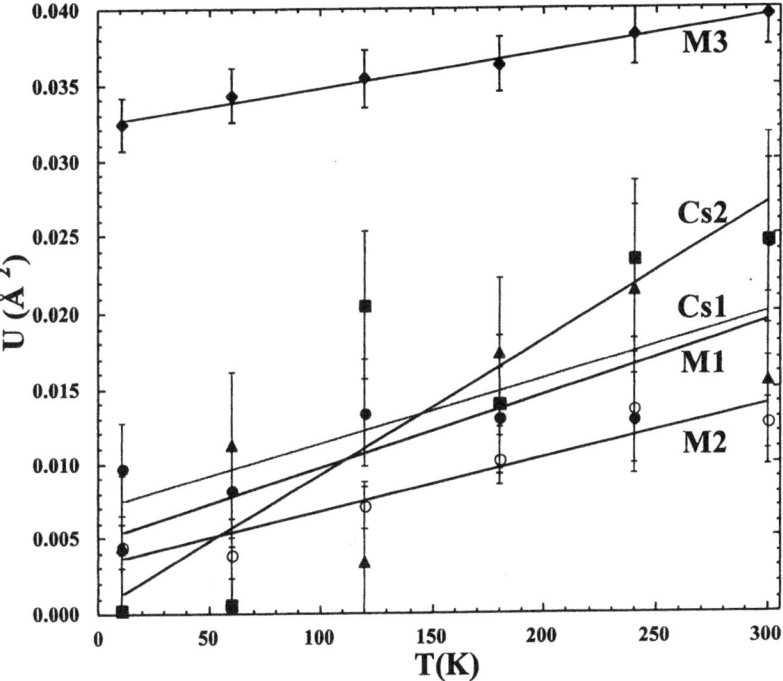

FIG. 11. Temperature dependence of the isotropic atomic displacement parameter for Cs_8Sn_{44}. The Sn sites are labeled M in this figure. The largest $U(T)$ is observed for Sn(3). The weak temperature dependence indicates that the relatively large $U(T)$ is most likely due to static disorder induced by the two vacancies per unit cell on the Sn(1) site. From Nolas et al. (2000b), Fig. 3.

isotropic ADP of Cs(1) and those of the three Sn sites are similar in magnitude. That of Cs(2) is much larger, as is shown in Fig. 10. The temperature-dependent ADP data for Cs(2) indicates that most of the disorder in this site is likely due to the "rattling," or thermally induced dynamic, motion of the Cs atom in the tetrakaidecahedra. The positions of the framework (Zn, Sn) atoms show a minimal temperature dependence, which indicates a relatively stiff framework. This is consistent with a model whereby the Cs atoms "rattle" about in the relatively large tetrakaidecahedra "cage." The data for Cs_8Sn_{44}, as shown in Fig. 11, are in distinct contrast with this model. The ADPs are generally similar for both the Cs(2) and the Cs(1) sites, whereas the largest ADP occurs for the Sn(3) site. In addition, there is little temperature dependence to $U(T)$ for Sn(3). This implies a distortion, that is, a static, positional disorder, of this site created by the Sn vacancies on the 6c site. This is corroborated by the Mössbauer spectral data (Nolas et al., 2000b). Apparently in the defect type-I structure of Cs_8Sn_{44} the positional disorder of the framework Sn atoms ruins the rigidity of the polyhedra (Nolas et al., 2000b). In fact the temperature dependence of κ_g increases as T^{-1} with decreasing temperature for Cs_8Sn_{44} (Cohn et al., 1999). Cohn et al. (1999) therefore used this clathrate as a model of a clathrate without "rattlers." As shown in Figs. 9, 10, and 11, and as described earlier, the structural details have a very distinct effect on the thermal transport properties of these compounds.

Although temperature-dependent structural analysis has not been undertaken for any of the type-II clathrate structures at this point, extensive room-temperature powder Rietveld refinements and Si NMR spectroscopy of Na_xSi_{136} have been reported (Ramachandran et al., 1999c; Reny et al., 1998; Gryko et al., 1998). By analyzing a series of Si_{136} clathrates partially filled with Na, that is, $x < 24$, Reny et al. (1998) and Ramachandran et al. (1999c) observed that the removal of Na from the smaller dodecahedra occurs first compared to that from the larger hexakaidecahedra in forming the partially filled specimens. Large thermal parameters for Na in the 8b site, that is, the Na(1) site centered in the hexakaidecahedra, were also reported. In Na_xSi_{136} specimens the Na(1) ADP was reported to be much larger than those of the other four (Na(2) and three silicon) crystallographic sites. Although these results are undoubtedly affected by the disorder introduced by the fact that all the Na sites are not fully occupied, that is, the random distribution of vacancies on the 8b site, the large ADP for a $Na_{23}Si_{136}$ specimen, a specimen with a fully occupied Na(1) site (Ramachandran et al., 1999c), demonstrates that indeed there is plenty of room for the Na to "rattle" about inside their large cages.

3. VIBRATION FREQUENCIES

The phonon transport of heat dominates the thermal conductivity of clathrates. Thus, an understanding of the phonon spectrum is important. In particular, we want to understand κ_g as a function of T/Θ, where Θ is the Debye temperature. Dong and Sankey (1999) calculated the phonon dispersion relations for both type I and II germanium clathrates and Dong et al. (1999) calculated the phonon dispersion relation for Type-II silicon clathrates, as well as its dependence on pressure. The range of phonon frequencies are about the same for both types of germanium clathrates, the acoustic phonons ranging from 0 to about 60 cm^{-1} and the optical phonons ranging from 60 to 285 cm^{-1}. For silicon, the acoustic modes are below 100 cm^{-1} while the optical modes range from 121 to 487 cm^{-1}. We do not know of any phonon dispersion calculations made for tin clathrates.

In Table IV we show a comparison between the phonon frequencies (in wave numbers and in kelvins) for the different clathrate structures and their corresponding diamond structure. In the case of tin, we used the experimen-

TABLE IV

COMPARISON BETWEEN TOP-OF-THE-BAND PHONON FREQUENCIES FOR SILICON, GERMANIUM, AND TIN IN CLATHRATE FORM AND IN THE DIAMOND STRUCTURE[a]

	Silicon		Germanium		Tin	
	Diamond	Clathrate	Diamond	Clathrate	Diamond	Clathrate
Optical	745 K	701 K	433 K	410 K	287 K	272 K
	517 cm^{-1}	487 cm^{-1}	301 cm^{-1}	285 cm^{-1}	200 cm^{-1}	190 cm^{-1}
Acoustic	591 K	143 K	343 K	86 K	224 K	56 K
	411 cm^{-1}	100 cm^{-1}	238 cm^{-1}	60 cm^{-1}	156 cm^{-1}	39 cm^{-1}

[a] In the three cases the values for the diamond structure were determined by inelastic neutron scattering. The values for the clathrate structure of silicon (Dong et al., 1999) and germanium (Dong and Sankey, 1999) were calculated. The values for the tin clathrate were estimated by scaling down the respective frequencies for silicon and germanium.

tal values measured for α-Sn at 90 K (Price et al., 1971) and estimated the clathrate frequencies based on the ratio between clathrate and diamond frequencies for both Si and Ge structures. It is interesting to note that the ratio between equivalent frequencies (acoustic and optical, respectively) for Si and Ge clathrates compared to the diamond structure is practically the same, namely about 0.94 for the optical modes and 0.25 for acoustic modes.

According to Dong et al. (1999), in the case of germanium clathrates, the optical modes, with the exception of a few within the 100–150 cm^{-1} region, have very small group velocities, and thus their contribution to the heat transport is also very small. That means that, following a PGEC model, the best "rattlers" for these systems are the ones with frequencies between 100 and 150 wave numbers for the low-lying optic branches or below 60 cm^{-1} for the acoustic phonons. Schujman and Slack (2000) estimated the vibration frequencies for different ions inside Ge and Sn type-I clathrates. The results are shown in Table V. The cage radii are calculated to be R_d = 3.428 Å, R_{tb} = 3.979 Å, and R_{tax} = 3.595 Å for Ge clathrates, and R_d = 3.988 Å, R_{tb} = 4.144 Å, and R_{tax} = 4.592 Å for Sn clathrates, assuming that the germanium and tin atoms are hard spheres. R_d is the average radius of the dodecahedral cage, and for the tetrakaidecahedral cages, R_{tb} is the average distance from the center of the cage to the center of the atoms that form the "equator" of the cage, and R_{tax} is the average distance from the center of the cage to the center of the atoms that form the hexagons. For the type-II clathrates, the average radii of the ideal hexakaidecahedral cage are 3.9426, 4.1069, and 4.7110 Å for Si, Ge, and Sn, respectively. The lowest vibration frequencies are for the heaviest ions in the most oversized cages. Given the fact that in order to reach κ_{min} it may be necessary to scatter phonons with two different ranges of frequencies (i.e., between 100 and 150

TABLE V

ESTIMATION OF VIBRATION FREQUENCIES FOR SEVERAL IONS INSIDE CAGES OF Ge AND Sn CLATHRATES[a]

Ion	Mass (g)	Ionic radius (Å)	Type-I						Type-II	
			Ge clathrate			Sn clathrate			Ge	Sn
			W_d	W_{tax}	W_{tb}	W_d	W_{tax}	W_{tb}	W_h	W_h
Na^{1+}	22.9898	1.00	97	78	32	68	53	19	63	44
Rb^{1+}	85.4678	1.50	77	73	61	50	44	25	46	30
Sr^{2+}	87.62	1.18	81	68	34	55	45	19	51	35
Ba^{2+}	137.33	1.38	117	67	35	51	43	22	47	32
La^{3+}	138.9055	1.15	77	64	30	53	42	17	49	34
Pr^{3+}	140.9077	1.07	71	58	25	50	39	15	46	32
Hf^{4+}	178.49	0.77	58	45	15	42	31	10	39	28
Th^{4+}	232.0381	0.99	60	48	20	43	44	12	40	28

[a] For the type-I case we are assuming a composition A_8B_{46}, where A is the ion and B = Ge, Sn. All the frequencies are in wave numbers. W_d, W_{tax}, and W_{tb} indicate vibrations inside the dodecahedral cages, the tetrakaidecahedral cages in the direction perpendicular to the equator, and in the plane of the equator, respectively. For the type-II case we estimate the vibration frequencies for the 28-atom hexakaidecahedral cages of the Ge and Sn ideal structures.

cm^{-1} and below 60 cm^{-1} for germanium clathrates), it may be more effective to use a combination of different rattlers rather than only one atomic species. This is indeed the case, as described in Section IV of this review.

We can estimate the Debye temperature for the different types of ideal, *empty* clathrates, where in this case the word "ideal" is used in the sense described in the crystal structure section, as presented by Schujman et al. (2000), that is, a structure where all the bonds have the same length, which is equal to the average diameter of the atoms forming the network. Although approximate, these estimations will be useful for discussing the effectiveness of different local modes in scattering the acoustic phonons that transport the heat in these semiconducting crystals.

We relate the bulk modulus for the clathrate and diamond structure by use of the approximation:

$$B_c = B_d \left(\frac{V_d}{V_c}\right)^{2/3}, \qquad (2)$$

where the subscripts "c" and "d" refer to "clathrate" and "diamond" respectively, $V = \delta^3$ is the volume per atom for each structure, and B is the bulk modulus. The $\frac{2}{3}$ exponent in Eq. (2) is approximate and based on a dimensional argument. Experimentally it has been observed (Ramachan-

TABLE VI

ESTIMATION OF DEBYE TEMPERATURES FOR THE DIFFERENT
CLATHRATE STRUCTURES AND A COMPARISON WITH THE
CORRESPONDING DIAMOND STRUCTURE (ALL VALUES IN KELVINS)

Structure	Silicon	Germanium	Tin
Diamond	645	374	230
Type-I	628	366	217
Type-II	630	366	215

dran et al., 1999a) that in silicon type-II clathrates Eq. (2) gives calculated B_c values very close to the observed ones.

The speed of sound can be written as

$$v_{sound} \propto \sqrt{\frac{BV}{\bar{M}}} \qquad (3)$$

where $\bar{M}$ is the average atomic mass, and finally, the Debye temperature (Slack, 1979) will be proportional to

$$\Theta \propto \frac{v_{sound}}{\delta}. \qquad (4)$$

Then, the Debye temperature of the clathrate can be estimated from the known values for the diamond structure by using

$$\Theta_c = \Theta_d \frac{\delta_d}{\delta_c} \left(\frac{V_c}{V_d}\right)^{1/6}. \qquad (5)$$

In Table VI we show the values obtained for the different empty clathrate structures and a comparison with the ones measured for the corresponding diamond structure. The values are very similar. For comparison we note that the Debye temperature of "filled" type I germanium clathrates are ~270 K from heat capacity measurements and structural data (Sales et al., 1999; G. S. Nolas, unpublished results).

Using this simplified approach we may estimate the thermal conductivity of Cs_8Sn_{44}, what we can take as a model of a clathrate without "rattlers" (Cohn et al., 1999; Schujman et al., 2000). Following Slack (1979), we can estimate the thermal conductivity in a temperature range near $T = \Theta$ at a given temperature as

$$\kappa_g(T) = B\bar{M}\delta\Theta^3 n^{-2/3} T^{-1} \gamma^{-1}, \qquad (6)$$

where $B = 3.04$ W/(g cm^2 K), n is the number of atoms in the primitive unit cell, T is the absolute temperature, and γ is the Grüneisen parameter. We assume $\gamma \approx 1$, which is reasonable for tetrahedrally bonded crystals (Slack, 1979). We need a value for the Debye temperature of Cs$_8$Sn$_{44}$. From Eq. (4) we can extract a value proportional to the velocity of sound for the empty Sn clathrate. Assuming that the bulk modulus will be similar for both the empty and filled clathrate, using Eq. (3) we can estimate a ratio between the velocity of sound in both filled and empty clathrates. We can then write the Debye temperature for the filled clathrate as

$$\Theta_{FC} = \Theta_{EC} \frac{v_{FC}}{\delta_{FC}} \frac{\delta_{EC}}{v_{EC}}. \quad (7)$$

With these considerations, the Debye temperature we obtain for the filled Cs$_8$Sn$_{44}$ clathrate is $\Theta_{FC} = 199$ K, and then, applying Eq. (6) and using $n = 46$ and $T = 300$ K, we calculate a theoretical thermal conductivity $\kappa_g = 0.025$ W/cm K. The value that Cohn et al. (1999) measured at room temperature is 0.01 W/cm K, which is just over a factor of 2 smaller than our calculated value. Thus, this simple estimate corroborates the fact that Cs$_8$Sn$_{44}$ has a thermal conductivity consistent with a model of κ_g limited by simple three-phonon scattering. There is no need to invoke a "rattle" scattering of the phonons by a cesium local mode, as demonstrated by thermal conductivity (Cohn et al., 1999) and temperature-dependent neutron diffraction (Nolas et al., 2000b) data.

4. BAND STRUCTURE

There have been several published band structure calculations on C, Si, and Ge type I and II clathrates, but none as yet have appeared on Sn clathrates. Nesper et al. (1993) performed augmented plane-wave calculations on carbon in clathrate and other zeolite structures. Their calculations indicated that carbon clathrates are very low in energy, in fact, much lower than the fullerene molecule C$_{60}$. Sekkal et al. (1999) used molecular dynamics calculations based on a Lennard–Jones potential in simulating the structural and thermodynamic properties of C$_{46}$ and C$_{136}$. A particularly interesting aspect of their calculation indicated that C$_{46}$ is unstable near room temperature.

Adams et al. (1994) employed self-consistent plane-wave calculations within the local density approximation (LDA) in an investigation of Si$_{46}$ and Si$_{136}$ clathrates. These calculations showed that both Si clathrate types have indirect bandgaps that are 0.7 eV wider than that of diamond-structured Si, whereas the bandgap of clathrates formed by carbon is reduced compared to that of diamond. The total energy of Si clathrates is only 0.07 eV/atom above that of the ground state of diamond-Si, and the volume increase in the clathrate phase is 17% compared to the diamond phase. This work also investigated the structure and energies of type I and II carbon clathrates, resulting in similar volume expansion but with larger energy differences compared to diamond. Menon et al. (1997) employed a generalized tight-binding molecular-dynamics scheme; Kahn and Lu (1997) used the tight-bonding density-matrix method to study the structural and vibrational properties of Si_{46} and Si_{136}. Their results verified those of Adams et al. (1994) while also providing the vibrational modes and elastic moduli of these clathrates. The vibrational modes of Si_{136} were also calculated by Dong et al. (1999), who also investigated their pressure dependence.

Demkov et al. (1995) used the self-consistent plane-wave technique in investigating the effect of alkali metals inside the polyhedra of Si clathrates. The electronic properties of the Si_{136} clathrate change, and perhaps may be altered in a controllable fashion, by doping with alkali metals. The effect of doping on the band structure is stronger for less electropositive metals. In particular, doping with Na introduces a narrow band that comes from the lowest conduction band in pure Si_{136}. This band is half occupied at low Na concentration ($x = 4$ in Na_xSi_{136}) and fully occupied at higher concentrations ($x = 8$). For K doping the effect is less pronounced and the lowest conduction bands will have structure similar to that of pure Si_{136}.

Using LDA, Saito and Oshiyama (1995) investigated the electronic properties of Si_{46} and $Na_2Ba_6Si_{46}$, a compound that was shown to have a superconducting transition temperature of 4 K by Kawaji et al. (1995). The calculations on Si_{46} generally agreed with those previously reported. For $Na_2Ba_6Si_{46}$ the valence-electron density is low around the Na(1), $2d$, site but there is substantial charge around the Ba(2), $6c$, site indicating that the Ba state may be hybridized more strongly with the framework Si conduction band state than the Na state. This strong hybridization of the Ba state with Si_{46} results in a high Fermi-level density of states, which may be the reason for the observed superconductivity of this compound.

Blake et al. (1999) employed density functional calculations to study the structure and electronic properties of $Sr_8Ga_{16}Ge_{30}$. Their calculations indicated that the Sr atoms are weakly bound to the "cages" formed by the (Ga, Ge) framework and undergo large-amplitude motion. In addition, $Sr_8Ga_{16}Ge_{30}$ is a metal in which the largest contribution to the electronic conductivity comes from a band in which the electrons are located on the (Ga, Ge) framework, while bands originating from the Sr atoms contribute little to conduction. The Sr atoms vibrations are weakly coupled to the conduction electrons since there is little charge transfer between the Sr atoms and the framework atoms in this model.

Dong et al. (1999) investigated the electronic structure and vibrational modes of pure Ge_{46} and Ge_{136} using the density functional plane-wave pseudopotential method. The energies of these Ge clathrates are 0.04 to 0.05 eV/atom higher than that of diamond-structured Ge, while their volumes are 13 to 14% higher. Both type I and II Ge clathrates are semiconductors with LDA calculated gaps 1.21 and 0.75 eV wider than that of diamond-Ge, respectively; however, the authors estimate the true bandgaps to be near 2 eV, because LDA calculations normally underestimate these values. The phonon dispersion curves show that the acoustic modes are limited to less than 60 cm^{-1}. The Raman and infrared optic modes were also calculated. Zhao et al. (1999) studied the structure and electronic properties of Ge_{46} and K_8Ge_{46} employing first principles calculations within the local-density approximation. These calculations indicated that Ge_{46} is a stable structure with only a slightly higher energy compared to diamond-structured Ge. The calculated bandgap and expanded volume of Ge_{46} were in agreement with that obtained by Dong and Sankey (1999). K_8Ge_{46}, however, is metallic with a moderate density of states. The valence-band and DOS are similar to that of pure Ge_{46}, while the conduction bands are modified by K. Almost complete charge transfer appears to take place from the K sites to the Ge framework.

IV. Thermal Conductivity

1. GLASSLIKE CONDUCTION

The measurement and analysis of the thermal conductivity of clathrates has to date mainly been undertaken by Nolas and co-workers (Nolas, 1999a; Nolas et al., 1999a,b, 2000a,b,c,d). Uher et al. (1999) also measured the thermal conductivity, κ, of polycrystalline $Sr_8Ga_{16}Ge_{30}$ and $Ba_8Ga_{16}Ge_{30}$. One of the most interesting and important discoveries, in terms of thermoelectrics, was the magnitude and temperature dependence of κ_g of Ge clathrates. Figure 12 shows κ as a function of temperature from 0.6 to 300 K for a typical phase-pure polycrystalline $Sr_8Ga_{16}Ge_{30}$ specimen with large average grain size and high resistivity. Essentially $\kappa \cong \kappa_g$ for this sample, assuming the Wiedemann–Franz relation. The electronic κ_e is negligible. Figure 12 also shows κ for single crystal Ge (Glassbrenner and Slack, 1964), amorphous Ge (a-Ge) (Cahill et al., 1989b), and amorphous SiO_2 (a-SiO_2) (Cahill et al., 1992), as well as κ_{min} calculated for Ge. Note that κ_{min} for the germanium clathrate will be somewhat lower. We note that κ is lower than that of a-SiO_2 above 100 K, it is close to that of a-Ge at room temperature, and it exhibits a temperature dependence that is reminiscent of amorphous materials. The low-temperature (<1 K) data indicates a T^2 temperature dependence, as shown by the straight-line fit to the data in Fig. 12. Higher temperature data shows a minimum, or dip, in the 4 to 35 K range indicative

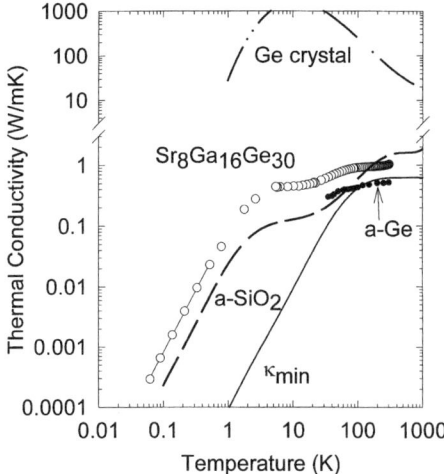

FIG. 12. Lattice thermal conductivity versus temperature for polycrystalline $Sr_8Ga_{16}Ge_{30}$, single crystal Ge, amorphous SiO_2, amorphous Ge, and κ_{min} for Ge. The straight-line fit to the $Sr_8Ga_{16}Ge_{30}$ data below 1 K produces a T^2-temperature dependence characteristic of glasses. From Nolas et al. (1999a), Fig. 1. Copyright 1998 IEEE.

of possible resonance scattering. The low-temperature data also has a T^2 dependence on temperature similar to that found for amorphous material, and similar to the T^2 dependence calculated for κ_{min} (Cahill et al., 1992; Slack, 1979). It is clear from this data that in the $Sr_8Ga_{16}Ge_{30}$ compound the traditional alloy phonon scattering, which predominantly scatters the highest frequency phonons, has been replaced by one or more much lower frequency scattering mechanisms. The highest frequency optic phonons in the clathrate structure have very low or zero group velocity and contribute little to the total thermal conductivity, while the low-frequency acoustic phonons have the highest group velocity and contribute most to κ. The scattering of these low-frequency acoustic phonons by the encaged ions results in low thermal conductivities.

2. CONNECTION WITH STRUCTURAL PROPERTIES

It is most instructive to employ the split-site or off-center, multiple well model, obtained from the structural refinements as described in the previous section, in evaluating the κ_g data, since mass fluctuation scattering or grain boundary scattering cannot explain the temperature dependence of κ_g of the Ge clathrate compounds. This was undertaken by Nolas and co-workers (Cohn et al., 1999; Nolas et al., 2000a,c; Chakoumakos et al., 1999) via a model that incorporates the dynamic as well as static disorder associated with the encapsulated atoms as required. The ADP data were also employed in determining the characteristic localized vibration frequencies for weakly bound atoms that "rattle" within their atomic "cages." This approach, which assumes that the "rattling" atoms act as harmonic oscillators, was shown to be successful in the case of skutterudite and other compounds by Sales et al. (1999). A detailed discussion of this approach can be found in Chapter 1 of the next volume in this series. The localized vibration of the "rattler" atom can be described by an Einstein oscillator model such that $U = k_B T / m(2\pi v)^2$, where U is the isotropic mean-square displacement, k_B is Boltzmann's constant, m is the mass of the "rattling" atoms under the assumption that their "cages" are relatively rigid, and v the frequency of vibration. The ADP data can then be used to estimate the "Einstein temperature" of these atoms, $T_E = hv/k_B$, where h is Planck's constant. Employing this approach we obtain $T_E = 85$ and 119 K for Sr(2) and Sr(1), respectively, from temperature-dependent neutron diffraction data on a polycrystalline specimen. Similar values to these were obtained for single crystal specimens from refinements of room temperature X-ray diffraction data (Nolas et al., 2000a).

Figure 13 shows the κ_g data for polycrystalline $Sr_8Ga_{16}Ge_{30}$ from 100 K down to 60 mK from Cohn et al. (1999), along with a fit employing the model described earlier (Nolas et al., 2000c). Data for a-SiO_2 and κ_{min} for Ge is also shown for comparison. The fit to the data incorporates structural observations in phenomenological fits of the κ_g data to a kinetic theory expression with a phonon mean free path that is a sum of terms representing tunnel system (TS), resonant, and Rayleigh (R) scattering (Cohn et al., 1999),

$$\kappa_g = \frac{v}{3}\int C(\omega)l(\omega)d\omega \tag{8}$$

$$l(\omega) = (l_{TS}^{-1} + l_{res}^{-1} + l_R^{-1})^{-1} + l_{min} \tag{9}$$

$$l_{TS}^{-1} = \frac{A\hbar\omega}{k_B}\tanh\left(\frac{\hbar\omega}{2k_BT}\right) + \frac{A}{2}\left(\frac{k_B}{\hbar\omega} + \frac{1}{BT^3}\right)^{-1} \tag{10}$$

$$l_{res}^{-1} = \sum_i \frac{C_i\omega^2 T^2}{(\omega_i^2 - \omega^2)^2 + \gamma_i\omega_i^2\omega^2} \tag{11}$$

$$l_R^{-1} = D\left(\frac{\hbar\omega}{k_B}\right)^4, \tag{12}$$

where ω_D is the Debye frequency, C is the heat capacity of the phonons, v is the average sound velocity, ω is the frequency, T the absolute temperature, $\hbar$ is Planck's constant divided by 2π, and γ is an average deformation potential. The lower limit on l is assumed to be a constant, l_{min}. The

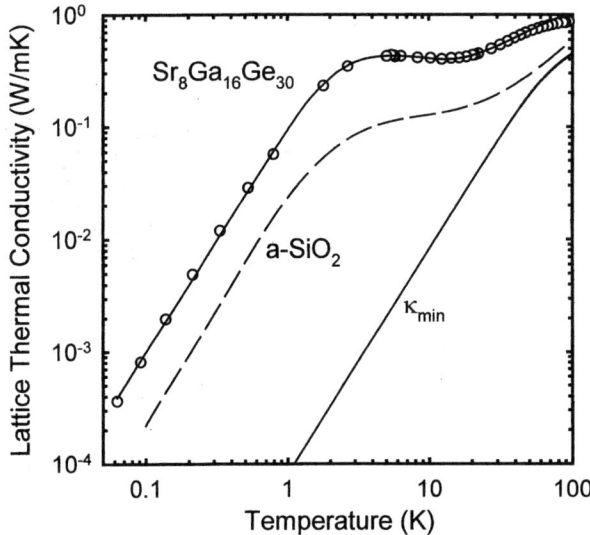

FIG. 13. Lattice thermal conductivity versus temperature from 60 mK to 100 K for $Sr_8Ga_{16}Ge_{30}$ with a fit (solid line) to the model discussed in the text. The fitting parameters are: $A = 9.0 \times 10^3$ m^{-1} K^{-2} s^{-2}, $B = 0.5 \times 10^{-3}$ K^{-2}, $C_1 = 2.09 \times 10^{-32}$ m^{-1} K^{-2} s^{-2}, $C_2 = 1.80 \times 10^{-31}$ m^{-1} K^{-2} s^{-2}, $D = 0.6$ m^{-1} K^{-1}, $\gamma_1 = 4$, $\gamma_2 = 0.8$, and $l_{min} = 5$ Å. The dashed line is for amorphous SiO_2 and the solid line for κ_{min} for Ge. From Cohn et al. (1999), Fig. 2.

constants A and B in Eq. (10) are related to microscopic variables describing the TS model (Graebner et al., 1986). Their ratio is given by $A/B = n(\hbar\omega)^2/\pi k_B$, with n the density of tunnel states per unit volume strongly coupled to phonons. The phenomenological resonance terms are of the form employed previously to describe phonon scattering in ionic crystals (Pohl, 1962; Walker and Pohl, 1963). Although the l_{res} term employed is not unique and the values of ω_1 and ω_2 are sensitive to the values of D (Cohn et al., 1999), good fits are achieved (solid curve in Fig. 13) using the T_E values obtained from the structural data, as described earlier, with $\Theta_D = 270$ K and $v = 2600$ m/s (Nolas et al., 2000a,c; Chakoumakos et al., 1999; Sales et al., 1999; G. S. Nolas, unpublished results). The values for the fitting parameters (see Fig. 13 caption) were similar to those obtained for κ_g on single crystal Ge clathrates (Nolas et al., 2000a). In addition, the TS parameters are comparable to those found for many amorphous solids (Graebner et al., 1986). As revealed by this semiquantitative analysis (Nolas et al., 2000a,c; Chakoumakos et al., 1999), at low temperatures the static disorder associated with the Sr positions in the tetrakaidecahedra, as described by the split-site model for the Sr(2) site, may be the origin of the TS associated with

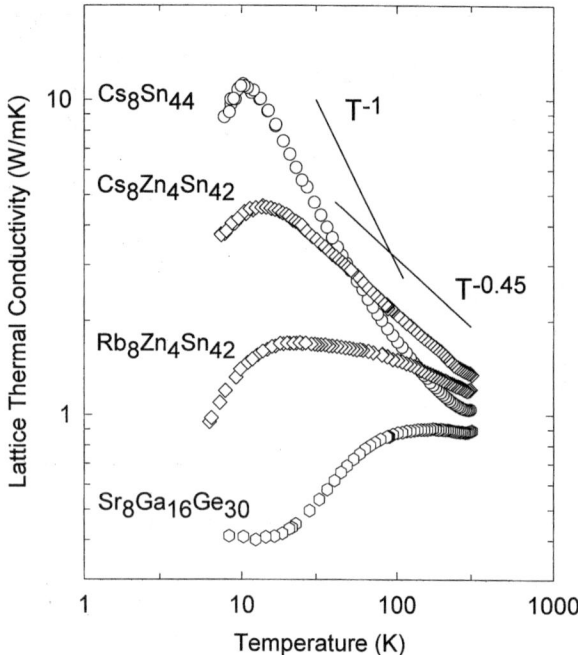

FIG. 14. Lattice thermal conductivity versus temperature for polycrystalline Cs_8Sn_{44}, $Cs_8Zn_4Sn_{42}$, $Rb_8Zn_4Sn_{42}$, and $Sr_8Ga_{16}Ge_{30}$.

a "freezing-out" of the "rattling" motion of Sr(2). Small energy barriers (≤ 1 K) presumably characterize the separation between the Sr(2) positions. The resonance dip in the data in the range $4 < T < 35$ K is associated with the dynamic, or "rattling," of the Sr atoms inside their atomic cages. As this material is one of only a few crystalline materials having glasslike thermal conductivity, other thermal and elastic properties at low temperatures are clearly of particular interest.

The correlation between the crystal structure and the thermal conductivity of semiconducting clathrates can be expanded to include other clathrate compounds as well as defect structured compounds, as described by Nolas et al. (1999a, 2000b,d; Nolas, 1999b). Figure 14 shows κ_g for large grain size and high-resistivity polycrystalline specimens of Cs_8Sn_{44}, $Cs_8Zn_4Sn_{42}$, $Rb_8Zn_4Sn_{42}$, and $Sr_8Ga_{16}Ge_{30}$ from 300 to 8 K (Nolas et al., 1998, 1999b, 2000b,d; Cohn et al., 1999). As seen in this figure, the Cs_8Sn_{44} compound exhibits a temperature-dependent κ_g that varies as T^{-1} and is typical of simple crystalline insulators dominated by phonon–phonon scattering. In the case of the $Cs_8Zn_4Sn_{42}$ compound, κ_g also increases with decreasing temperature and varies as $T^{-0.45}$, indicating that mass-fluctuation phonon scattering is present. In the case of $Cs_8Zn_4Sn_{42}$ the disorder created by the

Cs(2) "rattling" atoms in their rigid cages results in a strong phonon scattering in the optic phonon frequency range; this is not the case for Cs_8Sn_{44} (Nolas et al., 1999b, 2000b; Cohn et al., 1999). The additional bonding induced between Cs and Sn atoms neighboring the vacancies in Cs_8Sn_{44} apparently constrains the Cs(2) displacements so that they do not "rattle." This structural difference between the two compounds is the source of their differing thermal conductivities. The localized vibrations, or "rattling," of Cs(2) in $Cs_8Zn_4Sn_{42}$ probably couple to the optic phonons, producing resonant damping of these phonons. In the case of $Rb_8Zn_4Sn_{42}$ the Rb atom is smaller than the Cs atom and therefore is more free to "rattle" inside the polyhedra formed by the (Zn, Sn) framework. These localized vibrations may provide a more prominent phonon scattering effect in this compound, as indicated by the temperature dependence of κ_g for $Rb_8Zn_4Sn_{42}$ (Fig. 14). As noted by Nolas et al. (1998, 1999b, 2000b), the differences in grain-boundary scattering cannot explain the κ_g difference between these two compounds. Although the enhanced thermal motion of the Cs(2) or Rb(2) atoms in the Sn clathrates appears to diminish κ_g, the effect is not as great as that caused by Sr(2) motion in $Sr_8Ga_{16}Ge_{30}$, where the Sr vibration frequencies lie in the acoustic range. There the Sr(2) thermal ellipsoids are nearly an order of magnitude larger than those of the (Ga, Ge) framework atoms (Fig. 9). It may be that static as well as dynamic disorder is required in order to achieve glasslike thermal transport in these compounds.

Nolas and co-workers have investigated the thermal transport of several type-I clathrates. Figure 15 shows κ_g from room temperature to 6 K for five

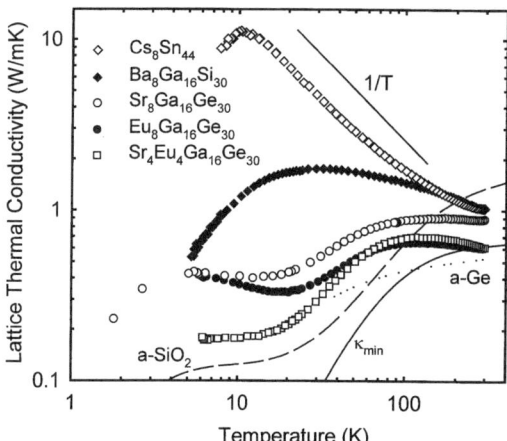

FIG. 15. Lattice thermal conductivity versus temperature for five representative polycrystalline type I clathrates. The dashed and dotted curves are for amorphous SiO_2 and amorphous Ge, respectively, and the solid curve is κ_{min} calculated for Ge. From Cohn et al. (1999), Fig. 2.

representative clathrate specimens (Cohn et al., 1999). From this figure a trend in the thermal transport emerges that is intimately related to the crystal structure of these compounds. The Ge clathrates exhibit a κ_g that is typical of amorphous solids. The values are well below that of a-SiO$_2$ and close to that of a-Ge near room temperature. In addition, the temperature dependence of the thermal conductivity of Ge clathrates is much like that of a-SiO$_2$. Europium (Eu^{2+}) is a smaller ion and has a larger ADP than Sr^{2+} in the (Ga, Ge) framework (Nolas et al., 2000a). In addition, Eu^{2+} is almost twice as massive as Sr^{2+} and therefore will have a larger effect on κ_g. In the case of Sr$_4$Eu$_4$Ga$_{16}$Ge$_{30}$ there are two different atoms in the voids of the crystal structure, which introduces six different resonant scattering frequencies (three for each ion). This compound exhibits the lowest κ_g values in the temperature range shown and tracks the temperature dependence of a-SiO$_2$ quite closely. The κ_g measurements on Sr$_4$Eu$_4$Ga$_{16}$Ge$_{30}$ clearly demonstrate that κ_g is further reduced compared to when the disorder of Sr or Eu is solely employed. We note than in the temperature range below 30 K there is also grain boundary scattering of the phonons; single crystal samples have somewhat higher κ_g values in this temperature range (Nolas et al., 2000a,c). The Ba$_8$Ga$_{16}$Si$_{30}$ sample has a relatively low κ_g; however, the temperature dependence is not similar to that of the Ge clathrates. Although Ba is much more massive than the elements that make up the host matrix (i.e., Ga or Si), a prerequisite for glasslike κ_g (Slack, 1995), the temperature dependence of κ_g is similar to that of a simple crystalline solid exhibiting mass-fluctuation scattering. This is due to the fact that Ba^{2+} is similar in size to the Si$_{20}$ and Si$_{24}$ cages, whereas Sr^{2+} or Eu^{2+} are smaller than the Ge cages. From X-ray diffraction data, it does not appear that Ga increases the average size of the cage very much. The case of Cs$_8$Sn$_{44}$ was extensively discussed previously.

V. Electronic Properties

1. Doping Variation

In the early work of Cros and co-workers (1970), the electronic properties of several of their Si clathrates were investigated. Figure 16 shows the Seebeck coefficient for four of their N-type specimens, three type-II Si-clathrates along with that for K$_7$Si$_{46}$, prepared by cold pressing and then sintering at 400°C (Cros et al., 1970). The Seebeck coefficient increases with decreasing Na content. This is in agreement with their magnetic susceptibility data and indicates a metal-to-insulator transition at $x \sim 11$ in Na$_x$Si$_{136}$. Their electronic conductivity data, however, were compromised by the quality of the specimens prepared for electronic measurements (Mott, 1973).

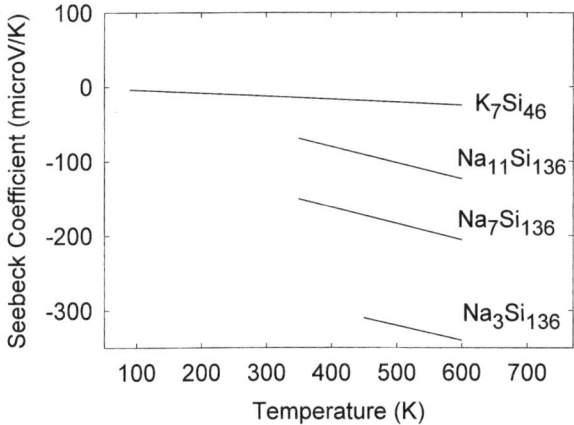

FIG. 16. Seebeck coefficient versus temperature for three type II clathrates plus K_7Si_{46} of type I. From Cros et al. (1970), Fig. 9.

A thorough investigation of the transport properties of the type II clathrate system has yet to be undertaken.

Figures 17 and 18 show resistivity, ρ, and absolute Seebeck coefficient, S, as functions of temperature from 300 to 5 K for three specimens with nominal compositions $Sr_8Ga_{16+x}Ge_{30-x}$, with x varied slightly in the three specimens (Nolas et al., 1998). The ability to vary the doping level of these semiconductor compounds by varying the chemical composition is one reason why they are of interest for thermoelectric applications. The Ga is randomly substituted for Ge in the structure (Chakoumakos et al., 1999) and is used to produce charge compensation for the divalent alkali-earth ion Sr^{2+} (Nolas et al., 1998). The doping level of this series of specimens was varied by changing the Ga-to-Ge ratio while maintaining a fixed Sr concentration. This is similar to doping diamond-structure Ge with As, for example. p-Type conduction was not obtained in the clathrate in this manner; however, p-type clathrates can be achieved by substituting Zn for Ge (G. S. Nolas, unpublished results). The absolute S decreases with increasing carrier concentration, as shown in Fig. 18. The absolute S also decreases with decreasing temperature, as expected in heavily doped semiconductors with negligible phonon drag. In the inset to Fig. 17 a plot of $\ln(\rho)$ versus $1/T$ is also shown for the specimen that had the lowest electron concentration. The straight line is a fit to the equation $\rho = \rho_o \exp(\Delta/T)$, where the activation energy for the donors is $E_a = 2\Delta$. From this fit we obtain $E_a = 15$ meV. This value is similar to that obtained employing optical absorption measurements and can be compared to 14.17 meV for shallow donors such as As in Ge. The intrinsic carrier mobility in the clathrate structure is at present unknown. The presence of

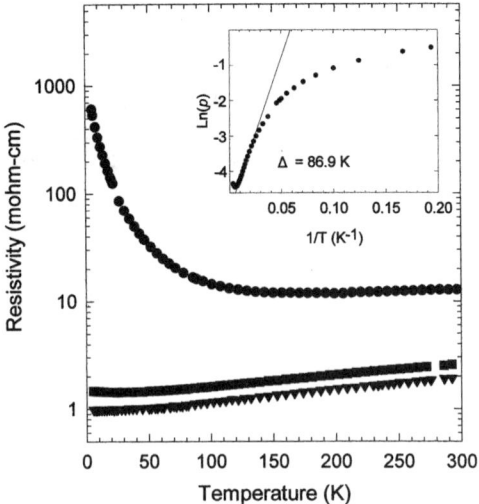

FIG. 17. Resistivity versus temperature from 5 to 300 K for three n-type $Sr_8Ga_{16}Ge_{30}$ specimens with different Ga-to-Ge ratios but with similar Sr concentrations resulting in varying carrier concentrations. The inset shows $\ln(\rho)$ versus $1/T$ for the specimen with the highest ρ, in ohm-cm. The straight line fit to the equation $\rho = \rho_0 \exp(\Delta/T)$ gives $\Delta = 86.9$ K. From Nolas et al. (1998), Fig. 1.

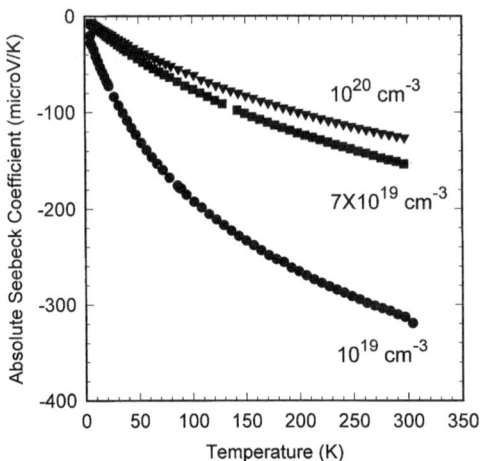

FIG. 18. Seebeck coefficient versus temperature for the three specimens from Fig. 17. The room temperature Hall carrier concentration of each sample is also shown. From Nolas et al. (1998), Fig. 2.

ionized electron donors such as Na^{+1} or Sr^{+2} decreases the mobility below its intrinsic value; by how much is not known. The Sr^{2+} ions may or may not cause very serious Coulomb scattering of the carriers (Iverson et al., 1999). This problem has been addressed by theoretical analyses (Blake et al., 1999).

2. METALS OR SEMICONDUCTORS

Nolas and co-workers (Nolas, 1999a; Nolas et al., 1998, 1999a,b, 2000b,d; Cohn et al., 1999) have measured the transport properties of several type I Si, Ge, and Sn-clathrate specimens. Compounds exhibited either semiconducting or metallic electronic properties depending on doping level and even stoichiometry. Figure 19 shows a compilation of five n-type polycrystalline

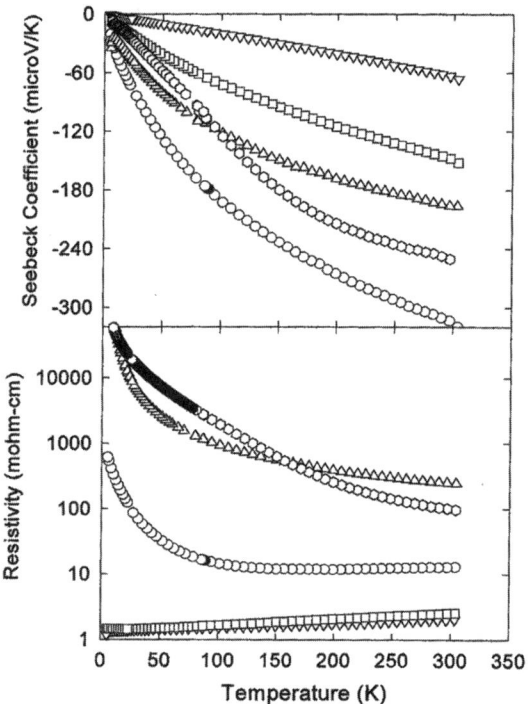

FIG. 19. Seebeck coefficient and electrical resistivity versus temperature for five representative n-type specimens illustrating the range of possible compositions and power factors (i.e., S^2/ρ) obtained by varying the carrier concentration; $Sr_8Ga_{16}Ge_{30}$ (data points represented by circles), $Rb_8Zn_4Sn_{42}$ (hexagons), $Cs_8Zn_4Sn_{34}Ge_8$ (up triangles), $Eu_8Ga_{16}Ge_{30}$ (squares), and $Ba_8Ga_{16}Si_{30}$ (down triangles).

compounds, illustrating the range and type of electronic properties associated with these compounds. This figure is intended to provide an idea of the variety of different Si, Ge, Sn, and mixed-crystal clathrates and their electronic properties. In general S increases with increasing temperature, typical of semiconductor behavior, with the magnitude dependent on the carrier concentration. In the case of Sn clathrates, very high S values and very high electrical resistivities have been measured. This is another feature of the differences between Sn clathrates and Si and Ge clathrates. More work is needed in order to understand the reason for this phenomena. n-Type conduction is most often obtained in clathrate compounds; however, p-type compounds have also been synthesized (G. S. Nolas, unpublished results). In certain compositions, such as $Sr_8Zn_8Ge_{38}$, both n- and p-type compounds can be synthesized (G. S. Nolas, unpublished results).

VI. Summary and Conclusions

The clathrate structure, Ge clathrates for example, can be thought of as a derivative of the four-coordinated diamond lattice structure of Ge. As described earlier, between the Ge–Ge atoms in the Ge diamond lattice there is enough space to hold small interstitial atoms such as H or He. However, this space is not large enough to hold Sr or Eu atoms. The presence of these "guests" induces a change in the Ge clathrate to a more open structure.

The fact that clathrate compounds can be synthesized to possess glasslike lattice thermal conductivity, the ability to vary the electronic properties by changing the doping level, and their relatively good electronic properties all indicate that this system is a PGEC system and therefore of interest for further research for thermoelectric applications. This is the only known system of compounds that have relatively good semiconductor properties in addition to exhibiting glasslike κ_g properties. As described in this chapter, the boride compounds also display semiconducting properties with glasslike κ_g (for the case of YB_{68}) although with poor mobility and lower Seebeck values. In all these compounds, however, the goal is to replace the traditional alloy phonon scattering, which predominantly scatters the highest frequency phonons, by a much lower frequency resonance and possibly disorder-type scattering. The highest frequency phonons in any crystal lattice have very low or zero group velocities and contribute little to the total thermal conductivity. The low frequency phonons have the highest group velocity and contribute most to κ_g. This is why the "caged impurities" produce such pronounced decreases in κ_g in clathrate and crypto-clathrate systems.

Thus far the highest room temperature dimensionless figures of merit, ZT, for n-type clathrate compounds range from 0.25 to 0.34. The ZT values increase with increasing temperature with $ZT > 1$ above 400°C. These values were obtained on polycrystalline type-I Ge clathrates. These results are very interesting for thermoelectrics applications and represent a high ZT value for basically unoptimized materials. The ZT values could presumably be increased by optimizing the doping level, by changing the rattle frequencies, or by employing single crystal samples instead of polycrystalline ones. The promising properties of these interesting materials suggest a new category of thermoelectric materials for future investigation.

Acknowledgments

George S. Nolas very gratefully acknowledges the support of Marlow Industries, Inc., and the U.S. Army Research Laboratory under contract DAAD17-99-C-0006. Glen A. Slack and Sandra B. Schujman very gratefully acknowledge the support of the Office of Naval Research under grant number 00014-94-1-0341.

References

G. B. Adams, M. O'Keeffe, A. A. Demkov, O. F. Sankey, and Y.-M. Huang, Wide-band-gap Si in open fourfold-coordinated clathrate structures, *Phys. Rev. B* 49, 8048 (1994).

M. I. Aivazov, T. I. Bryushkova, V. S. Mkrtchyan, and V. A. Rubanov, Thermal conductivity of LaB_6–EuB_6 solid solutions, *Teplofiz. Vis. Temp.* 17, 330 (1979) [*High Temp. USSR* 17, 277 (1979)].

P. A. Alekseev, A. S. Ivanov, K. A. Kikoin, A. S. Mischenko, A. N. Lazukov, A. Yu. Rumyantsev, and I. P. Sadikov, Lattice dynamics in fluctuation valence compound SmB_6, *Am. Inst. Phys. Conf. Proc.* 231, 318 (1991).

N. W. Ashcroft and N. D. Mermin, *Solid State Physics*. Saunders, Philadelphia, PA (1976).

N. P. Blake, L. Mollnitz, G. Kresse, and H. Metiu, Why clathrates are good thermoelectrics: A theoretical study of $Sr_8Ga_{16}Ge_{30}$, *J. Chem. Phys.* 111, 3133 (1999).

J. D. Bryan, V. I. Srdanov, G. Stucky, and D. Schmidt, Superconductivity in germanium clathrate $Ba_8Ga_{16}Ge_{30}$, *Phys. Rev. B* 60, 3064 (1999).

E. Busman, The Crystal Structure of KSi, RbSi, CsSi, KGe, RbGe, and CsGe, *Zeit. anorg. allgem. Chem.* 313, 90 (1961).

D. G. Cahill, H. E. Fisher, S. K. Watson, R. O. Pohl, and G. S. Slack, Thermal properties of boron and borides, *Phys. Rev. B*40, 3254 (1989a).

D. G. Cahill, H. E. Fisher, T. Klitsner, E. T. Swarts, and R. O. Pohl, Thermal conductivity of thin films: Measurements and understanding, *J. Vac. Sci. Technol. A* 7, 1259 (1989b).

D. G. Cahill, S. K. Watson, and R. O. Pohl, Lower limit to the thermal conductivity of disordered crystals, *Phys. Rev. B* 46, 6131 (1992) and references therein.

B. C. Chakoumakos, B. C. Sales, D. G. Mandrus, and G. S. Nolas, Structural disorder and thermal conductivity of the semiconducting clathrate $Sr_8Ga_{16}Ge_{30}$, *J. Alloys Comp.* 296, 801 (1999).

T. L. Chu, S. S. Chu, and R. L. Ray, Germanium arsenide iodide: A clathrate semiconductor, *J. Appl. Phys.* 53, 7102 (1982).

W. F. Claussen, Suggested structures of water in inert gas hydrates, *J. Chem. Phys.* 19, 259, (1951); A second water structure for inert gas hydrates, 19, 1425 (1951).

J. L. Cohn, G. S. Nolas, V. Fessatidis, T. H. Metcalf, and G. A. Slack, Glass-like heat conduction in high-mobility crystalline semiconductors, *Phys. Rev. Lett.* 82, 779 (1999).

J. G. Cook and M. J. Laubitz, The thermal conductivity of two clathrate hydrates, in *Proceedings of the 17th International Thermal Conductivity Conference*, p. 745. Plenum Press, New York, 1982.

C. Cros, Sur quelques nouveaux siliciures et germaniures alcalins à structure clathrate: étude cristallochimique et physique, Doctoral Thesis, University of Bordeaux, Bordeaux, France (1970).

C. Cros, M. Pouchard, and P. Hagenmuller, Sur deux nouvelles phases du systeme silicium-sodium, *C. R. Acad. Sc. Paris* 260, 4764 (1965).

C. Cros, M. Pouchard, P. Hagenmuller, and J. S. Kasper, Sur deux composes du potassium isotypes de l'hydrate de krypton, *Bull. Soc. Chim. Fr.* 7, 2737 (1968).

C. Cros, M. Pouchard, and P. Hagenmuller, Sur une nouvelle famille de clathrates mineraux isotypes des hydrates de gaz et de liquides. Interpretation des resultats obtenus, *J. Solid State Chem.* 2, 570 (1970).

C. Cros, M. Pouchard, and P. Hagenmuller, Sur deux nouvelles structures du silicium et du germanium de type clathrate, *Bull. Soc. Chim. Fr.* 2, 379 (1971).

A. Czybulka, B. Kuhl, and H.-U. Schujster, Neue ternäre Käfigverbindungen inder Systemen Barium-2B(3B)-Element-Germanium, *Z. anorg. allg. Chem.* 594, 23 (1991).

H. Davy, The elementary nature of chlorine, *Ann. Chim.* 79, 326 (1811).

A. A. Demkov, O. F. Sankey, K. E. Schmidt, G. B. Adams, and M. O'Keeffe, Theoretical investigation of alkali-metal doping in Si clathrates, *Phys. Rev. B* 50, 17001 (1995).

M. W. C. Dharma-Wardana, The thermal conductivity of the ice polymorphs and the ice clathrates, *J. Phys. Chem.* 87, 4185 (1983).

J. Dong and O. F. Sankey, Theoretical study of two expanded phases of crystalline germanium: Clathrate-I and clathrate-II, *J. Phys. Condens. Matter* 11, 6129 (1999).

J. Dong, O. F. Sankey, and G. Kern, Theoretical study of the vibrational modes and their pressure dependence in the pure clathrate-II silicon framework, *Phys. Rev. B* 60, 950 (1999).

P. Dünner, H. J. Heuvel, and M. Hörle, Absorber materials for control rod systems of fast breeder reactors, *J. Nuclear Mater.* 124, 185 (1984).

B. Einsenmann, H. Schäfer, and J. Zagler, Die Verbindungen $A_8^{II}B_{16}^{III}B_{30}^{IV}$ ($A^{II} \equiv$ Sr, Ba; $B^{III} \equiv$ Al, Ga; $B^{IV} \equiv$ Si, Ge, Sn) und ihre Käfigstrukturen, *J. Less-Common Met.* 118, 43 (1986).

J. Etourneau, J. P. Mercurio, and P. Hagenmuller, Chapter 9 in *Boron and Refractory Borides* (V. I. Matkovich, ed.). Springer Verlag, Berlin (1977).

M. Faraday, On hydrate of chlorine, *Quart. J. Sci. Lit. Arts.* 15, 71 (1823).

Franks, F. *Water, a Comprehensive Treatise*. Plenum Press, New York (1973).

J. Gallmeier, H. Schäffer, and A. Weiss, Eine Käfigstruktur als gemeinsames Bauprinzip der Verbindungen K_8E_{46} (E = Si, Ge, Sn), *Zeit. Naturforsch.* 24B, 665 (1969).

C. J. Glassbrenner and G. A. Slack, Thermal conductivity of silicon and germanium from 3 K to the melting point, *Phys. Rev.* 134, A1058 (1964).

J. E. Graebner, B. Golding, and L. C. Allen, Phonon localization in glasses, *Phys. Rev. B* 34, 5696 (1986).

J. Gryko, P. F. McMillan, R. F. Marzke, A. P. Dodokin, A. A. Demkov, and O. F. Sankey, Temperature-dependent ^{23}Na Knight shifts and sharply peaked structure in the electronic densities of state of Na–Si clathrates, *Phys. Rev. B* 57, 4172 (1998).

B. B. Iverson, A. E. C. Palmqvist, D. E. Cox, G. S. Nolas, G. D. Stucky, N. P. Blake, and H. Metiu, Why are clathrates good candidates for thermoelectric materials?, *J. Solid State Chem.* 149, 455 (1999).

D. Kahn and J. P. Lu, Structural properties and vibrational modes of Si_{46} and Si_{136} clathrates, *Phys. Rev. B* 56, 13898 (1997).

J. S. Kasper, P. Hagenmuller, M. Pouchard, and C. Cros, Clathrate structure of Silicon Na_8Si_{46} and Na_xSi_{136} ($x < 11$), *Science* 150, 1713 (1965).

H. Kawaji, H. Horie, S. Yamanaka, and M. Ishikawa, Superconductivity in the silicon clathrate compound $(Na, Ba)_xSi_{46}$, *Phys. Rev. Lett.* 74, 1427 (1995).

L. I. Kekelidze, I. A. Bairamashvili, V. N. Kovton, I. I. Petrov, and D. K. Tavartkiladze, The effect of structural defects on thermal conductivity of polycrystalline AlB_{12} and EuB_6, *Am. Inst. Phys. Conf. Proc.* 231, 371 (1991).

M. Korsukova, Vacancies and thermal vibrations of atoms in the crystal structure of rare-earth hexaborides, *Jap. J. Appl. Phys.*, Series 10, 15 (1993) (Proc. Eleventh Int. Symp. Boron and Borides).

R. Kröner, K. Peters, H. G. von Schnering, and R. Nesper, Crystal structure of the clathrates $K_8Ga_8Si_{38}$ and $K_8Ga_8Si_{38}$, *Z. Kristallogr.* 213, 667 (1998a).

R. Kröner, K. Peters, H. G. von Schnering, and R. Nesper, Crystal structure of the clathrates $Rb_8Al_8Ge_{38}$ and $Rb_8Al_8Sn_{38}$, *Z. Kristallogr.* 213, 669 (1998b).

R. Kröner, K. Peters, H. G. von Schnering, and R. Nesper, Crystal structure of the clathrates $Cs_8Ga_8Ge_{38}$ and $Cs_8Ga_8Sn_{38}$, *Z. Kristallogr.* 213, 671 (1998c).

R. Kröner, K. Peters, H. G. von Schnering, and R. Nesper, Crystal structure of the clathrates $K_8Al_8Ge_{38}$ and $K_8Al_8Sn_{38}$, *Z. Kristallogr.* 213, 675 (1998d).

B. Kuhl, A. Czybulka, and H.-U. Schuster, Neue ternäre Käfigverbindungen aus den systemen Barium-Indium/Zink/Cadmium-Germanium, *Z. Anorg. Allg. Chem.* 621, 1 (1995).

S. M. L'vov, V. F. Nemchenko, and Yu. B. Paderno, Thermal conductivity of the hexaborides of alkali metals and rare earth metals, *Doklady Akad. Nauk SSSR* 149, 1371 (1963).

Yu. M. Markov, G. N. Trokhina, E. E. Zernova, and V. V. Maksimovskii, Thermophysical properties of Y, Gd, and Sm hexaborides, *Neorgan. Mater.* 14, 79 (1978) [*Inorgan. Mater.* 14, 61 (1978)].

M. Menon, E. Richter, and K. R. Subbaswamy, Structural and vibrational properties of Si clathrates in a generalized tight-binding molecular-dynamics scheme, *Phys. Rev. B* 56, 12290 (1997).

J. P. Mercurio, J. Etourneau, R. Naslain, P. Hagenmuller, and J. B. Goodenough, Electrical and magnetic properties of solid solutions $La_xEu_{1-x}B_6$, *J. Solid State Chem.* 9, 37 (1974).

H. Misiorek, J. Mucha, A. Jezowski, Y. Paderno, and N. Shitsevalova, Thermal conductivity of rare-earth element dodecaborides, *J. Phys.—Cond. Matter* 7, 8927 (1995).

N. F. Mott, Properties of compounds of type Na_xSi_{46} and Na_xSi_{136}, *J. Solid State Chem.* 6, 348 (1973).

E. L Muetterties, *The Chemistry of Boron and Its Compounds*. Wiley, New York, (1967).

R. Nesper, K. Vogel, and P. Blochl, Hypothetical carbon modifications derived from zeolite frameworks, *Angew. Chem.* 32, 701 (1993).

G. S. Nolas, Semiconductor clathrates: A PGEC system with potential for thermoelectric applications, in *Thermoelectric Materials 1998—The Next Generation Materials for Small-Scale Refrigeration and Power Generation Applications* (T. M. Tritt, G. Mahan, H. B. Lyon, Jr., and M. G. Kanatzidis, eds.), pp. 435–442. Mater. Res. Soc. Symp. Proc., Vol. 545, Pittsburgh, PA (1999a).

G. S. Nolas, Semiconducting clathrates: A PGEC system with potential for thermoelectric applications, in *Thermoelectric Materials—The Next Generation Materials for Small-Scale Refrigeration and Power Generation Applications* (T. M. Tritt, G. Mahan, H. B. Lyon, Jr., and M. G. Kanatzidis, eds.), p. 435. Mater. Res. Soc. Symp. Proc. Vol. 545, Pittsburgh, PA (1999b).

G. S. Nolas, J. L. Cohn, G. A. Slack, and S. B. Schujman, Semiconducting Ge-clathrates: Promising candidates for thermoelectric applications, *Appl. Phys. Lett.* 73, 178 (1998).

G. S. Nolas, G. A. Slack, J. L. Cohn, and S. B. Schujman, The next generation of thermoelectric materials, in *Proceedings of the Eighteenth International Conference on Thermoelectrics*, p. 294. IEEE catalog #98TH8365, Piscataway, NJ (1999a).

G. S. Nolas, T. J. R. Weakley, and J. L. Cohn, Structural, chemical and transport properties of a new clathrate compound: $Cs_8Zn_4Sn_{42}$, *Chem. Mater.* 11, 2470 (1999b).

G. S. Nolas, D. T. Morelli, and T. M. Tritt, Skutterudites: A phonon-glass-electron crystal approach to advanced thermoelectric energy conversion applications, *Ann. Rev. Mater. Sci.* 29, 89 (1999c).

G. S. Nolas, T. J. R. Weakley, J. L. Cohn, and R. Sharma, Structural properties and thermal conductivity of crystalline Ge-clathrates, *Phys. Rev. B* 61, 3845 (2000a).

G. S. Nolas, B. C. Chakoumakos, B. Mahieu, G. J. Long, and T. J. R. Weakley, Structural characterization and thermal conductivity of type-I tin clathrates, *Chem. Mat.* (2000b).

G. S. Nolas, J. L. Cohn, B. C. Chakoumakos, and G. A. Slack, Glass-like heat conduction in crystalline semiconductors, in *Proceedings of the Twenty-fifth International Thermal Conductivity Conference* (C. Uher and D. T. Morelli, eds.), p. 122. Technomic Publishing Co., Lancaster, PA (2000c).

G. S. Nolas, J. L. Cohn, and E. Nelson, Transport properties of tin clathrates, in *Proceedings of the Nineteenth International Conference on Thermoelectrics* (2000d).

Yu. B. Paderno, E. S. Garf, T. Niemyskii, and I. Praeka, Electrophysical properties of fused hexaborides of the alkaline and rare-earth metals, *Poroshk. Metall. Acad. Nauk. Ukr. S. S. R.* 10, 55 (1969) [*Sov. Powd. Met. Met. Ceram.* 8, 821 (1969)].

L. Pauling and R. E. Marsh, The structure of chlorine hydrate, *Proc. Nat. Acad. Sci.* 38, 112 (1952).

R. O. Pohl, Thermal conductivity and phonon resonance scattering, *Phys. Rev. Lett.* 8, 481 (1962).

D. L. Price, J. M. Rowe, and R. M. Nicklow, Lattice dynamics of grey tin and indium antimonide, *Phys. Rev. B3*, 1268 (1971).

G. K. Ramachandran, P. F. McMillan, S. K. Deb, M. Somayazulu, J. Gryko, J. Dong and O. F. Sankey, High pressure phase transformation of the Guest-Free silicon clathrate Si_{136}, *J. Phys. Condens. Matter* 12, 4013 (1999a).

G. K. Ramachandran, P. F. McMillan, J. Diefenbacher, J. Gryko, J. Dong, and O. F. Sankey, ^{29}Si NMR study on the stoichiometry of the silicon clathrate Na_8Si_{46}, *Phys. Rev. B* 60, 12294 (1999b).

G. K. Ramachandran, J. Dong, J. Diefenbacher, J. Gryko, R. Marzke, O. Sankey, and P. McMillan, Synthesis and X-ray characterization of silicon clathrates, *J. Solid State Chem.* 145, 716 (1999c).

E. Reny, P. Gravereau, C. Cros, and M. Pouchard, Structural characterizations of the Na_xSi_{136} and Na_8Si_{46} silicon clathrates using the Rietveld method, *J. Mater. Chem.* 8, 2839 (1998).

R. G. Ross and P. Anderson, Clathrate and other solid phases in the tetrahydrofuran–water system: Thermal conductivity and heat capacity under pressure, *Can. J. Chem.* 60, 881 (1982).

R. G. Ross, P. Anderson, and G. Backström, Unusual PT dependence of thermal conductivity for a clathrate hydrate, *Nature (London)* 290, 322 (1981).

S. Saito and A. Oshiyama, Electronic Structure of Si_{46} and $Na_2Ba_6Si_{46}$, *Phys. Rev. B* 51, 2628 (1995).

B. C. Sales, B. C. Chakoumakos, D. Mandrus, and J. W. Sharp, Atomic displacement parameters and the lattice thermal conductivity of clathrate-like thermoelectric compounds, *J. Solid State Chem.* 146, 528 (1999).

R. Schneider, J. Geerk, and H. Rietschel, Electron tunnelling into a superconducting cluster compound: YB_6, *Europhys. Lett.* 4, 845 (1987).

S. B. Schujman and G. A. Slack, Getting rid of acoustic phonons, in *Proceedings of the Nineteenth International Conference on Thermoelectrics* (2000).

S. B. Schujman, G. S. Nolas, R. A. Young, C. Lind, A. P. Wilkinson, G. A. Slack, R. Patschke, M. G. Kanatzidis, M. Ulutagay, and S.-J. Hwu, Structural analysis of the $Sr_8Ga_{16}Ge_{30}$ clathrate compound, *J. Appl. Phys.* 87, 1529 (2000).

W. Sekkal, S. Ait Abderahmane, R. Terki, M. Certier, and H. Aourag, Molecular-dynamics simulation of carbon in the clathrate structure, *Mater. Sci. Eng.* B64, 123 (1999).

M. Sera, S. Kobayashi, M. Hiroi, and N. Kobayashi, Thermal conductivity of RB_6 single crystals, *Phys. Rev. B*54, 5207 (1996).

G. A. Slack, The thermal conductivity of nonmetallic solids, in *Solid State Physics*, Vol. 34, p. 1 (H. Ehrenreich, F. Seitz, and D. Turnbull, eds.). Academic Press, New York (1979).

G. A. Slack, Thermal conductivity of ice, *Phys. Rev. B* 22, 3065 (1980).

G. A. Slack, New materials and performance limits for thermoelectric cooling, in *CRC Handbook of Thermoelectrics* (D. M. Rowe, ed.), p. 407. CRC Press, Boca Raton, FL (1995).

G. A. Slack, Design concepts for improved thermoelectric materials, in *Thermoelectric Materials — New Directions and Approaches* (T. M. Tritt, M. G. Kanatzidis, H. B. Lyon, Jr., and G. D. Mahan, eds.), p. 47. Mat. Res. Soc. Symp. Proc. Vol. 478, Pittsburgh, PA (1997).

G. A. Slack, High thermal conductivity and low thermal conductivity solids, in *Proceedings of the Twenty-fifth International Thermal Conductivity Conference*, p. 308, Technomic Publishing Company, Lancaster, PA (2000).

G. A. Slack, D. W. Oliver, and F. H. Horn, Thermal conductivity of boron and some boron compounds, *Phys. Rev. B*4, 1714 (1971).

G. A. Slack, D. W. Oliver, G. D. Brower, and J. D. Young, Properties of melt-grown single crystals of YB_{68}, *J. Phys. Chem. Solids* 38, 45 (1977).

G. A. Slack, C. I. Hejna, M. F. Garbauskas, and J. S. Kasper, The crystal structure and density of β-rhombohedral boron, *J. Solid State Chem.* 76, 52 (1988a).

G. A. Slack, C. I. Hejna, M. F. Garbauskas, and J. S. Kasper, X-ray study of transition-metal dopants in β-boron, *J. Solid State Chem.* 76, 64 (1988b).

H. G. Smith, G. Dolling, S. Kunii, M. Kasaya, P. Liu, K. Takegahara, T. Kasuya, and T. Goto, Experimental study of lattice dynamics in LaB_6 and YbB_6, *Solid State Commun.* 53, 15 (1985).

E. Suess, G. Bohrmann, J. Greinert, and E. Lausch, Flammable ice, *Scientific American*, Nov. 1999, p. 76.

J. M. Tranquada, S. M. Heald, M. A. Pick, Z. Fisk, and J. L. Smith, Lattice dynamics of the heavy fermion compound UBe_{13}, *J. d. Phys. Colloq.* 47-C8, 937 (1986).

V. A. Trunov, A. L. Malyshev, D. Yu. Chernyshov, M. M. Korsukova, and N. N. Gurin, Thermal vibrations and static displacement of atoms in the crystal structure of Nd and Sm hexaborides, *Fiz. Tverd. Tela.* 36, 2687 (1994) [*Sov. Phys.-Solid State* 36 (1994)].

J. S. Tse, Localized oscillators and heat conduction in clathrate hydrates, *J. Inclusion Phenom.* 17, 259 (1994).

J. S. Tse and M. White, Origin of glassy crystalline behavior in the thermal properties of clathrate hydrates: a thermal conductivity study of tetrahydrofuran hydrate, *J. Phys. Chem.* 92, 5006 (1988).

C. Uher, J. Yang and S. Hu, Materials with open crystal structure as prospective novel thermoelectrics, in *Thermoelectric Materials 1998 — The Next Generation Materials for Small-Scale Refrigeration and Power Generation Applications* (T. M. Tritt, G. Mahan, H. B. Lyon, Jr., and M. G. Kanatzidis, eds.), p. 247. Mater. Res. Soc. Symp. Proc., Vol. 545, Pittsburgh, PA (1999).

A. Van Wieringen and N. Warmoltz, On the permeation of hydrogen and helium in single crystal silicon and germanium at elevated temperatures, *Physica* 22, 849 (1956).

H. G. von Schnering and H. Menke, Die partielle Substitution von Ge durch GaAs und GaSb in den Käfigverbindungen $Ge_{38}As_8J_8$ und $Ge_{38}Sb_8J_8$, *Z. anorg. allg. Chem.* 424, 108 (1978).

H. G. von Schnering, W. Carrillo-Cabrera, R. Kröner, E.-M. Peters, K. Peters, and R. Nesper, Crystal structure of the clathrates b-$Ba_8Ga_{16}Sn_{30}$, *Z. Kristallogr.* 213, 679 (1998a).

H. G. von Schnering, R. Kröner, M. Menke, K. Peters, and R. Nesper, Crystal structure of the clathrates $Rb_8Ga_8Sn_{38}$, $Rb_8Ga_8Ge_{38}$ and $Rb_8Ga_8Si_{38}$, *Z. Kristallogr.* 213, 677 (1998b).

H. G. von Schnering, H. Menke, R. Kröner, E.-M. Peters, K. Peters, and R. Nesper, Crystal Structure of the clathrates $Rb_8In_8Ge_{38}$ and $K_8In_8Ge_{38}$, *Z. Kristallogr.* 213, 673 (1998c).

M. Von Stackelberg, Solid gas hydrates, *Naturwiss.* 36, 27 (1949).

M. Von Stackelberg, O. Gotzen, J. Pietuchovsku, O. Witscher, H. Fruhbus, and W. Meinhold, Structure and formula of gas hydrates, *Fortschr. Mineral* 26, 122 (1947).

C. T. Walker and R. O. Pohl, Phonon scattering by point defects, *Phys. Rev.* 131, 1433 (1963).

H. Werheit, U. Kuhlmann, and T. Tanaka, Electronic transport and optical properties of YB_{66}, *Am. Inst. Phys. Conf. Proc.* 231, 125 (1991).

W. Westerhaus and H.-U. Schuster, Darstellung und Struktur weiterer ternärer Phasen mit modifizierter K_8Ge_{46}-Käfigstruktur, *Zeit. F. Naturforsch.* B 32, 1365 (1977).

J.-T. Zhao and J. D. Corbett, Zintl phases in alkali-metal-Tin systems: K_8Sn_{44} phases with a defect clathrate structure, *Inorg. Chem.* 33, 5721 (1994).

J. Zhao, A. Buldum, J. P. Lu, and C. Y. Fong, Structural and electronic properties of germanium clathrates Ge_{46} and K_8Ge_{46}, *Phys. Rev.* B 60, 14177 (1999).

Index

A

A–Bi–Se system, 75–82
Arc-plasma spraying, 119, 120, 130
Atomic displacement parameter (ADP), 154–156, 170, 273–278, 286, 290

B

Binary skutterudites
   magnetic susceptibility, 184–188
   structure and bonding, 141–149
   thermal conductivity, 214–219
Bipolar conduction, 13
Bismuth, 12–16
Bismuth–antimony alloys (BiSb)
   carrier mobility and density, 111–113
   crystal structure, 103–106
   electrical resistivity, 111, 120–123
   electronic structure, 106–111
   figure of merit, 127–129
   polycrystalline and powder metallurgy, synthesis of, 117–120
   polycrystalline samples, transport properties, 129–131
   research on, 101–103
   single crystals, synthesis of, 115–117
   single crystals, transport properties, 120–129
   thermal conductivity, 125–127
   thermoelectric power, 123–125
   transport properties, 120–131
β-$K_2Bi_8Se_{13}$, 74–75
Boltzmann's constant, 10, 286
Boltzmann transport theory, 53, 169
Boron/boron compounds, 261–265
Bridgman method, 14

C

Cage fillers, 164
Cage structure, clathrates and, 259–260
Carrier concentration and mobility, 28, 43–45
   of bismuth and antimony, 111–113
Chalcogenide materials, 58, 92
Chemical vapor deposition (CVD), 56
Chemical vapor transport method, 178
Clathrates, 21
   background information, 257–259
   band structure, 282–284
   crystal structure and bonding, 267–273
   doping, 290–293
   electronic resistivity, 290–294
   forms of ice, 256
   glasslike conduction, 284–285
   historic perspectives, 255–257
   methods for forming, 261
   modes of, 259–265
   origins of, 255–256
   rattling, 273–278, 286–289
   semiconducting or metallic properties, 293–294
   structural properties, 267–284
   summary and conclusions, 294–295
   synthesis methods, 265–267
   thermal conductivity, 284–290
   vibration frequencies, 278–282
Coefficient of performance (COP), 27
Comparative technique, 41

Crypto-clathrates, 259
Crystal structure
  of bismuth and antimony, 103–106
  of clathrates, 267–273
Curie law and constants, 186, 193
Curie–Weiss form, 188, 191, 193
Czochralski pulling technique, 117

D

Debye model, 222
Debye temperature, 278–282
Density functional theory, 157, 283
  plane-wave pseudopotential method, 283–284
Density of states (DOS), skutterudites, 167–170
Doping, 219–224, 290–293

E

Effective mass of carriers, 11
Einstein frequency, bare, 169–170, 171–172
Electrical conductivity, 2, 10
  thermoelectric measurements and, 31–36
Electrical contacts and contact effects, 30–31
Electrical resistivity
  of bismuth and antimony, 111, 120–123
  of bulk samples, 31–32
  of clathrates, 290–294
  of skutterudites, 196–214
  of thin disks or films, 33–36
Electron evaporation, skutterudites and, 178
Electronic structure
  of bismuth and antimony, 106–111
  of clathrates, 282–284
  of skutterudites, 156–167
Ellipsoidal nonparabolic (ENP) model, 108, 113
Ettingshausen effect, 8–9, 13, 45
Extrusion method, 14

F

Fermi–Dirac statistics, 10, 12

Figure of merit, 2, 5, 6, 7, 8, 11–12, 102
  of bismuth and antimony, 127–129
  limits on, 18–20
  two-dimensional, 21–22
Filled skutterudites
  magnetic susceptibility, 188–195
  structure and bonding, 149–156
  thermal conductivity, 229–245
Full-potential linearized augmented plane wave (FP-LAPW) method, 160

G

Germanium, 18

H

Hall coefficient, 28, 43–45, 111
Hall effect, 43
Hall factor, 44
Hall voltage, 44, 45
Harman technique (Z-meter), 45–47
High atomic weight rule, 12
Hole scattering, 128–129
Hydrothermal technique, 56–57

I

Iron disilicide, 17

J

Joule heating, 1, 30, 42

K

Kane model, two-band, 160, 161–162
Kelvin coefficient, 3
Kelvin effect, 2

L

Laser-flash thermal diffusivity method, 43
Lead and A–Bi–Se system, 75–82

Lennard–Jones potential, 282
Linearized-plane-wave (LAPW) method, 157–158, 163–164, 169
Linear muffin-tin orbital method in atomic sphere approximation (LMTO-ASA), 160, 164
Local density approximation (LDA), 157, 160, 163, 168, 282, 283, 284
Lorentz force, 44
Low-temperature solid state metathesis technique, 56, 178–179

M

Magnetic properties, skutterudites and, 180–195
Magnetic susceptibility
  binary skutterudites, 184–188
  filled skutterudites, 188–195
Mechanical alloying, 119
Melts, skutterudites from, 177–178
Molecular beam epitaxy (MBE), 56
Molten flux approach, 58–59, 178
Molten salts, 57
Montgomery technique, 36
Mössbauer spectroscopy, 181–184
Mott equation, 53, 55
Multiple well model, split-site or off-center, 285–290

N

National Institute of Standards and Technology (NIST), 29
Nernst effect, 8–9, 13, 45
Neutron scattering studies, inelastic, 173

O

Oftedal relation, 143

P

Peltier coefficient, 2–3, 27
Peltier effect, 1–2, 6–7, 27
  resistivity measurements and, 32
Phonon drag, 36
Phonon glass and electron crystal (PGEC), 20, 52, 140, 229–230, 259
Phonon modes and density of states, skutterudites, 167–170
Phonon–phonon Umklapp scattering, 216, 222
Planck's constant, 11, 286
Plane-wave technique, 283
Powder metallurgy, 14, 118–119
Power factor, 8, 19, 22, 27
Pulsed laser deposition, 178
Pyrex, 41
Pyroceram, 41

Q

Quantum wells, 54–55

R

Raman scattering studies, 171–173
Rattling, 57, 58
  clathrates, 273–278, 286–289
  skutterudites, 155, 170–174
Rayleigh scattering, 219, 286
Refrigeration, Peltier effect and, 6–7
Ridgi–Leduc effect, 45
Rietveld-refined neutron diffraction, 276–277

S

Seebeck coefficient, 2, 3, 27
  bismuth telluride and, 14–15
  in semiconductor thermoelements, 9–13
  thermoelectric measurements and, 36–39
Seebeck effect, 1–2
Selenides, 64–66
  doped samples, 70–73
  undoped samples, 67–70
Selenium, bismuth, 13–14
Semiconductor thermoelements, 9–13, 55
Shubnikov–de Haas oscillation amplitude, 162, 164
Silicon, 18
Sintering, 14, 179

Skutterudites, 21, 44
  band structure, 156–167
  binary, magnetic susceptibility, 184–188
  binary, structure and bonding, 141–149
  binary, thermal conductivity, 214–219
  conclusions, 246–247
  doping, effects of, 219–224
  electronic resistivity, 196–214
  filled, magnetic susceptibility, 188–195
  filled, structure and bonding, 149–156
  filled, thermal conductivity, 229–245
  magnetic properties, 180–195
  Mössbauer spectroscopy, 181–184
  origins and research on, 139–140
  phonon modes and density of states, 167–170
  rattling, 155, 170–174
  sample preparation aspects, 174–180
  solid solutions, 224–226
  ternary, 226–229
  thermal conductivity, 214–245
  transport properties, 196–245
  vibrational properties, 167–174
Solid solutions, 224–226
Solid-state chemistry, importance of, 55–57
Solvothermal technique, 56–57
Spin casting, 179
SQUID magnetometers, 180–181
Standards and samples, for thermoelectric measurements, 29–30
Steady-state method, 41
Stephan–Boltzmann constant, 41
Sulfides, 61–64

T

TAGS formulation, 18
Telluride systems
  antimony, 13–14
  $APb_2Bi_3Te_7$, 87–92
  bismuth, 12–16
  $CsBi_4Te_6$, 82–87
  lead, 17–18
  $RbBi_{3.66}Te_6$, 87–92
  tin, 18
Ternary skutterudites, 226–229
Thermal conductivity, 2, 11, 14
  of binary skutterudites, 214–219
  of bismuth and antimony, 125–127
  of clathrates, 284–290
  of filled skutterudites, 229–245
  lattice, 19, 20
  measurements, 39–43
Thermal parameter, 154
Thermal radiation, 22
Thermocouples, efficiency of, 4–5
Thermoelectric coefficients, 2–3
Thermoelectric effects
  development and applications of, 1–9
  limits on figure of merit, 18–20
  materials, 13–18
  new developments, 20–22
Thermoelectric materials
  See also under type of
  criteria for, 54–55
  description of, 13–18, 27–28
  discovery of, 57–59
  effects of errors in calculating, 28–29
  lead and A–Bi–Se system, 75–82
  outlook on, 96–98
  properties of, 28
  related to β-$K_2Bi_8Se_{13}$, 74–75
  research on, 25–27
  selenides, 64–73
  solid-state chemistry, importance of, 55–57
  sulfides, 61–64
  telluride systems, 82–92
  Zintl phases, 92–96
Thermoelectric materials, measurements of
  carrier concentration and mobility, 28, 43–45
  electrical conductivity, 31–36
  electrical contacts and contact effects, 30–31
  Hall coefficient, 28, 43–45
  resistivity of bulk samples, 31–32
  resistivity of thin disks or films, 33–36
  Seebeck coefficient, 36–39
  standards and samples, 29–30
  summary of, 47
  thermal conductivity, 39–43
  Z-meter (Harman technique), 45–47
Thermoelectric power, of bismuth and antimony, 123–125
Thermoelements, semiconductor, 9–13
Thomson coefficient, 3

Thomson effect, 2
3-ω technique, 41–43
Tight-binding molecular-dynamics, 283
Tight-bonding density-matrix method, 283
Transport properties
  of bismuth and antimony, 120–131
  of skutterudites, 196–245
Traveling heater method (THM), 116–117
Tunnel system (TS), 286–288

V

Vacuum diodes, 22
van der Pauw technique, 34–35, 45
van der Waals bonds, 14
Vergard's rule, 105–106
Vibrational properties, skutterudites and, 167–174

W

Wiedemann–Franz (WF) law, 9, 10, 52, 214, 284
Wilson–Sommerfeld ratio, 194

X

X-ray absorption near-edge structure (XANES), 165–167

Z

Zintl phases, 92–96
Z-meter (Harman technique), 45–47
Zone melting pass and multipass, 115–117
Zone-melted rods, grinding, 119, 130

Contents of Volumes in This Series

Volume 1 Physics of III–V Compounds

C. Hilsum, Some Key Features of III–V Compounds
F. Bassani, Methods of Band Calculations Applicable to III–V Compounds
E. O. Kane, The k-p Method
V. L. Bonch-Bruevich, Effect of Heavy Doping on the Semiconductor Band Structure
D. Long, Energy Band Structures of Mixed Crystals of III–V Compounds
L. M. Roth and P. N. Argyres, Magnetic Quantum Effects
S. M. Puri and T. H. Geballe, Thermomagnetic Effects in the Quantum Region
W. M. Becker, Band Characteristics near Principal Minima from Magnetoresistance
E. H. Putley, Freeze-Out Effects, Hot Electron Effects, and Submillimeter Photoconductivity in InSb
H. Weiss, Magnetoresistance
B. Ancker-Johnson, Plasma in Semiconductors and Semimetals

Volume 2 Physics of III–V Compounds

M. G. Holland, Thermal Conductivity
S. I. Novkova, Thermal Expansion
U. Piesbergen, Heat Capacity and Debye Temperatures
G. Giesecke, Lattice Constants
J. R. Drabble, Elastic Properties
A. U. Mac Rae and G. W. Gobeli, Low Energy Electron Diffraction Studies
R. Lee Mieher, Nuclear Magnetic Resonance
B. Goldstein, Electron Paramagnetic Resonance
T. S. Moss, Photoconduction in III–V Compounds
E. Antoncik and J. Tauc, Quantum Efficiency of the Internal Photoelectric Effect in InSb
G. W. Gobeli and I. G. Allen, Photoelectric Threshold and Work Function
P. S. Pershan, Nonlinear Optics in III–V Compounds
M. Gershenzon, Radiative Recombination in the III–V Compounds
F. Stern, Stimulated Emission in Semiconductors

Volume 3 Optical of Properties III–V Compounds

M. Hass, Lattice Reflection
W. G. Spitzer, Multiphonon Lattice Absorption
D. L. Stierwalt and R. F. Potter, Emittance Studies
H. R. Philipp and H. Ehrenveich, Ultraviolet Optical Properties
M. Cardona, Optical Absorption above the Fundamental Edge
E. J. Johnson, Absorption near the Fundamental Edge
J. O. Dimmock, Introduction to the Theory of Exciton States in Semiconductors
B. Lax and J. G. Mavroides, Interband Magnetooptical Effects
H. Y. Fan, Effects of Free Carries on Optical Properties
E. D. Palik and G. B. Wright, Free-Carrier Magnetooptical Effects
R. H. Bube, Photoelectronic Analysis
B. O. Seraphin and H. E. Bennett, Optical Constants

Volume 4 Physics of III–V Compounds

N. A. Goryunova, A. S. Borschevskii, and D. N. Tretiakov, Hardness
N. N. Sirota, Heats of Formation and Temperatures and Heats of Fusion of Compounds $A^{III}B^{V}$
D. L. Kendall, Diffusion
A. G. Chynoweth, Charge Multiplication Phenomena
R. W. Keyes, The Effects of Hydrostatic Pressure on the Properties of III–V Semiconductors
L. W. Aukerman, Radiation Effects
N. A. Goryunova, F. P. Kesamanly, and D. N. Nasledov, Phenomena in Solid Solutions
R. T. Bate, Electrical Properties of Nonuniform Crystals

Volume 5 Infrared Detectors

H. Levinstein, Characterization of Infrared Detectors
P. W. Kruse, Indium Antimonide Photoconductive and Photoelectromagnetic Detectors
M. B. Prince, Narrowband Self-Filtering Detectors
I. Melngalis and T. C. Harman, Single-Crystal Lead-Tin Chalcogenides
D. Long and J. L. Schmidt, Mercury-Cadmium Telluride and Closely Related Alloys
E. H. Putley, The Pyroelectric Detector
N. B. Stevens, Radiation Thermopiles
R. J. Keyes and T. M. Quist, Low Level Coherent and Incoherent Detection in the Infrared
M. C. Teich, Coherent Detection in the Infrared
F. R. Arams, E. W. Sard, B. J. Peyton, and F. P. Pace, Infrared Heterodyne Detection with Gigahertz IF Response
H. S. Sommers, Jr., Macrowave-Based Photoconductive Detector
R. Sehr and R. Zuleeg, Imaging and Display

Volume 6 Injection Phenomena

M. A. Lampert and R. B. Schilling, Current Injection in Solids: The Regional Approximation Method
R. Williams, Injection by Internal Photoemission
A. M. Barnett, Current Filament Formation

R. Baron and J. W. Mayer, Double Injection in Semiconductors
W. Ruppel, The Photoconductor-Metal Contact

Volume 7 Application and Devices
Part A

J. A. Copeland and S. Knight, Applications Utilizing Bulk Negative Resistance
F. A. Padovani, The Voltage-Current Characteristics of Metal-Semiconductor Contacts
P. L. Hower, W. W. Hooper, B. R. Cairns, R. D. Fairman, and D. A. Tremere, The GaAs Field-Effect Transistor
M. H. White, MOS Transistors
G. R. Antell, Gallium Arsenide Transistors
T. L. Tansley, Heterojunction Properties

Part B

T. Misawa, IMPATT Diodes
H. C. Okean, Tunnel Diodes
R. B. Campbell and Hung-Chi Chang, Silicon Junction Carbide Devices
R. E. Enstrom, H. Kressel, and L. Krassner, High-Temperature Power Rectifiers of $GaAs_{1-x}P_x$

Volume 8 Transport and Optical Phenomena

R. J. Stirn, Band Structure and Galvanomagnetic Effects in III–V Compounds with Indirect Band Gaps
R. W. Ure, Jr., Thermoelectric Effects in III–V Compounds
H. Piller, Faraday Rotation
H. Barry Bebb and E. W. Williams, Photoluminescence I: Theory
E. W. Williams and H. Barry Bebb, Photoluminescence II: Gallium Arsenide

Volume 9 Modulation Techniques

B. O. Seraphin, Electroreflectance
R. L. Aggarwal, Modulated Interband Magnetooptics
D. F. Blossey and Paul Handler, Electroabsorption
B. Batz, Thermal and Wavelength Modulation Spectroscopy
I. Balslev, Piezopptical Effects
D. E. Aspnes and N. Bottka, Electric-Field Effects on the Dielectric Function of Semiconductors and Insulators

Volume 10 Transport Phenomena

R. L. Rhode, Low-Field Electron Transport
J. D. Wiley, Mobility of Holes in III–V Compounds
C. M. Wolfe and G. E. Stillman, Apparent Mobility Enhancement in Inhomogeneous Crystals
R. L. Petersen, The Magnetophonon Effect

Volume 11 Solar Cells

H. J. Hovel, Introduction; Carrier Collection, Spectral Response, and Photocurrent; Solar Cell Electrical Characteristics; Efficiency; Thickness; Other Solar Cell Devices; Radiation Effects; Temperature and Intensity; Solar Cell Technology

Volume 12 Infrared Detectors (II)

W. L. Eiseman, J. D. Merriam, and R. F. Potter, Operational Characteristics of Infrared Photodetectors
P. R. Bratt, Impurity Germanium and Silicon Infrared Detectors
E. H. Putley, InSb Submillimeter Photoconductive Detectors
G. E. Stillman, C. M. Wolfe, and J. O. Dimmock, Far-Infrared Photoconductivity in High Purity GaAs
G. E. Stillman and C. M. Wolfe, Avalanche Photodiodes
P. L. Richards, The Josephson Junction as a Detector of Microwave and Far-Infrared Radiation
E. H. Putley, The Pyroelectric Detector — An Update

Volume 13 Cadmium Telluride

K. Zanio, Materials Preparations; Physics; Defects; Applications

Volume 14 Lasers, Junctions, Transport

N. Holonyak, Jr. and M. H. Lee, Photopumped III–V Semiconductor Lasers
H. Kressel and J. K. Butler, Heterojunction Laser Diodes
A Van der Ziel, Space-Charge-Limited Solid-State Diodes
P. J. Price, Monte Carlo Calculation of Electron Transport in Solids

Volume 15 Contacts, Junctions, Emitters

B. L. Sharma, Ohmic Contacts to III–V Compounds Semiconductors
A. Nussbaum, The Theory of Semiconducting Junctions
J. S. Escher, NEA Semiconductor Photoemitters

Volume 16 Defects, (HgCd)Se, (HgCd)Te

H. Kressel, The Effect of Crystal Defects on Optoelectronic Devices
C. R. Whitsett, J. G. Broerman, and C. J. Summers, Crystal Growth and Properties of $Hg_{1-x}Cd_xSe$ alloys
M. H. Weiler, Magnetooptical Properties of $Hg_{1-x}Cd_xTe$ Alloys
P. W. Kruse and J. G. Ready, Nonlinear Optical Effects in $Hg_{1-x}Cd_xTe$

Volume 17 CW Processing of Silicon and Other Semiconductors

J. F. Gibbons, Beam Processing of Silicon
A. Lietoila, R. B. Gold, J. F. Gibbons, and L. A. Christel, Temperature Distributions and Solid Phase Reaction Rates Produced by Scanning CW Beams

A. Leitoila and J. F. Gibbons, Applications of CW Beam Processing to Ion Implanted Crystalline Silicon
N. M. Johnson, Electronic Defects in CW Transient Thermal Processed Silicon
K. F. Lee, T. J. Stultz, and J. F. Gibbons, Beam Recrystallized Polycrystalline Silicon: Properties, Applications, and Techniques
T. Shibata, A. Wakita, T. W. Sigmon, and J. F. Gibbons, Metal-Silicon Reactions and Silicide
Y. I. Nissim and J. F. Gibbons, CW Beam Processing of Gallium Arsenide

Volume 18 Mercury Cadmium Telluride

P. W. Kruse, The Emergence of $(Hg_{1-x}Cd_x)Te$ as a Modern Infrared Sensitive Material
H. E. Hirsch, S. C. Liang, and A. G. White, Preparation of High-Purity Cadmium, Mercury, and Tellurium
W. F. H. Micklethwaite, The Crystal Growth of Cadmium Mercury Telluride
P. E. Petersen, Auger Recombination in Mercury Cadmium Telluride
R. M. Broudy and V. J. Mazurczyck, (HgCd)Te Photoconductive Detectors
M. B. Reine, A. K. Soad, and T. J. Tredwell, Photovoltaic Infrared Detectors
M. A. Kinch, Metal-Insulator-Semiconductor Infrared Detectors

Volume 19 Deep Levels, GaAs, Alloys, Photochemistry

G. F. Neumark and K. Kosai, Deep Levels in Wide Band-Gap III–V Semiconductors
D. C. Look, The Electrical and Photoelectronic Properties of Semi-Insulating GaAs
R. F. Brebrick, Ching-Hua Su, and Pok-Kai Liao, Associated Solution Model for Ga-In-Sb and Hg-Cd-Te
Y. Ya. Gurevich and Y. V. Pleskon, Photoelectrochemistry of Semiconductors

Volume 20 Semi-Insulating GaAs

R. N. Thomas, H. M. Hobgood, G. W. Eldridge, D. L. Barrett, T. T. Braggins, L. B. Ta, and S. K. Wang, High-Purity LEC Growth and Direct Implantation of GaAs for Monolithic Microwave Circuits
C. A. Stolte, Ion Implantation and Materials for GaAs Integrated Circuits
C. G. Kirkpatrick, R. T. Chen, D. E. Holmes, P. M. Asbeck, K. R. Elliott, R. D. Fairman, and J. R. Oliver, LEC GaAs for Integrated Circuit Applications
J. S. Blakemore and S. Rahimi, Models for Mid-Gap Centers in Gallium Arsenide

Volume 21 Hydrogenated Amorphous Silicon
Part A

J. I. Pankove, Introduction
M. Hirose, Glow Discharge; Chemical Vapor Deposition
Y. Uchida, di Glow Discharge
T. D. Moustakas, Sputtering
I. Yamada, Ionized-Cluster Beam Deposition
B. A. Scott, Homogeneous Chemical Vapor Deposition

F. J. Kampas, Chemical Reactions in Plasma Deposition
P. A. Longeway, Plasma Kinetics
H. A. Weakliem, Diagnostics of Silane Glow Discharges Using Probes and Mass Spectroscopy
L. Gluttman, Relation between the Atomic and the Electronic Structures
A. Chenevas-Paule, Experiment Determination of Structure
S. Minomura, Pressure Effects on the Local Atomic Structure
D. Adler, Defects and Density of Localized States

Part B

J. I. Pankove, Introduction
G. D. Cody, The Optical Absorption Edge of a-Si:H
N. M. Amer and W. B. Jackson, Optical Properties of Defect States in a-Si:H
P. J. Zanzucchi, The Vibrational Spectra of a-Si:H
Y. Hamakawa, Electroreflectance and Electroabsorption
J. S. Lannin, Raman Scattering of Amorphous Si, Ge, and Their Alloys
R. A. Street, Luminescence in a-Si:H
R. S. Crandall, Photoconductivity
J. Tauc, Time-Resolved Spectroscopy of Electronic Relaxation Processes
P. E. Vanier, IR-Induced Quenching and Enhancement of Photoconductivity and Photo luminescence
H. Schade, Irradiation-Induced Metastable Effects
L. Ley, Photoelectron Emission Studies

Part C

J. I. Pankove, Introduction
J. D. Cohen, Density of States from Junction Measurements in Hydrogenated Amorphous Silicon
P. C. Taylor, Magnetic Resonance Measurements in a-Si:H
K. Morigaki, Optically Detected Magnetic Resonance
J. Dresner, Carrier Mobility in a-Si:H
T. Tiedje, Information about band-Tail States from Time-of-Flight Experiments
A. R. Moore, Diffusion Length in Undoped a-Si:H
W. Beyer and J. Overhof, Doping Effects in a-Si:H
H. Fritzche, Electronic Properties of Surfaces in a-Si:H
C. R. Wronski, The Staebler-Wronski Effect
R. J. Nemanich, Schottky Barriers on a-Si:H
B. Abeles and T. Tiedje, Amorphous Semiconductor Superlattices

Part D

J. I. Pankove, Introduction
D. E. Carlson, Solar Cells
G. A. Swartz, Closed-Form Solution of I–V Characteristic for a a-Si:H Solar Cells
I. Shimizu, Electrophotography
S. Ishioka, Image Pickup Tubes

P. G. LeComber and W. E. Spear, The Development of the a-Si:H Field-Effect Transistor and Its Possible Applications
D. G. Ast, a-Si:H FET-Addressed LCD Panel
S. Kaneko, Solid-State Image Sensor
M. Matsumura, Charge-Coupled Devices
M. A. Bosch, Optical Recording
A. D'Amico and G. Fortunato, Ambient Sensors
H. Kukimoto, Amorphous Light-Emitting Devices
R. J. Phelan, Jr., Fast Detectors and Modulators
J. I. Pankove, Hybrid Structures
P. G. LeComber, A. E. Owen, W. E. Spear, J. Hajto, and W. K. Choi, Electronic Switching in Amorphous Silicon Junction Devices

Volume 22 Lightwave Communications Technology
Part A

K. Nakajima, The Liquid-Phase Epitaxial Growth of InGaAsP
W. T. Tsang, Molecular Beam Epitaxy for III–V Compound Semiconductors
G. B. Stringfellow, Organometallic Vapor-Phase Epitaxial Growth of III–V Semiconductors
G. Beuchet, Halide and Chloride Transport Vapor-Phase Deposition of InGaAsP and GaAs
M. Razeghi, Low-Pressure Metallo-Organic Chemical Vapor Deposition of $Ga_xIn_{1-x}As P_{1-y}$ Alloys
P. M. Petroff, Defects in III–V Compound Semiconductors

Part B

J. P. van der Ziel, Mode Locking of Semiconductor Lasers
K. Y. Lau and A. Yariv, High-Frequency Current Modulation of Semiconductor Injection Lasers
C. H. Henry, Special Properties of Semiconductor Lasers
Y. Suematsu, K. Kishino, S. Arai, and F. Koyama, Dynamic Single-Mode Semiconductor Lasers with a Distributed Reflector
W. T. Tsang, The Cleaved-Coupled-Cavity (C^3) Laser

Part C

R. J. Nelson and N. K. Dutta, Review of InGaAsP InP Laser Structures and Comparison of Their Performance
N. Chinone and M. Nakamura, Mode-Stabilized Semiconductor Lasers for 0.7–0.8- and 1.1–1.6-μm Regions
Y. Horikoshi, Semiconductor Lasers with Wavelengths Exceeding 2 μm
B. A. Dean and M. Dixon, The Functional Reliability of Semiconductor Lasers as Optical Transmitters
R. H. Saul, T. P. Lee, and C. A. Burus, Light-Emitting Device Design
C. L. Zipfel, Light-Emitting Diode-Reliability
T. P. Lee and T. Li, LED-Based Multimode Lightwave Systems
K. Ogawa, Semiconductor Noise-Mode Partition Noise

Part D

F. *Capasso*, The Physics of Avalanche Photodiodes
T. P. *Pearsall and M. A. Pollack*, Compound Semiconductor Photodiodes
T. *Kaneda*, Silicon and Germanium Avalanche Photodiodes
S. R. *Forrest*, Sensitivity of Avalanche Photodetector Receivers for High-Bit-Rate Long-Wavelength Optical Communication Systems
J. C. *Campbell*, Phototransistors for Lightwave Communications

Part E

S. *Wang*, Principles and Characteristics of Integrable Active and Passive Optical Devices
S. *Margalit and A. Yariv*, Integrated Electronic and Photonic Devices
T. *Mukai, Y. Yamamoto, and T. Kimura*, Optical Amplification by Semiconductor Lasers

Volume 23 Pulsed Laser Processing of Semiconductors

R. F. *Wood, C. W. White, and R. T. Young*, Laser Processing of Semiconductors: An Overview
C. W. *White*, Segregation, Solute Trapping, and Supersaturated Alloys
G. E. *Jellison, Jr.*, Optical and Electrical Properties of Pulsed Laser-Annealed Silicon
R. F. *Wood and G. E. Jellison, Jr.*, Melting Model of Pulsed Laser Processing
R. F. *Wood and F. W. Young, Jr.*, Nonequilibrium Solidification Following Pulsed Laser Melting
D. H. *Lowndes and G. E. Jellison, Jr.*, Time-Resolved Measurement During Pulsed Laser Irradiation of Silicon
D. M. *Zebner*, Surface Studies of Pulsed Laser Irradiated Semiconductors
D. H. *Lowndes*, Pulsed Beam Processing of Gallium Arsenide
R. B. *James*, Pulsed CO_2 Laser Annealing of Semiconductors
R. T. *Young and R. F. Wood*, Applications of Pulsed Laser Processing

Volume 24 Applications of Multiquantum Wells, Selective Doping, and Superlattices

C. *Weisbuch*, Fundamental Properties of III–V Semiconductor Two-Dimensional Quantized Structures: The Basis for Optical and Electronic Device Applications
H. *Morkoc and H. Unlu*, Factors Affecting the Performance of (Al, Ga)As/GaAs and (Al, Ga)As/InGaAs Modulation-Doped Field-Effect Transistors: Microwave and Digital Applications
N. T. *Linh*, Two-Dimensional Electron Gas FETs: Microwave Applications
M. *Abe et al.*, Ultra-High-Speed HEMT Integrated Circuits
D. S. *Chemla, D. A. B. Miller, and P. W. Smith*, Nonlinear Optical Properties of Multiple Quantum Well Structures for Optical Signal Processing
F. *Capasso*, Graded-Gap and Superlattice Devices by Band-Gap Engineering
W. T. *Tsang*, Quantum Confinement Heterostructure Semiconductor Lasers
G. C. *Osbourn et al.*, Principles and Applications of Semiconductor Strained-Layer Superlattices

Volume 25 Diluted Magnetic Semiconductors

W. Giriat and J. K. Furdyna, Crystal Structure, Composition, and Materials Preparation of Diluted Magnetic Semiconductors
W. M. Becker, Band Structure and Optical Properties of Wide-Gap $A^{II}_{1-x}Mn_xB_{IV}$ Alloys at Zero Magnetic Field
S. Oseroff and P. H. Keesom, Magnetic Properties: Macroscopic Studies
T. Giebultowicz and T. M. Holden, Neutron Scattering Studies of the Magnetic Structure and Dynamics of Diluted Magnetic Semiconductors
J. Kossut, Band Structure and Quantum Transport Phenomena in Narrow-Gap Diluted Magnetic Semiconductors
C. Riquaux, Magnetooptical Properties of Large-Gap Diluted Magnetic Semiconductors
J. A. Gaj, Magnetooptical Properties of Large-Gap Diluted Magnetic Semiconductors
J. Mycielski, Shallow Acceptors in Diluted Magnetic Semiconductors: Splitting, Boil-off, Giant Negative Magnetoresistance
A. K. Ramadas and R. Rodriquez, Raman Scattering in Diluted Magnetic Semiconductors
P. A. Wolff, Theory of Bound Magnetic Polarons in Semimagnetic Semiconductors

Volume 26 III–V Compound Semiconductors and Semiconductor Properties of Superionic Materials

Z. Yuanxi, III–V Compounds
H. V. Winston, A. T. Hunter, H. Kimura, and R. E. Lee, InAs-Alloyed GaAs Substrates for Direct Implantation
P. K. Bhattacharya and S. Dhar, Deep Levels in III–V Compound Semiconductors Grown by MBE
Y. Ya. Gurevich and A. K. Ivanov-Shits, Semiconductor Properties of Supersonic Materials

Volume 27 High Conducting Quasi-One-Dimensional Organic Crystals

E. M. Conwell, Introduction to Highly Conducting Quasi-One-Dimensional Organic Crystals
I. A. Howard, A Reference Guide to the Conducting Quasi-One-Dimensional Organic Molecular Crystals
J. P. Pouquet, Structural Instabilities
E. M. Conwell, Transport Properties
C. S. Jacobsen, Optical Properties
J. C. Scott, Magnetic Properties
L. Zuppiroli, Irradiation Effects: Perfect Crystals and Real Crystals

Volume 28 Measurement of High-Speed Signals in Solid State Devices

J. Frey and D. Ioannou, Materials and Devices for High-Speed and Optoelectronic Applications
H. Schumacher and E. Strid, Electronic Wafer Probing Techniques
D. H. Auston, Picosecond Photoconductivity: High-Speed Measurements of Devices and Materials
J. A. Valdmanis, Electro-Optic Measurement Techniques for Picosecond Materials, Devices, and Integrated Circuits.
J. M. Wiesenfeld and R. K. Jain, Direct Optical Probing of Integrated Circuits and High-Speed Devices
G. Plows, Electron-Beam Probing
A. M. Weiner and R. B. Marcus, Photoemissive Probing

Volume 29 Very High Speed Integrated Circuits: Gallium Arsenide LSI

M. Kuzuhara and T. Nazaki, Active Layer Formation by Ion Implantation
H. Hasimoto, Focused Ion Beam Implantation Technology
T. Nozaki and A. Higashisaka, Device Fabrication Process Technology
M. Ino and T. Takada, GaAs LSI Circuit Design
M. Hirayama, M. Ohmori, and K. Yamasaki, GaAs LSI Fabrication and Performance

Volume 30 Very High Speed Integrated Circuits: Heterostructure

H. Watanabe, T. Mizutani, and A. Usui, Fundamentals of Epitaxial Growth and Atomic Layer Epitaxy
S. Hiyamizu, Characteristics of Two-Dimensional Electron Gas in III–V Compound Heterostructures Grown by MBE
T. Nakanisi, Metalorganic Vapor Phase Epitaxy for High-Quality Active Layers
T. Nimura, High Electron Mobility Transistor and LSI Applications
T. Sugeta and T. Ishibashi, Hetero-Bipolar Transistor and LSI Application
H. Matsueda, T. Tanaka, and M. Nakamura, Optoelectronic Integrated Circuits

Volume 31 Indium Phosphide: Crystal Growth and Characterization

J. P. Farges, Growth of Discoloration-free InP
M. J. McCollum and G. E. Stillman, High Purity InP Grown by Hydride Vapor Phase Epitaxy
T. Inada and T. Fukuda, Direct Synthesis and Growth of Indium Phosphide by the Liquid Phosphorous Encapsulated Czochralski Method
O. Oda, K. Katagiri, K. Shinohara, S. Katsura, Y. Takahashi, K. Kainosho, K. Kohiro, and R. Hirano, InP Crystal Growth, Substrate Preparation and Evaluation
K. Tada, M. Tatsumi, M. Morioka, T. Araki, and T. Kawase, InP Substrates: Production and Quality Control
M. Razeghi, LP-MOCVD Growth, Characterization, and Application of InP Material
T. A. Kennedy and P. J. Lin-Chung, Stoichiometric Defects in InP

Volme 32 Strained-Layer Superlattices: Physics

T. P. Pearsall, Strained-Layer Superlattices
F. H. Pollack, Effects of Homogeneous Strain on the Electronic and Vibrational Levels in Semiconductors
J. Y. Marzin, J. M. Gerárd, P. Voisin, and J. A. Brum, Optical Studies of Strained III–V Heterolayers
R. People and S. A. Jackson, Structurally Induced States from Strain and Confinement
M. Jaros, Microscopic Phenomena in Ordered Superlattices

Volume 33 Strained-Layer Superlattices: Materials Science and Technology

R. Hull and J. C. Bean, Principles and Concepts of Strained-Layer Epitaxy
W. J. Schaff, P. J. Tasker, M. C. Foisy, and L. F. Eastman, Device Applications of Strained-Layer Epitaxy

S. T. Picraux, B. L. Doyle, and J. Y. Tsao, Structure and Characterization of Strained-Layer Superlattices
E. Kasper and F. Schaffer, Group IV Compounds
D. L. Martin, Molecular Beam Epitaxy of IV–VI Compounds Heterojunction
R. L. Gunshor, L. A. Kolodziejski, A. V. Nurmikko, and N. Otsuka, Molecular Beam Epitaxy of II–VI Semiconductor Microstructures

Volume 34 Hydrogen in Semiconductors

J. I. Pankove and N. M. Johnson, Introduction to Hydrogen in Semiconductors
C. H. Seager, Hydrogenation Methods
J. I. Pankove, Hydrogenation of Defects in Crystalline Silicon
J. W. Corbett, P. Deák, U. V. Desnica, and S. J. Pearton, Hydrogen Passivation of Damage Centers in Semiconductors
S. J. Pearton, Neutralization of Deep Levels in Silicon
J. I. Pankove, Neutralization of Shallow Acceptors in Silicon
N. M. Johnson, Neutralization of Donor Dopants and Formation of Hydrogen-Induced Defects in n-Type Silicon
M. Stavola and S. J. Pearton, Vibrational Spectroscopy of Hydrogen-Related Defects in Silicon
A. D. Marwick, Hydrogen in Semiconductors: Ion Beam Techniques
C. Herring and N. M. Johnson, Hydrogen Migration and Solubility in Silicon
E. E. Haller, Hydrogen-Related Phenomena in Crystalline Germanium
J. Kakalios, Hydrogen Diffusion in Amorphous Silicon
J. Chevalier, B. Clerjaud, and B. Pajot, Neutralization of Defects and Dopants in III–V Semiconductors
G. G. DeLeo and W. B. Fowler, Computational Studies of Hydrogen-Containing Complexes in Semiconductors
R. F. Kiefl and T. L. Estle, Muonium in Semiconductors
C. G. Van de Walle, Theory of Isolated Interstitial Hydrogen and Muonium in Crystalline Semiconductors

Volume 35 Nanostructured Systems

M. Reed, Introduction
H. van Houten, C. W. J. Beenakker, and B. J. van Wees, Quantum Point Contacts
G. Timp, When Does a Wire Become an Electron Waveguide?
M. Büttiker, The Quantum Hall Effects in Open Conductors
W. Hansen, J. P. Kotthaus, and U. Merkt, Electrons in Laterally Periodic Nanostructures

Volume 36 The Spectroscopy of Semiconductors

D. Heiman, Spectroscopy of Semiconductors at Low Temperatures and High Magnetic Fields
A. V. Nurmikko, Transient Spectroscopy by Ultrashort Laser Pulse Techniques
A. K. Ramdas and S. Rodriguez, Piezospectroscopy of Semiconductors
O. J. Glembocki and B. V. Shanabrook, Photoreflectance Spectroscopy of Microstructures
D. G. Seiler, C. L. Littler, and M. H. Wiler, One- and Two-Photon Magneto-Optical Spectroscopy of InSb and $Hg_{1-x}Cd_xTe$

Volume 37 The Mechanical Properties of Semiconductors

A.-B. Chen, A. Sher and W. T. Yost, Elastic Constants and Related Properties of Semiconductor Compounds and Their Alloys
D. R. Clarke, Fracture of Silicon and Other Semiconductors
H. Siethoff, The Plasticity of Elemental and Compound Semiconductors
S. Guruswamy, K. T. Faber and J. P. Hirth, Mechanical Behavior of Compound Semiconductors
S. Mahajan, Deformation Behavior of Compound Semiconductors
J. P. Hirth, Injection of Dislocations into Strained Multilayer Structures
D. Kendall, C. B. Fleddermann, and K. J. Malloy, Critical Technologies for the Micromachining of Silicon
I. Matsuba and K. Mokuya, Processing and Semiconductor Thermoelastic Behavior

Volume 38 Imperfections in III/V Materials

U. Scherz and M. Scheffler, Density-Functional Theory of sp-Bonded Defects in III/V Semiconductors
M. Kaminska and E. R. Weber, El2 Defect in GaAs
D. C. Look, Defects Relevant for Compensation in Semi-Insulating GaAs
R. C. Newman, Local Vibrational Mode Spectroscopy of Defects in III/V Compounds
A. M. Hennel, Transition Metals in III/V Compounds
K. J. Malloy and K. Khachaturyan, DX and Related Defects in Semiconductors
V. Swaminathan and A. S. Jordan, Dislocations in III/V Compounds
K. W. Nauka, Deep Level Defects in the Epitaxial III/V Materials

Volume 39 Minority Carriers in III–V Semiconductors: Physics and Applications

N. K. Dutta, Radiative Transitions in GaAs and Other III–V Compounds
R. K. Ahrenkiel, Minority-Carrier Lifetime in III–V Semiconductors
T. Furuta, High Field Minority Electron Transport in p-GaAs
M. S. Lundstrom, Minority-Carrier Transport in III–V Semiconductors
R. A. Abram, Effects of Heavy Doping and High Excitation on the Band Structure of GaAs
D. Yevick and W. Bardyszewski, An Introduction to Non-Equilibrium Many-Body Analyses of Optical Processes in III–V Semiconductors

Volume 40 Epitaxial Microstructures

E. F. Schubert, Delta-Doping of Semiconductors: Electronic, Optical, and Structural Properties of Materials and Devices
A. Gossard, M. Sundaram, and P. Hopkins, Wide Graded Potential Wells
P. Petroff, Direct Growth of Nanometer-Size Quantum Wire Superlattices
E. Kapon, Lateral Patterning of Quantum Well Heterostructures by Growth of Nonplanar Substrates
H. Temkin, D. Gershoni, and M. Panish, Optical Properties of $Ga_{1-x}In_xAs$/InP Quantum Wells

Volume 41 High Speed Heterostructure Devices

F. Capasso, F. Beltram, S. Sen, A. Pahlevi, and A. Y. Cho, Quantum Electron Devices: Physics and Applications
P. Solomon, D. J. Frank, S. L. Wright, and F. Canora, GaAs-Gate Semiconductor–Insulator–Semiconductor FET
M. H. Hashemi and U. K. Mishra, Unipolar InP-Based Transistors
R. Kiehl, Complementary Heterostructure FET Integrated Circuits
T. Ishibashi, GaAs-Based and InP-Based Heterostructure Bipolar Transistors
H. C. Liu and T. C. L. G. Sollner, High-Frequency-Tunneling Devices
H. Ohnishi, T. More, M. Takatsu, K. Imamura, and N. Yokoyama, Resonant-Tunneling Hot-Electron Transistors and Circuits

Volume 42 Oxygen in Silicon

F. Shimura, Introduction to Oxygen in Silicon
W. Lin, The Incorporation of Oxygen into Silicon Crystals
T. J. Schaffner and D. K. Schroder, Characterization Techniques for Oxygen in Silicon
W. M. Bullis, Oxygen Concentration Measurement
S. M. Hu, Intrinsic Point Defects in Silicon
B. Pajot, Some Atomic Configurations of Oxygen
J. Michel and L. C. Kimerling, Electical Properties of Oxygen in Silicon
R. C. Newman and R. Jones, Diffusion of Oxygen in Silicon
T. Y. Tan and W. J. Taylor, Mechanisms of Oxygen Precipitation: Some Quantitative Aspects
M. Schrems, Simulation of Oxygen Precipitation
K. Simino and I. Yonenaga, Oxygen Effect on Mechanical Properties
W. Bergholz, Grown-in and Process-Induced Effects
F. Shimura, Intrinsic/Internal Gettering
H. Tsuya, Oxygen Effect on Electronic Device Performance

Volume 43 Semiconductors for Room Temperature Nuclear Detector Applications

R. B. James and T. E. Schlesinger, Introduction and Overview
L. S. Darken and C. E. Cox, High-Purity Germanium Detectors
A. Burger, D. Nason, L. Van den Berg, and M. Schieber, Growth of Mercuric Iodide
X. J. Bao, T. E. Schlesinger, and R. B. James, Electrical Properties of Mercuric Iodide
X. J. Bao, R. B. James, and T. E. Schlesinger, Optical Properties of Red Mercuric Iodide
M. Hage-Ali and P. Siffert, Growth Methods of CdTe Nuclear Detector Materials
M. Hage-Ali and P Siffert, Characterization of CdTe Nuclear Detector Materials
M. Hage-Ali and P. Siffert, CdTe Nuclear Detectors and Applications
R. B. James, T. E. Schlesinger, J. Lund, and M. Schieber, $Cd_{1-x}Zn_xTe$ Spectrometers for Gamma and X-Ray Applications
D. S. McGregor, J. E. Kammeraad, Gallium Arsenide Radiation Detectors and Spectrometers
J. C. Lund, F. Olschner, and A. Burger, Lead Iodide
M. R. Squillante, and K. S. Shah, Other Materials: Status and Prospects
V. M. Gerrish, Characterization and Quantification of Detector Performance
J. S. Iwanczyk and B. E. Patt, Electronics for X-ray and Gamma Ray Spectrometers
M. Schieber, R. B. James, and T. E. Schlesinger, Summary and Remaining Issues for Room Temperature Radiation Spectrometers

Volume 44 II–IV Blue/Green Light Emitters: Device Physics and Epitaxial Growth

J. Han and R. L. Gunshor, MBE Growth and Electrical Properties of Wide Bandgap ZnSe-based II–VI Semiconductors
S. Fujita and S. Fujita, Growth and Characterization of ZnSe-based II–VI Semiconductors by MOVPE
E. Ho and L. A. Kolodziejski, Gaseous Source UHV Epitaxy Technologies for Wide Bandgap II–VI Semiconductors
C. G. Van de Walle, Doping of Wide-Band-Gap II–VI Compounds — Theory
R. Cingolani, Optical Properties of Excitons in ZnSe-Based Quantum Well Heterostructures
A. Ishibashi and A. V. Nurmikko, II–VI Diode Lasers: A Current View of Device Performance and Issues
S. Guha and J. Petruzello, Defects and Degradation in Wide-Gap II–VI-based Structures and Light Emitting Devices

Volume 45 Effect of Disorder and Defects in Ion-Implanted Semiconductors: Electrical and Physiochemical Characterization

H. Ryssel, Ion Implantation into Semiconductors: Historical Perspectives
You-Nian Wang and Teng-Cai Ma, Electronic Stopping Power for Energetic Ions in Solids
S. T. Nakagawa, Solid Effect on the Electronic Stopping of Crystalline Target and Application to Range Estimation
G. Müller, S. Kalbitzer and G. N. Greaves, Ion Beams in Amorphous Semiconductor Research
J. Boussey-Said, Sheet and Spreading Resistance Analysis of Ion Implanted and Annealed Semiconductors
M. L. Polignano and G. Queirolo, Studies of the Stripping Hall Effect in Ion-Implanted Silicon
J. Stoemenos, Transmission Electron Microscopy Analyses
R. Nipoti and M. Servidori, Rutherford Backscattering Studies of Ion Implanted Semiconductors
P. Zaumseil, X-ray Diffraction Techniques

Volume 46 Effect of Disorder and Defects in Ion-Implanted Semiconductors: Optical and Photothermal Characterization

M. Fried, T. Lohner and J. Gyulai, Ellipsometric Analysis
A. Seas and C. Christofides, Transmission and Reflection Spectroscopy on Ion Implanted Semiconductors
A. Othonos and C. Christofides, Photoluminescence and Raman Scattering of Ion Implanted Semiconductors. Influence of Annealing
C. Christofides, Photomodulated Thermoreflectance Investigation of Implanted Wafers. Annealing Kinetics of Defects
U. Zammit, Photothermal Deflection Spectroscopy Characterization of Ion-Implanted and Annealed Silicon Films
A. Mandelis, A. Budiman and M. Vargas, Photothermal Deep-Level Transient Spectroscopy of Impurities and Defects in Semiconductors
R. Kalish and S. Charbonneau, Ion Implantation into Quantum-Well Structures
A. M. Myasnikov and N. N. Gerasimenko, Ion Implantation and Thermal Annealing of III-V Compound Semiconducting Systems: Some Problems of III-V Narrow Gap Semiconductors

Volume 47 Uncooled Infrared Imaging Arrays and Systems

R. G. Buser and M. P. Tompsett, Historical Overview
P. W. Kruse, Principles of Uncooled Infrared Focal Plane Arrays
R. A. Wood, Monolithic Silicon Microbolometer Arrays
C. M. Hanson, Hybrid Pyroelectric-Ferroelectric Bolometer Arrays
D. L. Polla and J. R. Choi, Monolithic Pyroelectric Bolometer Arrays
N. Teranishi, Thermoelectric Uncooled Infrared Focal Plane Arrays
M. F. Tompsett, Pyroelectric Vidicon
T. W. Kenny, Tunneling Infrared Sensors
J. R. Vig, R. L. Filler and Y. Kim, Application of Quartz Microresonators to Uncooled Infrared Imaging Arrays
P. W. Kruse, Application of Uncooled Monolithic Thermoelectric Linear Arrays to Imaging Radiometers

Volume 48 High Brightness Light Emitting Diodes

G. B. Stringfellow, Materials Issues in High-Brightness Light-Emitting Diodes
M. G. Craford, Overview of Device issues in High-Brightness Light-Emitting Diodes
F. M. Steranka, AlGaAs Red Light Emitting Diodes
C. H. Chen, S. A. Stockman, M. J. Peanasky, and C. P. Kuo, OMVPE Growth of AlGaInP for High Efficiency Visible Light-Emitting Diodes
F. A. Kish and R. M. Fletcher, AlGaInP Light-Emitting Diodes
M. W. Hodapp, Applications for High Brightness Light-Emitting Diodes
I. Akasaki and H. Amano, Organometallic Vapor Epitaxy of GaN for High Brightness Blue Light Emitting Diodes
S. Nakamura, Group III-V Nitride Based Ultraviolet-Blue-Green-Yellow Light-Emitting Diodes and Laser Diodes

Volume 49 Light Emission in Silicon: from Physics to Devices

D. J. Lockwood, Light Emission in Silicon
G. Abstreiter, Band Gaps and Light Emission in Si/SiGe Atomic Layer Structures
T. G. Brown and D. G. Hall, Radiative Isoelectronic Impurities in Silicon and Silicon-Germanium Alloys and Superlattices
J. Michel, L. V. C. Assali, M. T. Morse, and L. C. Kimerling, Erbium in Silicon
Y. Kanemitsu, Silicon and Germanium Nanoparticles
P. M. Fauchet, Porous Silicon: Photoluminescence and Electroluminescent Devices
C. Delerue, G. Allan, and M. Lannoo, Theory of Radiative and Nonradiative Processes in Silicon Nanocrystallites
L. Brus, Silicon Polymers and Nanocrystals

Volume 50 Gallium Nitride (GaN)

J. I. Pankove and T. D. Moustakas, Introduction
S. P. DenBaars and S. Keller, Metalorganic Chemical Vapor Deposition (MOCVD) of Group III Nitrides
W. A. Bryden and T. J. Kistenmacher, Growth of Group III-A Nitrides by Reactive Sputtering
N. Newman, Thermochemistry of III-N Semiconductors
S. J. Pearton and R. J. Shul, Etching of III Nitrides

S. M. Bedair, Indium-based Nitride Compounds
A. Trampert, O. Brandt, and K. H. Ploog, Crystal Structure of Group III Nitrides
H. Morkoc, F. Hamdani, and A. Salvador, Electronic and Optical Properties of III V Nitride based Quantum Wells and Superlattices
K. Doverspike and J. I. Pankove, Doping in the III-Nitrides
T. Suski and P. Perlin, High Pressure Studies of Defects and Impurities in Gallium Nitride
B. Monemar, Optical Properties of GaN
W. R. L. Lambrecht, Band Structure of the Group III Nitrides
N. E. Christensen and P. Perlin, Phonons and Phase Transitions in GaN
S. Nakamura, Applications of LEDs and LDs
I. Akasaki and H. Amano, Lasers
J. A. Cooper, Jr., Nonvolatile Random Access Memories in Wide Bandgap Semiconductors

Volume 51A Identification of Defects in Semiconductors

G. D. Watkins, EPR and ENDOR Studies of Defects in Semiconductors
J.-M. Spaeth, Magneto-Optical and Electrical Detection of Paramagnetic Resonance in Semiconductors
T. A. Kennedy and E. R. Glaser, Magnetic Resonance of Epitaxial Layers Detected by Photoluminescence
K. H. Chow, B. Hitti, and R. F. Kiefl, μSR on Muonium in Semiconductors and Its Relation to Hydrogen
K. Saarinen, P. Hautojärvi, and C. Corbel, Positron Annihilation Spectroscopy of Defects in Semiconductors
R. Jones and P. R. Briddon, The Ab Initio Cluster Method and the Dynamics of Defects in Semiconductors

Volume 51B Identification of Defects in Semiconductors

G. Davies, Optical Measurements of Point Defects
P. M. Mooney, Defect Identification Using Capacitance Spectroscopy
M. Stavola, Vibrational Spectroscopy of Light Element Impurities in Semiconductors
P. Schwander, W. D. Rau, C. Kisielowski, M. Gribelyuk, and A. Ourmazd, Defect Processes in Semiconductors Studied at the Atomic Level by Transmission Electron Microscopy
N. D. Jager and E. R. Weber, Scanning Tunneling Microscopy of Defects in Semiconductors

Volume 52 SiC Materials and Devices

K. Järrendahl and R. F. Davis, Materials Properties and Characterization of SiC
V. A. Dmitriev and M. G. Spencer, SiC Fabrication Technology: Growth and Doping
V. Saxena and A. J. Steckl, Building Blocks for SiC Devices: Ohmic Contacts, Schottky Contacts, and p-n Junctions
M. S. Shur, SiC Transistors
C. D. Brandt, R. C. Clarke, R. R. Siergiej, J. B. Casady, A. W. Morse, S. Sriram, and A. K. Agarwal, SiC for Applications in High-Power Electronics
R. J. Trew, SiC Microwave Devices

J. Edmond, H. Kong, G. Negley, M. Leonard, K. Doverspike, W. Weeks, A. Suvorov, D. Waltz, and C. Carter, Jr., SiC-Based UV Photodiodes and Light-Emitting Diodes
H. Morkoç, Beyond Silicon Carbide! III–V Nitride-Based Heterostructures and Devices

Volume 53 Cumulative Subject and Author Index Including Tables of Contents for Volume 1–50

Volume 54 High Pressure in Semiconductor Physics I

W. Paul, High Pressure in Semiconductor Physics: A Historical Overview
N. E. Christensen, Electronic Structure Calculations for Semiconductors under Pressure
R. J. Neimes and M. I. McMahon, Structural Transitions in the Group IV, III-V and II-VI Semiconductors Under Pressure
A. R. Goni and K. Syassen, Optical Properties of Semiconductors Under Pressure
P. Trautman, M. Baj, and J. M. Baranowski, Hydrostatic Pressure and Uniaxial Stress in Investigations of the EL2 Defect in GaAs
M. Li and P. Y. Yu, High-Pressure Study of DX Centers Using Capacitance Techniques
T. Suski, Spatial Correlations of Impurity Charges in Doped Semiconductors
N. Kuroda, Pressure Effects on the Electronic Properties of Diluted Magnetic Semiconductors

Volume 55 High Pressure in Semiconductor Physics II

D. K. Maude and J. C. Portal, Parallel Transport in Low-Dimensional Semiconductor Structures
P. C. Klipstein, Tunneling Under Pressure: High-Pressure Studies of Vertical Transport in Semiconductor Heterostructures
E. Anastassakis and M. Cardona, Phonons, Strains, and Pressure in Semiconductors
F. H. Pollak, Effects of External Uniaxial Stress on the Optical Properties of Semiconductors and Semiconductor Microstructures
A. R. Adams, M. Silver, and J. Allam, Semiconductor Optoelectronic Devices
S. Porowski and I. Grzegory, The Application of High Nitrogen Pressure in the Physics and Technology of III-N Compounds
M. Yousuf, Diamond Anvil Cells in High Pressure Studies of Semiconductors

Volume 56 Germanium Silicon: Physics and Materials

J. C. Bean, Growth Techniques and Procedures
D. E. Savage, F. Liu, V. Zielasek, and M. G. Lagally, Fundamental Crystal Growth Mechanisms
R. Hull, Misfit Strain Accommodation in SiGe Heterostructures
M. J. Shaw and M. Jaros, Fundamental Physics of Strained Layer GeSi: Quo Vadis?
F. Cerdeira, Optical Properties
S. A. Ringel and P. N. Grillot, Electronic Properties and Deep Levels in Germanium-Silicon
J. C. Campbell, Optoelectronics in Silicon and Germanium Silicon
K. Eberl, K. Brunner, and O. G. Schmidt, $Si_{1-y}C_y$ and $Si_{1-x-y}Ge_xC_y$ Alloy Layers

Volume 57 Gallium Nitride (GaN) II

R. J. Molnar, Hydride Vapor Phase Epitaxial Growth of III-V Nitrides
T. D. Moustakas, Growth of III-V Nitrides by Molecular Beam Epitaxy
Z. Liliental-Weber, Defects in Bulk GaN and Homoepitaxial Layers
C. G. Van de Walle and N. M. Johnson, Hydrogen in III-V Nitrides
W. Götz and N. M. Johnson, Characterization of Dopants and Deep Level Defects in Gallium Nitride
B. Gil, Stress Effects on Optical Properties
C. Kisielowski, Strain in GaN Thin Films and Heterostructures
J. A. Miragliotta and D. K. Wickenden, Nonlinear Optical Properties of Gallium Nitride
B. K. Meyer, Magnetic Resonance Investigations on Group III-Nitrides
M. S. Shur and M. Asif Khan, GaN and AlGaN Ultraviolet Detectors
C. H. Qiu, J. I. Pankove, and C. Rossington, III-V Nitride-Based X-ray Detectors

Volume 58 Nonlinear Optics in Semiconductors I

A. Kost, Resonant Optical Nonlinearities in Semiconductors
E. Garmire, Optical Nonlinearities in Semiconductors Enhanced by Carrier Transport
D. S. Chemla, Ultrafast Transient Nonlinear Optical Processes in Semiconductors
M. Sheik-Bahae and E. W. Van Stryland, Optical Nonlinearities in the Transparency Region of Bulk Semiconductors
J. E. Millerd, M. Ziari, and A. Partovi, Photorefractivity in Semiconductors

Volume 59 Nonlinear Optics in Semiconductors II

J. B. Khurgin, Second Order Nonlinearities and Optical Rectification
K. L. Hall, E. R. Thoen, and E. P. Ippen, Nonlinearities in Active Media
E. Hanamura, Optical Responses of Quantum Wires/Dots and Microcavities
U. Keller, Semiconductor Nonlinearities for Solid-State Laser Modelocking and Q-Switching
A. Miller, Transient Grating Studies of Carrier Diffusion and Mobility in Semiconductors

Volume 60 Self-Assembled InGaAs/GaAs Quantum Dots

Mitsuru Sugawara, Theoretical Bases of the Optical Properties of Semiconductor Quantum Nano-Structures
Yoshiaki Nakata, Yoshihiro Sugiyama, and Mitsuru Sugawara, Molecular Beam Epitaxial Growth of Self-Assembled InAs/GaAs Quantum Dots
Kohki Mukai, Mitsuru Sugawara, Mitsuru Egawa, and Nobuyuki Ohtsuka, Metalorganic Vapor Phase Epitaxial Growth of Self-Assembled InGaAs/GaAs Quantum Dots Emitting at 1.3 μm
Kohki Mukai and Mitsuru Sugawara, Optical Characterization of Quantum Dots
Kohki Mukai and Mitsuru Sugawara, The Photon Bottleneck Effect in Quantum Dots
Hajime Shoji, Self-Assembled Quantum Dot Lasers
Hiroshi Ishikawa, Applications of Quantum Dot to Optical Devices
Mitsuru Sugawara, Kohki Mukai, Hiroshi Ishikawa, Koji Otsubo, and Yoshiaki Nakata, The Latest News

Volume 61 Hydrogen in Semiconductors II

Norbert H. Nickel, Introduction to Hydrogen in Semiconductors II
Noble M. Johnson and Chris G. Van de Walle, Isolated Monatomic Hydrogen in Silicon
Yurij V. Gorelkinskii, Electron Paramagnetic Resonance Studies of Hydrogen and Hydrogen-Related Defects in Crystalline Silicon
Norbert H. Nickel, Hydrogen in Polycrystalline Silicon
Wolfhard Beyer, Hydrogen Phenomena in Hydrogenated Amorphous Silicon
Chris G. Van de Walle, Hydrogen Interactions with Polycrystalline and Amorphous Silicon—Theory
Karen M. McNamara Rutledge, Hydrogen in Polycrystalline CVD Diamond
Roger L. Lichti, Dynamics of Muonium Diffusion, Site Changes and Charge-State Transitions
Matthew D. McCluskey and Eugene E. Haller, Hydrogen in III-V and II-VI Semiconductors
S. J. Pearton and J. W. Lee, The Properties of Hydrogen in GaN and Related Alloys
Jörg Neugebauer and Chris G. Van de Walle, Theory of Hydrogen in GaN

Volume 62 Intersubband Transitions in Quantum Wells: Physics and Device Applications I

Manfred Helm, The Basic Physics of Intersubband Transitions
Jerome Faist, Carlo Sirtori, Federico Capasso, Loren N. Pfeiffer, Ken W. West, Deborah L. Sivco, and Alfred Y. Cho, Quantum Interference Effects in Intersubband Transitions
H. C. Liu, Quantum Well Infrared Photodetector Physics and Novel Devices
S. D. Gunapala and S. V. Bandara, Quantum Well Infrared Photodetector (QWIP) Focal Plane Arrays

Volume 63 Chemical Mechanical Polishing in Si Processing

Frank B. Kaufman, Introduction
Thomas Bibby and Karey Holland, Equipment
John P. Bare, Facilitization
Duane S. Boning and Okumu Ouma, Modeling and Simulation
Shin Hwa Li, Bruce Tredinnick, and Mel Hoffman, Consumables I: Slurry
Lee M. Cook, CMP Consumables II: Pad
François Tardif, Post-CMP Clean
Shin Hwa Li, Tara Chhatpar, and Frederic Robert, CMP Metrology
Shin Hwa Li, Visun Bucha, and Kyle Wooldridge, Applications and CMP-Related Process Problems

Volume 64 Electroluminescence I

M. G. Craford, S. A. Stockman, M. J. Peanasky, and F. A. Kish, Visible Light-Emitting Diodes
H. Chui, N. F. Gardner, P. N. Grillot, J. W. Huang, M. R. Krames, and S. A. Maranowski, High-Efficiency AlGaInP Light-Emitting Diodes
R. S. Kern, W. Götz, C. H. Chen, H. Liu, R. M. Fletcher, and C. P. Kuo, High-Brightness Nitride-Based Visible-Light-Emitting Diodes
Yoshiharu Sato, Organic LED System Considerations
V. Bulović, P. E. Burrows, and S. R. Forrest, Molecular Organic Light-Emitting Devices

Volume 65 Electroluminescence II

V. Bulović and S. R. Forrest, Polymeric and Molecular Organic Light Emitting Devices: A Comparison
Regina Mueller-Mach and Gerd O. Mueller, Thin Film Electroluminescence
Markku Leskelä, Wei-Min Li, and Mikko Ritala, Materials in Thin Film Electroluminescent Devices
Kristiaan Neyts, Microcavities for Electroluminescent Devices

Volume 66 Intersubband Transitions in Quantum Wells: Physics and Device Applications II

Jerome Faist, Federico Capasso, Carlo Sirtori, Deborah L. Sivco, and Alfred Y. Cho, Quantum Cascade Lasers
Federico Capasso, Carlo Sirtori, D. L. Sivco, and A. Y. Cho, Nonlinear Optics in Coupled-Quantum-Well Quasi-Molecules
Karl Unterrainer, Photon-Assisted Tunneling in Semiconductor Quantum Structures
P. Haring Bolivar, T. Dekorsy, and H. Kurz, Optically Excited Bloch Oscillations — Fundamentals and Application Perspectives

Volume 67 Ultrafast Physical Processes in Semiconductors

Alfred Leitenstorfer and Alfred Laubereau, Ultrafast Electron–Phonon Interactions in Semiconductors: Quantum Kinetic Memory Effects
Christoph Lienau and Thomas Elsaesser, Spatially and Temporally Resolved Near-Field Scanning Optical Microscopy Studies of Semiconductor Quantum Wires
K. T. Tsen, Ultrafast Dynamics in Wide Bandgap Wurtzite GaN
J. Paul Callan, Albert M.-T. Kim, Christopher A. D. Roeser, and Eriz Mazur, Ultrafast Dynamics and Phase Changes in Highly Excited GaAs
Hartmut Haug, Quantum Kinetics for Femtosecond Spectroscopy in Semiconductors
T. Meier and S. W. Koch, Coulomb Correlation Signatures in the Excitonic Optical Nonlinearities of Semiconductors
Roland E. Allen, Traian Dumitrică, and Ben Torralva, Electronic and Structural Response of Materials to Fast, Intense Laser Pulses
E. Gornik and R. Kersting, Coherent THz Emission in Semiconductors

Volume 68 Isotope Effects in Solid State Physics

Vladimir G. Plekhanov: Elastic Properties; Thermal Properties; Vibrational Properties; Raman Spectra of Isotopically Mixed Crystals; Excitons in LiH Crystals; Exciton–Phonon Interaction; Isotopic Effect in the Emission Spectrum of Polaritons; Isotopic Disordering of Crystal Lattices; Future Developments and Applications; Conclusions

ISBN 0-12-752178-X